Lecture Notes in Computer Science 12152

More information about this series at http://www.springer.com/series/7407

Ryszard Janicki · Natalia Sidorova ·
Thomas Chatain (Eds.)

Application and Theory
of Petri Nets
and Concurrency

41st International Conference, PETRI NETS 2020
Paris, France, June 24–25, 2020
Proceedings

 Springer

Editors
Ryszard Janicki ⓘ
McMaster University
Hamilton, ON, Canada

Natalia Sidorova
Eindhoven University of Technology
Eindhoven, The Netherlands

Thomas Chatain ⓘ
LSV, CNRS & ENS Paris-Saclay
Cachan, France

ISSN 0302-9743   ISSN 1611-3349 (electronic)
Lecture Notes in Computer Science
ISBN 978-3-030-51830-1   ISBN 978-3-030-51831-8 (eBook)
https://doi.org/10.1007/978-3-030-51831-8

LNCS Sublibrary: SL1 – Theoretical Computer Science and General Issues

This Springer imprint is published by the registered company Springer Nature Switzerland AG
The registered company address is: Gewerbestrasse 11, 6330 Cham, Switzerland

Preface

This volume constitutes the proceedings of the 41st International Conference on Application and Theory of Petri Nets and Concurrency (Petri Nets 2020). This series of conferences serves as an annual meeting place to discuss progress in the field of Petrinets and related models of concurrency. These conferences provide a forum for researchers to present and discuss both applications and theoretical developments in this area. Novel tools and substantial enhancements to existing tools can also be presented.

Petri Nets 2020 includes a section devoted to Application of Concurrency to System Design, which until this year was a separate event.

The event was organized by the LoVe (Logics and Verification) team of the computer science laboratory LIPN (Laboratoire d'Informatique de Paris Nord), University Sorbonne Paris Nord, and CNRS, jointly with members of the Paris region MeFoSy-LoMa group (Méthodes Formelles pour les Systèmes Logiciels et Matériels). The conference was supposed to take place in the chosen area of Campus Condorcet, the new international research campus in humanities and social sciences in Paris.

Unfortunately, because of the coronavirus epidemic, the event was organized as a virtual conference. We would like to express our deepest thanks to the Organizing Committee chaired by Laure Petrucci and Ètienne André for the time and effort invested in the organization of this event.

This year, 56 papers were submitted to Petri Nets 2020 by authors from 21 different countries. Each paper was reviewed by three reviewers. The discussion phase and final selection process by the Program Committee (PC) were supported by the EasyChair conference system. From 44 regular papers and 12 tool papers, the PC selected 23 papers for presentation: 17 regular papers and 6 tool papers. After the conference, some of these authors were invited to submit an extended version of their contribution for consideration in a special issue of a journal.

We thank the PC members and other reviewers for their careful and timely evaluation of the submissions and the fruitful constructive discussions that resulted in the final selection of papers. The Springer LNCS team (notably Anna Kramer and Aliaksandr Birukou) provided excellent and welcome support in the preparation of this volume.

Due to a virtual format of the event the keynote presentations have been postponed to the 2021 edition of this conference. Alongside Petri Nets 2020, the following workshops and events took place: the 10th edition of the Model Checking Contest (MCC 2020), the Workshop on Algorithms and Theories for the Analysis of Event

Data (ATAED 2020), and the Workshop on Petri Netsand Software Engineering (PNSE 2020). All the above workshops and events were also delivered in virtual format.

We hope you enjoy reading the contributions in this LNCS volume.

June 2020   Ryszard Janicki
                                                      Natalia Sidorova
                                                      Thomas Chatain

Organization

Program Committee

Elvio Gilberto Amparore	Università di Torino, Italy
Paolo Baldan	Università di Padova, Italy
Benoit Barbot	Université Paris Est Créteil, France
Didier Buchs	University of Geneva, Switzerland
Josep Carmona	Universitat Politèecnica de Catalunya, Spain
Thomas Chatain	ENS Paris-Saclay, France
Isabel Demongodin	LSIS - UMR CNRS, France
Jörg Desel	Fernuniversität in Hagen, Germany
Raymond Devillers	Université Libre de Bruxelles, Bergium
Susanna Donatelli	Università di Torino, Italy
Javier Esparza	Technical University of Munich, Germany
David Frutos Escrig	Universidad Complutense de Madrid, Spain
Stefan Haar	Inria Saclay, LSV, ENS Cachan, France
Xudong He	Florida International University, USA
Loic Helouet	Inria, France
Petr Jancar	Palacky University, Czech Republic
Ryszard Janicki	McMaster University, Canada
Ekkart Kindler	Technical University of Denmark, Denmark
Jetty Kleijn	Leiden University, The Netherlands
Lars Kristensen	Western Norway University of Applied Sciences, Norway
Michael Köhler-Bußmeier	University of Applied Science at Hamburg, Germany
Irina Lomazova	National Research University Higher School of Economics, Russia
Robert Lorenz	Augsburg University, Germany
Roland Meyer	TU Braunschweig, Germany
Lukasz Mikulski	Nicolaus Copernicus University, Poland
Andrew Miner	Iowa State University, USA
Andrey Mokhov	Newcastle University, UK
Claire Pagetti	ONERA, IRIT-ENSEEIHT, France
Pierre-Alain Reynier	Aix-Marseille Université, France
Olivier H. Roux	École Centrale de Nantes, France
Arnaud Sangnier	Université Paris Diderot, France
Natalia Sidorova	Technische Universiteit Eindhoven, The Netherlands
Boudewijn Van Dongen	Technische Universiteit Eindhoven, The Netherlands
Karsten Wolf	Universität Rostock, Germany
Alex Yakovlev	Newcastle University, UK
Wlodek Zuberek	Memorial University, Canada

Additional Reviewers

Althoff, Matthias
Ballarini, Paolo
Bashkin, Vladimir
Basu, Samik
Baudru, Nicolas
Becker, Mike
Bergenthum, Robin
Biswal, Shruti
Boltenhagen, Mathilde
Brenner, Leonardo
Brunel, Julien
Carvalho, Rafael V.
Chini, Peter
Claviere, Arthur
Coet, Aurélien
Delahaye, Benoit
Delfieu, David
Ding, Junhua
Dong, Zhijiang
Geeraerts, Gilles
Haas, Thomas
Hoogeboom, Hendrik Jan
Jezequel, Loig
Knapik, Michał
Kot, Martin
Kurpiewski, Damian
Küpper, Sebastian
Laarman, Alfons
Leroux, Jérôme
Lime, Didier

Lukaynov, Georgy
Meggendorfer, Tobias
Metzger, Johannes
Meyer, Philipp J.
Mitsyuk, Alexey A.
Montali, Marco
Morard, Damien
Niwinski, Damian
Nowicki, Marek
Oualhadj, Youssouf
Outrata, Jan
Padoan, Tommaso
Pekergin, Nihal
Petrak, Lisa
Przymus, Piotr
Racordon, Dimitri
Randour, Mickael
Rivkin, Andrey
Rosa-Velardo, Fernando
Roux, Pierre
Sawa, Zdeněk
Seidner, Charlotte
Sensfelder, Nathanael
Stachtiari, Emmanouela
Valero, Valentin
van der Wall, Sören
Verbeek, Eric
Weil-Kennedy, Chana
Wolff, Sebastian
Yakovlev, Alex

Contents

Application of Concurrency to System Design

Automatic Decomposition of Petri Nets into Automata Networks – A Synthetic Account

Pierre Bouvier[1]([⊠]), Hubert Garavel[1]([⊠]), and Hernán Ponce-de-León[2]([⊠])

[1] Univ. Grenoble Alpes, Inria, CNRS, Grenoble INP, LIG, 38000 Grenoble, France
{pierre.bouvier,hubert.garavel}@inria.fr
[2] Research Institute CODE, Bundeswehr University Munich, Munich, Germany
hernan.ponce@unibw.de

Abstract. This article revisits the problem of decomposing a Petri net into a network of automata, a problem that has been around since the early 70s. We reformulate this problem as the transformation of an ordinary, one-safe Petri net into a flat, unit-safe NUPN (Nested-Unit Petri Net) and define a quality criterion based on the number of bits required for the structural encoding of markings. We propose various transformation methods, all of which we implemented in a tool chain that combines NUPN tools with third-party software, such as SAT solvers, SMT solvers, and tools for graph colouring and finding maximal cliques. We perform an extensive evaluation of these methods on a collection of more than 12,000 nets from diverse sources, including nets whose marking graph is too large for being explored exhaustively.

1 Introduction

The present article addresses the *decomposition problem* for Petri nets. Precisely, we study the automatic transformation of a (low-level) Petri net into an *automata network*, i.e., a set of sequential components (such as finite-state machines) that execute asynchronously, synchronize with each other, and exhibit the same global behaviour as the original Petri net. This problem is of practical interest for at least two reasons: (i) Petri nets are expressive, but poorly structured; decomposition is a means to restructure them automatically, making them more modular and, hopefully, easier to understand and reason about; (ii) automata networks contain structural information that formal verification algorithms may exploit to increase efficiency using, e.g., logarithmic encodings of reachable markings, easier detection of independent transitions for partial-order and stubborn-set reduction methods, and divide-and-conquer strategies for compositional verification.

To a large extent, we reformulate the decomposition problem in terms of *Nested-Unit Petri Nets* (NUPNs) [8], a modern extension of Petri nets, in which places can be grouped into *units* that express sequential components. Units can be recursively nested to reflect both the concurrent and the hierarchical nature of complex systems. This model of computation, originally developed

© Springer Nature Switzerland AG 2020
R. Janicki et al. (Eds.): PETRI NETS 2020, LNCS 12152, pp. 3–23, 2020.
https://doi.org/10.1007/978-3-030-51831-8_1

for translating process calculi to Petri nets, increases the efficiency of formal verification [1, 8, Sect. 6]. It has been so far implemented in thirteen software tools [8, Sect. 7] and adopted for the benchmarks of the Model Checking Contest and the parallel problems of the Rigorous Examination of Reactive Systems. Notice that certain NUPN features, such as the hierarchical nesting of units to an arbitrary depth, are not exploited in the present article.

Related Work. The decomposition of a Petri net into sequential processes has been studied since the early 70s at least [12], and gave rise to a significant body of academic literature. On the theoretical side, one can mention the decomposition of elementary nets into a set of concurrent communicating sequential components [23, Sect. 4.3–4.4], the decomposition of live and bounded free–choice nets into S-components or T-components [6, Chap. 5], and the distribution of a Petri net to geographical locations [5, 11]. On the algorithmic side, one can mention decomposition methods that compute a coverage of a net by strongly connected state machines, e.g., [12], approaches based on invariants and semiflows, e.g., [3, 21, 24], and approaches based on reachability analysis, some using concurrency graphs on which decomposition can be expressed as a graph colouring algorithm [27], others using hypergraphs, which seem more compact that concurrency graphs [28]. On the software implementation side, one can mention the Diane tool [19], which seems no longer accessible today, and the Hippo tool, which is developed at the University of Zielona Góra (Poland) and available on-line through a dedicated web portal[1]. Further references to related work are given throughout the next sections.

Outline. The present paper describes a state-of-the-art approach based on (partial) reachability analysis for translating ordinary, safe Petri nets into automata networks. This approach has been fully implemented in a software tool chain, and successfully applied to thousands of examples. The remainder of this article is organized as follows. Section 2 states the decomposition problem by precisely defining which kind of Petri nets are taken as input, which kind of automata networks are produced as output, and which quality criterion should guide the decomposition. Section 3 defines some key concepts used for decomposition, namely the concurrency relation, the concurrency matrix, and the concurrency graph. Section 4 provides means to efficiently search for solutions. The four next sections present various decomposition approaches based on graph colouring (Sect. 5), maximal-clique algorithms (Sect. 6), SAT solving (Sect. 7), and SMT solving (Sect. 8). Section 9 discusses the experimental results obtained by these various approaches and draws a comparison with the Hippo tool. Finally, Sect. 10 concludes the article.

---

[1] http://www.hippo.iee.uz.zgora.pl.

2 Problem Statement

2.1 Basic Definitions

We briefly recall the usual definitions of Petri nets and refer the reader to classical surveys for a more detailed presentation of Petri nets.

Definition 1. *A (marked) Petri Net is a 4-tuple* (P, T, F, M_0) *where:*

1. *P is a finite, non-empty set; the elements of P are called* places.
2. *T is a finite set such that* $P \cap T = \varnothing$*; the elements of T are called* transitions.
3. *F is a subset of* $(P \times T) \cup (T \times P)$*; the elements of F are called* arcs.
4. *M_0 is a non-empty subset of P; M_0 is called the* initial marking.

Notice that the above definition only covers *ordinary* nets (i.e., it assumes all arc weights are equal to one). Also, it only considers *safe* nets (i.e., each place contains at most one token), which enables the initial marking to be defined as a subset of P, rather than a function $P \to \mathbb{N}$ as in the usual definition of P/T nets. We now recall the classical firing rules for ordinary safe nets.

Definition 2. *Let* (P, T, F, M_0) *be a Petri Net.*

- *A* marking *M is defined as a set of places* $(M \subseteq P)$*. Each place belonging to a marking M is said to be* marked *or, also, to* possess *a token.*
- *The* pre-set *of a transition t is the set of places* ${}^{\bullet}t \stackrel{\text{def}}{=} \{p \in P \mid (p, t) \in F\}$*.*
- *The* post-set *of a transition t is the set of places* $t^{\bullet} \stackrel{\text{def}}{=} \{p \in P \mid (t, p) \in F\}$*.*
- *A transition t is* enabled *in some marking M iff* ${}^{\bullet}t \subseteq M$*.*
- *A transition t can* fire *from some marking M_1 to another marking M_2 iff t is enabled in M_1 and $M_2 = (M_1 \setminus {}^{\bullet}t) \cup t^{\bullet}$, which we note $M_1 \stackrel{t}{\longrightarrow} M_2$.*
- *A marking M is* reachable *from the initial marking M_0 iff $M = M_0$ or there exist $n \geq 1$ transitions $t_1, t_2, ..., t_n$ and $(n-1)$ markings $M_1, M_2, ..., M_{n-1}$ such that $M_0 \stackrel{t_1}{\longrightarrow} M_1 \stackrel{t_2}{\longrightarrow} M_2 ... M_{n-1} \stackrel{t_n}{\longrightarrow} M$.*
- *A place p is* dead *if it exists no reachable marking containing p.*
- *A transition t is* dead *if it exists no reachable marking in which t is enabled.*

We now recall the basic definition of a NUPN, referring the interested reader to [8] for a complete presentation of this model of computation.

Definition 3. *A (marked)* Nested-Unit Petri Net *(acronym: NUPN) is a 8-tuple* $(P, T, F, M_0, U, u_0, \sqsubseteq, \text{unit})$ *where* (P, T, F, M_0) *is a Petri net, and where:*

5. *U is a finite, non-empty set such that* $U \cap T = U \cap P = \varnothing$*; the elements of U are called* units.
6. *u_0 is an element of U; u_0 is called the* root *unit.*
7. *$\sqsubseteq$ is a binary relation over U such that $(U, \sqsupseteq)$ is a tree with a single root u_0, where $(\forall u_1, u_2 \in U)\ u_1 \sqsupseteq u_2 \stackrel{\text{def}}{=} u_2 \sqsubseteq u_1$; intuitively[2], $u_1 \sqsubseteq u_2$ expresses that unit u_1 is transitively nested in or equal to unit u_2.*

---

[2] $\sqsubseteq$ is reflexive, antisymmetric, transitive, and u_0 is the greatest element of U for $\sqsubseteq$.

8. unit *is a function* $P \to U$ *such that* $(\forall u \in U \setminus \{u_0\})$ $(\exists p \in P)$ unit$(p) = u$;
   intuitively, unit$(p) = u$ *expresses that unit* u *directly contains place* p.

Because NUPNs merely extend Petri nets by grouping places into units, they do not modify the Petri-net firing rules for transitions: all the concepts of Definition 2 for Petri nets also apply to NUPNs, so that Petri-net properties are preserved when NUPN information is added. Finally, we recall a few NUPN concepts to be used throughout this article; additional information can be found in [8], where such concepts (especially, unit safeness) are defined in a more general manner.

Definition 4. *Let* $N = (P, T, F, M_0, U, u_0, \sqsubseteq, \text{unit})$ *be a NUPN.*

- *The* local places *of a unit* u *are the set:* places$(u) \stackrel{\text{def}}{=} \{p \in P \mid \text{unit}(p) = u\}$.
- *A* void *unit is a unit having no local place.*[3]
- *A* leaf *unit is a minimal element of* $(U, \sqsubseteq)$, *i.e., a unit having no nested unit.*
- *The* height *of* N *is the length of the longest chain* $u_n \sqsubseteq \ldots \sqsubseteq u_1 \sqsubseteq u_0$ *of nested units, not counting the root unit* u_0 *if it is void.*
- *The* width *of* N *is the number of its leaf units.*
- *A* flat *NUPN is such that its height is equal to one.*
- *A* trivial *NUPN is such that its width is equal to the number of places.*[4]
- *A flat NUPN is* unit-safe *iff any reachable marking contains at most one local place of each unit.*

2.2   Input Formalism

From Definition 1 and 2, the Petri nets accepted as input by our methods must be ordinary, safe, and have at least one initially marked place[5]. Such restrictions are quite natural, given that our goal is the translation of a Petri net to a network of automata, which are also ordinary and safe by essence (this will be further discussed in Sect. 2.3). Unlike, e.g., [21], we do not handle bounded nets that are not safe, given that any bounded net can be converted to a safe net by duplicating (triplicating, etc.) some of its places.

Contrary to other approaches, we do not lay down additional restrictions on the Petri nets accepted as input. For instance:

- We do not require them to be free choice and *well formed* (i.e., live and safe) as in [12], nor free choice, live, and bounded as in [18].
- We do not require them to be *state machine coverable* [26, Def. 16.2.2 (180)], *state machine decomposable* [12, Chap. 5], nor *state machine allocatable* [13] (see [4, Sect. 7.1] for a discussion of the two latter concepts).
- We do not require them to be *pure* (i.e., free from self-loop transitions) as in [27], where Petri nets are represented using their incidence matrix, a data structure that cannot describe those transitions t such that ${}^\bullet t \cap t^\bullet \neq \varnothing$.
- We do not require them to be connected or strongly connected, and accept the presence of dead places and/or dead transitions.

---

[3] From item 8 of Definition 3, only the root unit u_0 may be void.
[4] Each unit has a single local place, except the root unit, which has either zero or one.
[5] However, they can have no transition.

Implementation and Experimentation. Our tool chain accepts as input a ".pnml" file containing a Petri net represented in the PNML standard format [14]. This file is then converted to a ".nupn" file, written in a concise and human-readable textual format[6] [8, Annex A] for storing NUPNs. This conversion is performed using PNML2NUPN[7], a translator developed at LIP6 (Paris, France), which discards those PNML features that are irrelevant to our problem (e.g., colored, timed, or graphical attributes,) to produce a P/T net. All NUPNs generated by PNML2NUPN are trivial, meaning that the translator follows the scheme given in [8, proof of Prop. 11] by putting each place in a separate unit; thus, the translator makes no attempt at discovering concurrency in PNML models.

Our tool chain also accepts a file directly written in the ".nupn" format, rather than being generated from PNML. Such a NUPN model may be trivial or already have a decomposition, either as a network of automata (in the case of flat non-trivial models) or as a hierarchy of nested concurrent units (in the case of non-flat models). We treat non-trivial models like trivial ones by purposely ignoring (most of) the information about the structure of non-trivial models.

The next step is to make sure that each NUPN model to be decomposed is safe. Many NUPN models generated from higher-level specification languages are unit safe by construction, and thus safe, which is indicated by a pragma "!unit_safe" present in the ".nupn" file. In absence of this pragma, one must check whether the underlying Petri net is safe, which is a PSPACE-complete problem. Various tools that can handle ".nupn" files are available for such purpose, e.g., CÆSAR.BDD[8] or CÆSAR.SDD[9].

To perform experiments, we used a collection of 12,728 models in ".nupn" format. This collection, which has been patiently built at INRIA Grenoble since 2013, gathers models derived from "realistic" specifications (i.e., written in high-level languages by humans rather than randomly generated, many of which developed for industrial problems). It also contains all ordinary, safe models from the former PetriWeb collection[10] and from the Model Checking Context benchmarks[11]. To our knowledge, our collection is the largest ever reported in the scientific literature on Petri nets.

A statistical survey confirms the diversity of our collection of models. Table 1 gives the percentage of models that satisfy or not some usual (structural and behavioural) properties of Petri nets, as well as topological properties of NUPNs; the answer is unknown for some large models that CÆSAR.BDD (with option "-mcc") could not process entirely. Table 2 gives numerical information about

---

[6] http://cadp.inria.fr/man/nupn.html.

[7] http://pnml.lip6.fr/pnml2nupn.

[8] http://cadp.inria.fr/man/caesar.bdd.html (see option "-check").

[9] http://github.com/ahamez/caesar.sdd (see option "--check").

[10] http://pnrepository.lip6.fr/pweb/models/all/browser.html.

[11] http://mcc.lip6.fr/models.php.

Table 1. Structural, behavioural, and topological properties of our collection

Property	Yes	No	Unknown	Property	Yes	No	Unknown
Pure	62.5%	37.4%	0.1%	Conservative	16.7%	83.3%	
Free-choice	42.4%	57.6%		Sub-conservative	29.8%	70.2%	
Extended free-choice	43.9%	56.0%	0.1%	Dead places	15.7%	80.0%	4.3%
Marked graph	3.6%	96.4%		Dead transitions	15.8%	80.7%	3.5%
State machine	12.5%	87.5%		Trivial	11.3%	88.7%	
Connected	94.1%	5.9%		Flat, Non trivial	25.0%	75.0%	
Strongly connected	13.9%	86.1%		Non flat	63.7%	36.3%	

Table 2. Numerical properties of our collection

Feature	Min value	Max value	Average	Median	Std deviation
#places	1	131,216	221.1	14	2,389
#transitions	0	16,967,720	9,399.5	19	274,364
#arcs	0	146,528,584	74,627.1	50	2,173,948
Arc density	0%	100%	14.9%	9.7%	0.2
#units	1	50,001	86.7	6	1,125
Height	1	2,891	4.1	2	43
Width	1	50,000	82.3	4	1,122

the size of the Petri nets (number of places, transitions, and arcs, as well as arc density[12]) and the size of the NUPNs (number of units, height, and width).

2.3   Output Formalism

Our goal is to translate ordinary, safe Petri nets into automata networks; however, many automata-based formalisms have been proposed in the literature, most of which are candidate targets for our translation, but differ in subtle details. A thorough survey was given in [4], yet newer proposals have been made since then. In this subsection, we address this issue by precisely stating which constraints a suitable output formalism should satisfy.

Our first constraint is that each automaton should be sequential (contrary to, e.g., [5], where concurrent transitions can be assigned to the same location), that our automata networks should be flat (contrary to, e.g., [7,16] and [20], where Petri nets are translated to hierarchical models in which processes can have nested sub-processes), and that the semantics of automata networks should be aligned with the usual interpretation of Petri nets, i.e., the states of each automaton should reflect Petri net places, and the global state of the

---

[12] We define arc density as the number of arcs divided by twice the product of the number of places and the number of transitions, i.e., the amount of memory needed to store the arc relation as a pair of place×transition matrices.

automata network should be the union of all the local states of its component automata. The transitions of each automaton should also behave as Petri net transitions, meaning that synchronization between several automata should be achieved using the Petri-net firing rules (see Definition 2) rather than alternative mechanisms, such as synchronization states in *synchronized automata* [22], asynchronous message passing in *reactive automata* [2], or FIFO buffers in *communicating automata* [10].

Our second constraint goes further by requiring a one-to-one mapping between the places of the Petri net and the states of the automata network, and between each transition of the Petri net and the corresponding (possibly synchronized) transition(s) of the automata network. Consequently, all structural and behavioural properties of the Petri net should be preserved by decomposition; in particular, the graph of reachable markings for the Petri net should be isomorphic (modulo some renaming of places and transitions) to the global state space of the automata network.

Our third constraint demands that the sets of local states of all automata in a network are pairwise disjoint, thus forbidding the possibility of having shared states between two or more automata. Such a criterion draws a clear separation line between the various models proposed in the literature. For instance, *open nets* [19] rely on shared places for input/output communications between components. Also, among the models surveyed in [4, Sect. 6–7], six models (*synchronized state machines, state machine decomposable nets, state machine allocatable nets, proper nets, strict free choice nets,* and *medium composable nets*) allow shared states, while two models (*superposed automata nets* and *basic modular Petri nets*) forbid shared states. We opt for the latter approach, which provides a sound model of concurrency (concurrent automata are likely to be executed on different processors and, thus, should not have shared states) and which is mathematically simpler (the local states of all automata form a partition of the set of states). It is worth mentioning that the decomposition approach described in [27] first generates a decomposition with shared states, but later gets rid of these by introducing auxiliary states (noted "NOP"); while we agree with the goal of having pairwise disjoint local state sets, we cannot reuse the same approach, as the addition of extra states in the automata network would violate the one-to-one mapping required by our second constraint.

Our fourth constraint is that a suitable output formalism should be general enough to support, without undue restrictions, all input Petri nets that are ordinary and safe. For instance, among the eight aforementioned models of [4, Sect. 6–7], three models (*state machine decomposable nets, state machine allocatable nets,* and *strict free choice nets*) require each automaton to be strongly connected; such a restriction obviously hinders decomposition, as a very simple net with two places and a transition between these places cannot be expressed using a single strongly connected automaton. Another frequent, yet questionable, restriction is the requirement that automata should be state machines, i.e., always have a token in any global state; all the eight aforementioned models of [4, Sect. 6–7] have this restriction, which, we believe, is unsuitable: a very simple net with a single place and a transition going out of this place cannot be represented as a state machine, unless an extra state is added to the post-set of the transition, thus violating our second constraint; more generally, decomposition into state machines assumes that the input Petri net is conservative (i.e., each transition has the same number of input and output places), where Table 1 indicates that only 16.7% of models satisfy this condition in practice; one must therefore consider a more flexible model, in which automata can be started and halted dynamically, meaning that each automaton does not necessary have a token in the initial state, and that it may lose its token in the course of its execution.

Eventually, the suitable output formalism that matches the four above constraints turns out to be a *flat, unit-safe NUPN*, i.e., nothing else than the input Petri net model augmented with a partition of the set of places into *units*, each featuring an automaton, such that, in any reachable marking, at most one place of each unit has a token. If there is more than one unit, an additional (root) unit will be created, which is void and encapsulates all other (leaf) units.

A key advantage of this output formalism is that decomposition can be seen as an operation within the NUPN domain, taking as input a trivial, unit-safe NUPN and producing as output a flat, unit-safe NUPN. This makes definitions simpler, as all places, transitions, and arcs of the initial Petri net are kept unchanged, and all structural and behavioural properties are preserved by the translation. In this article, we have so far carefully avoided the term of "state machines", which implies conservativeness, preferring the more vague term "automata network". In the sequel, we switch to the precise NUPN terminology, referring to "automata" as "(leaf) units" and "local states" as "places".

Implementation and Experimentation. Concretely, our tool chain produces as output a ".nupn" file containing the result of the decomposition. To a large extent, this file is identical to the input ".nupn" file, but contains new information about units. Finally, the output ".nupn" file can easily be translated to standard PNML format using the CÆSAR.BDD tool (with option "-pnml"); the information about units produced by the decomposition is retained and stored in the ".pnml" file using a "toolspecific" section[13] [8, Annex B].

2.4   Existence and Multiplicity of Solutions

Our decomposition problem, as stated above, consists in finding an appropriate set of units (namely, a partition of the set of places) to convert an ordinary, safe Petri net into an "isomorphic" flat, unit-safe NUPN.

Contrary to other decomposition approaches (starting from [12]) that may have no solution for certain classes of input nets, our problem always has at least one solution: the trivial NUPN corresponding to the input Petri net (see [8, proof of Prop. 11] for a formal definition) is flat (because its height is one) and it is unit-safe (because its underlying Petri net is safe [8, Prop. 7]).

---

[13] http://mcc.lip6.fr/nupn.php.

There may exist several solutions for the decomposition problem. For instance, given a valid solution containing a unit with several places, splitting this unit into two separate units also produces a valid solution; also, if this unit contains a dead place, moving this dead place to any other leaf unit also produces a valid solution. In any valid solution, the number of leaf units belongs to an interval $[Min, Max]$, where:

- Min is the largest value, for any reachable marking M, of $card(M)$, i.e., the number of tokens in M. For instance, if the initial marking M_0 contains n places, then any valid solution must have at least n leaf units. A valid solution may have more leaf units than Min; by example, a net with 4 places $p_0, ..., p_3$ (only p_0 is marked initially) and 3 transitions $t_1, ..., t_3$ such that ${}^\bullet t_1 = {}^\bullet t_2 = {}^\bullet t_3 = \{p_0\}$, $t_1{}^\bullet = \{p_2, p_3\}$, $t_2{}^\bullet = \{p_1, p_3\}$, and $t_3{}^\bullet = \{p_1, p_2\}$ has at most 2 tokens but at least 3 leaf units in any valid decomposition.
- Max is the number of places (this follows from item 8 of Definition 3). The upper bound Max is reached by, and only by, the trivial solution. Approaches to obtain an upper bound smaller than Max are discussed in Sect. 4 below.

Implementation and Experimentation. The CÆSAR.BDD tool (with option "-min-concurrency") can quickly compute (an under-approximation of) the value of Min without exploring the reachable marking graph entirely.

2.5   Criteria for Optimal Solutions

Since the decomposition problem usually admits multiple solutions, the next question is to select an "optimal" solution. For instance, the trivial solution is valid, but uninteresting. Various criteria can be used to compare solutions. From a theoretical point of view, and especially for the purpose of formal verification, a suitable criterion is to minimize the number of bits needed to represent any reachable marking. There are different ways of encoding the reachable markings of a safe Petri net. The least compact encoding consists in having one bit per place. The most compact encoding consists in first exploring all reachable markings, then encoding them using $\lceil \log_2 n \rceil$ bits, where n is the number of reachable markings; of course, this encoding is unrealistic, since exploration cannot be done without already having an encoding.

Between these two extremes, one should select an encoding that can be computed without exploring reachable markings and faithfully measures the compactness of each solution. For such purpose, [8, Sect. 6] lists five encodings for unit-safe NUPNs and evaluates their compactness. We base our approach on the most compact encoding of [8], noted (b), which we further enhance by detecting those units that never get a token or never lose their token.

Definition 5. *Let* $N = (P, T, F, M_0, U, u_0, \sqsubseteq, \mathsf{unit})$ *be a NUPN.*

- *Let the* projection *of a marking M on a unit u be:* $M \rhd u \overset{\text{def}}{=} M \cap \mathsf{places}(u)$.
- *A unit u is* idle *if has no token in the initial marking and no transition puts a token in this unit, i.e.,* $(M_0 \rhd u = \varnothing) \wedge (\forall t \in T) (t^\bullet \rhd u \neq \varnothing \Rightarrow {}^\bullet t \rhd u \neq \varnothing)$.

– *A unit u is permanent if has a token in the initial marking and no transition takes the token away from this unit, i.e., $(M_0 \rhd u \neq \varnothing) \wedge (\forall t \in T)$ ($^\bullet t \rhd u \neq \varnothing \Rightarrow t^\bullet \rhd u \neq \varnothing$).*

Notice that, in a flat NUPN, the root unit is always idle, since it is void. Using the encoding (b) of [8, Sect. 6], the state of each unit having n local places can be encoded using $\lceil \log_2(n+1) \rceil$ bits. This number of bits can be further reduced if the unit is idle or permanent.

Definition 6. *Let $N = (P, T, F, M_0, U, u_0, \sqsubseteq, \mathsf{unit})$ be a NUPN. The number of bits needed to represent the markings of N is defined as $\Sigma_{u \in U} \, \nu(u)$, where $\nu(u) = 0$ if u is idle, $\nu(u) = \lceil \log_2(card\,(\mathsf{places}\,(u))) \rceil$ if u is permanent, or $\nu(u) = \lceil \log_2(card\,(\mathsf{places}\,(u)) + 1) \rceil$ otherwise.*

We finally base our comparison criterion upon Definition 6: the smaller the number of bits, the better the decomposition.

Implementation and Experimentation. Given a NUPN obtained by decomposition, its number of bits can be computed using CÆSAR.BDD (option "-bits"). The determination of idle and permanent units is done in linear time, with a simple iteration that examines the pre-set and post-set of each transition.

3   Concurrent Places

3.1   Concurrency Relation

All the decomposition methods presented in this article are based on a *concurrency relation* defined over places, which appears in many scientific publications under various names: *coexistency defined by markings* [15, Sect. 9], *concurrency graph* [27], or *concurrency relation* [9,17,18,25], etc. Mostly identical, these definitions sometimes differ in details, such as the kind of Petri nets considered, or the handling of reflexivity, i.e., whether a place is concurrent with itself or not. We adopt the following definition:

Definition 7. *Let $N = (P, T, F, M_0)$ be a Petri net. Two places p_1 and p_2 are concurrent, noted "$p_1 \| p_2$", iff there exists a reachable marking M such that $p_1 \in M$ and $p_2 \in M$.*

This relation is most relevant to the decomposition problem, as it generates weak positive constraints (if two places are not concurrent, they *may* belong the same unit in the output NUPN) and strong negative constraints (if two places are concurrent, they *must not* belong to the same unit, i.e., $(p_1 \neq p_2) \wedge (p_1 \| p_2) \Rightarrow \mathsf{unit}\,(p_1) \neq \mathsf{unit}\,(p_2)$—otherwise, the output NUPN would not be unit safe).

For certain classes of Petri nets (namely, extended free choice nets that are bounded and live), there exists a polynomial algorithm for computing the concurrency relation [18]. We have not used this algorithm so far, since, in our collection (see Table 1), less than a half of the models are extended free choice, and most models are not strongly connected, thus not live.

3.2   Concurrency Matrix

We concretely represent the concurrency relation as a matrix indexed by places. The cells of this matrix are equal to "1" if the corresponding two places are concurrent, or to "0" if they are not. The matrix is computed using reachability analysis, but the set of reachable markings may be too large to be explored entirely. In such case, certain cells of the matrix may be undefined, a situation we record by giving them an unknown value noted ".". A matrix will be said to be *incomplete* if it contains at least one unknown value, or *complete* otherwise.

Implementation and Experimentation. In our tool chain for decomposition, the computation of the concurrent matrix is performed by CÆSAR.BDD (with option "-concurrent-places"). Several practical issues are faced.

For a large input model, the matrix can get huge (e.g., several gigabytes). This problem is addressed in two ways: (i) since the concurrency relation is symmetric, only the lower half of the matrix is represented; (ii) each line of the matrix is compacted using a run-length compression algorithm.

To fight algorithmic complexity when generating the matrix of a large input model, CÆSAR.BDD combines various techniques. Initially, the entire matrix is initialized to unknown values. Then, a symbolic exploration (using BDDs) of the reachable markings is undertaken, possibly with a user-specified timeout limitation; during this exploration, all the places reached and all the transitions fired are marked as not dead. If the exploration terminates within the allowed time, then all markings are known, so that the matrix is complete. Otherwise, only a subset of all markings is known, which enables one to write "1" in all the matrix cells corresponding to places found present together in some reached marking; then, various techniques are used to decrease the remaining number of unknown values, stopping as soon as this number drops to zero:

- If the input model is non-trivial and contains the pragma "!unit_safe", than all pairs of places in the same unit (resp., in two transitively nested units) are declared to be non-concurrent.
- If the diagonal of the matrix contains unknown values, an auxiliary explicit-state algorithm that computes a subset of dead places is run.
- If a place is dead, then it is not concurrent with any other place.
- If a transition t (dead or not) has a single input place p, then p is not concurrent with any output place of t different from p (otherwise, the net would not be safe).

Although the generation of concurrency matrices is a CPU-intensive operation, we found it to succeed for most models of our collection: 99.9% of all the matrices have been generated; the 12 matrices that could not be generated correspond to large NUPNs, all of which have more than 5556 places, 7868 transitions, and 18,200 arcs. Moreover, 97.3 of all the matrices are complete; the 350 incomplete matrices correspond to large NUPNs, all of which have more than 120 places, 144 transitions, and 720 arcs.

In the sequel, we proceed with the concurrency matrix, which we abstract away by replacing all unknown values by the value "1", thus assuming that, by default (i.e., in absence of positive information), individual places are live and pairs of places are concurrent. Such pessimistic assumptions are essential to the correctness of our decomposition methods: if the matrix is incomplete, the decomposed NUPN will still be unit-safe, although perhaps suboptimal.

3.3   Concurrency and Sequentiality Graphs

Derived from the (abstracted) concurrency matrix, we introduce two definitions, which have been already given by other authors, e.g., [25,27], etc.

Definition 8. *Given a net and its concurrency matrix, the* concurrency graph *is an undirected graph, whose vertices correspond to the places of the net, and such that it exists an edge between two different vertices iff the value of the matrix cell for the corresponding places is equal to "1". There are no self-loops in the concurrency graph.*

The decomposition problem can be formulated as a graph colouring problem on the concurrency graph, where leaf units play the role of *colours*. To ensure unit safeness, any two concurrent places must be put into different units; this amounts to the problem of assigning different colors to any two vertices connected by an edge in the concurrency graph. Then, for each colour, a unit is created, which contains all the vertices (i.e., places) having the same colour.

The concurrency graph is derived from the relation "$\|$"; one can also consider its complement graph (up to self-loops), which is derived from the relation "$\not\|$":

Definition 9. *Given a net and its concurrency matrix, the* sequentiality graph *is an undirected graph, whose vertices correspond to the places of the net, and such that it exists an edge between two different vertices iff the value of the matrix cell for the corresponding places is equal to "0". There are no self-loops in the sequentiality graph.*

4   Solution Search

Bits vs Units. In Sect. 2.5, we have selected the number of bits for encoding reachable markings (see Definition 6) as the most sensible criterion to finely measure the quality of a decomposition. However, this criterion is not easy to express in methods based on graph theory or SAT/SMT solving, because it involves the transcendental function "$\log_2$" and the continuous-discrete conversion function "$\lceil . \rceil$". For this reason, our approaches do not directly focus on reducing the number of bits specified in Definition 6, but target instead another goal that is much simpler to implement: reducing the number of units in the output NUPN. Unfortunately, reducing the number of units does not always coincide with reducing the number of bits. Consider a network with 10 places: a decomposition in 2 permanent units with 5 places each will require 6 bits, whereas another decomposition in 3 permanent units with, respectively, 2, 2, and 6 places will require 5 bits only[14]. However, a statistical analysis and our experiments show that a search oriented towards reducing the number of units also reduces, on average, the number of bits for encoding reachable markings.

Dichotomy vs Linear Search. We have seen above in Sect. 2.4 that all the solutions of the decomposition problem have their number of leaf units n within an interval $[Min, Max]$. Having implemented dichotomy, ascending linear search (i.e., starting from Min and incrementing n until a solution is found), and descending linear search (starting from Max and decrementing n until no solution is found), we observed that the latter is usually more efficient. A key advantage of the latter approach is that it always produces a valid (yet perhaps sub-optimal) solution, even if the search is halted due to some user-specified timeout limitation. Whenever possible, we accelerate the convergence by asking the solver whether it exists a solution having at most n leaf units (rather than a solution having exactly n leaf units). If the solver finds a solution having m leaf units, with $m < n - 1$, the next iteration will use m rather than $n - 1$.

Upper Bound Reduction. Having opted for descending linear search, we now present three approaches for reducing the value of the upper bound Max (i.e., the number of places). This boosts efficiency by restricting the search space, without excluding relevant solutions (the lower bound Min defined in Sect. 2.4 is kept unchanged):

1. As mentioned in Sect. 2.4, dead places, if present, can be put into any leaf unit of the output NUPN, still preserving the unit-safeness property. With respect to our quality criterion based on the number of bits, it would not be wise to create one extra unit to contain all the dead places nor, even worse, one extra unit per dead place. Instead, our approach is to put dead places into "normal" units, taking advantage of *free slots*, i.e., unused bit values in logarithmic encoding. For instance, if a non-permanent unit has n places, there are $m = \lceil \log_2(n+1) \rceil - (n+1)$ free slots, meaning, if m is not zero, that m dead places can be added to this unit without increasing its cost in bits. If there are less free slots than dead places, then we augment the number of free slots by adding one bit (or even more) to the unit having already the largest number of places. Thus, in presence of D dead places, we start by putting these places apart, discarding from the concurrency matrix the corresponding lines and columns, all of which contain only cells equal to "0". We then perform a descending linear search starting from $Max - D$ rather than Max. Finally, we distribute the dead places into the resulting units as explained above.
2. After discarding dead places, one can still start the descending linear search from a lower value than $Max - D$. Let Sum_p be the sum, in the concurrency

---

[14] The same observation holds for non-permanent units, e.g., $4 + 5$ vs $7 + 1 + 1$ places.

matrix[15], of all cell values on the line[16] corresponding to place p. Let Sum be the maximum, for all places p, of Sum_p. Clearly, $Min \leq Sum \leq Max - D$. One can start the search from Sum (rather than $Max - D$) since no solution has more than Sum leaf units. Indeed, Sum is equal to $\Delta + 1$, where Δ is the maximum degree of the concurrency graph[17] (the increment "+1" corresponds to the diagonal cell, whose value is always "1") and, from Brooks' theorem (extended to possibly non-connected graphs), the number of colours needed for the concurrency graph is at most $\Delta + 1$, so there exists at least one solution having at most Sum leaf units.

3. If the concurrency matrix was initially complete, and if the input NUPN contains the "!unit_safe" pragma, let W be the width of the input NUPN. If this NUPN is not flat, one can modify it by keeping only its root unit and leaf units, and by moving, for each non-leaf unit u, the local places of u into any leaf unit nested in u; clearly, the modified NUPN is flat, unit safe[18], and has also width W. So, there exists at least one solution with W leaf units. Thus, the descending linear search can be started from $min(W, Sum - D)$ rather than $Sum - D$, excluding the potential solutions having more leaf units than the input NUPN.

5 Methods Based on Graph Colouring

As stated in Sect. 3.3, the decomposition problem can be expressed as a colouring problem on the concurrency graph (see Definition 8). Thus, the simplest method to perform decomposition is to invoke a graph coloring tool, keeping in mind that graph colouring is an NP-complete problem. Moreover, if the concurrency matrix was initially complete, the minimal number of colours (i.e., the chromatic number of the graph) gives a decomposition with the smallest number of leaf units.

Implementation and Experimentation. We decided to use the Color6 software[19], which is developed at Université de Picardie Jules Verne (France) and is one of the most recent tools for graph colouring. Since Color6 does not necessarily compute an optimal solution (i.e., with the chromatic number) but instead returns a solution having at most a number of colors specified by the user, the tool must be invoked repeatedly, using the linear decreasing search strategy described in 4.

We developed Bourne-shell and Awk scripts that convert the concurrency graph into standard DIMACS format, invoke Color6 iteratively using descending linear search, parse the output of Color6, and finally assign places to units upon

---

[15] With or without dead places—this does not change the result since, given that $M_0 \neq \varnothing$, there is always at least one place that is not dead.

[16] Or column, since the concurrency matrix is symmetric.

[17] With or without dead places.

[18] Based upon the general definition of unit safeness [8, Sect. 3] for non-flat NUPNs.

[19] https://home.mis.u-picardie.fr/~cli/EnglishPage.html.

completion of the iterations. The results obtained are presented, as for all other decomposition methods, in Sect. 9.

We also experimented with dichotomy and ascending linear search, which we found, on average, 8% and 18% slower than descending linear search; indeed, Color6 often takes much more time to conclude there is no solution (i.e., when given a number of colors smaller than the chromatic number) than to find a solution when it exists.

6   Methods Based on Maximal Cliques

There is another (and, up to our knowledge, novel) approach to the decomposition problem. In the sequentiality graph (see Definition 9), a *clique* is a set of vertices that are pairwise connected (i.e., a complete subgraph), meaning that a clique corresponds to a potential unit, i.e., a set of places having at most one token in any reachable marking.

Colouring the concurrency graph is equivalent to finding, in the sequentiality graph, a *minimal set* of cliques that covers all vertices. Instead, we use a software tool that computes a *maximal clique*, i.e., one single clique containing as many vertices as possible. Although this problem is also NP-complete, it is usually faster to solve in practice than graph colouring.

The tool is invoked repeatedly: at each iteration, a maximal clique is found, from which, a unit is created; the sequentiality graph is then simplified by removing the vertices and edges of the clique, and the next iteration takes place until the sequentiality graph becomes empty. Hence, neither dichotomy nor linear searches apply to the approach described in the present section.

Implementation and Experimentation. We experimented with four maximum-clique tools: MaxCliqueDyn[20], BBMC[21], MaxCliquePara[22]—all developed at Institut Jožef Stefan in Ljubljana (Slovenia), and MoMC[23] developed at Université de Picardie Jules Verne (France). We developed Bourne shell and Awk scripts that generate sequentiality graphs in the DIMACS format, remove cliques from these graphs, invoke the tools, and extract maximal cliques from their output.

7   Methods Based on SAT Solving

We now present methods that encode the constraints of the concurrency matrix as propositional logic formulas passed to a SAT solver, contrary to the methods of Sect. 5 and 6, in which these constraints are expressed using graphs. We believe that the use of SAT solving for the decomposition problem is a novel approach. For efficiency, we generate formulas in Conjunctive Normal Form (CNF), a restriction (natively accepted by most SAT-solvers) of propositional logic.

---

[20] http://insilab.org/maxclique/.

[21] http://commsys.ijs.si/~matjaz/maxclique/BBMC.

[22] http://commsys.ijs.si/~matjaz/maxclique/MaxCliquePara.

[23] https://home.mis.u-picardie.fr/~cli/EnglishPage.html.

When applying decreasing linear search (see Sect. 4), one must produce a formula asking whether it exists a decomposition having at most n units. For each place p and each unit u, we create a propositional variable x_{pu} that is true iff place p belongs to unit u. We then add constraints over these variables: (i) For each unit u and each two places p and p' such that $\#p < \#p'$, where $\#p$ is a bijection from places names to the interval $[1, \text{card}(P)]$, if the cell (p, p') of the concurrency matrix contains the value "1", we add the constraint $\neg x_{pu} \vee \neg x_{p'u}$ to express that two concurrent places cannot be in the same unit; (ii) For each place p, we could add the constraint $\bigvee_u x_{pu}$ to express that p belongs to at least one unit, but this constraint is too loose and allows $n!$ similar solutions, just by permuting unit names; we thus replace the previous constraint by a stricter one that breaks the symmetry between units: for each place p, we add the refined constraint $\bigvee_{1 \le \#u \le min(\#p,n)} x_{pu}$, where $\#u$ is a bijection from unit names to the interval $[1, n]$.

Notice that no constraint requires that each place belongs to one single unit, as such a constraint would generate large formulas (quadratic in the number of units) when using the CNF fragment only. Thus, the SAT solver may compute an overly general answer, which includes invalid decompositions in which a place belongs to several units. We then refine this answer by assigning each of these places to a single unit, also trying to minimize the number of bits corresponding to the chosen valid decomposition. Our approach takes inspiration from the first-fit-decreasing bin-packing algorithm: places are first sorted by increasing numbers of units to which they can belong according to the SAT-solver; then, each place is put into the unit having the most free slots (see Sect. 4) and, in case of equality, the most local places.

Implementation and Experimentation. We experimented with two SAT solvers: MiniSat, which was chosen for its popularity, and CaDiCaL, which solved the most problems during the SAT Race 2019 competition. We developed Python scripts that generate formulas in the DIMACS-CNF format, invoke a SAT solver, parse its outputs, and assign places to units.

8   Methods Based on SMT Solving

We also considered SMT solvers, which accept formulas in richer logics than SAT solvers, namely (fragments of) first-order logic. We encoded the decomposition problem into five (quantifier-free) logic fragments: BV, DT, UFDT, IDL, and UFIDL, which we define below. As in Sect. 7, we perform descending linear search, generating formulas from the concurrency matrix for a given number n of units.

BV corresponds to *quantifier-free bit-vector* logic, which supports fixed-size boolean vectors and logical, relational, and arithmetical operators on these vectors. Our encoding creates, for each place p, a bit vector b_p of length n such that $b_p[u]$ is true iff place p can belong to unit u. Constraints are then added in the same way as for SAT solving, shifting from a quadratic set of propositional variables to a linear set of bit vectors.

DT corresponds to *quantifier-free data-type* logic, which supports the definition of algebraic data types. Our encoding defines an enumerated type *Unit*, which contains one value per unit; it creates also, for each place p, one variable x_p of type *Unit*. Then, constraints are added in the same way as for SAT solving, replacing each propositional variable x_{pu} by the predicate $x_p = u$, with the difference that our encoding implicitly warrants that each place is assigned to one, and only one, unit.

UFDT corresponds to *quantifier-free uninterpreted-function data-type* logic. Our encoding is based on that of DT but, instead of the x_p variables, defines both an enumerated type *Place*, which contains one value per place, and an uninterpreted function u : *Place* → *Unit*, each occurrence of x_p being replaced with u(p) in the constraints.

IDL corresponds to *quantifier-free integer-difference* logic, which supports integer variables and arithmetic constraints on the difference between two variables. Our encoding is based on that of DT but declares the variables x_p with the integer type instead of *Unit*. Since integers are unbounded, each variable x_p whose place number $\#p$ is greater or equal than n must be constrained by adding $x_p \in \{1, ..., n\}$, since x_p is not subject to a symmetry-break constraint.

UFIDL corresponds to *quantifier-free uninterpreted-function integer-difference* logic. Our encoding is based on that of IDL, with the same changes as for evolving from DT to UFDT. The additional constraint for each place p not subject to symmetry breaking is u$(x_p) \in \{1, ..., n\}$.

Implementation and Experimentation. We experimented with four SMT solvers: Z3 and CVC4, which are general enough to support all the aforementioned logic fragments, as well as Boolector and Yices, which support fewer fragments but were among the two fastest solvers in their respective "single query tracks" of the SMT-COMP 2019 competition. We developed Python scripts to generate formulas in the standard SMT-LIB 2 format, invoke the SMT solvers, analyze their output, and produce the decomposition.

We have 14 combinations (logic fragment, solver), to which we add a 15th combination by processing the IDL fragment with the linear-optimization capabilities of Z3: this is done by enriching the IDL formula with a (Z3-specific) directive "`min`" that asks Z3 to compute a solution with the smallest number of units; thus, no iteration is required for this approach, which we note "z3opt".

9   Experiment Results

Validation of Results. We checked each output NUPN systematically to ensure that: (i) it is syntactically and semantically correct, by running CÆSAR.BDD (with option "`-check`"); (ii) it is presumably unit safe, by checking that none of its units contains two places declared to be concurrent in the concurrency matrix; if the concurrency matrix was initially complete, this guarantees unit safeness; (iii) if the concurrency matrix was initially incomplete, we also check that the output NUPN is presumably unit safe, by exploring (part of) its marking

Table 3. Comparative results of our 22 decomposition methods

Decompos. method	Success	Failures	Timeouts	Total time (HH:MM:SS)	Bit ratio Total	Mean	Unit ratio Total	Mean
col-color6	97.8%	13	338	14:20:01	76.4%	60.0%	71.9%	28.3%
clq-bbmc	99.1%	15	170	17:48:34	66.2%	57.7%	61.7%	27.8%
clq-maxcl	99.2%	15	157	17:11:48	68.1%	57.8%	63.5%	28.1%
clq-mcqd	99.1%	15	178	18:26:56	69.9%	57.8%	65.5%	27.9%
clq-momc	98.7%	16	213	19:00:01	73.7%	58.3%	69.0%	28.5%
sat-cadical	96.2%	12	697	26:57:30	73.1%	59.9%	69.4%	29.5%
sat-minisat	95.5%	12	795	31:59:00	77.9%	61.1%	74.3%	30.1%
smt-bv-boolector	95.1%	12	728	29:34:34	77.4%	60.5%	74.0%	30.4%
smt-bv-cvc4	94.2%	12	855	35:04:50	78.7%	61.4%	75.5%	31.3%
smt-bv-yices	95.4%	12	685	26:05:16	73.7%	61.3%	70.0%	30.1%
smt-bv-z3	94.0%	12	881	37:19:22	78.3%	61.7%	75.1%	31.4%
smt-dt-cvc4	92.9%	12	1104	48:50:21	81.9%	62.7%	79.0%	32.4%
smt-dt-z3	92.7%	12	1100	41:12:21	83.3%	62.4%	80.6%	32.7%
smt-idl-cvc4	92.5%	12	1220	47:59:45	84.6%	63.6%	81.7%	33.5%
smt-idl-yices	93.2%	12	1082	40:52:12	83.7%	63.0%	80.7%	32.2%
smt-idl-z3	95.0%	12	848	33:30:06	80.0%	61.3%	76.5%	30.6%
smt-idl-z3opt	87.3%	12	1921	73:01:03	87.2%	65.9%	85.1%	37.2%
smt-ufdt-cvc4	92.4%	12	1167	44:39:34	84.2%	63.5%	81.5%	33.1%
smt-ufdt-z3	92.9%	12	1078	40:36:39	82.9%	62.4%	80.2%	32.6%
smt-ufidl-cvc4	90.8%	12	1468	57:03:43	85.7%	64.7%	83.1%	34.6%
smt-ufidl-yices	94.9%	12	823	30:41:15	77.5%	60.9%	74.1%	30.6%
smt-ufidl-z3	94.0%	12	952	36:46:32	80.7%	61.8%	77.5%	31.4%

graph, still using CÆSAR.BDD with option "-check", setting a timeout of one minute for this exploration; (iv) the input and the output NUPNs have the same structural and behavioural properties, including those of Table 1 and 2, but the three last lines devoted to units in both tables; this is done by comparing the outputs of CÆSAR.BDD (with option "-mcc") under a timeout of one minute.

Presentation of Results. Table 3 summarize the experimental results for all our decomposition methods listed in column 1 of this table. All the methods use the concurrency matrices, which have been pre-computed (when possible) for each input NUPN model of our collection. Each method was tried on each model, with a timeout of two minutes per model. If a method failed (because, e.g., a model was too large, its concurrency matrix absent, the constraints too complex for the solver, etc.), the trivial solution was taken as the result. If a timeout occurred for a method using descending linear search, the last computed intermediate solution was retained. We produced a large number of output NUPNs (potentially 12,728 input NUPNs × 22 decomposition methods, actually 265,121 output NUPNs). Column 2 (*successes*) gives the percentage of models decomposed by the corresponding method, whether a timeout occurred or not. Column 3 (*failures*) gives the number of models that the method failed to decompose, without occurrence of a timeout. Column 4 (*timeouts*) give the number of models for which a timeout occurred. Column 5 (*total time*) gives the time spent by the method on all models, excluding matrix pre-computation and validation of results. Column 6 (*bit ratio*) measures the quality of the decomposition by giving the quotient of the number of bits in the decomposed model divided by the number of bits in the input model[24]; the "*total*" sub-column gives the quotient between the sums of bits for all models, while the "*mean*" sub-column gives the average of the quotients obtained for each model. Column 7 (*unit ratio*) does the same as Column 6 for the number of (leaf) units instead of bits.

Many observations can be drawn from Table 3. In a nutshell, the maximal-cliques methods perform best: they are among the fastest approaches, with the fewest timeouts, and produce the most compact decompositions—possibly because they manage to handle large models for which the potential gains in bits and units are the most important. Interestingly, our results contradict the claim that BDD size is smaller when units have a *balanced* number of places [25]; one rather observes that the most compact encodings are achieved using maximal cliques, which produce fundamentally *imbalanced* decompositions by computing very large cliques during the first iterations and very small ones at the end.

Comparison with Hippo. We compared our methods with the three decomposition methods ("comparability graphs", "heuristic invariants", and "hypergraph colouring") of Hippo, on all the 223 safe Petri nets available (after removing duplicates) from the Hippo web portal. Our tool chain revealed that several of these nets were incorrect and that the two latter decomposition methods could produce invalid results; we reported these issues, later solved by the Hippo team.

Pursuing our assessment, we measured that the three methods of Hippo took, respectively, 65 seconds, 25 min, and 1 hour 53 min to process the collection of benchmarks. The Hippo portal uses a timeout of nearly 22-minutes, since, for five of the models, [28] reports that decomposition using hypergraphs takes more than one hour. Actually, the three methods could process, respectively, 70.5%, 92.5%, and 91.1% of the collection. In contrast, our tool chain took only 16 seconds to process 100% of the collection, using the Color6 tool on a 9-year old laptop. The decomposed models produced by our tool chain are also more compact, with an average bit ratio of 60.6%, compared to 80.6%, 92.2%, and 76.21% for the three methods of Hippo, respectively.

10   Conclusion

For the decomposition problem, which is nearly 50-year old, we presented various methods to automatically translate an ordinary, safe Petri net into a flat NUPN preserving all the structural and behavioural properties of the input model. For

---

[24] If an input model is non-trivial, it is replaced by the equivalent trivial NUPN.

such purpose, the concepts and results of the NUPN theory [8] have been found remarkably well-adapted.

Our methods have been implemented in a complete tool chain that accepts and produces models in standard PNML format. The tool chain takes advantage of recent advances in graph algorithms and SAT/SMT solvers, and is modular, allowing certain components to be replaced by more efficient ones. The tool chain is helpful to automatically restructure "legacy" Petri nets and "upgrade" them to NUPNs by inferring their (hidden or lost) concurrent structure.

Decomposition is a difficult problem, and several steps in our tool chain are NP-complete or PSPACE-complete; thus, there will always exist arbitrarily large Petri nets that cannot be decomposed. Yet, this is asymptotic complexity, which does not infirm the high success rate (from 87.3% to 99.2%) of our methods assessed on a large collection of 12,728 models from multiple, diverse origins.

References

1. Amparore, E.G., Beccuti, M., Donatelli, S.: Gradient-based variable ordering of decision diagrams for systems with structural units. In: D'Souza, D., Narayan Kumar, K. (eds.) ATVA 2017. LNCS, vol. 10482, pp. 184–200. Springer, Cham (2017). https://doi.org/10.1007/978-3-319-68167-2_13

2. Badouel, E., Caillaud, B., Darondeau, P.: Distributing finite automata through Petri net synthesis. Formal Asp. Comput. **13**(6), 447–470 (2002)

3. Balaguer, S., Chatain, T., Haar, S.: A concurrency-preserving translation from time Petri nets to networks of timed automata. Formal Methods Syst. Des. **40**(3), 330–355 (2012)

4. Bernardinello, L., De Cindio, F.: A survey of basic net models and modular net classes. In: Rozenberg, G. (ed.) Advances in Petri Nets 1992. LNCS, vol. 609, pp. 304–351. Springer, Heidelberg (1992). https://doi.org/10.1007/3-540-55610-9_177

5. Best, E., Darondeau, P.: Petri net distributability. In: Clarke, E., Virbitskaite, I., Voronkov, A. (eds.) PSI 2011. LNCS, vol. 7162, pp. 1–18. Springer, Heidelberg (2012). https://doi.org/10.1007/978-3-642-29709-0_1

6. Desel, J., Esparza, J.: Free Choice Petri nets. Cambridge University Press, Cambridge (1995)

7. Eshuis, R.: Translating safe Petri nets to statecharts in a structure-preserving way. In: Cavalcanti, A., Dams, D.R. (eds.) FM 2009. LNCS, vol. 5850, pp. 239–255. Springer, Heidelberg (2009). https://doi.org/10.1007/978-3-642-05089-3_16

8. Garavel, H.: Nested-unit Petri nets. J. Log. Algebraic Method Program. **104**, 60–85 (2019)

9. Garavel, H., Serwe, W.: State space reduction for process algebra specifications. Theor. Comput. Sci. **351**(2), 131–145 (2006)

10. Genest, B., Kuske, D., Muscholl, A.: On communicating automata with bounded channels. Fundam. Inform. **80**(1–3), 147–167 (2007)

11. van Glabbeek, R., Goltz, U., Schicke-Uffmann, J.-W.: On distributability of Petri nets. In: Birkedal, L. (ed.) FoSSaCS 2012. LNCS, vol. 7213, pp. 331–345. Springer, Heidelberg (2012). https://doi.org/10.1007/978-3-642-28729-9_22

12. Hack, M.: Analysis of Production Schemata by Petri Nets, Master thesis (computer science), MIT, Cambridge, MA, USA (1972)

13. Hack, M.: Extended State-Machine Allocatable Nets (ESMA), an Extension of Free Choice Petri Net Results. Technical report, 78–1, MIT (1974)
14. ISO/IEC: High-level Petri Nets - Part 2: Transfer Format. IS 15909–2:2011 (2011)
15. Janicki, R.: Nets, Sequential components and concurrency relations. Theor. Comput. Sci. **29**, 87–121 (1984)
16. Karatkevich, A., Andrzejewski, G.: Hierarchical decomposition of Petri nets for digital microsystems design. In: Modern Problems of Radio Engineering, Telecommunications, and Computer Science, pp. 518–521. IEEE (2006)
17. Kovalyov, A.V.: Concurrency relations and the safety problem for Petri nets. In: Jensen, K. (ed.) ICATPN 1992. LNCS, vol. 616, pp. 299–309. Springer, Heidelberg (1992). https://doi.org/10.1007/3-540-55676-1_17
18. Kovalyov, A., Esparza, J.: A polynomial algorithm to compute the concurrency relation of free-choice signal transition graphs. In: Workshop on Discrete Event Systems, pp. 1–6 (1996)
19. Mennicke, S., Oanea, O., Wolf, K.: Decomposition into open nets. In: Algorithmen und Werkzeuge für Petrinetze, pp. 29–34. CEUR-WS.org (2009)
20. Munoz-Gama, J., Carmona, J., van der Aalst, W.M.P.: Hierarchical conformance checking of process models based on event logs. In: Colom, J.-M., Desel, J. (eds.) PETRI NETS 2013. LNCS, vol. 7927, pp. 291–310. Springer, Heidelberg (2013). https://doi.org/10.1007/978-3-642-38697-8_16
21. Pastor, E., Cortadella, J., Peña, M.A.: Structural methods to improve the symbolic analysis of Petri nets. In: Donatelli, S., Kleijn, J. (eds.) ICATPN 1999. LNCS, vol. 1639, pp. 26–45. Springer, Heidelberg (1999). https://doi.org/10.1007/3-540-48745-X_3
22. Petit, A.: Distribution and synchronized automata. Theor. Comput. Sci. **76**(2–3), 285–308 (1990)
23. Rozenberg, G., Engelfriet, J.: Elementary net systems. In: Reisig, W., Rozenberg, G. (eds.) ACPN 1996. LNCS, vol. 1491, pp. 12–121. Springer, Heidelberg (1998). https://doi.org/10.1007/3-540-65306-6_14
24. Schmidt, K.: Using Petri net invariants in state space construction. In: Garavel, H., Hatcliff, J. (eds.) TACAS 2003. LNCS, vol. 2619, pp. 473–488. Springer, Heidelberg (2003). https://doi.org/10.1007/3-540-36577-X_35
25. Semenov, A., Yakovlev, A.: Combining partial orders and symbolic traversal for efficient verification of asynchronous circuits. In: CHDL. IEEE (1995)
26. Starke, P.H.: Analyse von Petri-Netz-Modellen. Teubner, Leitfäden und Monographien der Informatik (1990)
27. Wiśniewski, R., Karatkevich, A., Adamski, M., Costa, A., Gomes, L.: Prototyping of concurrent control systems with application of Petri nets and comparability graphs. IEEE Trans. Control Syst. Technol. **26**(2), 575–586 (2018)
28. Wiśniewski, R., Wiśniewska, M., Jarnut, M.: C-exact hypergraphs in concurrency and sequentiality analyses of cyber-physical systems specified by safe Petri nets. IEEE Access **7**, 13510–13522 (2019)

Data Centric Workflows
for Crowdsourcing

Pierre Bourhis[1], Loïc Hélouët[2(✉)], Zoltan Miklos[3], and Rituraj Singh[3]

[1] CNRS, Univ. Lille, Lille, France
pierre.bourhis@univ-lille.fr
[2] INRIA Rennes, Rennes, France
loic.helouet@inria.fr
[3] Univ. Rennes, Rennes, France
{zoltan.miklos,rituraj.singh}@irisa.fr

Abstract. Crowdsourcing consists in hiring workers on internet to perform large amounts of simple, independent and replicated work units, before assembling the returned results. A challenge to solve intricate problems is to define orchestrations of tasks, and allow higher-order answers where workers can suggest *a process to obtain data* rather than a *plain answer*. Another challenge is to guarantee that an orchestration with correct input data terminates, and produces correct output data. This work proposes *complex workflows*, a data-centric model for crowdsourcing based on orchestration of concurrent tasks and higher order schemes. We consider termination (whether some/all runs of a complex workflow terminate) and correctness (whether some/all runs of a workflow terminate with data satisfying FO requirements). We show that existential termination/correctness are undecidable in general excepted for specifications with bounded recursion. However, universal termination/correctness are decidable when constraints on inputs are specified in a decidable fragment of FO, and are at least in $co-2EXPTIME$.

1 Introduction

Crowdsourcing leverages intelligence of crowd to realize tasks where human skills still outperform machines [26]. It was successful in contributive science initiatives, such as CRUK's Trailblazer [3], Galaxy Zoo [5], etc. Most often, a crowdsourcing project consists in deploying a huge number of tasks that can be handled by humans in a reasonable amount of time. Generally, work units are simple micro-tasks, that are independent, cheap, repetitive, and take a few minutes to an hour to complete. They can be labeling of images, writing scientific blogs, etc. The requester publishes the tasks on a platform with a small incentive (a few cents, reputation gain, goodies ...), and waits for the participation from the crowd. However, many projects, and in particular scientific workflows, can be seen as a coordination of high-level composite tasks. As noted by [39], composite tasks are not or poorly supported by crowdsourcing platforms. Crowdsourcing markets such as Amazon Mechanical Turk [1], Foule Factory [4], CrowdFlower

© Springer Nature Switzerland AG 2020
R. Janicki et al. (Eds.): PETRI NETS 2020, LNCS 12152, pp. 24–45, 2020.
https://doi.org/10.1007/978-3-030-51831-8_2

[2], etc. already propose interfaces and APIs to access crowds, but the specification of crowd based complex processes is still in its infancy. Some works propose solutions for data acquisition and management or deployment of *workflows*, mainly at the level of micro-tasks [18,30]. Crowdforge uses Map-Reduce techniques to solve complex tasks [24]. Turkit [31] builds on an imperative language embedding calls to services of a crowdsourcing platform, which requires programming skills to design complex orchestrations. Turkomatic [27] and [42] implement a Price, Divide and Solve (PDS) approach: crowd workers divide tasks into orchestrations of subtasks, up to the level of micro-tasks. However, current PDS approaches ask clients to monitor workflows executions. In this setting, clients cannot use PDS crowdsourcing solutions as high-level services.

The next stage of crowdsourcing is hence to design more involved processes still relying on the crowd. A first challenge is to fill the gap between a high-level process that a requester wants to realize, and its implementation in terms of micro-tasks composition. Moving from one description level to the other is not easy, and we advocate the use of expertise of the crowd for such refinement. This can be achieved with higher-order answers, allowing a knowledgeable worker to return an orchestration of simpler tasks instead of a crisp answer to a question. A second challenge in this setting is to give guarantees on the termination (whether some/all runs terminate) and correctness of the overall process (whether the output data returned by a crowdsourced process meet some requirements w.r.t input data).

This paper proposes a data-centric model called *complex workflows*, that define orchestrations of tasks with higher-order schemes, and correctness requirements on input and output data of the overall processes. Complex workflows are mainly orchestrations of data transformation tasks, that start from initial input datasets that are explicitly given as crisp data or represented symbolically with a logical formula. Workers can contribute to micro-tasks (input data, add tags ...) or refine tasks at runtime. Some easy tasks (simple SQL queries or record updates) are automated. Input/Output (I/O) requirements are specified with a fragment of FO. We address the question of termination (whether all/some runs terminate) and correctness (whether all/some runs terminate and meet the I/O requirements). Due to higher-order, complex workflows are Turing complete, and hence existence of a terminating run is undecidable. However, one can decide whether **all** runs of a complex workflow terminate as soon as initial data is fixed or specified in an FO fragment where satisfiability is decidable. Existential termination becomes decidable with the same constraints on initial data as soon as recursion in higher-order rewritings is bounded. We then show that existential correctness is undecidable, and universal correctness is decidable if constraints on data are expressed in a decidable fragment of FO. The complexity of termination and correctness depends mainly on the length of runs, and on the complexity of satisfiability for the FO fragment used to specify initial data and I/O requirements. It is at least in (co-)2EXPTIME for the simplest fragments of FO, but can increase to an n-fold complexity for FO fragments that use both existential and universal quantification.

Several formal models have been proposed for orchestration of tasks and verification in data-centric systems or business processes. Workflow nets [40,41], a variant of Petri nets, address the control part of business processes, but data is not central. Orchestration models such as ORC [23] or BPEL [37] can define business processes, but cannot handle dynamic orchestrations and are not made to reason on data. [10] proposes a verification scheme for a workflow model depicting business processes with external calls, but addresses operational semantics issues (whether some "call pattern" can occur) rather than correctness of the computed data. Several variants of Petri nets with data have been proposed. Obviously, colored Petri nets [21] have a sufficient expressive power to model complex workflows. But they are defined at a low detail level, and encoding higher-order necessarily have to be done through complex color manipulations. Complex workflows separate data, orchestration, and data transformations. Several variants of PNs where transitions manipulate global variables exist (see for instance [15]), but they cannot model evolution of unbounded datasets. Other variants of PNs with tokens that carry data have also been proposed in [28]. Termination and boundedness are decidable for this model, but with non-elementary complexity, even when equality is the only predicate used. The closest Petri net variant to complex workflows are Structured Data nets [9], in which tokens carry structured data. Higher order was introduced in *nested* Petri nets [32] a model where places contain sub-nets. Interestingly, termination of nested nets is decidable, showing that higher-order does not imply Turing completeness. Data-centric models and their correctness have also been considered. Guarded Active XML [7] (GAXML) is a model where guarded services are embedded in structured data. Though GAXML does not address explicitly crowdsourcing nor higher-order, the rewritings performed during service calls can be seen as a form of *task* refinement. Restrictions on recursion in GAXML allows for verification of Tree LTL (a variant of LTL where propositions are statements on data). More recently, [8] has proposed a model for collaborative workflows where peers have a local view of a global instance, and collaborate via local updates. With some restrictions, PLTL-FO (LTL-FO with past operators) is decidable. Business artifacts were originally developed by IBM [36], and verification mechanisms for LTL-FO (LTL formula embedding FO statements over universally quantified variables) were proposed in [14,25]. LTL-FO is decidable for subclasses of artifacts with data dependencies and arithmetic. [16] considers systems composed of peers that communicate asynchronously. LTL-FO is decidable for systems with bounded queues. Business artifacts allow for data inputs during the lifetime of an artifact, but are static orchestrations of guarded tasks, described as legal relations on datasets before and after execution of a task. Further, LTL-FO verification focuses mainly on dynamics of systems (termination, reachability) but does not address correctness. [19] considers data-centric dynamic systems (DCDS), i.e. relational databases equipped with guarded actions that can modify their contents, and call external services. Fragments of FO μ-calculus are decidable for DCDS with bounded recursion.

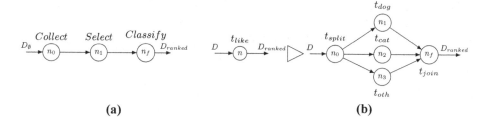

Fig. 1. a) A simple actor popularity poll, and b) a rule example.

2   Motivation

Our objectives are to provide tools to develop applications in which human actors execute tasks or propose solutions to realize a complex task, and check the correctness of the designed services. The envisioned scenario is the following: a client provides a coarse grain workflow depicting important phases of a complex task to process data, and a description of the expected output. The tasks can be completed in several ways, but cannot be fully automated. It is up to a pool of crowdworkers to complete them, possibly after refining difficult tasks up to the level of an orchestration of easy or automated micro-tasks.

The crowdsourcing model presented in Sect. 4 is currently used to design a real example, the SPIPOLL initiative [6]. The objective of SPIPOLL is to study populations of pollinating insects. The standard processes used by SPIPOLL call for experience of large pools of workers to collect, sort and tag large databases containing insects pictures. The features provided by our model fit particularly with the needs of SPIPOLL, as tagging pictures of rare insects is not a simple annotation task, and usually calls for complex validation processes.

Let us illustrate the needs and main features of complex workflows, refinement, and human interactions on a simpler example. A client (for instance a newspaper) wants to rank the most popular actors of the moment, in the categories *comedy*, *drama* and *action* movies. The task is sent to a crowdsourcing platform as a high-level process decomposed into three sequential phases: first a collection of the most popular actors, then a selection of the 50 most cited names, followed by a classification of these actors in comedy/drama/action category. The ranking ends with a vote for each category, that asks contributors to associate a score to each name. The client does not input data to the system, but has some requirements on the output: it is a dataset D_{ranked} with relational schema $r_{out}(name, cites, category, score)$, where *name* is a key, *cites* is an integer that gives the number of cites of an actor, *category* ranges over $\{drama, comedy, action\}$ and *score* is a rational number between 0 and 10. Further, every actor appearing in the final database should have a *score* and a number of *cites* greater than 0. Obviously, there are several ways to collect actors names, several ways to associate a category tag, to vote, etc. However, the clients needs are defined in terms of an orchestration of high-level tasks, without information on how the crowd will complete them, and without any termination guarantee. The orchestration is depicted in Fig. 1-a. It starts from an empty input dataset $D_\emptyset$, and returns an actor popularity dataset D_{ranked}.

In the rest of the paper, we formalize complex workflows and their semantics (Sect. 4). We then address termination (Sect. 5) and correctness (Sect. 6). More precisely, given a complex workflow CW, we consider the following problems:

Universal Termination: Does every run of CW terminate?

Existential Termination: Is there an input for which at least one run of CW terminates?

Universal Correctness: For a given I/O requirement $\psi^{in,out}$ relating input and output data, does every run of CW terminate and satisfy $\psi^{in,out}$?

Existential Correctness: For a given I/O requirement $\psi^{in,out}$, is there a particular input and at least one run of CW that terminates and satisfies $\psi^{in,out}$?

3   Preliminaries

We use a standard relational model [13], i.e., data is organized in *datasets*, that follow *relational schemas*. We assume finite set of domains **dom** = $dom_1, \ldots, dom_s$, a finite set of attribute names **att** and a finite set of relation names **relnames**. Each attribute $a_i \in$ **att** is associated with a domain $dom(a_i) \in$ **dom**. A *relational schema* (or table) is a pair $rs = (rn, A)$, where rn is a relation name and $A \subseteq$ **att** denotes a finite set of attributes. Intuitively, attributes in A are column names in a table, and rn the table name. The *arity* of rs is the size of its attributes set. A *record* of a relational schema $rs = (rn, A)$ is tuple $rn(v_1, \ldots v_{|A|})$ where $v_i \in dom(a_i)$ (it is a row of the table), and a *dataset* with relational schema rs is a multiset of records of rs. A *database schema* DB is a non-empty finite set of tables, and an instance over a database DB maps each table in DB to a dataset.

We address properties of datasets $D_1, \ldots D_k$ with *First Order logic* (FO) and predicates on records. For simplicity and efficiency, we will consider predicates defined as sets of linear inequalities over reals and integer variables. We use FO to define relations between the inputs and outputs of a task, and requirements on possible values of inputs and outputs of a complex workflow.

Definition 1 (First Order). *A* First Order formula *(in prenex normal form) over a set of variables* $\overrightarrow{X} = x_1, \ldots, x_n$ *is a formula of the form* $\varphi ::= \alpha(\overrightarrow{X}).\psi(\overrightarrow{X})$ *where* $\alpha(\overrightarrow{X})$ *is an alternation of quantifiers and variables of* $\overrightarrow{X}$, *i.e. sentences of the form* $\forall x_1 \exists x_2, \ldots$ *called the* prefix *of* φ *and* $\psi(\overrightarrow{X})$ *is a quantifier free formula called the* matrix *of* φ. $\psi(\overrightarrow{X})$ *is a boolean combinations of atoms of the form* $rn_i(x_1, \ldots x_k) \in D_i$, $P_j(x_1, \ldots x_n)$, *where* $rn_i(x_1, \ldots x_k)$*'s are relational statements, and* $P_j(x_1, \ldots x_n)$*'s are boolean function on* $x_1, \ldots x_n$ *called* predicates.

In the rest of the paper, we consider variables that either have finite domains or real valued domains, and predicates specified by simple linear inequalities of dimension 2, and in particular equality of variables, i.e. statements of the form $x_i = x_j$. One can decide in NP if a set of linear inequalities has a valid assignment, and in polynomial time [22] if variables values are reals. Let $\overrightarrow{X_1} = \{x_1, \ldots x_k\} \subseteq \overrightarrow{X}$. We write $\forall \overrightarrow{X_1}$ instead of $\forall x_1.\forall x_2 \ldots \forall x_k$, and $\exists \overrightarrow{X_1}$ instead of $\exists x_1.\exists x_2 \ldots \exists x_k$. We use w.l.o.g. formulas of the form $\forall \overrightarrow{X_1} \exists \overrightarrow{X_2} \ldots \psi(X)$ or $\exists \overrightarrow{X_1} \forall \overrightarrow{X_2} \ldots \psi(X)$, where $\psi(X)$ is a quantifier free matrix, and for every $i \neq j$, $\overrightarrow{X_i} \cap \overrightarrow{X_j} = \emptyset$. Every set of variables $\overrightarrow{X_i}$ is called a *block*. By allowing blocks of arbitrary size, and in particular empty blocks, this notation captures all FO formulas in prenex normal form. We denote by $\varphi_{[t_1/t_2]}$ the formula obtained by replacing every instance of term t_1 in φ by term t_2. Each variable x_i in $\overrightarrow{X}$ has a domain $Dom(x_i)$. A variable assignment is a function μ that associates a value d_x from $Dom(x)$ to each variable $x \in \overrightarrow{X}$.

Definition 2 (Satisfiability). *A variable free formula is* satisfiable *iff it evaluates to true. A formula of the form $\exists x, \phi$ is* satisfiable *iff there exists a value $d_x \in Dom(x)$ such that $\phi_{[x/d_x]}$ is satisfiable. A formula of the form $\forall x, \phi$ is satisfiable iff, for every value $d_x \in Dom(x)$, $\phi_{[x/d_x]}$ is satisfiable.*

It is well known that satisfiability of an FO formula is undecidable in general, but it is decidable for several fragments. The *universal fragment* (resp. existential fragment) of FO is the set of formulas of the form $\forall \overrightarrow{X}.\varphi$ (resp. $\exists \overrightarrow{Y}.\varphi$) where φ is quantifier free. We denote by $\forall$FO the universal fragment of FO and by $\exists$FO the existential fragment. Checking satisfiability of the existential/universal fragment of FO can be done non-deterministically in polynomial time. In our setting, where atoms in FO formula are relational statements and linear inequalities, satisfiability of formulas with only universal or existential quantifiers is decidable and (co)-NP-complete. We refer interested readers to [12] for a proof.

One needs not restrict to existential or universal fragments of FO to get decidability of satisfiability. A well known decidable fragment is (FO^2), that uses only two variables [35]. However, this fragment forbids atoms of arity greater than 2, which is a severe limitation when addressing properties of datasets. The BS fragment of FO is the set of formulas of the form $\exists \overrightarrow{Y_1}.\forall \overrightarrow{X_2}.\psi$, where ψ is quantifier free, may contain predicates, but no equality. The *Bernays-Schonfinkel-Ramsey* fragment of FO [11] (BSR-FO for short) extends the BS fragment by allowing equalities in the matrix ψ. Satisfiability of a formula in the BS or BSR fragment of FO is NEXPTIME-complete (w.r.t. the size of the formula) [29].

Recent results [38] exhibited a new fragment, called the *separated fragment* of FO, defined as follows: Let $Vars(A)$ be the set of variables appearing in an atom A. We say that two sets of variables $Y, Z \subseteq X$ are separated in a quantifier free formula $\phi(X)$ iff for every atom At of $\phi(X)$, $Vars(At) \cap Y = \emptyset$ or $Vars(At) \cap Z = \emptyset$. A formula in the *Separated Fragment* of FO (SF-FO for short) is a formula of the form $\forall \vec{X}_1.\exists \vec{Y}_2.\ldots.\forall \vec{X}_n.\exists \vec{Y}_n\phi$, where $\vec{X}_1 \cdots \cup \vec{X}_n$ and $\vec{Y}_1 \cdots \cup \vec{Y}_n$ are separated. The *SF* fragment is powerful and subsumes the *Monadic Fragment* [33] (where predicates can only be unary) and the BSR fragment. Every separated formula can be rewritten into an equivalent BSR formula (which yields decidability of satisfiability for SF formulas), but at the cost of an n-fold exponential blowup in the size of the original formula. Satisfiability of a separated formula φ is hence decidable [38], but with a complexity in $O(2^{\downarrow n|\varphi|})$.

A recent extension of FO^2 called FO^2BD allows atoms of arbitrary arity, but only formulas over sets of variables where at most two variables have unbounded domain. Interestingly, FO^2BD formulas are closed under computation of weakest preconditions for a set of simple SQL operations [20]. We show in Sect. 4 that conditions needed for a non-terminating execution of a complex workflow are in $\forall FO$, and that $\forall$FO, $\exists$FO, BSR-FO, SF-FO are closed under precondition calculus.

4   Complex Workflows

This section formalizes the notion of *complex workflow*, and gives their semantics through operational rules. This model is inspired by artifacts systems [14], but uses higher-order constructs (task decomposition), and relies on human actors (the *crowdworkers*) to complete tasks. We assume a *client* willing to use the power of crowdsourcing to realize a *complex task* that needs human contribution to collect, annotate, or organize data. This complex task is specified as a workflow that orchestrates elementary tasks or other complex tasks. The client can input data to the system (i.e. enter a dataset D_{in}), and may have a priori knowledge on the relation between the contents of his input and the plausible outputs returned after completion of the complex workflow. This scenario fits several types of applications such as opinion polls, citizen participation, etc. High-level answers of workers are seen as workflow refinements, and elementary tasks realizations are simple operations that transform data. During an execution of a complex workflow, we consider that each worker is engaged in the execution of at most one task.

Tasks: A task t is a work unit designed to transform input data into output data. It can be a high-level description submitted by a client, a very basic atomic task that can be easily realized by a single worker (e.g. tagging images), a task that can be fully automated, or a complex task that requires an orchestration of sub-tasks to reach its objective. We define a set of tasks $\mathcal{T} = \mathcal{T}_{ac} \uplus \mathcal{T}_{cx} \uplus \mathcal{T}_{aut}$ where $\mathcal{T}_{ac}$ is a set of *atomic tasks* that can be completed in one step by a worker, $\mathcal{T}_{cx}$ is a set of *complex tasks* which need to be refined as an orchestration of smaller sub-tasks to produce an output, and $\mathcal{T}_{aut}$ is a set of automated tasks performed by a machine (e.g. simple SQL queries: selection, union, projection ...). Tasks in $\mathcal{T}_{ac}$ and $\mathcal{T}_{aut}$ cannot be refined, and tasks in $\mathcal{T}_{aut}$ do not require contribution of a worker.

Definition 3 (Workflow). *A* workflow *is a labeled acyclic graph* $W = (N, \rightarrow, \lambda)$ *where N is a finite set of* nodes, *modeling occurrences of tasks,* $\rightarrow \subseteq N \times N$ *is a precedence relation, and* $\lambda \colon N \rightarrow \mathcal{T}$ *associates a task name to each node. A node $n \in N$ is a* source *iff it has no predecessor, and a* sink *iff it has no successor.*

In the rest of the paper, we fix a finite set of tasks $\mathcal{T}$, and denote by $\mathcal{W}$ the set of all possible workflows over $\mathcal{T}$. Intuitively, if $(n_1, n_2) \in \longrightarrow$, then an occurrence of task named $\lambda(n_1)$ represented by n_1 must be completed before an occurrence of task named $\lambda(n_2)$ represented by n_2, and the data computed by n_1 is used as an input for n_2. We denote by $min(W)$ the set of *sources* of W, by $succ(n_i)$ the set of successors of a node n_i, and by $pred(n_i)$ its predecessors. The *size* of W is the number of nodes in N and is denoted $|W|$. We assume that when a task in a workflow has several predecessors, its role is to aggregate data provided by preceding tasks, and when a task has several successors, its role is to distribute excerpts from its input datasets to its successors. With this convention, one can model situations where a large database is split into smaller datasets of reasonable sizes that are then processed independently. We denote by $W \backslash \{n_i\}$ the restriction of W to $N \backslash \{n_i\}$, i.e. a workflow from which node n_i is removed along with all edges which origins or goals are node n_i. We assume some well-formedness properties of workflows: Every workflow has a single *sink* node n_f. Informally, we can think of n_f as the task that returns the dataset computed during the execution of the workflow. There exists a path from every node n_i of W to the *sink* n_f. The property prevents from launching tasks which results are never used to build an answer to a client. We define a *higher-order* answer as a *refinement* of a node n in a workflow by another workflow. Intuitively, n is simply replaced by W' in W.

Definition 4 (Refinement). *Let $W = (N, \longrightarrow, \lambda)$ be a workflow, $W' = (N', \longrightarrow', \lambda')$ be a workflow with a unique source node $n'_{src} = min(W')$, a unique sink node n'_f such that $N \cap N' = \emptyset$. The refinement of $n \in N$ by W' in W is the workflow $W_{[n/W']} = (N_{[n/W']}, \longrightarrow_{[n/W']}, \lambda_{[n/W']})$, where $N_{[n/W']} = (N \backslash \{n\}) \cup N'$, $\lambda_{[n/W']}(n) = \lambda(n)$ if $n \in N, \lambda'(n)$ otherwise, and $\longrightarrow_{[n/W']} = \longrightarrow' \cup \{(n_1, n_2) \in \longrightarrow | n_1 \neq n \wedge n_2 \neq n\} \cup \{(n_1, n'_{src}) \mid (n_1, n) \in \longrightarrow\} \cup \{(n'_f, n_2) \mid (n, n_2) \in \longrightarrow\}$*

Definition 5. *A* Complex Workflow *is a tuple $CW = (W_0, \mathcal{T}, \mathcal{U}, sk, \mathcal{R})$ where $\mathcal{T}$ is a set of tasks, $\mathcal{U}$ a finite set of workers, $\mathcal{R} \subseteq \mathcal{T} \times 2^{\mathcal{W}}$ is a set of rewriting rules with $\mathcal{W}$ a finite set of workflows, and $sk \subseteq (\mathcal{U} \times \mathcal{R}) \cup (\mathcal{U} \times \mathcal{T}_{ac})$ defines workers competences. W_0 is an initial workflow, that contains a single source node n_{init} and a single sink node n_f.*

We assume that in every rule $(t, W) \in \mathcal{R}$, the labeling λ of W is injective. This results in no loss of generality, as one can create copies of a task for each node in W, but simplifies proofs and notations afterwards. Further, W has a unique source node $src(W)$. The relation sk specifies which workers have the right to perform or refine a particular task. This encodes a very simple competence model. We refer to [34] for further studies on competences and more elaborated competence models.

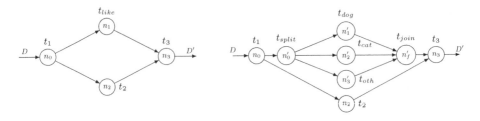

Fig. 2. A refinement of node n_1 using rule (t_{like}, W_{like}) of Fig. 1-b.

Let us consider the example of Fig. 2-left: a workflow contains a complex task t_{like} which objective is to rank large collections of images of different animals with a score between 0 and 10. The relational schema for the dataset D used as input for t_{like} is a collection of records of the form $Picdata(nb, name, kind)$ where nb is a key, $name$ is an identifier for a picture, $kind$ denotes the species obtained from former annotation of data by crowdworkers. Let us assume that a worker u knows how to handle task t_{like} (i.e. $(u, t_{like}) \in sk$), and wants to divide dataset D into three disjoint datasets containing pictures of cats, dogs, and other animals, rank them separately, and aggregate the results. This is captured by the rule $R = (t_{like}, W_{like})$ of Fig. 1-b, where node n_0 is an occurrence of an automated tasks that splits an input dataset into datasets containing pictures of dogs, cats, and other animals, n_1, n_2, n_3 represent tagging tasks for the respective animal kinds, and node n_f is an automated task that aggregates the results obtained after realization of preceding tasks. A higher-order answer of worker u is to apply rule R to refine node n_1 in the original workflow with W_{like}. The result is shown in Fig. 2-right.

It remains to show how automated and simple tasks are realized and process their input data to produce output data. At every node n representing a task $t = \lambda(n)$, the relational schemas of all input (resp. output) datasets are known and denoted $rs_1^{in}, \ldots rs_k^{in}$ (resp. $rs_1^{out}, \ldots rs_k^{out}$). We denote by $\mathcal{D}^{in} = D_1^{in}, \ldots D_k^{in}$ the set of datasets provided by predecessors of t, and by $\mathcal{D}^{out} = D_1^{out} \ldots D_q^{out}$ the set of output datasets computed by task t. During a run of a complex workflow, we allow tasks executions **only for nodes which inputs are not empty**. The contents of every D_i^{out} is the result of one of the operations below:

SQL-LIKE OPERATIONS: We allow standard SQL operations:

- **Selection:** For a given input dataset D_i^{in} with schema $rn(x_1, \ldots, x_n)$ and a predicate $P(x_1, \ldots x_n)$, compute $D_j^{out} = \{rn(x_1, \ldots x_n) \mid rn(x_1, \ldots x_n) \in D_i^{in} \wedge P(x_1, \ldots x_n)\}$.
- **Projection:** For a given input dataset D_i^{in} with schema $rn(x_1, \ldots x_n)$ and an index $k \in 1..n$ compute $D_j^{out} = \{rn(x_1, \ldots x_{k-1}, x_{k+1}, \ldots x_n) \mid rn(x_1, \ldots x_n) \in D_i^{in}\}$.
- **Insertion/deletion:** For a given input dataset D_i^{in} and a fixed tuple $rn(a_1, \ldots a_n)$, compute $D_j^{out} = D_i^{in} \cup \{rn(a_1, \ldots a_n)\}$ (resp. $D_j^{out} = D_i^{in} \setminus \{rn(a_1, \ldots a_n)\}$)

- **Union:** For two input dataset D_i^{in}, D_k^{in} with schema $rn(x_1, \ldots x_n)$, compute $D_j^{out} = D_i^{in} \cup D_k^{in}$
- **Join:** For two input dataset D_i^{in}, D_k^{in} with schema $rn_i(x_1, \ldots x_n)$ and $rn_k(y_1, \ldots y_q)$, for a chosen pair of indices ind_i, ind_k compute $D_j^{out} = \{rn'(x_1, \ldots x_n, y_1, \ldots y_{ind_k-1}, y_{ind_k+1}, \ldots y_q) \mid x_{ind_i} = y_{ind_k} \wedge rn_i(x_1, \ldots x_n) \in D_i^{in} \wedge rn_k(y_1, \ldots y_q) \in D_k^{in}\}$
- **Difference:** For two input dataset D_i^{in}, D_k^{in} with the same schema $rn(x_1, \ldots x_n)$, compute $D_j^{out} = D_i^{in} \setminus D_k^{in}$

WORKERS OPERATIONS: These are elementary tasks performed by workers to modify the datasets computed so far. These operations may perform non-deterministic choices among possible outputs.

- **Field addition:** Given an input dataset D_i^{in} with schema $rn(x_1, \ldots x_n)$, a predicate $P(.)$, compute a dataset D_j^{out} with schema $rn'(x_1, \ldots x_n, x_{n+1})$ such that every tuple $rn(a_1, \ldots, a_n) \in D_i^{in}$ is extended to a tuple $rn(a_1, \ldots, a_n, a_{n+1}) \in D_j^{out}$ such that $P(a_1, \ldots a_{n+1})$ holds. Note that the value of field x_{n+1} can be chosen **non-deterministically** for each record.
- **Record input:** Given an input dataset D_i^{in} with schema $rn(x_1, \ldots x_n)$, and a predicate P, compute a dataset D_j^{out} on the same schema $rn(x_1, \ldots x_n)$ with an additional record $rn(a_1, \ldots, a_n)$ such that $P(a_1, \ldots, a_n)$ holds. Note that the value of fields $x_1, \ldots x_{n+1}$ can be non-deterministically chosen. Intuitively, P defines the set of possible entries in a dataset.
- **Field update:** For each record $rn(a_1 \ldots a_n)$ of D_i^{in}, compute a record $rn(b_1, \ldots b_n)$ in D_j^{out} such that some linear arithmetic predicate $P(a_1, a_n, b_1, \ldots b_n)$ holds. Again, any value for $b_1, \ldots b_n$ that satisfies P can be chosen.

RECORD TO RECORD ARITHMETIC OPERATIONS: For each record $rn_i(a_1 \ldots a_n)$ of D_i^{in}, compute a record in D_j^{out} of the form $rn_j(b_1, \ldots b_q)$ such that each $b_k, k \in 1..q$ is a linear combination of $a_1 \ldots a_n$.

These operations can easily define tasks that split a database D_1^{in} in two datasets D_1^{out}, D_2^{out}, one containing records that satisfy some predicate P and the other one records that do not satisfy P. Similarly, one can describe input of record from a worker, operations that assemble datasets originating from different threads ... To summarize, when a node n with associated task t with I predecessors is executed, we can slightly abuse our notations and write $D_k^{out} = f_{k,t}(D_1^{in}, \ldots D_I^{in})$ when the task performs a deterministic calculus (SQL-based operations or record to record calculus), and $D_k^{out} \in F_{k,t}(D_1^{in}, \ldots D_I^{in})$ when the tasks involves non-deterministic choice of a worker. We now give the operational semantics of complex workflows.

An **execution** starts from the initial workflow W_0 (the initial high-level description provided by a requester). Executing a complex workflow consists in realizing all its tasks following the order given by the dependency relation $\longrightarrow$ in the orchestration, possibly after some refinement steps. At each step of an execution, the remaining part of the workflow to execute, the assignments of tasks to workers and the data input to tasks are memorized in a *configuration*. Execution steps consist in updating configurations according to operational rules. They assign a task to a competent worker, execute an atomic or automated task (i.e. produce output data from input data), or refine a complex task. Executions end when the remaining workflow to execute contains only the final node n_f.

A *worker assignment* for a workflow $W = (N, \longrightarrow, \lambda)$ is a partial map $\mathbf{wa}$: $N \to \mathcal{U}$ that assigns a worker to a node in the workflow. Let $\mathbf{wa}(n) = w_i$. If $\lambda(n)$ is a complex task, then there exists a rule $r = (\lambda(n), W) \in \mathcal{R}$ such that $(w_i, r) \in sk$ (worker w_i knows how to refine task $\lambda(n)$). Similarly, if $\lambda(n)$ is an atomic task, then $(w_i, \lambda(n)) \in sk$ (worker w_i has the competences needed to realize $\lambda(n)$). We furthermore require map $\mathbf{wa}$ to be injective, i.e. a worker is involved in at most one task. We say that $w_i \in \mathcal{U}$ is free if $w_i \notin \mathbf{wa}(N)$. If $\mathbf{wa}(n)$ is not defined, and w_i is a free worker, $\mathbf{wa} \cup \{(n, w_i)\}$ is the map that assigns node n to worker w_i, and is unchanged for other nodes. Similarly, $\mathbf{wa} \setminus \{n\}$ is the restriction of $\mathbf{wa}$ to $N \setminus \{n\}$.

A *data assignment* for a workflow W is a function $\mathbf{Dass} : N \to (DB \uplus \{\emptyset\})^*$, that maps nodes in W to sequence of input datasets. For a node with k predecessors $n_1, \ldots n_k$, we have $\mathbf{Dass}(n) = D_1 \ldots D_k$. A dataset D_i can be empty if n_i has not been executed yet, and hence has produced no data. $\mathbf{Dass}(n)_{[i/X]}$ is the sequence obtained by replacement of D_i by X in $\mathbf{Dass}(n)$.

Definition 6 (Configuration). *A configuration of a complex workflow is a triple $C = (W, \mathbf{wa}, \mathbf{Dass})$ where W is a workflow depicting remaining tasks that have to be completed, $\mathbf{wa}$ is a worker assignment, and $\mathbf{Dass}$ is a data assignment.*

A complex workflow execution starts from the *initial configuration* $C_0 = (W_0, \mathbf{wa}_0, \mathbf{Dass}_0)$, where $\mathbf{wa}_0$ is the empty map, $\mathbf{Dass}_0$ associates dataset D_{in} provided by client to n_{init} and sequences of empty datasets to all other nodes of W_0. A *final configuration* is a configuration $C_f = (W_f, \mathbf{wa}_f, \mathbf{Dass}_f)$ such that W_f contains only node n_f, $\mathbf{wa}_f$ is the empty map, and $\mathbf{Dass}_f(n_f)$ represents the dataset that was assembled during the execution of all nodes preceding n_f and has to be returned to the client. The intuitive understanding of this type of configuration is that n_f needs not be executed, and simply terminates the workflow by returning final output data. Note that due to data assignment, there can be more than one final configuration, and we denote by $\mathcal{C}_f$ the set of all final configurations.

We define the operational semantics of a complex workflow with 4 rules that transform a configuration $C = (W, \mathbf{wa}, \mathbf{Dass})$ in a successor configuration $C' = (W', \mathbf{wa}', \mathbf{Dass}')$. Rule 1 defines task assignments to free workers, Rule 2 defines the execution of an atomic task by a worker, Rule 3 defines the execution of an automated task, and Rule 4 formalizes refinement. We only give an intuitive description of semantic rules and refer reader to [12] for a complete operational semantics.

Rule 1 (WORKER ASSIGNMENT): A worker $u \in \mathcal{U}$ is assigned to a node n. The rule applies if u is free, has the skills required by $t = \lambda(n)$, if t is not an automated task ($t \notin \mathcal{T}_{aut}$) and if node n is not already assigned to a worker. Note that a worker can be assigned to a node even if it does not have input data yet, and is not yet executable. This rule only changes the worker assignment part in a configuration.

Rule 2 (ATOMIC TASK COMPLETION): An atomic task $t = \lambda(n)$ can be executed if node n is *minimal* in current workflow W, it is assigned to a worker $u = \mathbf{wa}(n)$ and its input data $\mathbf{Dass}(n)$ does not contain an empty dataset. Upon completion of task t, worker u publishes the produced data $\mathcal{D}^{out}$ to the succeeding nodes of n in the workflow and becomes available. This rule modifies the workflow part (node n is removed), the worker assignment, and the data assignment (new data is produced and made available to successors of n).

Rule 3 (AUTOMATED TASK COMPLETION): An automated task $t=\lambda(n)$ can be executed if node n is minimal in the workflow and its input data do not contain an empty dataset. The difference with atomic tasks completion is that n is not assigned a worker, and the produced outputs are a deterministic function of task inputs. Rule 3 modifies the workflow part (node n is removed), and the data assignment.

Rule 4 (COMPLEX TASK REFINEMENT): The refinement of a node n with $t = \lambda(n) \in \mathcal{T}_{cx}$ by worker $u = \mathbf{wa}(n)$ uses a refinement rule R_t such that $(u, t) \in sk$, and $R_t = (t, W_t)$. Rule 4 refines node n with workflow $W_t = (N_s, \longrightarrow_s, \lambda_s)$ (see Definition 4). Data originally used as input by n become inputs of the source node of W_t. All other newly inserted nodes have empty input datasets. This rule changes the workflow part of configurations and data assignment accordingly.

We say that there exists a *move* from a configuration C to a configuration C', or equivalently that C' is a successor of configuration C and write $C \rightsquigarrow C'$ whenever there exists a rule that transforms C into C'.

Definition 7 (Run). *A run $\rho = C_0.C_1 \ldots C_k$ of a complex workflow is a finite sequence of configurations such that C_0 is an initial configuration, and for every $i \in 1 \ldots k$, $C_{i-1} \rightsquigarrow C_i$. A run is maximal if C_k has no successor. A maximal run is terminated iff C_k is a final configuration, and it is deadlocked otherwise.*

In the rest of the paper, we denote by $\mathcal{R}uns(CW, D_{in})$ the set of maximal runs originating from *initial configuration $C_0 = (W_0, \mathbf{wa}_0, \mathbf{Dass}_0)$* (where $\mathbf{Dass}_0$ associates dataset D_{in} to node n_{init}). We denote by $\mathcal{R}each(CW, D_{in})$ the set of configurations that can be reached from C_0. Along a run, the size of datasets in use can grow, and the size of the workflow can also increase, due to refinement of tasks. Hence, $\mathcal{R}each(CW, D_{in})$ and $\mathcal{R}uns(CW, D_{in})$ need not be finite. Indeed, complex tasks and their refinement can encode unbounded recursive schemes in which workflow parts or datasets grow up to arbitrary sizes. Even when $\mathcal{R}each(CW, D_{in})$ is finite, a complex workflow may exhibit infinite cyclic behaviors. Hence, without restriction, complex workflows define transitions systems of arbitrary size, with growing data or workflow components, and with cycles.

In Sect. 5, we give an algorithm to check termination of a complex workflow with bounded recursion. Roughly speaking, this algorithms searches a reachable configuration C_{bad} in which emptiness of a dataset D could stop an execution. Once such a configuration is met, it remains to show that the statement $D = \emptyset$ is compatible with the operations performed by the run (insertion, projections, unions of datasets ...) before reaching C_{bad}. This is done by computing backward the weakest preconditions ensuring $D = \emptyset$ along the followed run, and checking their satisfiability. Weakest precondition were introduced in [17] to prove correctness of programs, and used recently to verify web applications with embedded SQL [20].

Definition 8. *Let $C \rightsquigarrow C'$ be a move. Let m be the nature of this move (an automated task realization, a worker assignment, a refinement ...). We denote by $wp[m]\psi$ the weakest precondition required on C such that ψ holds in C' after move m.*

Moves create dependencies among input and output datasets of a task, that are captured as FO properties. Further, the weakest precondition $wp[m]\psi$ of an FO property ψ is also an FO property.

Proposition 1. *Let CW be a complex workflow, r be the maximal arity of relational schemas in CW and ψ be an FO formula. Then for any move m of CW, $wp[m]\psi$ is an effectively computable FO formula, and is of size in $O(r.|\psi|)$.*

A complete proof is available in [12]. We show in Sect. 5 that many properties of complex workflows need restrictions on recursion and forbid SQL difference to be decidable. Proposition 1 does not need such assumption. Weakest preconditions calculus is rather syntactic and is effective for any FO formula and any move. Now, if automated tasks use SQL difference, universal quantifiers can be introduced in existential blocks, leading to weakest preconditions in an FO fragment where satisfiability is undecidable. Interestingly, if automated tasks do not use SQL difference, a weakest precondition is mainly obtained by syntactic replacement of relational statements and changes of variables. It can increase the number of variables, but it does not change the number of quantifier blocks nor their ordering, and does not introduce new quantifiers when replacing an atom. We hence easily obtain:

Corollary 1. *The existential, universal, BSR and SF fragments of FO are closed under calculus of a weakest precondition if tasks do not use SQL difference.*

5   Termination of Complex Workflows

Complex workflows use the knowledge and skills of crowd workers to complete a task starting from input data provided by a client. However, a workflow may never reach a final configuration because data cannot be processed properly by the workflow, or because infinite recursive schemes appear during the execution.

Definition 9 (Deadlock, Termination). *Let CW be a complex workflow, D_{in} be an initial dataset, $\mathcal{D}_{in}$ be a set of datasets. CW terminates existentially on input D_{in} iff there exists a run in $\mathcal{R}uns(CW, D_{in})$ that is terminated. CW terminates universally on input D_{in} iff all runs in $\mathcal{R}uns(CW, D_{in})$ are terminated. Similarly, CW terminates universally on input set $\mathcal{D}_{in}$ iff CW terminates universally on every input $D_{in} \in \mathcal{D}_{in}$, and CW terminates existentially on $\mathcal{D}_{in}$ iff some run of CW terminates for an input $D_{in} \in \mathcal{D}_{in}$.*

When addressing termination for a set of inputs, we describe $\mathcal{D}_{in}$ symbolically with a decidable fragment of FO ($\forall$FO, $\exists$FO, BSR, or SF-FO). Complex workflows are Turing powerful (we show an encoding of a counter machine in [12]). So we get:

Theorem 1. *Existential termination of complex workflows is undecidable.*

An interesting point is that undecidability does not rely on arguments based on the precise contents of datasets (that are ignored by semantic rules). Indeed, execution of tasks only requires non-empty input datasets. Undecidability holds as soon as higher order operations (semantic rule 4) are used.

Universal termination is somehow an easier problem than existential termination. We show in this section that it is indeed decidable for many cases and in particular when the datasets used as inputs of a complex workflow are explicitly given or are specified in a decidable fragment of *FO*. We proceed in several steps. We first define symbolic configurations, i.e. descriptions of the workflow part of configurations decorated with relational schemas depicting data available as input of tasks. We define a successor relation for symbolic configurations. We then identify the class of non-recursive complex workflows, in which the length of executions is bounded by some value $K_{\mathcal{T}_{cx}}$. We show that for a given finite symbolic executions ρ^S, and a description of inputs, one can check whether there exists an execution ρ that coincides with ρ_S. This proof builds on the effectiveness of calculus of weakest preconditions along a particular run (see Proposition 1). Then, showing that a complex workflow does not terminate amounts to proving that it is not recursion-free, or that it has a finite symbolic run which preconditions allow a deadlock.

Definition 10 (Symbolic configuration). *Let $CW = (W_0, \mathcal{T}, \mathcal{U}, sk, \mathcal{R})$ be a complex workflow with database schema DB. A symbolic configuration of CW is a triple $C^S = (W, Ass, Dass^S)$ where $W = (N, \rightarrow, \lambda)$ is a workflow, $Ass : N \rightarrow \mathcal{U}$ assigns workers to nodes, and $Dass^S : N \rightarrow (DB)^*$ associates a list of relational schemas to nodes of the workflow.*

Symbolic configuration describe the status of workflow execution as in standard configurations (see Definition 6) but leaves the data part underspecified. For every node n that is minimal in W, the meaning of $\mathbf{Dass}^S(n) = rs_1, \ldots rs_k$ is that task attached to node n takes as inputs datasets $D_1 \ldots D_k$ where each D_i conforms to relational schema rs_i. For a given symbolic configuration, we can find all rules that apply (there is only a finite number of worker assignments or task executions), and compute successor symbolic configurations. This construction is detailed in [12].

Definition 11 (Deadlocks, Potential deadlocks). *A symbolic configuration* $C^S = (W, Ass, Dass^S)$ *is* final *if* W *consists of a single node* n_f. *It is a* deadlock *if it has no successor. It is a* potential deadlock *iff a task execution can occur from this node, i.e. there exists* $n, \in \min(W)$ *such that* $\lambda(n)$ *is an automated or atomic task.*

A deadlocked symbolic configuration represents a situation where progress is blocked due to shortage of competent workers to execute tasks. A potential deadlock is a symbolic configuration C^S where empty datasets may stop an execution. This deadlock is *potential* because $Dass^S$ does not indicate whether a particular dataset D_i is empty. We show in this section that one can decide whether a potential deadlock situation in C^S represents a real and reachable deadlock, by considering how the contents of dataset D_i is forged along the execution leading to C^S.

Definition 12 (Symbolic run). *A symbolic run is a sequence* $\rho^S = C_0^S \xrightarrow{m_1} C_1^S \xrightarrow{m_2} \dots \xrightarrow{m_k} C_k^S$ *where each* C_i^S *is a symbolic configuration,* C_{i+1} *is a successor of* C_i, *and* $C_0^S = (W_0, Ass_0, \mathbf{Dass}^S)$ *where* W_0, Ass_0 *have the same meaning as for configurations, and* $Dass_0^S$ *associates to the minimal node* n_0 *in* W_0 *the relational schema of* $Dass_0(n_0)$.

One can associate to every execution of a complex workflow $\rho = C_0 \xrightarrow{m_1} C_1 \dots C_k$ a symbolic execution $\rho^S = C_0^S \xrightarrow{m_1} C_1^S \xrightarrow{m_2} \dots \xrightarrow{m_k} C_k^S$ called its *signature* by replacing data assignment in each $C_i = (W_i, Ass_i, \mathbf{Dass}_i)$ by a function from each node n to the relational schemas of the datasets in $\mathbf{Dass}_i(n)$. It is not true, however, that every symbolic execution is the signature of an execution of CW, as some moves might not be allowed when a dataset is empty (this can occur for instance when datasets are split, or after a selection). A natural question when considering a symbolic execution ρ^S is whether it is the signature of an actual run of CW. The proposition below shows that the decidability of this question depends on assumptions on input datasets.

Proposition 2. *Let* CW *be a complex workflow,* D_{in} *be a dataset,* $\mathcal{D}_{in}$ *be an FO formula with* n_{in} *variables, and* $\rho^S = C_0^S \dots C_i^S$ *be a symbolic run. If tasks do not use SQL difference, then deciding if there exists a run* ρ *with input dataset* D_{in} *and signature* ρ^S *is in* $2EXPTIME$. *Checking if there exists a run* ρ *of* CW *and an input dataset* D_{in} *that satisfies* $\mathcal{D}_{in}$ *with signature* ρ^S *is undecidable in general. If tasks do not use SQL difference, then it is in*

- $2EXPTIME$ *if* $\mathcal{D}_{in}$ *is in* $\exists FO$ *and* $3EXPTIME$ *if* $\mathcal{D}_{in}$ *is in* $\forall FO$ *or BSR-FO.*
- $n_{in} - foldEXPTIME$ *where* n_{in} *is the size of the formula describing* $\mathcal{D}_{in}$ *if* $\mathcal{D}_{in}$ *is in SF-FO*

A proof of this Proposition is given in [12]. It does not yet give an algorithm to check termination, as a complex workflows may have an infinite set of runs and runs of unbounded length. However, semantic rules 1, 2, 3 can be used a bounded number of times from a configuration (they decrease the number of available users or remaining tasks in W). Unboundedness then comes from the refinement rule 4, that implements recursive rewriting schemes.

Definition 13. *Let t be a complex task. We denote by $Rep(t)$ the task names that can appear when refining task t, i.e. $Rep(t) = \{t' \mid \exists u, W, (u, t) \in sk \wedge (t, W) \in \mathcal{R} \wedge t' \in \lambda(N_W)\}$. The rewriting graph of a complex workflow $CW = (W_0, \mathcal{T}, \mathcal{U}, sk, \mathcal{R})$ is the graph $RG(CW) = (\mathcal{T}_{cx}, \longrightarrow_R)$ where $(t_1, t_2) \in \longrightarrow_R$ iff $t_2 \in Rep(t_1)$. Then CW is recursion-free if there is no cycle of $RG(CW)$ that is accessible from a task appearing in W_0.*

When a complex workflow is not recursion free, then some executions may exhibit infinite behaviors in which some task t_i is refined infinitely often. Such an infinite rewriting loop can contain a deadlock. In this case, the complex workflow does not terminate. If this infinite rewriting loop does not contain deadlocks, the complex workflow execution will never reach a final configuration in which all tasks have been executed. Hence, complex workflows that are not recursion-free do not terminate universally (see the proof of Proposition 3 for further explanations). Recursion-freeness is decidable and in $O(|\mathcal{T}_{cx}^2| + |\mathcal{R}|)$ (see [12] for details). Further, letting d denote the maximal number of complex tasks appearing in a rule, there exists a bound $K_{\mathcal{T}_{cx}} \leq d^{|\mathcal{T}_{cx}|}$ on the size of W in a configuration, and the length of a (symbolic) execution of CW is in $O(3.K_{\mathcal{T}_{cx}})$ (see [12]). We now characterize complex workflows that terminate universally.

Proposition 3. *A complex workflow terminates universally if and only if: i) it is recursion free, ii) it has no (symbolic) deadlocked execution, iii) there exists no run with signature $C_0^S \dots C_i^S$ where C_i^S is a potential deadlock, with $D_k = \emptyset$ for some $D_k \in \mathbf{Dass}(n_j)$ and for some minimal node n_j of W_i.*

Condition i) can be easily verified (see [12] for a proof). If i) holds, then there is a bound $3.K_{\mathcal{T}_{cx}}$ on the length of all symbolic executions, and checking condition ii) is an exploration of reachable symbolic configurations to find deadlocks. It requires $\log(3.K_{\mathcal{T}_{cx}}).K_{\mathcal{T}_{cx}}$ space, i.e. can be done in EXPTIME (w.r.t. $K_{\mathcal{T}_{cx}}$). Checking condition iii) is more involved. First, it requires finding a potential deadlock, i.e. a reachable symbolic execution C_i^S with a minimal node n_j representing a task to execute. Then one has to check if there exists an actual run $\rho = C_0 \dots C_i$ with signature $C_0^S \dots C_i^S$ such that one of the input datasets D_k in sequence $\mathbf{Dass}_i(n_j)$ is empty. Let $rs_j = (rn_k, A_k)$ be the relational schema of D_k, with $A_j = \{a_1, \dots, a_{|A_j|}\}$. Then, emptiness of D_k can be encoded as a universal FO formula of the form $\psi_i := \forall x_1, \dots x_{|A_j|}, rn_k(x_1, \dots x_{|A_j|}) \notin D_k$. Then all weakest preconditions needed to reach C_i with $D_k = \emptyset$ can be computed iteratively, and remain in the universal fragment of FO (see Proposition 1). If a precondition ψ_j computed this way (at step $j < i$) is unsatisfiable, then there is no actual run with signature ρ^S such that D_k is an empty dataset. If one ends weakest precondition calculus on configuration C_0^S with a condition ψ_0 that is satisfiable, and $D_{in} \models \psi_0$, then such a run exists and $iii)$ does not hold. Similarly with inputs described by $\mathcal{D}_{in}$, if $\mathcal{D}_{in} \wedge \psi_0$ is satisfiable, then condition $iii)$ does not hold. Note that this supposes that $\mathcal{D}_{in}$ is written in a decidable FO fragment.

Theorem 2. *Let CW be a complex workflow, in which tasks do not use SQL difference. Let D_{in} be an input dataset, and $\mathcal{D}_{in}$ be an FO formula. Universal termination of CW on input D_{in} is in $co-2EXPTIME$. Universal termination on inputs that satisfy $\mathcal{D}_{in}$ is undecidable in general. It is in*

- *$co - 2EXPTIME$ (in K, the length of runs) if $\mathcal{D}_{in}$ is in $\forall FO$, $co - 3EXPTIME$ if $\mathcal{D}_{in}$ is $\exists FO$ or $BSR\text{-}FO$*
- *$co - n_{in}\text{-}fold\text{-}EXPTIME$, where $n_{in} = |\mathcal{D}_{in}| + 2^K$ if $\mathcal{D}_{in}$ is in $SF\text{-}FO$.*

One can notice that the algorithm to check universal termination of CWs guesses a path, and is hence non-deterministic. In the worst case, one may have to explore *all* symbolic executions of size at most $3 \cdot K_{\mathcal{T}_{cx}}$.

Undecidability of existential termination has several consequences: As complex workflows are Turing complete, automatic verification of properties such as reachability, coverability, boundedness of datasets, or more involved properties written in logics such as LTL FO [14] are also undecidable. However, the counter machine encoding in the proof of Theorem 1 uses recursive rewriting schemes. A natural question is then whether existential termination is decidable when all runs have a length bounded by some integer $K_{\max}$. Indeed, when such a bound exists it is sufficient to find a terminating witness, i.e. find a signature ρ^S of size $K_{\max}$ ending on a final configuration non-deterministically, compute weakest preconditions $\psi_{|\rho^S|}, \ldots, \psi_0$ and check their satisfiability. The length of ψ_0 is in $O(r^{K_{\max}})$, and as one has to verify non-emptiness of datasets used as inputs of tasks to allow execution of a run with signature ρ^S, the preconditions are in the existential fragment of FO, yielding a satisfiability of preconditions in $O(2^{r^{K_{\max}}})$.

Theorem 3. *Let CW be complex workflow in which tasks do not use SQL difference, and which runs are of length $\leq K_{\max}$. Let D_{in} be a dataset, and $\mathcal{D}_{in}$ a FO formula. One can decide in $2 - EXPTIME$ (in $K_{\max}$) whether CW terminates existentially on input D_{in}. If $\mathcal{D}_{in}$ is in $\exists FO$, termination of CW is also in $2 - EXPTIME$. It is in $3 - EXPTIME$ when $\mathcal{D}_{in}$ is in $\forall FO$ or $BSR\text{-}FO$*

Recursion-free CWs have runs of length bounded by $3.K_{\mathcal{T}_{cx}}$, so existential termination is decidable as soon as automated tasks do not use SQL difference (which makes alternations of quantifiers appear in weakest preconditions). The bound $K_{\mathcal{T}_{cx}}$ can be exponential (see proof of Proposition 2), but in practice, refinements are supposed to transform a complex task into an orchestration of simpler subtasks, and one can expect $K_{\mathcal{T}_{cx}}$ to be a simple polynomial in the number of complex tasks. Another way to bound recursion in a CW is to limit the number of refinements that can occur during an execution. For instance, if each complex task can be decomposed at most k times, the bound on length of runs becomes $K_{\max} = 3 \cdot k \cdot n^2 + 3 \cdot |W_0|$.

6   Correctness of Complex Workflows

Complex workflows provide a service to a client, that inputs some data (a dataset D_{in}) to a complex task, and expects some answer, returned as a dataset D_{out}. A positive answer to a termination question means that the process specified by a complex workflow does not deadlock in some/all executions. Yet, the returned data can still be incorrect. We assume the client sees the crowdsourcing platform as a black box, and simply asks for the realization of a complex task that needs specific competences. However, the client may have requirements on the type of output returned for a particular input. We express this constraint with a FO formula $\psi^{in,out}$ relating inputs and outputs, and extend the notions of existential and universal termination to capture the fact that a complex workflow implements client's needs if some/all runs terminate, and in addition fulfill requirements $\psi^{in,out}$. This is called *correctness*.

Definition 14. *A constraint on inputs and outputs is an FO formula* $\psi^{in,out} ::= \psi_E^{in,out} \wedge \psi_A^{in,out} \wedge \psi_{AE}^{in,out} \wedge \psi_{EA}^{in,out}$, *where*

- $\psi_E^{in,out}$ *is a conjunction of* $\exists FO$ *formulas addressing the contents of the input/output dataset, of the from* $\exists x, y, z, rn(x, y, z) \in D_{in} \wedge P(x, y, z)$, *where* $P(.)$ *is a predicate,*
- $\psi_A^{in,out}$ *is a conjunction of* $\forall FO$ *formulas constraining all tuples of the input/output dataset, of the form* $\forall x, y, z, rn(x, y, z) \in D_{in} \Rightarrow P(x, y, z)$
- $\psi_{AE}^{in,out}$ *is a conjunction of formulas relating the contents of inputs and outputs, of the form* $\forall x, y, z, rn(x, y, z) \in D_{in} \Rightarrow \exists(u, v, t), \varphi(x, y, z, u, v, t)$, *where* φ *is a predicate.*
- $\psi_{EA}^{in,out}$ *is a conjunction of formulas relating the contents of inputs and outputs, of the form* $\exists x, y, z, rn(x, y, z) \in D_{in}, \forall(u, v, t), \varphi(x, y, z, u, v, t)$

The ψ_{AE} part of the I/O constraint can be used to require that every record in an input dataset is tagged in the output. The ψ_{EA} part can be used to specify that the output is a particular record selected from the input dataset (to require correctness of a workflow that implements a vote).

Definition 15 (Correctness). *Let CW be a complex workflow, $\mathcal{D}_{in}$ be a set of input datasets, and $\psi^{in,out}$ be a constraint given by a client. A run in $\mathcal{R}uns(CW, D_{in})$ is correct if it ends in a final configuration and returns a dataset D_{out} such that $D_{in}, D_{out} \models \psi^{in,out}$. CW is existentially correct with inputs $\mathcal{D}_{in}$ iff there exists a correct run $\mathcal{R}uns(CW, D_{in})$ for some $D_{in} \in \mathcal{D}_{in}$. CW is universally correct with inputs $\mathcal{D}_{in}$ iff all runs in $\mathcal{R}uns(CW, D_{in})$ are correct for every $D_{in} \in \mathcal{D}_{in}$.*

In general, termination does not guarantee correctness. A terminated run starting from an input dataset D_{in} may return a dataset D_{out} such that pair D_{in}, D_{out} does not comply with constraint $\psi^{in,out}$ imposed by the client. For instance, a run may terminate with an empty dataset while the client required at least one answer. Similarly, a client may ask all records in the input dataset to appear with an additional tag in the output. If any input record is missing, the output will be considered as incorrect. As for termination, correctness can be handled through symbolic manipulation of datasets, but has to consider constraints that go beyond emptiness of datasets. Weakest preconditions can be effectively computed (Proposition 1): one derives successive formulas $\psi_i^{in,out}, \ldots \psi_0^{in,out}$ between D_{in}, D_{out} and datasets in use at step $i, \ldots 0$ of a run. However, the $\psi_{AE}^{in,out}$ part of formulas is already in an undecidable fragment of FO, so even universal termination is undecidable in general, and even when a bound on the length of runs is known. It becomes decidable only with some restrictions on the fragment of FO used to write $\psi^{in,out}$.

Theorem 4. *Existential and universal correctness of CW are undecidable, even when runs are of bounded length K. If tasks do not use SQL difference, and $\psi^{in,out}$ is in a decidable fragment of FO, then*

- *existential correctness is decidable for CWs with runs of bounded length, and is respectively in $2EXPTIME$, $3EXPTIME$ and 2^K-fold-$EXPTIME$ for the $\exists FO, \forall FO, BSR,$ and SF fragments.*
- *universal correctness is decidable and is respectively in $co - 2EXPTIME$, $co - 3EXPTIME$ and co-2^K-fold-$EXPTIME$ for the $\forall FO, \exists FO, BSR,$ and SF fragments.*

Proof. (Sketch). First, a CW that does not terminate (existentially or universally) cannot be correct, and setting $\psi^{in,out} :: = true$ we get a termination question. So existential correctness is undecidable for any class of input/output constraint. Then, if $\psi^{in,out}$ is in a decidable fragment of FO, and operations do not use SQL difference, then weakest preconditions preserve this fragment, and we can apply the algorithm used for Theorem 2 and 3, starting from precondition $\psi_{|\rho^S|} = \psi^{in,out}$. □

Restricting constraints to $\exists FO, \forall FO, BSR,$ or SF-FO can be seen as a limitation. However, $\exists FO$ can already express non-emptiness properties: $\exists x_1, \ldots, \exists x_k, \; rn(x_1, \ldots, x_k) \in D_{out}$ says that the output should contain at least one record. Now, to impose that every input is processed correctly, one needs a formula of the form $\psi_{in,out}^{valid} :: = \forall x_1, \ldots, x_k, rn(x_1, \ldots x_k) \in D_{in} \implies \exists y_1, \ldots, y_q, rn(x_1, \ldots x_k, y_1, \ldots y_q) \in D_{out} \wedge P(x_1, \ldots x_k, y_1, \ldots y_q)$, that asks that every input in D_{in} appears in the output, and $P()$ describes correct outputs. Clearly, $\psi_{in,out}^{valid}$ is not in the separated fragment of FO. We can decide correctness for formulas $\psi_{AE}^{in,out}$ of the form $\forall \overrightarrow{X_1} \exists \overrightarrow{Y_2}, \varphi$ as soon as every atom in φ that is not separated contains only existential variables that take values from a finite domain. Then $\psi_{AE}^{in,out}$ can be transformed into an $\forall FO$ formula which matrix is a boolean combination of separated atoms.

7   Conclusion

We have proposed *complex workflows*, a model for crowdsourcing applications which enables intricate data centric processes built on higher order schemes and studied their termination and correctness. Unsurprisingly, termination of a complex workflow is undecidable. Now the question of whether *all* runs terminate can be answered when the initial data is specified in a fragment of FO for which satisfiability is decidable. Similar remarks apply to correctness. Table 1 below summarizes the complexities of termination and correctness for static complex workflows (without higher order answer) or with bounded recursion, and for generic workflows with higher order. We consider complexity of termination and correctness for different decidable FO fragments. The (co)-2EXPTIME bound for the fragments with the lowest complexity mainly comes from the exponential size of the formula depicting preconditions that must hold at initial configuration (the EXPTIME complexity is in the maximal length of runs). This can be see as an untractable complexity, but one can expect depth of recursion to be quite low, or even enforce such a depth.

Table 1. Complexity of termination and correctness ($EXPT$ stands for EXPTIME).

Workflow Type	FO Fragment (for $\mathcal{D}_{in}$ or $\psi^{in,out}$)	Problems & Complexity (no SQL diff.)			
		Existential Termination	Universal Termination	Existential Correctness	Universal Correctness
Static, Recursive Bounded	FO	Undecidable	Undecidable	Undecidable	Undecidable
	$\exists^*$ ($\forall^*$ if univ. PB)	$2EXPT$	$co-2EXPT$	$2EXPT$	$co-2EXP$
	BSR,$\forall^*$($\exists^*$ if univ. PB)	$3EXPT$	$co-3EXPT$	$3EXPT$	$co-3EXPT$
	SF	$n_{in}-foldEXPT$	$co-n_{in}$-fold-$EXPT$	$2^{K_{Tcx}}$-fold-$EXPT$	co-$2^{K_{Tcx}}$-fold-$EXPT$
Recursive Unbounded	FO	Undecidable	Undecidable	Undecidable	Undecidable
	$\exists^*$ ($\forall^*$ if univ. PB)	Undecidable	$co-2EXPT$	Undecidable	$co-2EXPT$
	BSR,$\forall^*$($\exists^*$ if univ. PB)	Undecidable	$co-3EXPT(K)$	Undecidable	$co-3EXPT$
	SF	Undecidable	$co-n_{in}-foldEXPT$	Undecidable	co-$2^{K_{Tcx}}$-fold-$EXPT$

Several questions remain open: So far, we do not know whether the complexity bounds are sharp. Beyond complexity issues, crowdsourcing relies heavily on incentives to make sure that a task progresses. Placing appropriate incentives to optimize the overall cost of a complex workflow and ensure progress in an important topic. Similarly, a crux in crowdsourcing is monitoring, in particular to propose tasks to the most competent workers. Cost optimization and monitoring can be addressed as a kind of quantitative game. Other research directions deal with the representation and management of imprecision. So far, there is no measure of trust nor plausibility on values input by workers during a complex workflow execution. Equipping domains with such measures is a way to provide control techniques targeting improvement of trust in answers returned by a complex workflow, and trade-offs between performance and accuracy of answers.

References

1. Amazon's Mechanical Turk. http://www.mturk.com
2. Crowdflower. http://www.crowdflower.com
3. Cruk's trailblazer. http://www.cancerresearchuk.org
4. Foule factory. http://www.foulefactory.com
5. Galaxy zoo. http://zoo1.galaxyzoo.org
6. Spipoll. http://www.spipoll.org
7. Abiteboul, S., Segoufin, L., Vianu, V.: Static analysis of active XML systems. Trans. Database Syst. **34**(4), 23:1–23:44 (2009)
8. Abiteboul, S., Vianu, V.: Collaborative data-driven workflows: think global, act local. In: Proceedings of PODS 2013, pp. 91–102. ACM (2013)
9. Badouel, E., Hélouët, L., Morvan, C.: Petri nets with structured data. Fundam. Inform. **146**(1), 35–82 (2016)
10. Beeri, C., Eyal, A., Kamenkovich, S., Milo, T.: Querying business processes. In: Proceedings of VLDB 2006, pp. 343–354. ACM (2006)
11. Bernays, P., Schönfinkel, M.: Zum entscheidungsproblem der mathematischen logik. Math. Ann. **99**(1), 342–372 (1928)
12. Bourhis, O., Hélouët, L., Miklos, Z., Singh, R.: Data centric workflows for crowdsourcing. Technical report, Univ. Rennes, INRIA, (2019). https://hal.inria.fr/hal-02508838
13. Codd, E.F.: Relational completeness of data base sublanguages. Database Systems, pp. 65–98 (1972)
14. Damaggio, E., Deutsch, A., Vianu, V.: Artifact systems with data dependencies and arithmetic. Trans. Database Syst. **37**(3), 22 (2012)
15. de Leoni, M., Felli, P., Montali, M.: A holistic approach for soundness verification of decision-aware process models. In: Trujillo, J.C., et al. (eds.) ER 2018. LNCS, vol. 11157, pp. 219–235. Springer, Cham (2018). https://doi.org/10.1007/978-3-030-00847-5_17
16. Deutsch, A., Sui, L., Vianu, V., Zhou, D.: Verification of communicating data-driven web services. In: Proceedings of PODS 2006, pp. 90–99. ACM (2006)
17. Dijkstra, E.W.: Guarded commands, nondeterminacy and formal derivation of program. Commun. ACM **18**(8), 453–457 (1975)
18. Garcia-Molina, H., Joglekar, M., Marcus, A., Parameswaran, A., Verroios, V.: Challenges in data crowdsourcing. Trans. Knowl. Data Eng. **28**(4), 901–911 (2016)
19. Hariri, B.B., Calvanese, D., De Giacomo, G., Deutsch, A., Montali, M.: Verification of relational data-centric dynamic systems with external services. In: Proceedings of PODS 2013, pp. 163–174 (2013)
20. Itzhaky, S., et al.: On the automated verification of web applications with embedded SQL. In: Proceedings of ICDT 2017, LIPIcs, vol. 68, pp. 16:1–16:18 (2017)
21. Jensen, K.: Coloured Petri nets: a high level language for system design and analysis. In: Rozenberg, G. (ed.) ICATPN 1989. LNCS, vol. 483, pp. 342–416. Springer, Heidelberg (1991). https://doi.org/10.1007/3-540-53863-1_31
22. Khashiyan, L.G.: Polynomial algorithms in linear programming. U.S.S.R. Comput. Math. Math. Phys. **20**, 51–68 (1980)
23. Kitchin, D., Cook, W.R., Misra, J.: A language for task orchestration and its semantic properties. In: Baier, C., Hermanns, H. (eds.) CONCUR 2006. LNCS, vol. 4137, pp. 477–491. Springer, Heidelberg (2006). https://doi.org/10.1007/11817949_32

24. Kittur, A., Smus, B., Khamkar, S., Kraut, R.E.: Crowdforge: crowdsourcing complex work. In: Proceedings of UIST 2011, pp. 43–52. ACM (2011)
25. Koutsos, A., Vianu, V.: Process-centric views of data-driven business artifacts. J. Comput. Syst. Sci. **86**, 82–107 (2017)
26. Kucherbaev, P., Daniel, F., Tranquillini, S., Marchese, M.: Crowdsourcing processes: a survey of approaches and opportunities. IEEE Internet Comput. **20**(2), 50–56 (2016)
27. Kulkarni, A., Can, M., Hartmann, B.: Collaboratively crowdsourcing workflows with turkomatic. In: Proceedings of CSCW 2012, pp. 1003–1012. ACM (2012)
28. Lazic, R., Newcomb, T., Ouaknine, J., Roscoe, A.W., Worrell, J.: Nets with tokens which carry data. Fundam. Inform. **88**(3), 251–274 (2008)
29. Lewis, H.R.: Complexity results for classes of quantificational formulas. J. Comput. Syst. Sci. **21**(3), 317–353 (1980)
30. Li, G., Wang, J., Zheng, Y., Franklin, M.J.: Crowdsourced data management: a survey. Trans. Knowl. Data Eng. **28**(9), 2296–2319 (2016)
31. Little, G., Chilton, L.B., Goldman, M., Miller, R.C.: Turkit: tools for iterative tasks on mechanical turk. In: Proceedings of HCOMP 2009, pp. 29–30. ACM (2009)
32. Lomazova, I.A., Schnoebelen, P.: Some decidability results for nested Petri nets. In: Bjøner, D., Broy, M., Zamulin, A.V. (eds.) PSI 1999. LNCS, vol. 1755, pp. 208–220. Springer, Heidelberg (2000). https://doi.org/10.1007/3-540-46562-6_18
33. Löwenheim, L.: Über möglichkeiten im relativkalkül. Math. Ann. **76**(4), 447–470 (1915)
34. Mavridis, P., Gross-Amblard, D., Miklós, Z.: Using hierarchical skills for optimized task assignment in knowledge-intensive crowdsourcing. In: Proceedings of WWW 2016, pp. 843–853. ACM (2016)
35. Mortimer, M.: On languages with two variables. Math. Log. Q. **21**(1), 135–140 (1975)
36. Nigam, A., Caswell, N.S.: Business artifacts: an approach to operational specification. IBM Syst. J. **42**(3), 428–445 (2003)
37. OASIS. Web Services Business Process Execution Language. Technical report, OASIS (2007). http://docs.oasis-open.org/wsbpel/2.0/OS/wsbpel-v2.0-OS.pdf
38. Sturm, T., Voigt, M., Weidenbach, C.: Deciding first-order satisfiability when universal and existential variables are separated. In: Proceedings of LICS 2016, pp. 86–95 (2016)
39. Tranquillini, S., Daniel, F., Kucherbaev, P., Casati, F.: Modeling, enacting, and integrating custom crowdsourcing processes. TWEB **9**(2), 7:1–7:43 (2015)
40. van der Aalst, W.M.P., Dumas, M., Gottschalk, F., ter Hofstede, A.H.M., La Rosa, M., Mendling, J.: Correctness-Preserving configuration of business process models. In: Fiadeiro, J.L., Inverardi, P. (eds.) FASE 2008. LNCS, vol. 4961, pp. 46–61. Springer, Heidelberg (2008). https://doi.org/10.1007/978-3-540-78743-3_4
41. Van Der Aalst, W.M.P., et al.: Soundness of workflow nets: classification, decidability, and analysis. Formal Asp. Comput. **23**(3), 333–363 (2011)
42. Zheng, Q., Wang, W., Yu, Y., Pan, M., Shi, X.: Crowdsourcing complex task automatically by workflow technology. In: Cao, J., Liu, J. (eds.) MIPaC 2016. CCIS, vol. 686, pp. 17–30. Springer, Singapore (2017). https://doi.org/10.1007/978-981-10-3996-6_2

Synthesis for Multi-weighted Games with Branching-Time Winning Conditions

Isabella Kaufmann, Kim Guldstrand Larsen, and Jiří Srba[✉]

Department of Computer Science, Aalborg University,
Selma Lagerlofs Vej 300, 9220 Aalborg East, Denmark
srba@cs.aau.dk

Abstract. We investigate the synthesis problem in a quantitative game-theoretic setting with branching-time objectives. The objectives are given in a recursive modal logic with semantics defined over a multi-weighted extension of a Kripke structure where each transition is annotated with multiple nonnegative weights representing quantitative resources such as discrete time, energy and cost. The objectives may express bounds on the accumulation of each resource both in a global scope and in a local scope (on subformulae) utilizing a reset operator. We show that both the model checking problem as well as the synthesis problem are decidable and that the model checking problem is EXPTIME-complete, while the synthesis problem is in 2-EXPTIME and is NEXPTIME-hard. Furthermore, we encode both problems to the calculation of maximal fixed points on dependency graphs, thus achieving on-the-fly algorithms with the possibility of early termination.

1 Introduction

Formal verification is used to ensure that a system model M conforms to a given specification φ, denoted by $M \vDash \varphi$. It relies on having/creating a model of the system and formalising the system's specification in some (temporal) logic. We can distinguish the following verification problems. *Satisfiability*: given a specification φ, does there exist a model M s.t. $M \vDash \varphi$? *Model checking*: given a specification φ and a model M, is it the case that $M \vDash \varphi$? *Synthesis*: given a specification φ and a partial design D, can we construct a controller strategy σ s.t. for the restricted design $D\!\restriction\!\sigma$ we have $D\!\restriction\!\sigma \vDash \varphi$?

We consider the synthesis problem where we are given a specification and some initial (unfinished) design [26]. The goal is then to finish the design s.t. the resulting system satisfies the given specification. The design can model both the behaviour we can control as well as the uncontrollable behaviour that results from the influence of some external environment. In some sense, the synthesis problem can be seen as a generalization of both model checking and satisfiability, as the initial design can express a high degree of freedom for the controller (allowing us to check for satisfiability) or none at all (model checking).

We study the synthesis problem in a *multi-weighted* setting with nonnegative weights. This allows to reason about resources such as discrete time, energy and

© Springer Nature Switzerland AG 2020
R. Janicki et al. (Eds.): PETRI NETS 2020, LNCS 12152, pp. 46–66, 2020.
https://doi.org/10.1007/978-3-030-51831-8_3

(Tasks completed, Time units, Excess heat)

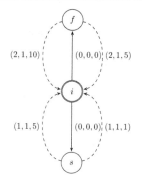

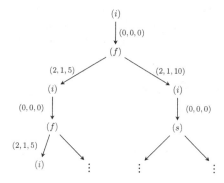

(a) Scheduling game graph   **(b)** Unfolded game according to strategy

Fig. 1. Task scheduling in a heat sensitive environment and the result of applying a strategy to the choices of using fast/slow processor

cost. In contrast to other works, we investigate the problem for specifications expressed in a *branching-time* recursive modal logic where the accumulated cost of an execution in the system can be bounded both globally and locally. The branching nature of the logic allows us to express both possibility (the existential case), certainty (the universal case) or a mixture of both.

To argue for the relevance of the problem, we present a simple motivating example. Consider a processor working on a set of tasks. Depending on a set of outside factors such as the state of the hardware, the duration of the sessions etc. it will vary how long a task takes to finish and how much excess heat it produces. We model a simple version of this scenario in a game graph illustrated in Fig. 1a. In the initial state (i) the processor is idle and does not produce any heat nor complete any tasks. From this state there are two possible actions modelled as transitions annotated with nonnegative weights, representing the number of completed tasks, the time units spent and the amount of excess heat produced. Both of these are controllable (illustrated as solid arrows) and represent a choice we have in either using the regular slow setting (s) or to utilize over-clocking to get a faster result (f). From each of these states there are two arrows leading back to the idle state. These represent the influence of the environment (and are therefore illustrated as dashed arrows) and differ in the produced excess heat.

We see that as a side result of over-clocking, the processor produces more heat than the regular setting (relative to the spent time units). However, we also see that the number of tasks completed in a single time unit is doubled.

We can now use this model to consider optimistic objectives such as "can we possibly (if the environment cooperates i.e. the arriving tasks are computationally easy) complete 4 tasks within 2 time units?". We can also ensure certain worst case invariant properties like "can we ensure that there is never produced more than 15 units of excess heat within 2 time units?". This specific property keeps us from overheating. We are able to express the conjunction of these properties as the branching time specification allows us to reason about both a single branch (possibility) and all branches (certainty).

For these types of specifications, where we are concerned with the use of resources, the correct choice for the controller in any given state of the game will depend on the values of these resources and the possible future involvement of the environment. In our example it is necessary to at least alternate between over-clocking and regular settings to ensure that we never overheat. However this in itself does not give us a productive enough schedule in the best case. To have the possibility of completing 4 tasks within 2 time units, we must begin with over-clocking the processor as the regular settings will use the entire time span on 2 tasks. In the event that the environment picks the transition $f \xrightarrow{(2,1,5)} i$ we can safely pick the fast option again as our strategy. Otherwise, we must choose the slow option not to risk overheating. If we unfold the game according to this informally presented strategy we get a structure like the multi-weighted Kripke structure illustrated in Fig. 1b on which we can verify our objectives.

Our Contributions. We present a recursive modal logic with semantics defined over a nonnegative multi-weighted extension of Kripke structures. We consider this formalism in a game theoretic setting and investigate both the model checking problem (essentially a single player with no choice) and the synthesis problem (two players representing a controller and an uncontrollable environment). For model checking we show that the problem is EXPTIME-complete.

Synthesis in a branching-time setting is challenged by the fact that it is not compositional in the structure of the formula. Consider the game presented in Fig. 1a with a winning condition stating that we are able to reach s before completing the first task (i.e. the first component is equal to 0) and at the same time we are able to reach f before completing the first task. Separately, each subformula has a winning strategy as we can reach either s or f before completing the first task. However, when these two formula are considered in conjunction, there is no winning strategy. As a result, all subformulae in a conjunction must be kept together as we require a uniform control strategy choice for all of them. To deal with this complication, we provide a translation to a suitable weight-annotated normal form and show that the synthesis problem can then be solved in 2-EXPTIME by reducing it to the calculation of the maximal fixed-point assignment on a dependency graph [23]. Last, we provide an NEXPTIME lower-bound for the synthesis problem.

Related Work. One reason to use quantities is to argue about the performance of a system [1,3,4] while another reason is to model and optimize quantitative aspects of a system such as the use of resources. In this paper we consider the latter approach and study a quantitative version of branching-time logic for specifying winning conditions in a weighted game. An obvious choice for such a logic is some variant of weighted CTL. Indeed this type of logic has previously been considered for model checking [2,5,13,15,23] where (some subset) of CTL is extended to express bounds on the accumulated cost of the resources. We have taken inspiration in these logics but have chosen to look at a more general type of logic, namely a modal logic with recursive definitions. Our syntax is inspired by the weighted version of the alternation-free modal μ-calculus presented in [22] which was studied in the context of satisfiability checking.

In regards to synthesis, the problem was for expressive logics like μ-calculus and its sub-logics such as CTL, LTL and CTL* [9,18–20,24,25,28]. However, the synthesis problem is still open for many quantitative extensions of these logics, and to our knowledge there is no work on quantitative branching-time logics in this area. Synthesis has been studied in a weighted setting in the form of energy, mean-payoff and weighted parity games [6,10,11,17,17]. Objectives that allow one to reason (in conjunction) about several complex quantitative requirements on a system has only recently received attention [27].

The main novelty of our work is that we consider a branching-time logic for expressing the specifications, allowing us to capture, at the same time, the quantitative goals for both the optimistic and pessimistic scenarios as demonstrated in our introductory example. Similar objectives are pursued in [8].

2   Preliminaries

An n-Weighted Kripke Structure (n-WKS) is a tuple $K = (S, s_0, \mathcal{AP}, L, T)$ where S is a set of states, $s_0 \in S$ is the initial state, $\mathcal{AP}$ is a set of atomic propositions, $L : S \to \mathcal{P}(\mathcal{AP})$ is a labelling function and $T \subseteq S \times \mathbb{N}^n \times S$ is a transition relation, with a weight vector of n dimensions, where for all $s \in S$ there is some outgoing transition $(s, \bar{c}, s') \in T$. When $(s, \bar{c}, s') \in T$, where $s, s' \in S$ and $\bar{c} \in \mathbb{N}^n$, we write $s \xrightarrow{\bar{c}} s'$. We define the set of outgoing transitions from a state $s \in S$ as $out(s) = \{s \xrightarrow{\bar{c}} s' \in T\}$. For the remainder of the section we fix an n-WKS $K = (S, s_0, \mathcal{AP}, L, T)$ where K is $finite$ i.e. S is a finite set of states and T is a finite transition relation and $\mathcal{AP}$ is a finite set of atomic propositions.

Let $\overline{w} \in \mathbb{N}^n$ be a vector of dimension n. We denote the ith component of $\overline{w}$ by $\overline{w}[i]$, where $1 \leq i \leq n$. To set the ith component of $\overline{w}$ to a specific value $k \in \mathbb{N}$ we write $\overline{w}[i \to k]$ and to set multiple components to a specific value $k \in \mathbb{N}$ we write $\overline{w}[I \to k]$ where $I \subseteq \{1, \ldots, n\}$. Given $\overline{w}, \overline{w}' \in \mathbb{N}^n$, we write $\overline{w} \leq \overline{w}'$ iff $\overline{w}[i] \leq \overline{w}'[i]$ for all $1 \leq i \leq n$.

A run in K is a sequence of states and transitions $\rho = s_0 \xrightarrow{\bar{c}_0} s_1 \xrightarrow{\bar{c}_1} s_2 \xrightarrow{\bar{c}_2} \ldots$ where $s_i \xrightarrow{\bar{c}_i} s_{i+1} \in T$ for all $i \geq 0$. Given a position $i \in \mathbb{N}$ along ρ, let $\rho(i) = s_i$, and $last(\rho)$ be the state at the last position along ρ, if ρ is finite. We also define the concatenation operator $\circ$, s.t. if $(\rho = s_0 \xrightarrow{\bar{c}_0} s_1 \xrightarrow{\bar{c}_1} \ldots \xrightarrow{\bar{c}_{m-1}} s_m)$ then $\rho \circ (s_m \xrightarrow{\bar{c}_m} s_{m+1}) = (s_0 \xrightarrow{\bar{c}_0} s_1 \xrightarrow{\bar{c}_1} s_2 \ldots s_m \xrightarrow{\bar{c}_m} s_{m+1})$. We denote the set of all runs ρ in K of the form $(\rho = s_0 \xrightarrow{\bar{c}_0} s_1 \xrightarrow{\bar{c}_1} \ldots)$ as Π_K. Furthermore, we denote the set of all finite runs ρ in K of the form $(\rho = s_0 \xrightarrow{\bar{c}_0} \ldots \xrightarrow{\bar{c}_{m-1}} s_m)$ as Π_K^{fin}.

Given a run $(\rho = s_0 \xrightarrow{\bar{c}_0} s_1 \xrightarrow{\bar{c}_1} \ldots) \in \Pi_K$, the cost of ρ at position $i \in \mathbb{N}$ is then defined as: $cost_\rho(i) = 0^n$ if $i = 0$ and $cost_\rho(i) = \sum_{j=0}^{i-1} \bar{c}_j$ otherwise. If ρ is finite, we denote $cost(\rho)$ as the cost of the last position along ρ. Lastly, we define a state and cost pair $(s, \overline{w}) \in S \times \mathbb{N}^n$ as a configuration. The set of all configurations in K is denoted $\mathcal{C}_K$.

3   RML and the Model Checking Problem

We can now define our logic.

Definition 1 (RML equation system). *Let $\mathcal{AP}$ be a set of atomic proposi-tions and $\mathcal{V}ar = \{X_0, \ldots, X_m\}$ be a finite set of variables. A Recursive Modal Logic (RML) equations system is a function $\mathcal{E} : \mathcal{V}ar \to \mathcal{F}_{\mathcal{V}ar}$, denoted by $\mathcal{E} = [X_0 = \varphi_0, \ldots, X_m = \varphi_m]$, where $\mathcal{F}_{\mathcal{V}ar}$ is the set of all RML formulae given by:*

$$\varphi \ := \quad \varphi_1 \wedge \varphi_2 \mid \varphi_1 \vee \varphi_2 \mid \beta$$
$$\beta \ := \quad \text{TRUE} \mid \text{FALSE} \mid a \mid \neg a \mid e \bowtie c \mid reset\ R\ in\ EX(X) \mid reset\ R\ in\ AX(X)$$
$$e \ := \quad \#i \mid c \mid e_1 \oplus e_2$$

where $a \in \mathcal{AP}$, $\bowtie \in \{>, \geq, =, \leq, <\}$, $1 \leq i \leq n$ is the index of a vector compo-nent, $R \subseteq \{1, \ldots, n\}$ is a set of indexes of vector components, $c \in \mathbb{N}$, $\oplus \in \{+, \cdot\}$ and $X \in \mathcal{V}ar$ is a variable.

Given a formula $reset\ R\ in\ EX(X)$ (or $reset\ R\ in\ AX(X)$) where $R = \emptyset$ we relieve the notation slightly and simply write $EX(X)$ (or $AX(X)$) instead. We denote the set of all subformulae in a formula φ as $Sub(\varphi)$. Furthermore, we refer to a formula given by the syntactical category β as a basic formula and a formula given by e as an expression.

Remark 1. We limit ourselves to bounds of the form $(e \bowtie c)$ as more expressive bounds of the form $(e_1 \bowtie e_2)$ allows us to simulate a two-counter Minsky machine and thus make the model checking problem undecidable [14].

Given n-WKS K, a variable $X \in \mathcal{V}ar$ is evaluated in an environment $\epsilon : \mathcal{V}ar \to 2^{\mathcal{C}_K}$ assigning a set of configurations $(s, \overline{w}) \in \mathcal{C}_K$ to a variable X. We denote the set of all environments as Env and assume the ordering s.t. for $\epsilon, \epsilon' \in Env$ we have $\epsilon \subseteq \epsilon'$ if $\epsilon(X) \subseteq \epsilon'(X)$ for all $X \in \mathcal{V}ar$. Formally the semantics of an RML formula is a function $\mathcal{F}_{\mathcal{V}ar} \times Env \to 2^{\mathcal{C}_K}$ mapping a RML formula and an environment to a set of configurations. The semantics for a formula φ is thus defined based on an environment ϵ as follows:

$$[\![\varphi_1 \vee \varphi_2]\!]_\epsilon = [\![\varphi_1]\!]_\epsilon \cup [\![\varphi_2]\!]_\epsilon \qquad\qquad [\![\varphi_1 \wedge \varphi_2]\!]_\epsilon = [\![\varphi_1]\!]_\epsilon \cap [\![\varphi_2]\!]_\epsilon$$

and the semantics for a basic formula β is defined as follows:

$$[\![\text{TRUE}]\!]_\epsilon = \mathcal{C}_K \quad [\![\text{FALSE}]\!]_\epsilon = \emptyset$$

$$[\![a]\!]_\epsilon = \{(s,\overline{w}) \in \mathcal{C}_K | a \in L(s)\} \quad [\![\neg a]\!]_\epsilon = \{(s,\overline{w}) \in \mathcal{C}_K | a \notin L(s)\}$$

$$[\![e \bowtie c]\!]_\epsilon = \{(s,\overline{w}) \in \mathcal{C}_K | eval_{\overline{w}}(e) \bowtie c\}$$

$$[\![reset\ R\ in\ EX(X)]\!]_\epsilon = \left\{(s,\overline{w}) \in \mathcal{C}_K \left| \begin{array}{l} \text{there is } (s,\overline{c},s') \in T \text{ s.t.} \\ (s',(\overline{w}[R \to 0] + \overline{c})) \in \epsilon(X) \end{array} \right.\right\}$$

$$[\![reset\ R\ in\ AX(X)]\!]_\epsilon = \left\{(s,\overline{w}) \in \mathcal{C}_K \left| \begin{array}{l} \text{for all } (s,\overline{c},s') \in T \text{ we have} \\ (s',(\overline{w}[R \to 0] + \overline{c})) \in \epsilon(X) \end{array} \right.\right\}$$

where $eval_{\overline{w}}(c) = c$, $eval_{\overline{w}}(\#i) = \overline{w}[i]$ and $eval_{\overline{w}}(e_1 \oplus e_2) = eval_{\overline{w}}(e_1) \oplus eval_{\overline{w}}(e_2)$.

The semantics of an RML equation system is defined by the function $\mathcal{F} : Env \to Env$ where for all $\mathcal{E}(X) = \varphi$ we have $\mathcal{F}(\epsilon)(X) = [\![\varphi]\!]_\epsilon$. By the semantics, we have that $\mathcal{F}$ is monotonic, given the complete lattice formed by $(2^{\mathcal{C}_K}, \subseteq)$. Hence by Knaster-Tarski's theorem we have that there exists a unique maximal fixed point defined as

$$\nu\mathcal{E} = \bigcup\{\epsilon \in Env | \epsilon \subseteq \mathcal{F}(\epsilon)\}.$$

Let K be an n-WKS and $\mathcal{E}$ be an RML equation system. If $(s,\overline{w}) \in [\![\varphi]\!]_{\nu\mathcal{E}}$ we write $(s,\overline{w}) \vDash_K \varphi$. When the n-WKS K is obvious from context, we omit it and simply write $(s,\overline{w}) \vDash \varphi$.

Definition 2. (Model checking problem). *The model checking problem asks, given an n-WKS $K = (S, s_0, \mathcal{AP}, L, T)$, a configuration $(s,\overline{w}) \in \mathcal{C}_K$, an RML equation system $\mathcal{E}$ and an RML formula φ in $\mathcal{E}$ as input, whether $(s,\overline{w}) \vDash_K \varphi$.*

In the remainder of the paper, we assume, unless otherwise indicated, that the model checking problem for an n-WKS K and an RML equation system $\mathcal{E}$ is the question of whether $(s_0, 0^n) \vDash_K \mathcal{E}(X_0)$ where X_0 is the first variable in $\mathcal{E}$.

Remark 2. In our logic, we can encode some instances of reachability, specifically cost-bounded reachability, in an n-WKS where cost in divergent i.e. there are no infinite runs where the cost-component does not increase. Consider a formula that specifies that we can fulfil in all paths some property φ before the first cost-component reaches 10. In a weighted CTL-style syntax, this will be written as $AF_{\#1 \leq 10}\ \varphi$. This is a cost-bounded reachability property and we can encode it as the following equation system $\mathcal{E} = [X = ((AX(X) \vee \varphi) \wedge (\#1 \leq 10))]$. As the environment $\nu\mathcal{E}$ is a maximum fixed point, the satisfaction of this formula may be in general witnessed also by an infinite run satisfying $(\#1 \leq 10)$ but never satisfying φ. However, as all cycles are strictly increasing in the first component, this is not possible and the only way to satisfy the specification is to eventually reach φ within the bound $(\#1 \leq 10)$.

Consider now the following properties used in our motivating example.

1. It is always the case that the produced excess heat never exceeds 15 units within 2 time units.

2. There is a possibility of completing 4 tasks within 2 time units.

We now formalise the conjunction of these objectives as an RML equation system $\mathcal{E}$. Recall that the first component counts the number of completed tasks while the second component measures the time units and the third measures the produced excess heat. First, we express the bound on the excess heat in the formula $p_H = (\#2 > 2 \vee \#3 \leq 15)$ which states that if less than 2 units of time have gone by then the amount of excess heat is not allowed to be larger than 15 units. The first equation enforces p_H in the current state and it ensures that it is always satisfied in the future.

$$\mathcal{E}(X_H) = p_H \wedge AX(X_H) \wedge reset \ \{2,3\} \ in \ AX(X_H)$$

We add the last conjunction $reset \ \{2,3\} \ in \ AX(X_H)$ so that we do not stop checking the property the first time component two exceeded 2 units. By resetting the components every time we take a step, we ensure that this invariant is checked with "fresh values" no matter the cost of any previous steps. Next, we express the possibility of completing 4 tasks within 2 time units by $p_T = (\#1 \geq 4 \wedge \#2 \leq 2)$. We define the second equation

$$\mathcal{E}(X_T) = p_T \vee (EX(X_T) \wedge \#2 \leq 2)$$

where the condition $\#2 \leq 2$ in the second part of the equation ensures that only finite runs which satisfy p_T at some point are considered (see Remark 2). The final property is now formulated as the conjunction of X_H and X_T.

3.1   Finite Representation

Given a finite n-WKS and an RML equation system over the finite set of variables $Var = \{X_0, \ldots, X_m\}$, we only need to consider a finite number of configurations when evaluating a variable X in $\mathcal{E}$.

First, we fix a finite n-WKS $K = (S, s_0, \mathcal{AP}, L, T)$ and an RML equation system $\mathcal{E}$. We define the set of all subformulae in the equation system $\mathcal{E}$ as

$$Sub(\mathcal{E}) = \bigcup_{i \in \{0, \ldots, m\}} Sub(\mathcal{E}(X_i)).$$

Second, let $gb = \max\{c | (e \bowtie c) \in Sub(\mathcal{E})\}$ be the largest bound of $\mathcal{E}$. (Note that if there are no expressions of the type $e \bowtie c$ then the weights can be simply ignored.) This gives an upper-bound to the value of the cost vectors we need to consider. We say that the constant gb is *derived from* $\mathcal{E}$ and based on it we define a function Cut used to limit the number of vectors we need to represent.

Definition 3 (Cut). *Let gb be the constant derived from $\mathcal{E}$. The function Cut : $\mathbb{N}^n \to \mathbb{N}^n$ is then defined for all $1 \leq i \leq n$:*

$$Cut(\overline{w})[i] = \begin{cases} \overline{w}[i] & if \ \overline{w}[i] \leq gb \\ gb + 1 & otherwise. \end{cases}$$

Based on the finite set of configurations $\mathcal{C}_K^{cut} = \{(s, Cut(\overline{w}))|(s, \overline{w}) \in \mathcal{C}_K\}$ we can define a new cut environment. The new semantics is a straightforward extension. The main changes are made to the rules concerning the bounds and the next operators:

$$[\![e \bowtie c]\!]_{\epsilon_{cut}}^{cut} = \{(s, \overline{w}) \in \mathcal{C}_K^{cut}|eval_{Cut(\overline{w})}(e) \bowtie c\}$$

$$[\![reset\ R\ in\ EX(X)]\!]_{\epsilon_{cut}}^{cut} = \left\{(s, Cut(\overline{w}))\ \middle|\ \begin{array}{c} \text{there is some } (s, \overline{c}, s') \in T \text{ s.t.} \\ (s', Cut(\overline{w}[R \to 0] + \overline{c})) \in \epsilon_{cut}(X) \end{array}\right\}$$

$$[\![reset\ R\ in\ AX(X)]\!]_{\epsilon_{cut}}^{cut} = \left\{(s, Cut(\overline{w}))\ \middle|\ \begin{array}{c} \text{for all } (s, \overline{c}, s') \in T \text{ we have} \\ (s', Cut(\overline{w}[R \to 0] + \overline{c})) \in \epsilon_{cut}(X) \end{array}\right\}.$$

The semantics for an RML equation system is then defined as the regular semantics but with cut environments:

$$\nu\mathcal{E}_{cut} = \bigcup\{\epsilon_{cut} \in Env_{cut}|\epsilon_{cut} \subseteq \mathcal{F}(\epsilon_{cut})\}.$$

If $(s, \overline{w}) \in [\![\varphi]\!]_{\nu\mathcal{E}_{cut}}^{cut}$ then we write $(s, \overline{w}) \vDash_K^{cut} \varphi$.

Lemma 1 (Equivalence of the cut semantics). *Let K be an n-WKS and $\mathcal{E}$ be an RML equation system. Given a configuration $(s, \overline{w}) \in \mathcal{C}_K$ we have $(s, \overline{w}) \vDash_K \varphi$ iff $(s, Cut(\overline{w})) \vDash_K^{cut} \varphi$.*

Proof (Sketch). We show by structural induction on the formula φ that $(s, \overline{w}) \in [\![\varphi]\!]_{\epsilon}$ iff $(s, \overline{w}) \in [\![\varphi]\!]_{\epsilon_{cut}}^{cut}$ whenever we cut with a constant gb derived from $\mathcal{E}$. This means that in any environment ϵ we have $\epsilon \subseteq \mathcal{F}(\epsilon)$ iff $\epsilon_{cut} \subseteq \mathcal{F}(\epsilon_{cut})$. $\qquad\square$

Lemma 2 (Hardness of Model Checking). *The model checking problem is EXPTIME-hard, already for an n-WKS K with a single weight.*

Proof. We show that the problem is EXPTIME-hard by reduction from countdown games that are EXPTIME-complete [16]. A countdown game (Q, R) consists of a finite set of states Q and a finite transition relation $R \subseteq Q \times \mathbb{N} \times Q$. We write transitions as $(q, k, q') \in R$ and say that the duration of the transition is k. A configuration in the game is a pair (q, c) where $q \in Q$ and $c \in \mathbb{N}$. Given the configuration (q, c) the rules of the game are defined as follows:

○ if $c = 0$ then player 1 wins, else
○ if for all transitions $(q, k, q') \in R$ we have $k > c$ and $c > 0$ player 2 wins,
○ otherwise there exists some transition $(q, k, q') \in R$ s.t. $k \leq c$. Player 1 must choose such a duration k while player 2 chooses a target state q' s.t. $(q, k, q') \in R$. The new configuration is $(q', c - k)$. We repeat this until a winner is found.

We now reduce the problem of deciding which player is the winner of the countdown game (Q, R) given the configuration (q, c) to deciding the model checking problem for an n-WKS and an RML equation system. We create the 1-WKS $K = (S, s_0, \mathcal{AP}, L, T)$ from the countdown game (Q, R) as follows: $S = Q \cup \{s_q^k|\exists(q, k, q') \in R\}$ with the initial state $s_0 = q_0$, $\mathcal{AP} = \emptyset$ and the set of transitions are defined as $T = \{(q, k, s_q^k)|\exists(q, k, q') \in R \text{ and } \exists s_q^k \in$

$S\} \cup \{(s_q^k, 0, q') | \exists s_q^k \in S \text{ and } \exists (q, k, q') \in R\}$. To enforce the rules, we create the following RML equation system:

$$\mathcal{E} = \begin{bmatrix} X_0 = ((\#1 = c) \vee (EX(X_1) \wedge \#1 < c)) \\ X_1 = AX(X_0) \end{bmatrix}.$$

Now we can observe that $(s_0, 0) \vDash_K \mathcal{E}(X_0)$ iff player 1 has a winning strategy in the corresponding countdown game given the initial configuration (q_0, c).   □

Theorem 1 (Model Checking Complexity). *Given an n-WKS and an RML equation system, the model checking problem is EXPTIME-complete.*

Proof. The upper-bound follows from Lemma 1 and noticing that the fixed-point computation (based on the cut semantics) runs in exponential time; there are at most exponentially many cut configurations and it takes at most an exponential number of rounds to reach the fixed point. The lower-bound is by Lemma 2.   □

4   Game Theoretic Framework

In this section we introduce multi-weighted two-player games, where one player acts as the controller and one player act as the uncontrollable environment. We end the section by expressing the synthesis problem as the problem of finding a winning strategy for the controller in such a game.

Definition 4 (n-Weighted Game Graph). *An n-Weighted Game Graph (n-WGG) is a tuple $\mathcal{G} = (S, s_0, \mathcal{AP}, L, T_c, T_u)$ where T_c and T_u are disjoint sets and $K_\mathcal{G} = (S, s_0, \mathcal{AP}, L, T_c \cup T_u)$ is the underlying n-WKS.*

We fix an n-WGG $\mathcal{G} = (S, s_0, \mathcal{AP}, L, T_c, T_u)$ for the remainder of the section. The set of transitions T_c are owned by the controller while the set T_u is owned by the uncontrollable environment. When $(s, \overline{c}, s') \in T_c$, we write $s \xrightarrow{\overline{c}} s'$ and when $(s, \overline{c}, s') \in T_u$ we write $s \xdashrightarrow{\overline{c}} s'$. When s has some outgoing controllable transition we write $s \rightarrow$ (or $s \dashrightarrow$ for uncontrollable transitions). Similarly, if there are no outgoing transitions we write $s \nrightarrow$ (or $s \ndashrightarrow$ for uncontrollable transitions).

Definition 5 (Game). *An n-Weighted Game (n-WG) is a tuple $(\mathcal{G}, \mathcal{E})$ where $\mathcal{G}$ is an n-WGG and the winning condition is an RML equation system $\mathcal{E}$.*

In an n-WG the controller's actions are based on a *strategy* that specifies which transition to take at a given position. Formally a strategy σ is a function that, given the history of the game (the current branch), outputs the controller's next move. Recall that $K_\mathcal{G}$ is the underlying n-WKS of an n-WGG $\mathcal{G}$ and $\Pi_{K_\mathcal{G}}^{fin}$ is the set of all finite runs in $K_\mathcal{G}$.

Definition 6 (Strategy). *A strategy for the controller is a function* $\sigma : \Pi_{K_\mathcal{G}}^{fin} \to T_c \cup \{\text{NIL}\}$ *mapping a finite run* ρ *to a transition s.t.*

$$\sigma(\rho) = \begin{cases} last(\rho) \xrightarrow{\bar{c}} s' & \text{if } last(\rho) \to \\ \text{NIL} & \text{otherwise} \end{cases}$$

where NIL *is the choice to do nothing which is only allowed if there is no controllable transition to choose.*

For the uncontrollable actions, we are forced to consider all possible options as the winning condition allows us to reason about branching properties. Based on the strategy's choices and all choices available to the environment, we unfold the game into an n-WKS on which we can verify the objective.

Definition 7 (Strategy restricted n-WKS). *Given a game graph* $\mathcal{G}$ *and a strategy* σ, *we define* $\mathcal{G}{\restriction}\sigma = (S', s_0, \mathcal{AP}, L', T_c{\restriction}\sigma \cup T_u')$ *as the n-WKS resulting from restricting the game graph under the strategy* σ *s.t.:*

- $S' = \Pi_{K_G}^{fin}$
- $L'(\rho) = L(last(\rho))$
- $T_c'{\restriction}\sigma = \left\{ (\rho, \bar{c}, (\rho \circ \sigma(\rho))) \,\middle|\, \sigma(\rho) = (last(\rho) \xrightarrow{\bar{c}} s') \right\}$
- $T_u' = \{(\rho, \bar{c}, \rho \circ (last(\rho) \dashrightarrow^{\bar{c}} s')) \,\middle|\, (last(\rho) \dashrightarrow^{\bar{c}} s')\}$

A strategy σ is *winning* for the game $(\mathcal{G}, \mathcal{E})$ iff $(s_0, 0^n) \vDash_{\mathcal{G}{\restriction}\sigma} \mathcal{E}(X_0)$.

Definition 8 (Synthesis Problem). *Given the n-WG $(\mathcal{G}, \mathcal{E})$, the synthesis problem is to decide if there is a strategy* σ *s.t.* $(s_0, 0^n) \vDash_{\mathcal{G}{\restriction}\sigma} \mathcal{E}(X_0)$.

We return to the motivating example of a processor completing a set of tasks introduced earlier. Let $\mathcal{G}$ be the n-WGG presented in Fig. 1a and $\mathcal{E}$ be the formalised winning condition presented in Sect. 3. As stated in the introduction, one winning strategy is to initially choose the fastest option $i \xrightarrow{(0,0,0)} f$ and repeat this choice whenever the preceding uncontrollable move did not generate more than 5 units of excess heat. In all other situations we choose the safe slow option. However, we notice that there is a simpler alternative which still satisfies the winning condition.

After the first two turns, we no longer need to consider the optimistic objective which aims to finish 4 units of work in 2 time units, as 2 time units have already passed. We can therefore focus solely on keeping the excess heat down and can simply always choose the slow option after that. We formally define this simpler strategy σ below:

$$\sigma(\rho) = \begin{cases} i \xrightarrow{(0,0,0)} f & \text{if } last(\rho) = i \text{ and } cost_\rho[2] \leq 5 \\ i \xrightarrow{(0,0,0)} s & \text{if } last(\rho) = i \text{ and } cost_\rho[2] > 5 \\ \text{NIL} & \text{otherwise.} \end{cases}$$

The result of applying σ to the game graph $\mathcal{G}$ is the strategy restricted game graph $\mathcal{G}\!\restriction\!\sigma$ illustrated in Fig. 2. As the second property only expresses a possibility it is enough that our first move might lead to completing the tasks and this defined strategy is indeed winning.If we are to formally define the strategy σ_{alt}, which is the first option discussed, we need to know some of the history of the play in order to choose alternating transitions from the state i. In fact, we need to look at the last four transitions to ensure a safe pick. In the remainder of this paper, we carefully consider the memory requirements of winning strategies for RML winning conditions (we encode the memory needed as the remainder of the winning condition which is still to be satisfied).

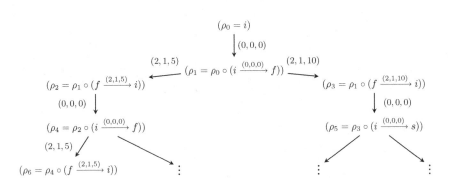

Fig. 2. Strategy restricted game graph $\mathcal{G}\!\restriction\!\sigma$ for $\mathcal{G}$ from Fig. 1a

5   Dependency Graphs

To solve the synthesis problem we shall propose a reduction to the problem of calculating the maximal fixed-point assignment of a dependency graph. A dependency graph [23] is a pair $D = (V, H)$ where V is a finite set of nodes and $H : V \times \mathcal{P}(V)$ is a finite set of hyper-edges. Given a hyper-edge $h = (v, T) \in H$ we say that $v \in V$ is the source node and $T \subseteq V$ is the set of target nodes.

An assignment of a dependency graph D is a function $A : V \to \{0, 1\}$ that assigns value 0 (false) or 1 (true) to each node in the graph. We also assume a component-wise ordering $\sqsubseteq$ of assignments s.t. $A_1 \sqsubseteq A_2$ whenever $A_1(v) \leq A_2(v)$ for all $v \in V$. The set of all assignments is denoted by $\mathcal{A}$ and clearly $(\mathcal{A}, \sqsubseteq)$ is a complete lattice. A (post) fixed-point assignment of D is an assignment A s.t. for any $v \in V$ if for all $(v, T) \in H$ there exists an $u \in T$ s.t. $A(u) = 0$ then we have that $A(v) = 0$. This is formalized by the monotonic function $f : \mathcal{A} \to \mathcal{A}$ defined as

$$f(A)(v) = \bigvee_{(v,T)\in H} \bigwedge_{(u\in T)} A(u)$$

where, by convention, conjunction over the empty set is true while disjunction over the empty set is false. By Knaster-Tarski's theorem we know that there

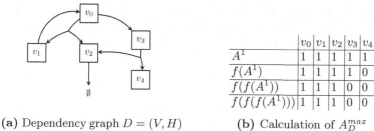

(a) Dependency graph $D = (V, H)$

	v_0	v_1	v_2	v_3	v_4
A^1	1	1	1	1	1
$f(A^1)$	1	1	1	1	0
$f(f(A^1))$	1	1	1	0	0
$f(f(f(A^1)))$	1	1	1	0	0

(b) Calculation of A_D^{max}

Fig. 3. Dependency graph D and the fixed point assignment A^{max}

exists a unique maximum fixed-point assignment of D, denoted A_D^{max}. When the dependency graph D is clear from context, we simply denote the fixed-point assignment as A^{max}. To compute the maximum fixed-point assignment, we apply f repeatedly on the assignment, starting with $f(A^1)$ where A^1 is the initial assignment defined s.t. $A^1(v) = 1$ for all $v \in V$, as shown in Fig. 3.

Theorem 2 ([23]). *There is a linear time (on-the-fly) algorithm to compute the maximal fixed point of a dependency graph.*

6   On-the-Fly Synthesis Algorithm for n-WG

We shall now present our encoding of the synthesis problem to the problem of calculating the maximal fixed-point assignment of a dependency graph. The initial idea is that given a game $(\mathcal{G}, \mathcal{E})$, we construct a dependency graph $D_{(\mathcal{G}, \mathcal{E})}$ with nodes of the form $\langle \varphi, (s, \overline{w}) \rangle$ such that $A^{max}(\langle \varphi, (s, \overline{w}) \rangle) = 1$ iff there exists a strategy σ where $(s, \overline{w}) \vDash_{\mathcal{G} \restriction \sigma} \varphi$. We are referring to this as an on-the-fly algorithm as we do not necessarily need to calculate the maximal fixed-point assignment of the entire dependency graph to terminate.

The first challenge in the encoding is to keep track of which parts of the winning condition must be considered together (i.e. cannot be looked at compositionally). We can analyse disjunction compositionally by decomposing it to the individual disjuncts. However, for conjunction, we must consider the whole formula together as the controller's choice must be done in agreement with all subformulae. As a consequence of keeping all conjuncts together, the reset operator may force us to evaluate the same subformula for (possibly several) different cost vectors at any given point.

This technical challenge is solved by annotating the basic formulae with weight vectors under which they must be evaluated. Then we transform the weight annotated formulae into a disjunctive normal form such that we can separate the disjuncts (as they can be solved independently) and then for a given disjunct select one controllable transition. To create such an annotated normal form, we now propose a more succinct notation for evaluating a formula with respect to multiple different cost vectors.

Definition 9 (Weighted basic formula). *We define the function $\varphi[\overline{w}]$ as the operation of pushing the cost vector $\overline{w}$ through an RML formula φ s.t. each basic formula is prefixed with the cost vector:*

$$\varphi[\overline{w}] = \begin{cases} \varphi_1[\overline{w}] \wedge \varphi_2[\overline{w}] & \text{if } \varphi = \varphi_1 \wedge \varphi_2 \\ \varphi_1[\overline{w}] \vee \varphi_2[\overline{w}] & \text{if } \varphi = \varphi_1 \vee \varphi_2 \\ (\overline{w} : \varphi) & \text{if } \varphi = \beta. \end{cases}$$

The result is a positive Boolean combination of weighted basic formulae for which we write $s \vDash (\overline{w} : \beta)$ whenever $(s, \overline{w}) \vDash \beta$.

Notice that for propositions of the form a, $\neg a$, TRUE and FALSE the cost vector $\overline{w}$ does not impact the satisfiability and can be ignored such that instead of $s \vDash (\overline{w} : a)$ we simply consider $s \vDash a$. For $(e \bowtie c)$ this corresponds to $eval_{\overline{w}}(e) \bowtie c$ which evaluates to either TRUE or FALSE and can be replaced by its truth value.

We use the function dnf to transforms the conjunction $\bigwedge_{m \in M} \varphi_m[\overline{w}_m]$ into an equivalent formula in disjunctive normal form $dnf(\bigwedge_{m \in M}(\varphi_m[\overline{w}_m])) = \bigvee_{i \in I} \psi_i$ where each ψ_i is of the form:

$$\bigwedge_{\ell \in L} p_\ell \wedge \bigwedge_{j \in J} \overline{w}_j : reset\ R_j\ in\ EX(X_j) \wedge \bigwedge_{k \in K} \overline{w}_k : reset\ R_k\ in\ AX(X_k) \qquad (1)$$

where $p := a|\neg a|\text{TRUE}|\text{FALSE}$.

For the remainder of the section, we fix a conjunction of weighted basic formulae ψ such that L, J or K refer to the indices used in Eq. (1). Additionally, we fix an n-WGG $\mathcal{G} = (S, s_0, \mathcal{AP}, L, T_c, T_u)$. Based on a strategy choice $s \xrightarrow{\overline{c}} s' \in T_c$ we define the set of final transitions as $finalOut(s \xrightarrow{\overline{c}} s') = \{(s \xrightarrow{\overline{c}_u} s_u) \in T_u\} \cup \{(s \xrightarrow{\overline{c}} s')\}$.

6.1 Determining a Winning Move

Given a conjunction of weighted basic formulae ψ, a strategy choice $s \xrightarrow{\overline{c}} s'$ is winning if all propositions and bounds are fulfilled i.e. $s \vDash \bigwedge_{\ell \in L} p_\ell$ and (i) every $(\overline{w}_j : reset\ R_j\ in\ EX(X_j))$ subformula is satisfied by some final outgoing transition, and (ii) all $(\overline{w}_k : reset\ R_k\ in\ AX(X_k))$ subformulae are satisfied by all final outgoing transitions.

The second challenge in the encoding is thus to determine how the formula should envolve in the different branches of the resulting computation tree i.e. which existential subformula should be satisfied by a given final transition. We frame this choice as a mapping from existential subformulae to final transitions $\alpha : J \to finalOut(s \xrightarrow{\overline{c}} s')$. Given such a mapping we can formally define the remaining winning condition for each final transition $(s, \overline{c}', s'') \in finalOut(s \xrightarrow{\overline{c}} s')$ in the *move* function where $move(\psi, \alpha, (s, \overline{c}', s'')) = $ FALSE if $\exists \ell \in L$ s.t. $s \nvDash p_\ell$ and otherwise $move(\psi, \alpha, (s, \overline{c}', s'')) =$

$$\bigwedge_{\substack{j \in J\ s.t. \\ \alpha(j) = (s, \overline{c}', s'')}} \mathcal{E}(X_j)[(Cut(\overline{w}_j[R_j \to 0] + \overline{c}')] \wedge \bigwedge_{k \in K} \mathcal{E}(X_k)[(Cut(\overline{w}_k[R_k \to 0] + \overline{c}')] \ .$$

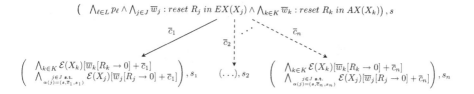

Fig. 4. Successor states assigned their remaining formula

In Fig. 4 we show an illustration of the final outgoing transitions as well as the remaining formula based on the move function.

6.2 Dependency Graph Encoding

We are now ready to present the encoding of the synthesis problem to the computation of the maximal fixed-point assignment on a dependency graph. The encoding consists of two types of nodes. Either $\langle \psi, s \rangle$ which represent the synthesis problem from the state s and winning condition ψ, or intermediate nodes with additional information used to explore possible strategies and mappings.

Given the game $(\mathcal{G}, \mathcal{E})$ and the initial state s_0, we construct the dependency graph $D_{(\mathcal{G}, \mathcal{E})}$ with the root node $\langle \mathcal{E}(X_0)[0^n], s_0 \rangle$. The outgoing hyper-edges are defined in Fig. 5. Notice that the nodes representing synthesis problems are illustrated as square boxes, while intermediate nodes are illustrated as ellipses. Below a brief description of the encoding is given:

1. Given a conjunction $\bigwedge_{m \in M}(\mathcal{E}(X_m)[\overline{w}_m])$, we first put it into disjunctive normal form (Fig. 5a),
2. and then the resulting formula $\bigvee_{i \in I} \psi_i$ is split into conjunctions of weighted basic formulae (Fig. 5b).
3. The propositions TRUE (Fig. 5c) and FALSE (Fig. 5d) are handled in the obvious way (note that FALSE has no outgoing edges and evaluates to 0).
4. For a conjunction of weighted basic formulae ψ, we create hyper-edges to intermediate nodes which explore each possible strategy (Fig. 5e).
5. From the intermediate node, we create a hyper-edge with a target node for each outgoing transition. Each target node records the last state and updates the remaining formula based on the move function (Fig. 5f). Goto step 1.
6. If there are no outgoing controllable transitions, we combine steps 4 and 5 and explore all mappings and advance the formula by one step (Fig. 5g).

Given a dependency graph $D_{(\mathcal{G}, \mathcal{E})}$ with the maximal fixed-point assignment A^{max}, we have that if $A^{max}(\langle \mathcal{E}(X_0)[0^n], s_0 \rangle) = 1$ we can extract a winning strategy from the initial state s_0 by evaluating how the node got that assignment. Essentially for a node $\langle s, \psi \rangle$ we follow the assignment through either Fig. 5e and Fig. 5f, or Fig. 5g and extract the strategy from the intermediate node (if there is one). We generate the runs used for the next strategy choices based on the

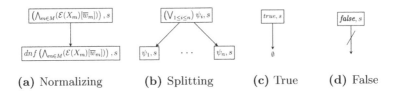

(a) Normalizing   **(b)** Splitting   **(c)** True   **(d)** False

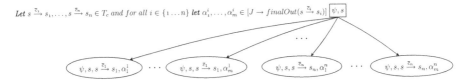

(e) Choosing a controllable transition (strategy choice) and a mapping

(f) Advancing the formula, given a strategy choice and a mapping, when $s \to$

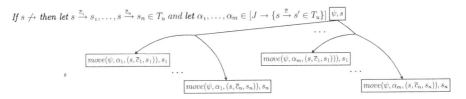

(g) Choosing a mapping and advancing the formula when $s \not\to$

Fig. 5. Encoding of the synthesis problem

final transitions. If there is no remaining winning condition to satisfy (or no way of satisfying it in the future) we go to Fig. 5c or 5d.

Lemma 3 (Correctness of the encoding). *Let* $(\mathcal{G}, \mathcal{E})$ *be an* n-*WG. There is a strategy* σ *such that* $s_0 \vDash_{\mathcal{G} \upharpoonright \sigma} \mathcal{E}(X_0)[0^n]$ *iff* $A^{max}_{D(\mathcal{G},\mathcal{E})}(\langle \mathcal{E}(X_0)[0^n], s_0 \rangle) = 1$.

Theorem 3. *The synthesis problem for* n-*WGs is in 2-EXPTIME.*

Proof. To show that the problem is in 2-EXPTIME, we first notice that given a game $(\mathcal{G}, \mathcal{E})$, the number of nodes in the dependency graph $D_{(\mathcal{G},\mathcal{E})}$ is doubly exponential in the size of $K_{\mathcal{G}}$. This is because there is an exponential number of weighted basic formulae and the nodes can contain an arbitrary conjunction (subset of) weighted basic formulae. As finding the maximal fixed point of a dependency graph can be then done in linear time in the size of the dependency graph (Theorem 2), this gives us a doubly exponential algorithm. □

6.3 Hardness of the Synthesis Problem

We shall now argue that the synthesis problem is NEXPTIME-hard. This is proved by a reduction from Succinct Hamiltonian path problem.

Definition 10 (Succinct Hamiltonian). *Let $n, r \in \mathbb{N}$ be natural numbers and let $\varphi(x_1, x_2, \ldots, x_{2n+r})$ be a Boolean formula over the variables $x_1, \ldots, x_{2n+r}$. This input defines a directed graph $G = (V, E)$ where $V = \{0, 1\}^n$ and $(v, v') \in E$ iff there exists $y \in \{0, 1\}^r$ such that $\varphi(v, v', y) = \text{TRUE}$. The Succinct Hamiltonian problem is to decide if there is a Hamiltonian path in G (containing each vertex exactly once).*

Lemma 4. *The Succinct Hamiltonian problem is NEXPTIME-complete.*

Proof. In [12] the succinct Hamiltonian path problem is proved NEXPTIME-complete where the edge relation is defined by a Boolean circuit. The Boolean formula representation is obtained by employing the Tseytin transformation. □

Theorem 4. *The synthesis problem for n-WGs is NEXPTIME-hard.*

Proof. By polynomial time reduction from Succinct Hamiltonian problem. Let $n, r \in \mathbb{N}$ and $\varphi(x_1, x_2, \ldots, x_{2n+r})$ be an instance of the problem. We construct an $(3n + r + 1)$-WG $\mathcal{G}$ and a formula such that the answer to the synthesis problem is positive iff there is a Hamiltonian path. The cost-vectors that we use are of the form

$$\overline{w} = [\underbrace{1, \ldots, n}_{u_1}, \underbrace{n+1, \ldots, 2n}_{u_2}, \underbrace{2n+1, \ldots, 3n}_{u_3}, \underbrace{3n+1, \ldots, 3n+r}_{y}, \underbrace{3n+r+1}_{\text{counter}}]$$

where (u_1, u_2) or (u_2, u_1) represents the current edge that we are exploring (the direction of the edge depends on the phase of the edge generation process), u_3 stores the encoding of a node that must appear in the path, y represents the internal variables for the evaluation of φ and the last weight coordinate is a counter where we store the number of processed edges. Whenever we write u_1 we refer to the first n components of the cost-vector, for u_2 we refer to the next n components etc. The constructed game graph $\mathcal{G}$ together with the winning condition is given in Fig. 6 such that the notation $\#i++$ represents a vector where at position i the weight value is 1 and at all other positions the value is 0. Notice that all transitions in the game graph are controllable. Let us now argue about the correctness of the reduction.

First, let us assume that there is a Hamiltonian path in G, i.e. a sequence $v_1, v_2, \ldots, v_{2^n} \in V$ where every node from V is part of the sequence and for all i, $1 \leq i < 2^n$, there is $y \in \{0, 1\}^r$ such that $\varphi(v_i, v_{i+1}, y) = \text{TRUE}$. We shall define a winning strategy σ for the constructed game graph with a winning condition

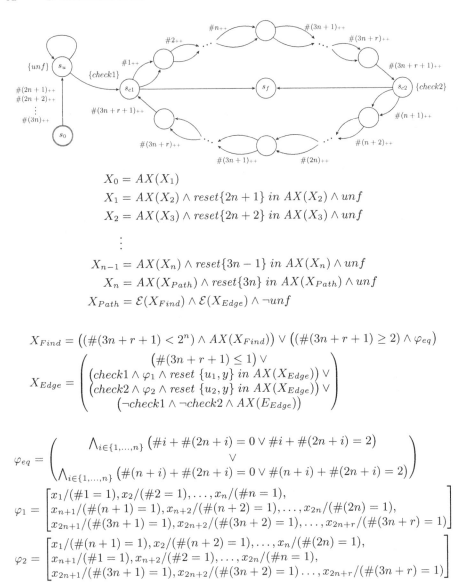

$$X_0 = AX(X_1)$$
$$X_1 = AX(X_2) \wedge reset\{2n + 1\} \ in \ AX(X_2) \wedge unf$$
$$X_2 = AX(X_3) \wedge reset\{2n + 2\} \ in \ AX(X_3) \wedge unf$$

$$\vdots$$

$$X_{n-1} = AX(X_n) \wedge reset\{3n - 1\} \ in \ AX(X_n) \wedge unf$$
$$X_n = AX(X_{Path}) \wedge reset\{3n\} \ in \ AX(X_{Path}) \wedge unf$$
$$X_{Path} = \mathcal{E}(X_{Find}) \wedge \mathcal{E}(X_{Edge}) \wedge \neg unf$$

$$X_{Find} = \left((\#(3n + r + 1) < 2^n) \wedge AX(X_{Find}) \right) \vee \left((\#(3n + r + 1) \geq 2) \wedge \varphi_{eq} \right)$$

$$X_{Edge} = \begin{pmatrix} (\#(3n + r + 1) \leq 1) \vee \\ (check1 \wedge \varphi_1 \wedge reset \{u_1, y\} \ in \ AX(X_{Edge})) \vee \\ (check2 \wedge \varphi_2 \wedge reset \{u_2, y\} \ in \ AX(X_{Edge})) \vee \\ (\neg check1 \wedge \neg check2 \wedge AX(E_{Edge})) \end{pmatrix}$$

$$\varphi_{eq} = \begin{pmatrix} \bigwedge_{i \in \{1,\ldots,n\}} \left(\#i + \#(2n + i) = 0 \vee \#i + \#(2n + i) = 2 \right) \\ \vee \\ \bigwedge_{i \in \{1,\ldots,n\}} \left(\#(n + i) + \#(2n + i) = 0 \vee \#(n + i) + \#(2n + i) = 2 \right) \end{pmatrix}$$

$$\varphi_1 = \begin{bmatrix} x_1/(\#1 = 1), x_2/(\#2 = 1), \ldots, x_n/(\#n = 1), \\ x_{n+1}/(\#(n + 1) = 1), x_{n+2}/(\#(n + 2) = 1), \ldots, x_{2n}/(\#(2n) = 1), \\ x_{2n+1}/(\#(3n + 1) = 1), x_{2n+2}/(\#(3n + 2) = 1), \ldots, x_{2n+r}/(\#(3n + r) = 1) \end{bmatrix}$$

$$\varphi_2 = \begin{bmatrix} x_1/(\#(n + 1) = 1), x_2/(\#(n + 2) = 1), \ldots, x_n/(\#(2n) = 1), \\ x_{n+1}/(\#1 = 1), x_{n+2}/(\#2 = 1), \ldots, x_{2n}/(\#n = 1), \\ x_{2n+1}/(\#(3n + 1) = 1), x_{2n+2}/(\#(3n + 2) = 1) \ldots, x_{2n+r}/(\#(3n + r) = 1) \end{bmatrix}$$

Fig. 6. Game graph and winning condition

defined by the variable X_0. The strategy σ moves from the initial state to s_u that satisfies the proposition unf and then loops n times in s_u. Afterwards, it proceeds to s_{c_1} as this is the only way to satisfy the equations for variables X_0 to X_n. From this point, we define a strategy based on the Hamiltonian path $v_1, v_2, \ldots, v_{2n}$. First, we move from s_{c_1} to s_{c_2} and bit by bit select the appropriate edges so that the encoding of node v_1 stored in the first n bits of the weight vector. Then we select the encoding for the y part of the weight vector and increase the counter value to 1. We repeat the process in the symmetric lower part of the loop and store the encoding of the node v_2 in the second part of the weight vector, making sure that the y part is selected so that $\varphi(v_1, v_2, y) = \text{TRUE}$. Then the control strategy is defined so that the node v_3 is stored instead of the node v_1 and an appropriate vector y is generated so that $\varphi(v_2, v_3, y) = \text{TRUE}$. Next, the node v_4 is generated instead of the node v_2 with the corresponding vector y and so on until the whole Hamiltonian path is traversed in this way, until we move to the state s_f. It remains to show that the strategy σ satisfies the variable X_0.

To argue that this is a winning strategy we consider both the invariant X_{Edge} and X_{Find}, activated by X_{Path} in the state s_{c_1}. The use of *reset* in the equations from X_0 to X_n created 2^n active versions of the remaining winning condition X_{Path} by creating a new cost-prefix for every Boolean combination of the components in u_3. Hence every node in the graph has an active version of X_{Find} trying to satisfy the subformula φ_{eq} that proves that the vector in the third weight component (u_3) appears within less than 2^n steps at least once in the generated sequence of nodes. As the generated path is Hamiltonian, this is indeed the case and X_{Find} is hence satisfied. Next we argue for the satisfiability of the invariant X_{Find}. The satisfiability of this variable is never affected by u_3 and so we can ignore the multiple active versions. In the strategy σ we are generating nodes according to the Hamiltonian path and every time we are in s_{c_1} that satisfies the proposition *check*1 we must guarantee we have a valid edge between u_1 and u_2 (required by φ_1) and similarly we must satisfy φ_2 whenever we are in s_{c_2}. Again, as the generated sequence is a path in the graph, this is the case.

Let us now assume that there is a winning strategy σ for $(\mathcal{G}, \mathcal{E})$. We shall argue that the sequence of nodes generated in u_1 and u_2 defines a Hamiltonian path. As σ is winning it must include a move to s_u, n loops on s_u and finally a move to s_{c1}. This is needed to satisfy the equations from X_0 to X_n. Given this sequence of moves, we have that from s_{c1} the strategy must enforce 2^n active versions of X_{Edge} and X_{Find}. To satisfy all active versions of X_{Find} all nodes must be represented in either u_1 or u_2 before the counter reaches 2^n. As there are exactly 2^n nodes, this implies that there cannot be a repetition of any node in the first 2^n nodes generated by σ. To satisfy X_{Edge} the only choices are to either continuously satisfy the second or third clause or move to s_f where the last clause trivially holds. If the strategy choice is the move to s_f before all nodes in V have been represented in either u_1 or u_2, not all active versions of X_{Find} will be satisfied. Hence this would not be a winning strategy. The only alternative is that σ satisfies the second or third clause 2^n times before moving to s_f. By definition of X_{Edge} this implies that σ generates 2^n nodes which form a valid sequence of edges (to satisfy the checks performed by φ_1 and φ_2). Hence the existence of a winning strategy implies the existence of a Hamiltonian path. □

7 Conclusion

We presented a recursive modal logic with semantics defined over a nonnegative multi-weighted extension of Kripke structures. We proved that the model checking problem is EXPTIME-complete and EXPTIME-hard even for a single weight. We then introduced the synthesis problem given multi-weighted two-player games (with a controller and an environment as players) with objectives formulated in the recursive modal logic and with the game-arena given by a multi-weighted Kripke structure. We proved that the synthesis problem is in 2-EXPTIME and is NEXPTIME-hard. The containment result is achieved by a (doubly exponential) reduction to dependency graph that allows for on-the-fly algorithms for finding the maximum fixed-point value of a given node. It required a nontrivial treatment of conjunctive subformulae, as in the branching-time setting we have to guarantee that uniform choices of controllable transitions are made across all formulae in a conjunction.

Acknowledgements. The research leading to these results has received funding from the project DiCyPS funded by the Innovation Fund Denmark, ERC Advanced Grant LASSO and the QASNET DFF project.

References

1. Almagor, S., Boker, U., Kupferman, O.: Formally reasoning about quality. J. ACM **24**(1–24), 56 (2016)
2. Bauer, S.S., Juhl, L., Larsen, K.G., Srba, J., Legay, A.: A logic for accumulated-weight reasoning on multiweighted modal automata. In: TASE, pp. 77–84 (2012)
3. Bloem, R., Chatterjee, K., Henzinger, T.A., Jobstmann, B.: Better quality in synthesis through quantitative objectives. In: Bouajjani, A., Maler, O. (eds.) CAV 2009. LNCS, vol. 5643, pp. 140–156. Springer, Heidelberg (2009). https://doi.org/10.1007/978-3-642-02658-4_14
4. Bloem, R., et al.: Synthesizing robust systems. Acta Inform. **51**(3–4), 193–220 (2013). https://doi.org/10.1007/s00236-013-0191-5
5. Boker, U., Chatterjee, K., Henzinger, T.A., Kupferman, O.: Temporal specifications with accumulative values. TOCL **27**, 1–27:25 (2014)
6. Bouyer, P., Markey, N., Randour, M., Larsen, K.G., Laursen, S.: Average-energy games. Acta Inform. **55**, 91–127 (2018)
7. Brázdil, T., Chatterjee, K., Kučera, A., Novotný, P.: Efficient controller synthesis for consumption games with multiple resource types. In: Madhusudan, P., Seshia, S.A. (eds.) CAV 2012. LNCS, vol. 7358, pp. 23–38. Springer, Heidelberg (2012). https://doi.org/10.1007/978-3-642-31424-7_8
8. Bruyère, V., Filiot, E., Randour, M., Raskin, J.-F.: Meet your expectations with guarantees: beyond worst-case synthesis in quantitative games. Inf. Comput. **254**, 259–295 (2017)
9. Buchi, J.R., Landweber, L.H.: Solving sequential conditions by finite-state strategies. In: Mac Lane, S., Siefkes, D. (eds.) The Collected Works of J. Richard Büchi, pp. 525–541. Springer, New York (1990). https://doi.org/10.1007/978-1-4613-8928-6_29

10. Chatterjee, K., Randour, M., Raskin, J.-F.: Strategy synthesis for multi-dimensional quantitative objectives. Acta Inform. **51**(3–4), 129–163 (2014)
11. Chatterjee, K., Doyen, L., Henzinger, T.A., Raskin, J.-F.: Generalized mean-payo and energy games. In: FSTTCS 2010. LiPIcs (2010)
12. Galperin, H., Wigderson, A.: Succinct representations of graphs. Inf. Control **56**, 183–198 (1983)
13. Jensen, J.F., Larsen, K.G., Srba, J., Oestergaard, L.K.: Efficient model checking of weighted CTL with upper-bound constraints. STTT **18**(4), 409–426 (2016)
14. Jensen, L.S., Kaufmann, I., Nielsen, S.M.: Symbolic synthesis of non-negative multi-weighted games with temporal objectives. Master thesis, Department of Computer Science, Aalborg University, Denmark (2017)
15. Jensen, L.S., Kaufmann, I., Larsen, K.G., Nielsen, S.M., Srba, J.: Model checking and synthesis for branching multi-weighted logics. JLAMP **105**, 28–46 (2019)
16. Jurdziński, M., Laroussinie, F., Sproston, J.: Model checking probabilistic timed automata with one or two clocks. In: Grumberg, O., Huth, M. (eds.) TACAS 2007. LNCS, vol. 4424, pp. 170–184. Springer, Heidelberg (2007). https://doi.org/10.1007/978-3-540-71209-1_15
17. Jurdziński, M., Lazić, R., Schmitz, S.: Fixed-dimensional energy games are in pseudo-polynomial time. In: Halldórsson, M.M., Iwama, K., Kobayashi, N., Speckmann, B. (eds.) ICALP 2015. LNCS, vol. 9135, pp. 260–272. Springer, Heidelberg (2015). https://doi.org/10.1007/978-3-662-47666-6_21
18. Kupferman, O., Vardi, M.Y.: μ–calculus synthesis. In: Nielsen, M., Rovan, B. (eds.) MFCS 2000. LNCS, vol. 1893, pp. 497–507. Springer, Heidelberg (2000). https://doi.org/10.1007/3-540-44612-5_45
19. Kupferman, O., Madhusudan, P., Thiagarajan, P.S., Vardi, M.Y.: Open systems in reactive environments: control and synthesis. In: Palamidessi, C. (ed.) CONCUR 2000. LNCS, vol. 1877, pp. 92–107. Springer, Heidelberg (2000). https://doi.org/10.1007/3-540-44618-4_9
20. Kupfermant, O., Vardi, M.Y.: Synthesis with incomplete information. In: Barringer, H., Fisher, M., Gabbay, D., Gough, G. (eds.) Advances in Temporal Logic, vol. 16, pp. 109–127. Springer, Netherlands (2000). https://doi.org/10.1007/978-94-015-9586-5_6
21. Laroussinie, F., Markey, N., Schnoebelen, P.: Efficient timed model checking for discrete-time systems. Theor. Comput. Sci. **353**, 249–271 (2006)
22. Larsen, K.G., Mardare, R., Xue, B.: Alternation-free weighted mu-calculus: decidability and completeness. Electron. Notes Theor. Comput. Sci. **319**, 289–313 (2015)
23. Liu, X., Smolka, S.A.: Simple linear-time algorithms for minimal fixed points. In: Larsen, K.G., Skyum, S., Winskel, G. (eds.) ICALP 1998. LNCS, vol. 1443, pp. 53–66. Springer, Heidelberg (1998). https://doi.org/10.1007/BFb0055040
24. Pnueli, A., Rosner, R.: On the synthesis of a reactive module. In: Proceedings of the 16th ACM SIGPLAN-SIGACT Symposium on Principles of Programming Languages. POPL 1989, pp. 179–190. ACM (1989)
25. Pnueli, A., Rosner, R.: On the synthesis of an asynchronous reactive module. In: Ausiello, G., Dezani-Ciancaglini, M., Della Rocca, S.R. (eds.) ICALP 1989. LNCS, vol. 372, pp. 652–671. Springer, Heidelberg (1989). https://doi.org/10.1007/BFb0035790
26. Ramadge, P.J.G., Wonham, W.M.: The control of discrete event systems. In: Proceedings of the IEEE, pp. 81–98 (1989)

27. Le Roux, S., Pauly, A., Randour, M.: Extending finite-memory determinacy by boolean combination of winning conditions. In: FSTTCS 2018, LIPIcs. vol. 122, pp. 38:1–38:20 (2018)
28. Vardi, M.Y.: An automata-theoretic approach to fair realizability and synthesis. In: Wolper, P. (ed.) CAV 1995. LNCS, vol. 939, pp. 267–278. Springer, Heidelberg (1995). https://doi.org/10.1007/3-540-60045-0_56

Languages and Synthesis

On the High Complexity of Petri Nets ω-Languages

Olivier Finkel[(✉)] [iD]

Institut de Mathématiques de Jussieu - Paris Rive Gauche
CNRS et Université Paris 7, Paris, France
Olivier.Finkel@math.univ-paris-diderot.fr

Abstract. We prove that ω-languages of (non-deterministic) Petri nets and ω-languages of (non-deterministic) Turing machines have the same topological complexity: the Borel and Wadge hierarchies of the class of ω-languages of (non-deterministic) Petri nets are equal to the Borel and Wadge hierarchies of the class of ω-languages of (non-deterministic) Turing machines. We also show that it is highly undecidable to determine the topological complexity of a Petri net ω-language. Moreover, we infer from the proofs of the above results that the equivalence and the inclusion problems for ω-languages of Petri nets are Π_2^1-complete, hence also highly undecidable.

Keywords: Automata and formal languages · Petri nets · Infinite words · Logic in computer science · Cantor topology · Borel hierarchy · Wadge hierarchy · Wadge degrees · Highly undecidable properties

1 Introduction

In the sixties, Büchi was the first to study acceptance of infinite words by finite automata with the now called Büchi acceptance condition, in order to prove the decidability of the monadic second order theory of one successor over the integers. Since then there has been a lot of work on regular ω-languages, accepted by Büchi automata, or by some other variants of automata over infinite words, like Muller or Rabin automata, see [27,36,37]. The acceptance of infinite words by other finite machines, like pushdown automata, counter automata, Petri nets, Turing machines, …, with various acceptance conditions, has also been studied, see [3,8,35,36,38].

The Cantor topology is a very natural topology on the set Σ^ω of infinite words over a finite alphabet Σ which is induced by the prefix metric. Then a way to study the complexity of languages of infinite words accepted by finite machines is to study their topological complexity and firstly to locate them with regard to the Borel and the projective hierarchies [8,24,36,37].

Every ω-language accepted by a deterministic Büchi automaton is a $\mathbf{\Pi}_2^0$-set. On the other hand, it follows from Mac Naughton's Theorem that every regular ω-language is accepted by a deterministic Muller automaton, and thus is a

© Springer Nature Switzerland AG 2020
R. Janicki et al. (Eds.): PETRI NETS 2020, LNCS 12152, pp. 69–88, 2020.
https://doi.org/10.1007/978-3-030-51831-8_4

boolean combination of ω-languages accepted by deterministic Büchi automata. Therefore every regular ω-language is a $\mathbf{\Delta}_3^0$-set. Moreover Landweber proved that the Borel complexity of any ω-language accepted by a Muller or Büchi automaton can be effectively computed (see [23,27]). In a similar way, every ω-language accepted by a deterministic Muller Turing machine, and thus also by any Muller deterministic finite machine is a $\mathbf{\Delta}_3^0$-set, [8,36].

On the other hand, the Wadge hierarchy is a great refinement of the Borel hierarchy, firstly defined by Wadge via reductions by continuous functions [39]. The trace of the Wadge hierarchy on the ω-regular languages is called the Wagner hierarchy. It has been completely described by Klaus Wagner in [40]. Its length is the ordinal ω^ω. Wagner gave an automaton-like characterization of this hierarchy, based on the notions of chain and superchain, together with an algorithm to compute the Wadge (Wagner) degree of any given ω-regular language, see also [29,31–33].

The Wadge hierarchy of deterministic context-free ω-languages was determined by Duparc in [5,6]. Its length is the ordinal $\omega^{(\omega^2)}$. We do not know yet whether this hierarchy is decidable or not. But the Wadge hierarchy induced by deterministic partially blind 1-counter automata was described in an effective way in [10], and other partial decidability results were obtained in [11]. Then, it was proved in [13] that the Wadge hierarchy of 1-counter or context-free ω-languages and the Wadge hierarchy of effective analytic sets (which form the class of all the ω-languages accepted by non-deterministic Turing machines) are equal. Moreover similar results hold about the Wadge hierarchy of infinitary rational relations accepted by 2-tape Büchi automata, [14]. Finally, the Wadge hierarchy of ω-languages of deterministic Turing machines was determined by Selivanov in [30].

We consider in this paper acceptance of infinite words by Petri nets. Petri nets are used for the description of distributed systems [9,20,28], and form a very important mathematical model in Concurrency Theory that has been developed for general concurrent computation. In the context of Automata Theory, Petri nets may be defined as (partially) blind multicounter automata, as explained in [8,19,38]. First, one can distinguish between the places of a given Petri net by dividing them into the bounded ones (the number of tokens in such a place at any time is uniformly bounded) and the unbounded ones. Then each unbounded place may be seen as a partially blind counter, and the tokens in the bounded places determine the state of the partially blind multicounter automaton that is equivalent to the initial Petri net. The transitions of the Petri net may then be seen as the finite control of the partially blind multicounter automaton and the labels of these transitions are then the input symbols. The infinite behavior of Petri nets was first studied by Valk [38] and by Carstensen in the case of deterministic Petri nets [1].

On one side, the topological complexity of ω-languages of *deterministic* Petri nets is completely determined. They are $\mathbf{\Delta}_3^0$-sets and their Wadge hierarchy has been determined by Duparc, Finkel and Ressayre in [7]; its length is the ordinal ω^{ω^2}. On the other side, Finkel and Skrzypczak proved in [18] that there exist

Σ_3^0-complete, hence non Δ_3^0, ω-languages accepted by *non-deterministic* one-partially-blind-counter Büchi automata. The existence of a Σ_1^1-complete, hence non Borel, ω-language accepted by a Petri net was independently proved by Finkel and Skrzypczak in [17,34]. Moreover, Skrzypczak has proved in [34] that one blind counter is sufficient. In this paper, we fill the gap between Σ_3^0 and Σ_1^1 for Petri nets ω-languages. Notice that ω-languages accepted by (non-blind) one-counter Büchi automata have the same topological complexity as ω-languages of Turing machines, [13], but the non-blindness of the counter, i.e. the ability to use the zero-test of the counter, was essential in the proof of this result.

Using a simulation of a given real time 1-counter (with zero-test) Büchi automaton $\mathcal{A}$ accepting ω-words x over the alphabet Σ by a real time 4-blind-counter Büchi automaton $\mathcal{B}$ reading some special codes $h(x)$ of the words x, we prove here that ω-languages of *non-deterministic* Petri nets and effective analytic sets have the same topological complexity: the Borel and Wadge hierarchies of the class of ω-languages of Petri nets are equal to the Borel and Wadge hierarchies of the class of effective analytic sets. In particular, for each non-null recursive ordinal $\alpha < \omega_1^{\mathrm{CK}}$ there exist some Σ_α^0-complete and some Π_α^0-complete ω-languages of Petri nets, and the supremum of the set of Borel ranks of ω-languages of Petri nets is the ordinal γ_2^1, which is strictly greater than the first non-recursive ordinal ω_1^{CK}.

Notice that the topological equivalences we get in this paper are different from the language theoretical equivalences studied by Carstensen and Valk.

We also show that it is highly undecidable to determine the topological complexity of a Petri net ω-language. Moreover, we infer from the proofs of the above results that the equivalence and the inclusion problems for ω-languages of Petri nets are Π_2^1-complete, hence also highly undecidable.

The paper is organized as follows. In Sect. 2 we review the notions of (blind) counter automata and ω-languages. In Sect. 3 we recall notions of topology, and the Borel and Wadge hierarchies on a Cantor space. We prove our main results in Sect. 4. We show that the topological complexity of a Petri net ω-language is highly undecidable in Sect. 5. The equivalence and the inclusion problems for ω-languages of Petri nets are shown to be Π_2^1-complete in Sect. 6. Concluding remarks are given in Sect. 7.

2   Counter Automata

We assume the reader to be familiar with the theory of formal $(\omega$-)languages [27,36]. We recall the usual notations of formal language theory.

If Σ is a finite alphabet, a *non-empty finite word* over Σ is any sequence $x = a_1 \ldots a_k$, where $a_i \in \Sigma$ for $i = 1, \ldots, k$, and k is an integer ≥ 1. The *length* of x is k, denoted by $|x|$. The *empty word* is denoted by λ; its length is 0. $\Sigma^\star$ is the *set of finite words* (including the empty word) over Σ, and we denote $\Sigma^+ = \Sigma^\star \setminus \{\lambda\}$. A (finitary) *language* V over an alphabet Σ is a subset of $\Sigma^\star$.

The *first infinite ordinal* is ω. An *ω-word* over Σ is an ω-sequence $a_1 \ldots a_n \ldots$, where for all integers $i \geq 1$, $a_i \in \Sigma$. When $\sigma = a_1 \ldots a_n \ldots$ is an ω-word over Σ, we write $\sigma(n) = a_n$, $\sigma[n] = \sigma(1)\sigma(2)\ldots\sigma(n)$ for all $n \geq 1$ and $\sigma[0] = \lambda$.

The usual concatenation product of two finite words u and v is denoted $u \cdot v$ (and sometimes just uv). This product is extended to the product of a finite word u and an ω-word v: the infinite word $u \cdot v$ is then the ω-word such that:

$$(u \cdot v)(k) = u(k) \text{ if } k \leq |u| \text{ , and } (u \cdot v)(k) = v(k - |u|) \text{ if } k > |u|.$$

The *set of ω-words* over the alphabet Σ is denoted by Σ^ω. An *ω-language* V over an alphabet Σ is a subset of Σ^ω, and its complement (in Σ^ω) $\Sigma^\omega \setminus V$, denoted V^-.

The *prefix relation* is denoted $\sqsubseteq$: a finite word u is a *prefix* of a finite word v (respectively, an infinite word v), denoted $u \sqsubseteq v$, if and only if there exists a finite word w (respectively, an infinite word w), such that $v = u \cdot w$.

Let k be an integer ≥ 1. A k-counter machine has k *counters*, each of which containing a non-negative integer. The machine can test whether the content of a given counter is zero or not, but this is not possible if the counter is a blind (sometimes called partially blind, as in [19]) counter. This means that if a transition of the machine is enabled when the content of a blind counter is zero then the same transition is also enabled when the content of the same counter is a positive integer. And transitions depend on the letter read by the machine, the current state of the finite control, and the tests about the values of the counters. Notice that in the sequel we shall only consider real-time automata, i.e. λ-transitions are not allowed (but the general results of this paper will be easily extended to the case of non-real-time automata).

Formally a *real time k-counter machine* is a 4-tuple $\mathcal{M} = (K, \Sigma, \Delta, q_0)$, where K is a finite set of states, Σ is a finite input alphabet, $q_0 \in K$ is the initial state, and $\Delta \subseteq K \times \Sigma \times \{0,1\}^k \times K \times \{0,1,-1\}^k$ is the transition relation.

If the machine $\mathcal{M}$ is in state q and $c_i \in \mathbf{N}$ is the content of the ith counter $\mathcal{C}_i$ then the configuration (or global state) of $\mathcal{M}$ is the $(k+1)$-tuple $(q, c_1, \ldots, c_k)$.

For $a \in \Sigma$, $q, q' \in K$ and $(c_1, \ldots, c_k) \in \mathbf{N}^k$ such that $c_j = 0$ for $j \in E \subseteq \{1, \ldots, k\}$ and $c_j > 0$ for $j \notin E$, if $(q, a, i_1, \ldots, i_k, q', j_1, \ldots, j_k) \in \Delta$ where $i_j = 0$ for $j \in E$ and $i_j = 1$ for $j \notin E$, then we write:

$$a : (q, c_1, \ldots, c_k) \mapsto_{\mathcal{M}} (q', c_1 + j_1, \ldots, c_k + j_k).$$

Thus the transition relation must obviously satisfy: if $(q, a, i_1, \ldots, i_k, q', j_1, \ldots, j_k) \in \Delta$ and $i_m = 0$ for some $m \in \{1, \ldots, k\}$ then $j_m = 0$ or $j_m = 1$ (but j_m may not be equal to -1).

Moreover if the counters of $\mathcal{M}$ are blind, then, if $(q, a, i_1, \ldots, i_k, q', j_1, \ldots, j_k) \in \Delta$ holds, and $i_m = 0$ for some $m \in \{1, \ldots, k\}$ then $(q, a, i_1, \ldots, i_k, q', j_1, \ldots, j_k) \in \Delta$ also holds if $i_m = 1$ and the other integers are unchanged.

An ω-sequence of configurations $r = (q_i, c_1^i, \ldots c_k^i)_{i \geq 1}$ is called a run of $\mathcal{M}$ on an ω-word $\sigma = a_1 a_2 \ldots a_n \ldots$ over Σ iff:

(1) $(q_1, c_1^1, \ldots c_k^1) = (q_0, 0, \ldots, 0)$
(2) for each $i \geq 1$,

$$a_i : (q_i, c_1^i, \ldots c_k^i) \mapsto_{\mathcal{M}} (q_{i+1}, c_1^{i+1}, \ldots c_k^{i+1}).$$

For every such run r, $\text{In}(r)$ is the set of all states entered infinitely often during r.

Definition 1. *A Büchi k-counter automaton is a 5-tuple $\mathcal{M} = (K, \Sigma, \Delta, q_0, F)$, where $\mathcal{M}' = (K, \Sigma, \Delta, q_0)$ is a k-counter machine and $F \subseteq K$ is the set of accepting states. The ω-language accepted by $\mathcal{M}$ is:*
$$L(\mathcal{M}) = \{\sigma \in \Sigma^\omega \mid \text{ there exists a run } r \text{ of } \mathcal{M} \text{ on } \sigma \text{ such that } \text{In}(r) \cap F \neq \emptyset\}.$$

Definition 2. *A Muller k-counter automaton is a 5-tuple $\mathcal{M} = (K, \Sigma, \Delta, q_0, \mathcal{F})$, where $\mathcal{M}' = (K, \Sigma, \Delta, q_0)$ is a k-counter machine and $\mathcal{F} \subseteq 2^K$ is the set of accepting sets of states. The ω-language accepted by $\mathcal{M}$ is: $L(\mathcal{M}) = \{\sigma \in \Sigma^\omega \mid \text{ there exists a run } r \text{ of } \mathcal{M} \text{ on } \sigma \text{ such that } \text{In}(r) \in \mathcal{F}\}$.*

It is well known that an ω-language is accepted by a non-deterministic (real time) Büchi k-counter automaton iff it is accepted by a non-deterministic (real time) Muller k-counter automaton [8]. Notice that it cannot be shown without using the non determinism of automata and this result is no longer true in the deterministic case.

The class of ω-languages accepted by real time k-counter Büchi automata (respectively, real time k-blind-counter Büchi automata) is denoted **r-CL$(k)_\omega$**. (respectively, **r-BCL$(k)_\omega$**). (Notice that in previous papers, as in [13], the class **r-CL$(k)_\omega$** was denoted **r-BCL$(k)_\omega$** so we have slightly changed the notation in order to distinguish the different classes).

The class **CL$(1)_\omega$** is a strict subclass of the class **CFL$_\omega$** of context free ω-languages accepted by pushdown Büchi automata.

If we omit the counter of a real-time Büchi 1-counter automaton, then we simply get the notion of Büchi automaton. The class of ω-languages accepted by Büchi automata is the class of regular ω-languages.

3   Hierarchies in a Cantor Space

3.1   Borel Hierarchy and Analytic Sets

We assume the reader to be familiar with basic notions of topology which may be found in [24,25,27,36]. There is a natural metric on the set Σ^ω of infinite words over a finite alphabet Σ containing at least two letters which is called the *prefix metric* and is defined as follows. For $u, v \in \Sigma^\omega$ and $u \neq v$ let $\delta(u, v) = 2^{-l_{\text{pref}(u,v)}}$ where $l_{\text{pref}(u,v)}$ is the first integer n such that the $(n+1)^{st}$ letter of u is different from the $(n+1)^{st}$ letter of v. This metric induces on Σ^ω the usual Cantor topology in which the *open subsets* of Σ^ω are of the form $W \cdot \Sigma^\omega$, for $W \subseteq \Sigma^\star$. A set $L \subseteq \Sigma^\omega$ is a *closed set* iff its complement $\Sigma^\omega - L$ is an open set.

Define now the *Borel Hierarchy* of subsets of Σ^ω:

Definition 3. *For a non-null countable ordinal α, the classes $\mathbf{\Sigma}_\alpha^0$ and $\mathbf{\Pi}_\alpha^0$ of the Borel Hierarchy on the topological space Σ^ω are defined as follows: $\mathbf{\Sigma}_1^0$ is the class of open subsets of Σ^ω, $\mathbf{\Pi}_1^0$ is the class of closed subsets of Σ^ω, and for any countable ordinal $\alpha \geq 2$:*

$\mathbf{\Sigma}^0_\alpha$ is the class of countable unions of subsets of Σ^ω in $\bigcup_{\gamma<\alpha} \mathbf{\Pi}^0_\gamma$.

$\mathbf{\Pi}^0_\alpha$ is the class of countable intersections of subsets of Σ^ω in $\bigcup_{\gamma<\alpha} \mathbf{\Sigma}^0_\gamma$.

The class of *Borel sets* is $\mathbf{\Delta}^1_1 := \bigcup_{\xi<\omega_1} \mathbf{\Sigma}^0_\xi = \bigcup_{\xi<\omega_1} \mathbf{\Pi}^0_\xi$, where ω_1 is the first uncountable ordinal. There are also some subsets of Σ^ω which are not Borel. In particular the class of Borel subsets of Σ^ω is strictly included into the class $\mathbf{\Sigma}^1_1$ of *analytic sets* which are obtained by projection of Borel sets.

Definition 4. *A subset A of Σ^ω is in the class $\mathbf{\Sigma}^1_1$ of analytic sets iff there exists another finite set Y and a Borel subset B of $(\Sigma \times Y)^\omega$ such that $x \in A \leftrightarrow \exists y \in Y^\omega$ such that $(x, y) \in B$, where (x, y) is the infinite word over the alphabet $\Sigma \times Y$ such that $(x, y)(i) = (x(i), y(i))$ for each integer $i \geq 1$.*

We now define completeness with regard to reduction by continuous functions. For a countable ordinal $\alpha \geq 1$, a set $F \subseteq \Sigma^\omega$ is said to be a $\mathbf{\Sigma}^0_\alpha$ (respectively, $\mathbf{\Pi}^0_\alpha$, $\mathbf{\Sigma}^1_1$)-*complete set* iff for any set $E \subseteq Y^\omega$ (with Y a finite alphabet): $E \in \mathbf{\Sigma}^0_\alpha$ (respectively, $E \in \mathbf{\Pi}^0_\alpha$, $E \in \mathbf{\Sigma}^1_1$) iff there exists a continuous function $f : Y^\omega \to \Sigma^\omega$ such that $E = f^{-1}(F)$.

Let us now recall the definition of the arithmetical hierarchy of ω-languages, see for example [25,36]. Let Σ be a finite alphabet. An ω-language $L \subseteq \Sigma^\omega$ belongs to the class Σ_n iff there exists a recursive relation $R_L \subseteq (\mathbb{N})^{n-1} \times \Sigma^\star$ such that $L = \{\sigma \in \Sigma^\omega \mid \exists a_1 \ldots Q_n a_n \ (a_1, \ldots, a_{n-1}, \sigma[a_n + 1]) \in R_L\}$, where Q_i is one of the quantifiers $\forall$ or $\exists$ (not necessarily in an alternating order). An ω-language $L \subseteq \Sigma^\omega$ belongs to the class Π_n if and only if its complement $\Sigma^\omega - L$ belongs to the class Σ_n. The inclusion relations that hold between the classes Σ_n and Π_n are the same as for the corresponding classes of the Borel hierarchy and the classes Σ_n and Π_n are strictly included in the respective classes $\mathbf{\Sigma}^0_n$ and $\mathbf{\Pi}^0_n$ of the Borel hierarchy.

As in the case of the Borel hierarchy, projections of arithmetical sets (of the second Π-class) lead beyond the arithmetical hierarchy, to the analytical hierarchy of ω-languages. The first class of the analytical hierarchy of ω-languages is the (lightface) class Σ^1_1 of effective analytic sets. An ω-language $L \subseteq \Sigma^\omega$ belongs to the class Σ^1_1 if and only if there exists a recursive relation $R_L \subseteq (\mathbb{N}) \times \{0,1\}^\star \times \Sigma^\star$ such that: $L = \{\sigma \in \Sigma^\omega \mid \exists \tau (\tau \in \{0,1\}^\omega \wedge \forall n \exists m ((n, \tau[m], \sigma[m]) \in R_L))\}$. Thus an ω-language $L \subseteq \Sigma^\omega$ is in the class Σ^1_1 iff it is the projection of an ω-language over the alphabet $\{0,1\} \times \Sigma$ which is in the class Π_2.

Kechris, Marker and Sami proved in [22] that the supremum of the set of Borel ranks of (lightface) Π^1_1, so also of (lightface) Σ^1_1, sets is the ordinal γ^1_2. This ordinal is precisely defined in [22]. It holds that $\omega^{\mathrm{CK}}_1 < \gamma^1_2$, where ω^{CK}_1 is the first non-recursive ordinal, called the Church-Kleene ordinal.

Notice that it seems still unknown whether *every* non null ordinal $\gamma < \gamma^1_2$ is the Borel rank of a (lightface) Π^1_1 (or Σ^1_1) set. On the other hand it is known that for every ordinal $\gamma < \omega^{\mathrm{CK}}_1$ there exist some $\mathbf{\Sigma}^0_\gamma$-complete and $\mathbf{\Pi}^0_\gamma$-complete sets in the class Δ^1_1.

Recall that a Büchi Turing machine is just a Turing machine working on infinite inputs with a Büchi-like acceptance condition, and that the class of ω-languages accepted by Büchi Turing machines is the class Σ^1_1 [3,36].

3.2   Wadge Hierarchy

We now introduce the Wadge hierarchy, which is a great refinement of the Borel hierarchy defined via reductions by continuous functions, [4,39].

Definition 5 (Wadge [39]). *Let X, Y be two finite alphabets. For $L \subseteq X^\omega$ and $L' \subseteq Y^\omega$, L is said to be Wadge reducible to L' ($L \leq_W L'$) iff there exists a continuous function $f : X^\omega \to Y^\omega$, such that $L = f^{-1}(L')$. L and L' are Wadge equivalent iff $L \leq_W L'$ and $L' \leq_W L$. This will be denoted by $L \equiv_W L'$. And we shall say that $L <_W L'$ iff $L \leq_W L'$ but not $L' \leq_W L$.*

A set $L \subseteq X^\omega$ is said to be self dual iff $L \equiv_W L^-$, and otherwise it is said to be non self dual.

The relation $\leq_W$ is reflexive and transitive, and $\equiv_W$ is an equivalence relation. The *equivalence classes* of $\equiv_W$ are called *Wadge degrees*. The Wadge hierarchy WH is the class of Borel subsets of a set X^ω, where X is a finite set, equipped with $\leq_W$ and with $\equiv_W$.

For $L \subseteq X^\omega$ and $L' \subseteq Y^\omega$, if $L \leq_W L'$ and $L = f^{-1}(L')$ where f is a continuous function from X^ω into Y^ω, then f is called a continuous reduction of L to L'. Intuitively it means that L is less complicated than L' because to check whether $x \in L$ it suffices to check whether $f(x) \in L'$ where f is a continuous function. Hence the Wadge degree of an ω-language is a measure of its topological complexity.

Notice that in the above definition, we consider that a subset $L \subseteq X^\omega$ is given together with the alphabet X.

We can now define the *Wadge class* of a set L:

Definition 6. *Let L be a subset of X^ω. The Wadge class of L is :*
$$[L] = \{L' \mid L' \subseteq Y^\omega \text{ for a finite alphabet } Y \text{ and } L' \leq_W L\}.$$

Recall that each *Borel class* $\mathbf{\Sigma}^0_\alpha$ and $\mathbf{\Pi}^0_\alpha$ is a *Wadge class*. A set $L \subseteq X^\omega$ is a $\mathbf{\Sigma}^0_\alpha$ (respectively $\mathbf{\Pi}^0_\alpha$)-*complete set* iff for any set $L' \subseteq Y^\omega$, L' is in $\mathbf{\Sigma}^0_\alpha$ (respectively $\mathbf{\Pi}^0_\alpha$) iff $L' \leq_W L$.

There is a close relationship between Wadge reducibility and games which we now introduce.

Definition 7. *Let $L \subseteq X^\omega$ and $L' \subseteq Y^\omega$. The Wadge game $W(L, L')$ is a game with perfect information between two players, player 1 who is in charge of L and player 2 who is in charge of L'. Player 1 first writes a letter $a_1 \in X$, then player 2 writes a letter $b_1 \in Y$, then player 1 writes a letter $a_2 \in X$, and so on. The two players alternatively write letters a_n of X for player 1 and b_n of Y for player 2. After ω steps, the player 1 has written an ω-word $a \in X^\omega$ and the player 2 has written an ω-word $b \in Y^\omega$. The player 2 is allowed to skip, even infinitely often, provided he really writes an ω-word in ω steps. The player 2 wins the play iff $[a \in L \leftrightarrow b \in L']$, i.e. iff :*

$$[(a \in L \text{ and } b \in L') \text{ or } (a \notin L \text{ and } b \notin L' \text{ and } b \text{ is infinite})].$$

Recall that a strategy for player 1 is a function $\sigma : (Y \cup \{s\})^\star \to X$. And a strategy for player 2 is a function $f : X^+ \to Y \cup \{s\}$. The strategy σ is a winning strategy for player 1 iff he always wins a play when he uses the strategy σ, i.e. when the nth letter he writes is given by $a_n = \sigma(b_1 \cdots b_{n-1})$, where b_i is the letter written by player 2 at step i and $b_i = s$ if player 2 skips at step i. A winning strategy for player 2 is defined in a similar manner.

Martin's Theorem states that every Gale-Stewart game $G(X)$ (see [21]), with X a Borel set, is determined and this implies the following:

Theorem 8 (Wadge). *Let $L \subseteq X^\omega$ and $L' \subseteq Y^\omega$ be two Borel sets, where X and Y are finite alphabets. Then the Wadge game $W(L, L')$ is determined: one of the two players has a winning strategy. And $L \leq_W L'$ iff the player 2 has a winning strategy in the game $W(L, L')$.*

Theorem 9 (Wadge). *Up to the complement and $\equiv_W$, the class of Borel subsets of X^ω, for a finite alphabet X having at least two letters, is a well ordered hierarchy. There is an ordinal $|WH|$, called the length of the hierarchy, and a map d_W^0 from WH onto $|WH| - \{0\}$, such that for all $L, L' \subseteq X^\omega$:*
$$d_W^0 L < d_W^0 L' \leftrightarrow L <_W L' \text{ and}$$
$$d_W^0 L = d_W^0 L' \leftrightarrow [L \equiv_W L' \text{ or } L \equiv_W L'^-].$$

The Wadge hierarchy of Borel sets of **finite rank** has length $^1\varepsilon_0$ where $^1\varepsilon_0$ is the limit of the ordinals α_n defined by $\alpha_1 = \omega_1$ and $\alpha_{n+1} = \omega_1^{\alpha_n}$ for n a non negative integer, ω_1 being the first non countable ordinal. Then $^1\varepsilon_0$ is the first fixed point of the ordinal exponentiation of base ω_1. The length of the Wadge hierarchy of Borel sets in $\mathbf{\Delta}_\omega^0 = \mathbf{\Sigma}_\omega^0 \cap \mathbf{\Pi}_\omega^0$ is the ω_1^{th} fixed point of the ordinal exponentiation of base ω_1, which is a much larger ordinal. The length of the whole Wadge hierarchy of Borel sets is a huge ordinal, with regard to the ω_1^{th} fixed point of the ordinal exponentiation of base ω_1. It is described in [4,39] by the use of the Veblen functions.

4   Wadge Degrees of ω-Languages of Petri Nets

We are firstly going to prove the following result.

Theorem 10. *The Wadge hierarchy of the class* **r-BCL**$(4)_\omega$ *is equal to the Wadge hierarchy of the class* **r-CL**$(1)_\omega$.

In order to prove this result, we first define a coding of ω-words over a finite alphabet Σ by ω-words over the alphabet $\Sigma \cup \{A, B, 0\}$ where A, B and 0 are new letters not in Σ.

We shall code an ω-word $x \in \Sigma^\omega$ by the ω-word $h(x)$ defined by

$$h(x) = A0x(1)B0^2x(2)A \cdots B0^{2n}x(2n)A0^{2n+1}x(2n+1)B \cdots$$

This coding defines a mapping $h : \Sigma^\omega \to (\Sigma \cup \{A, B, 0\})^\omega$.

The function h is continuous because for all ω-words $x, y \in \Sigma^\omega$ and each positive integer n, it holds that $\delta(x, y) < 2^{-n} \to \delta(h(x), h(y)) < 2^{-n}$.

We now state the following lemma.

Lemma 11. *Let $\mathcal{A}$ be a real time 1-counter Büchi automaton accepting ω-words over the alphabet Σ. Then one can construct a real time 4-blind-counter Büchi automaton $\mathcal{B}$ reading words over the alphabet $\Gamma = \Sigma \cup \{A, B, 0\}$, such that $L(\mathcal{A})$ $= h^{-1}(L(\mathcal{B}))$, i.e. $\quad \forall x \in \Sigma^{\omega} \quad h(x) \in L(\mathcal{B}) \longleftrightarrow x \in L(\mathcal{A})$.*

Proof. Let $\mathcal{A} = (K, \Sigma, \Delta, q_0, F)$ be a real time 1-counter Büchi automaton accepting ω-words over the alphabet Σ. We are going to explain informally the behaviour of the 4-blind-counter Büchi automaton $\mathcal{B}$ when reading an ω-word of the form $h(x)$, even if we are going to see that $\mathcal{B}$ may also accept some infinite words which do not belong to the range of h. Recall that $h(x)$ is of the form

$$h(x) = A0x(1)B0^2x(2)A \cdots B0^{2n}x(2n)A0^{2n+1}x(2n+1)B \cdots$$

Notice that in particular every ω-word in $h(\Sigma^{\omega})$ is of the form:

$$y = A0^{n_1}x(1)B0^{n_2}x(2)A \cdots B0^{n_{2n}}x(2n)A0^{n_{2n+1}}x(2n+1)B \cdots$$

where for all $i \geq 1$, $n_i > 0$ is a positive integer, and $x(i) \in \Sigma$.

Moreover it is easy to see that the set of ω-words $y \in \Gamma^{\omega}$ which can be written in the above form is a regular ω-language $\mathcal{R} \subseteq \Gamma^{\omega}$, and thus we can assume, using a classical product construction (see for instance [27]), that the automaton $\mathcal{B}$ will only accept some ω-words of this form.

Now the reading by the automaton $\mathcal{B}$ of an ω-word of the above form

$$y = A0^{n_1}x(1)B0^{n_2}x(2)A \cdots B0^{n_{2n}}x(2n)A0^{n_{2n+1}}x(2n+1)B \cdots$$

will give a decomposition of the ω-word y of the following form:

$$y = Au_1v_1x(1)Bu_2v_2x(2)Au_3v_3x(3)B \cdots$$
$$\cdots Bu_{2n}v_{2n}x(2n)Au_{2n+1}v_{2n+1}x(2n+1)B \cdots$$

where, for all integers $i \geq 1$, $u_i, v_i \in 0^{\star}$, $x(i) \in \Sigma$, $|u_1| = 0$.

The automaton $\mathcal{B}$ will use its four *blind* counters, which we denote $\mathcal{C}_1, \mathcal{C}_2, \mathcal{C}_3, \mathcal{C}_4$, in the following way. Recall that the automaton $\mathcal{B}$ being non-deterministic, we do not describe the unique run of $\mathcal{B}$ on y, but the general case of a possible run.

At the beginning of the run, the value of each of the four counters is equal to zero. Then the counter $\mathcal{C}_1$ is increased of $|u_1|$ when reading u_1, i.e. the counter $\mathcal{C}_1$ is actually not increased since $|u_1| = 0$ and the finite control is here used to check this. Then the counter $\mathcal{C}_2$ is increased of 1 for each letter 0 of v_1 which is read until the automaton reads the letter $x(1)$ and then the letter B. Notice that at this time the values of the counters $\mathcal{C}_3$ and $\mathcal{C}_4$ are still equal to zero. Then the behaviour of the automaton $\mathcal{B}$ when reading the next segment $0^{n_2}x(2)A$ is as follows. The counters $\mathcal{C}_1$ is firstly decreased of 1 for each letter 0 read, when reading k_2 letters 0, where $k_2 \geq 0$ (notice that here $k_2 = 0$ because the value of the counter $\mathcal{C}_1$ being equal to zero, it cannot decrease under 0). Then the counter $\mathcal{C}_2$ is decreased of 1 for each letter 0 read, and next the automaton has to read one more letter 0, leaving unchanged the counters $\mathcal{C}_1$ and $\mathcal{C}_2$, before reading the letter $x(2)$. The end of the decreasing mode of $\mathcal{C}_1$ coincide with the beginning of the decreasing mode of $\mathcal{C}_2$, and this change may occur in a *non-deterministic way* (because the automaton $\mathcal{B}$ cannot check whether the value of $\mathcal{C}_1$ is equal to zero). Now we describe the behaviour of the counters $\mathcal{C}_3$ and $\mathcal{C}_4$ when reading the segment $0^{n_2}x(2)A$. Using its finite control, the automaton $\mathcal{B}$ has checked that $|u_1| = 0$, and then if there is a transition of the automaton $\mathcal{A}$ such that $x(1) : (q_0, |u_1|) \mapsto_\mathcal{A} (q_1, |u_1| + N_1)$ then the counter $\mathcal{C}_3$ is increased of 1 for each letter 0 read, during the reading of the $k_2 + N_1$ first letters 0 of 0^{n_2}, where k_2 is described above as the number of which the counter $\mathcal{C}_1$ has been decreased. This determines u_2 by $|u_2| = k_2 + N_1$ and then the counter $\mathcal{C}_4$ is increased by 1 for each letter 0 read until $\mathcal{B}$ reads $x(2)$, and this determines v_2. Notice that the automaton $\mathcal{B}$ keeps in its finite control the memory of the state q_1 of the automaton $\mathcal{A}$, and that, after having read the segment $0^{n_2} = u_2v_2$, the values of the counters $\mathcal{C}_3$ and $\mathcal{C}_4$ are respectively $|\mathcal{C}_3| = |u_2| = k_2 + N_1$ and $|\mathcal{C}_4| = |v_2| = n_2 - (|u_2|)$.

Now the run will continue. Notice that generally when reading a segment $B0^{n_{2n}}x(2n)A$ the counters $\mathcal{C}_1$ and $\mathcal{C}_2$ will successively decrease when reading the first $(n_{2n} - 1)$ letters 0 and then will remain unchanged when reading the last letter 0, and the counters $\mathcal{C}_3$ and $\mathcal{C}_4$ will successively increase, when reading the (n_{2n}) letters 0. Again the end of the decreasing mode of $\mathcal{C}_1$ coincide with the beginning of the decreasing mode of $\mathcal{C}_2$, and this change may occur in a *non-deterministic way*. But the automaton has kept in its finite control whether $|u_{2n-1}| = 0$ or not and also a state q_{2n-2} of the automaton $\mathcal{A}$. Now, if there is a transition of the automaton $\mathcal{A}$ such that $x(2n - 1) : (q_{2n-2}, |u_{2n-1}|) \mapsto_\mathcal{A} (q_{2n-1}, |u_{2n-1}| + N_{2n-1})$ for some integer $N_{2n-1} \in \{-1; 0; 1\}$, and the counter $\mathcal{C}_1$ is decreased of 1 for each letter 0 read, when reading k_{2n} first letters 0 of $0^{n_{2n}}$, then the counter $\mathcal{C}_3$ is increased of 1 for each letter 0 read, during the reading of the $k_{2n} + N_{2n-1}$ first letters 0 of $0^{n_{2n}}$, and next the counter $\mathcal{C}_4$ is increased by 1 for each letter 0 read until $\mathcal{B}$ reads $x(2n)$, and this determines v_{2n}. Then after having read the segment $0^{n_{2n}} = u_{2n}v_{2n}$, the values of the counters $\mathcal{C}_3$ and $\mathcal{C}_4$ have respectively increased of $|u_{2n}| = k_{2n} + N_{2n-1}$ and $|v_{2n}| = n_{2n} - |u_{2n}|$. Notice that one cannot ensure that, after the reading of $0^{n_{2n}} = u_{2n}v_{2n}$, the exact values of these counters are $|\mathcal{C}_3| = |u_{2n}| = k_{2n} + N_{2n-1}$ and $|\mathcal{C}_4| = |v_{2n}| = n_{2n} - |u_{2n}|$. Actually this is due to the fact that one cannot ensure that the values of $\mathcal{C}_3$ and $\mathcal{C}_4$ are equal to zero at the beginning of the reading of the segment $B0^{n_{2n}}x(2n)A$ although we will see this is true and important in the particular case of a word of the form $y = h(x)$.

The run will continue in a similar manner during the reading of the next segment $A0^{n_{2n+1}}x(2n + 1)B$, but here the role of the counters $\mathcal{C}_1$ and $\mathcal{C}_2$ on one side, and of the counters $\mathcal{C}_3$ and $\mathcal{C}_4$ on the other side, will be interchanged. More precisely the counters $\mathcal{C}_3$ and $\mathcal{C}_4$ will successively decrease when reading the first $(n_{2n+1} - 1)$ letters 0 and then will remain unchanged when reading the last letter 0, and the counters $\mathcal{C}_1$ and $\mathcal{C}_2$ will successively increase, when reading the (n_{2n+1}) letters 0. The end of the decreasing mode of $\mathcal{C}_3$ coincide with the beginning of the decreasing mode of $\mathcal{C}_4$, and this change may occur in a *non-deterministic way.* But the automaton has kept in its finite control whether $|u_{2n}| = 0$ or not and also a state q_{2n-1} of the automaton $\mathcal{A}$. Now, if there is a transition of the automaton $\mathcal{A}$ such that $x(2n) : (q_{2n-1}, |u_{2n}|) \mapsto_{\mathcal{A}} (q_{2n}, |u_{2n}| + N_{2n})$ for some integer $N_{2n} \in \{-1; 0, 1\}$, and the counter $\mathcal{C}_3$ is decreased of 1 for each letter 0 read, when reading k_{2n+1} first letters 0 of $0^{n_{2n+1}}$, then the counter $\mathcal{C}_1$ is increased of 1 for each letter 0 read, during the reading of the $k_{2n+1} + N_{2n}$ first letters 0 of $0^{n_{2n+1}}$, and next the counter $\mathcal{C}_2$ is increased by 1 for each letter 0 read until $\mathcal{B}$ reads $x(2n+1)$, and this determines v_{2n+1}. Then after having read the segment $0^{n_{2n+1}} = u_{2n+1}v_{2n+1}$, the values of the counters $\mathcal{C}_1$ and $\mathcal{C}_2$ have respectively increased of $|u_{2n+1}| = k_{2n+1} + N_{2n}$ and $|v_{2n+1}| = n_{2n+1} - |u_{2n+1}|$. Notice that again one cannot ensure that, after the reading of $0^{n_{2n+1}} = u_{2n+1}v_{2n+1}$, the exact values of these counters are $|\mathcal{C}_1| = |u_{2n+1}| = k_{2n+1} + N_{2n}$ and $|\mathcal{C}_2| = |v_{2n+1}| = n_{2n+1} - |u_{2n+1}|$. This is due to the fact that one cannot ensure that the values of $\mathcal{C}_1$ and $\mathcal{C}_2$ are equal to zero at the beginning of the reading of the segment $A0^{n_{2n+1}}x(2n+1)B$ although we will see this is true and important in the particular case of a word of the form $y = h(x)$.

The run then continues in the same way if it is possible and in particular if there is no blocking due to the fact that one of the counters of the automaton $\mathcal{B}$ would have a negative value.

Now an ω-word $y \in \mathcal{R} \subseteq \Gamma^\omega$ of the above form will be accepted by the automaton $\mathcal{B}$ if there is such an infinite run for which a final state $q_f \in F$ of the automaton $\mathcal{A}$ has been stored infinitely often in the finite control of $\mathcal{B}$ in the way which has just been described above.

We now consider the particular case of an ω-word of the form $y = h(x)$, for some $x \in \Sigma^\omega$. Let then

$$y = h(x) = A0x(1)B0^2x(2)A0^3x(3)B\cdots B0^{2n}x(2n)A0^{2n+1}x(2n+1)B\cdots$$

We are going to show that, if y is accepted by the automaton $\mathcal{B}$, then $x \in L(\mathcal{A})$. Let us consider a run of the automaton $\mathcal{B}$ on y as described above and which is an accepting run. We first show by induction on $n \geq 1$, that after having read an initial segment of the form $A0x(1)B0^2x(2)A\cdots A0^{2n-1}x(2n-1)B$, the values of the counters $\mathcal{C}_3$ and $\mathcal{C}_4$ are equal to zero, and the values of the counters $\mathcal{C}_1$ and $\mathcal{C}_2$ satisfy $|\mathcal{C}_1| + |\mathcal{C}_2| = 2n - 1$. And similarly after having read an initial segment of the form $A0x(1)B0^2x(2)A\cdots B0^{2n}x(2n)A$ the values of the counters $\mathcal{C}_1$ and $\mathcal{C}_2$ are equal to zero, and the values of the counters $\mathcal{C}_3$ and $\mathcal{C}_4$ satisfy $|\mathcal{C}_3| + |\mathcal{C}_4| = 2n$.

For $n = 1$, we have seen that after having read the initial segment $A0x(1)B$, the values of the counters $\mathcal{C}_1$ and $\mathcal{C}_2$ will be respectively 0 and $|v_1|$ and here $|v_1| = 1$ and thus $|\mathcal{C}_1| + |\mathcal{C}_2| = 1$. On the other hand the counters $\mathcal{C}_3$ and $\mathcal{C}_4$ have not yet increased so that the value of each of these counters is equal to zero. During the reading of the segment 0^2 of $0^2x(2)A$ the counters $\mathcal{C}_1$ and $\mathcal{C}_2$ successively decrease. But here $\mathcal{C}_1$ cannot decrease (with the above notations, it holds that $k_2 = 0$) so $\mathcal{C}_2$ must decrease of 1 because after the decreasing mode the automaton $\mathcal{B}$ must read a last letter 0 without decreasing the counters $\mathcal{C}_1$ and $\mathcal{C}_2$ and then the letter $x(2) \in \Sigma$. Thus after having read $0^2x(2)A$ the values of C_1 and C_2 are equal to zero. Moreover the counters C_3 and C_4 had their values equal to zero at the beginning of the reading of $0^2x(2)A$ and they successively increase during the reading of 0^2 and they remain unchanged during the reading of $x(2)A$ so that their values satisfy $|C_3| + |C_4| = 2$ after the reading of $0^2x(2)A$.

Assume now that for some integer $n > 1$ the claim is proved for all integers $k < n$ and let us prove it for the integer n. By induction hypothesis we know that at the beginning of the reading of the segment $A0^{2n-1}x(2n-1)B$ of y, the values of the counters C_1 and C_2 are equal to zero, and the values of the counters C_3 and C_4 satisfy $|C_3| + |C_4| = 2n - 2$. When reading the $(2n-2)$ first letters 0 of $A0^{2n-1}x(2n-1)B$ the counters C_3 and C_4 successively decrease and they must decrease completely because after there must remain only one letter 0 to be read by B before the letter $x(2n-1)$. Therefore after the reading of $A0^{2n-1}x(2n-1)B$ the values of the counters C_3 and C_4 are equal to zero. And since the values of the counters C_1 and C_2 are equal to zero before the reading of $0^{2n-1}x(2n-1)B$ and these counters successively increase during the reading of 0^{2n-1}, their values satisfy $|C_1| + |C_2| = 2n - 1$ after the reading of $A0^{2n-1}x(2n-1)B$. We can reason in a very similar manner for the reading of the next segment $B0^{2n}x(2n)A$, the role of the counters C_1 and C_2 on one side, and of the counters C_3 and C_4 on the other side, being simply interchanged. This ends the proof of the claim by induction on n.

It is now easy to see by induction that for each integer $n \geq 2$, it holds that $k_n = |u_{n-1}|$. Then, since with the above notations we have $|u_{n+1}| = k_{n+1} + N_n = |u_n| + N_n$, and there is a transition of the automaton $\mathcal{A}$ such that $x(n) : (q_{n-1}, |u_n|) \mapsto_\mathcal{A} (q_n, |u_n| + N_n)$ for $N_n \in \{-1; 0; 1\}$, it holds that $x(n) : (q_{n-1}, |u_n|) \mapsto_\mathcal{A} (q_n, |u_{n+1}|)$. Therefore the sequence $(q_i, |u_i|)_{i \geq 0}$ is an accepting run of the automaton $\mathcal{A}$ on the ω-word x and $x \in L(\mathcal{A})$. Notice that the state q_0 of the sequence $(q_i)_{i \geq 0}$ is also the initial state of $\mathcal{A}$.

Conversely, it is easy to see that if $x \in L(\mathcal{A})$ then there exists an accepting run of the automaton $\mathcal{B}$ on the ω-word $h(x)$ and $h(x) \in L(\mathcal{B})$. □

The above Lemma 11 shows that, given a real time 1-counter (with zero-test) Büchi automaton $\mathcal{A}$ accepting ω-words over the alphabet Σ, one can construct a real time 4-blind-counter Büchi automaton $\mathcal{B}$ which can simulate the 1-counter automaton $\mathcal{A}$ on the code $h(x)$ of the word x. On the other hand, we cannot describe precisely the ω-words which are accepted by $\mathcal{B}$ but are not in the set $h(\Sigma^\omega)$. However we can see that all these words have a special shape, as stated by the following lemma.

Lemma 12. *Let $\mathcal{A}$ be a real time 1-counter Büchi automaton accepting ω-words over the alphabet Σ, and let $\mathcal{B}$ be the real time 4-blind-counter Büchi automaton reading words over the alphabet $\Gamma = \Sigma \cup \{A, B, 0\}$ which is constructed in the proof of Lemma 11. Let $y \in L(\mathcal{B}) \setminus h(\Sigma^\omega)$ being of the following form*

$$y = A0^{n_1}x(1)B0^{n_2}x(2)A0^{n_3}x(3)B \cdots B0^{n_{2n}}x(2n)A0^{n_{2n+1}}x(2n+1)B \cdots$$

and let i_0 be the smallest integer i such that $n_i \neq i$. Then it holds that either $i_0 = 1$ or $n_{i_0} < i_0$.

Proof. Assume first that $y \in L(\mathcal{B}) \setminus h(\Sigma^\omega)$ is of the following form

$$y = A0^{n_1}x(1)B0^{n_2}x(2)A \cdots B0^{n_{2n}}x(2n)A0^{n_{2n+1}}x(2n+1)B \cdots$$

and that the smallest integer i such that $n_i \neq i$ is an even integer $i_0 > 1$. Consider an infinite accepting run of $\mathcal{B}$ on y. It follows from the proof of the above Lemma 11 that after the reading of the initial segment

$$A0^{n_1}x(1)B0^{n_2}x(2)A \cdots A0^{i_0-1}x(i_0 - 1)B$$

the values of the counters $\mathcal{C}_3$ and $\mathcal{C}_4$ are equal to zero, and the values of the counters $\mathcal{C}_1$ and $\mathcal{C}_2$ satisfy $|\mathcal{C}_1| + |\mathcal{C}_2| = i_0 - 1$. Thus since the two counters must successively decrease during the next $n_{i_0} - 1$ letters 0, it holds that $n_{i_0} - 1 \leq i_0 - 1$ because otherwise either $\mathcal{C}_1$ or $\mathcal{C}_2$ would block. Therefore $n_{i_0} < i_0$ since $n_{i_0} \neq i_0$ by definition of i_0. The reasoning is very similar in the case of an odd integer i_0, the role of the counters $\mathcal{C}_1$ and $\mathcal{C}_2$ on one side, and of the counters $\mathcal{C}_3$ and $\mathcal{C}_4$ on the other side, being simply interchanged. $\qquad\square$

Let $\mathcal{L} \subseteq \Gamma^\omega$ be the ω-language containing the ω-words over Γ which belong to one of the following ω-languages.

- $\mathcal{L}_1$ is the set of ω-words over the alphabet $\Sigma \cup \{A, B, 0\}$ which have not any initial segment in $A \cdot 0 \cdot \Sigma \cdot B$.
- $\mathcal{L}_2$ is the set of ω-words over the alphabet $\Sigma \cup \{A, B, 0\}$ which contain a segment of the form $B \cdot 0^n \cdot a \cdot A \cdot 0^m \cdot b$ or of the form $A \cdot 0^n \cdot a \cdot B \cdot 0^m \cdot b$ for some letters $a, b \in \Sigma$ and some positive integers $m \leq n$.

Lemma 13. *The ω-language $\mathcal{L}$ is accepted by a (non-deterministic) real-time 1-blind counter Büchi automaton.*

Proof. First, it is easy to see that $\mathcal{L}_1$ is in fact a regular ω-language, and thus it is also accepted by a real-time 1-blind counter Büchi automaton (even without active counter). On the other hand it is also easy to construct a real time 1-blind counter Büchi automaton accepting the ω-language $\mathcal{L}_2$. The class of ω-languages accepted by *non-deterministic* real time 1-blind counter Büchi automata being closed under finite union in an effective way, one can construct a real time 1-blind counter Büchi automaton accepting $\mathcal{L}$. $\qquad\square$

Lemma 14. *Let $\mathcal{A}$ be a real time 1-counter Büchi automaton accepting ω-words over the alphabet Σ. Then one can construct a real time 4-blind counter Büchi automaton $\mathcal{P}_\mathcal{A}$ such that $L(\mathcal{P}_\mathcal{A}) = h(L(\mathcal{A})) \cup \mathcal{L}$.*

Proof. Let $\mathcal{A}$ be a real time 1-counter Büchi automaton accepting ω-words over Σ. We have seen in the proof of Lemma 11 that one can construct a real time 4-blind counter Büchi automaton $\mathcal{B}$ reading words over the alphabet $\Gamma = \Sigma \cup \{A, B, 0\}$, such that $L(\mathcal{A}) = h^{-1}(L(\mathcal{B}))$, i.e. $\forall x \in \Sigma^\omega \quad h(x) \in L(\mathcal{B}) \longleftrightarrow x \in L(\mathcal{A})$. Moreover By Lemma 12 it holds that $L(\mathcal{B}) \setminus h(\Sigma^\omega) \subseteq \mathcal{L}$. and thus $h(L(\mathcal{A})) \cup \mathcal{L} = L(\mathcal{B}) \cup \mathcal{L}$. But By Lemma 13 the ω-language $\mathcal{L}$ is accepted by a (non-deterministic) real-time 1-blind counter Büchi automaton, hence also by a real-time 4-blind counter Büchi automaton. The class of ω-languages accepted by (non-deterministic) real-time 4-blind counter Büchi automata is closed under finite union in an effective way, and thus one can construct a real time 4-blind counter Büchi automaton $\mathcal{P}_A$ such that $L(\mathcal{P}_A) = h(L(\mathcal{A})) \cup \mathcal{L}$.   □

We are now going to prove that if $L(\mathcal{A}) \subseteq \Sigma^\omega$ is accepted by a real time 1-counter automaton $\mathcal{A}$ with a Büchi acceptance condition then $L(\mathcal{P}_A) = h(L(\mathcal{A})) \cup \mathcal{L}$ will have the same Wadge degree as the ω-language $L(\mathcal{A})$, except for some very simple cases.

We first notice that $h(\Sigma^\omega)$ is a closed subset of Γ^ω. Indeed it is the image of the compact set Σ^ω by the continuous function h, and thus it is a compact hence also closed subset of $\Gamma^\omega = (\Sigma \cup \{A, B, 0\})^\omega$. Thus its complement $h(\Sigma^\omega)^- = (\Sigma \cup \{A, B, 0\})^\omega - h(\Sigma^\omega)$ is an open subset of Γ^ω. Moreover the set $\mathcal{L}$ is an open subset of Γ^ω, as it can be easily seen from its definition and one can easily define, from the definition of the ω-language $\mathcal{L}$, a finitary language $V \subseteq \Gamma^\star$ such that $\mathcal{L} = V \cdot \Gamma^\omega$. We shall also denote $\mathcal{L}' = h(\Sigma^\omega)^- \setminus \mathcal{L}$ so that Γ^ω is the disjoint union $\Gamma^\omega = h(\Sigma^\omega) \cup \mathcal{L} \cup \mathcal{L}'$. Notice that $\mathcal{L}'$ is the difference of the two open sets $h(\Sigma^\omega)^-$ and $\mathcal{L}$.

We now wish to return to the proof of the above Theorem 10 stating that the Wadge hierarchy of the class **r-BCL**$(4)_\omega$ is equal to the Wadge hierarchy of the class **r-CL**$(1)_\omega$.

To prove this result we firstly consider non self dual Borel sets. We recall the definition of Wadge degrees introduced by Duparc in [4] and which is a slight modification of the previous one.

Definition 15

(a) $d_w(\emptyset) = d_w(\emptyset^-) = 1$
(b) $d_w(L) = sup\{d_w(L') + 1 \mid L' \text{ non self dual and } L' <_W L\}$ *(for either L self dual or not, $L >_W \emptyset$).*

Wadge and Duparc used the operation of sum of sets of infinite words which has as counterpart the ordinal addition over Wadge degrees.

Definition 16 (Wadge, see [4,39]). *Assume that $X \subseteq Y$ are two finite alphabets, $Y - X$ containing at least two elements, and that $\{X_+, X_-\}$ is a partition of $Y - X$ in two non empty sets. Let $L \subseteq X^\omega$ and $L' \subseteq Y^\omega$, then*
$$L' + L =_{df} L \cup \{u \cdot a \cdot \beta \mid u \in X^\star, (a \in X_+ \text{ and } \beta \in L') \text{ or } (a \in X_- \text{ and } \beta \in L'^-)\}$$

This operation is closely related to the *ordinal sum* as it is stated in the following:

Theorem 17 (Wadge, see [4,39]). *Let $X \subseteq Y$, $Y - X$ containing at least two elements, $L \subseteq X^\omega$ and $L' \subseteq Y^\omega$ be non self dual Borel sets. Then $(L + L')$ is a non self dual Borel set and $d_w(L' + L) = d_w(L') + d_w(L)$.*

A player in charge of a set $L' + L$ in a Wadge game is like a player in charge of the set L but who can, at any step of the play, erase his previous play and choose to be this time in charge of L' or of L'^-. Notice that he can do this only one time during a play.

The following lemma was proved in [13]. Notice that below the emptyset is considered as an ω-language over an alphabet Δ such that $\Delta - \Sigma$ contains at least two elements.

Lemma 18. *Let $L \subseteq \Sigma^\omega$ be a non self dual Borel set such that $d_w(L) \geq \omega$. Then it holds that $L \equiv_W \emptyset + L$.*

We can now prove the following lemma.

Lemma 19. *Let $L \subseteq \Sigma^\omega$ be a non self dual Borel set accepted by a real time 1-counter Büchi automaton $\mathcal{A}$. Then there is an ω-language L' accepted by a real time 4-blind counter Büchi automaton such that $L \equiv_W L'$.*

Proof. Recall first that there are regular ω-languages of every finite Wadge degree, [29,36]. These regular ω-languages are Boolean combinations of open sets, and they obviously belong to the class $\mathbf{r\text{-}BCL}(4)_\omega$ since every regular ω-language belongs to this class.

So we have only to consider the case of non self dual Borel sets of Wadge degrees greater than or equal to ω.

Let then $L = L(\mathcal{A}) \subseteq \Sigma^\omega$ be a non self dual Borel set, accepted by a real time 1-counter Büchi automaton $\mathcal{A}$, such that $d_w(L) \geq \omega$. By Lemma 14, $L(\mathcal{P}_\mathcal{A}) = h(L(\mathcal{A})) \cup \mathcal{L}$ is accepted by a a real time 4-blind counter Büchi automaton $\mathcal{P}_\mathcal{A}$, where the mapping $h : \Sigma^\omega \to (\Sigma \cup \{A, B, 0\})^\omega$ is defined, for $x \in \Sigma^\omega$, by:

$$h(x) = A0x(1)B0^2x(2)A0^3x(3)B \cdots B0^{2n}x(2n)A0^{2n+1}x(2n+1)B \cdots$$

We set $L' = L(\mathcal{P}_\mathcal{A})$ and we now prove that $L' \equiv_W L$.

Firstly, it is easy to see that the function h is a continuous reduction of L to L' and thus $L \leq_W L'$.

To prove that $L' \leq_W L$, it suffices to prove that $L' \leq_W \emptyset + (\emptyset + L)$ because Lemma 18 states that $\emptyset + L \equiv_W L$, and thus also $\emptyset + (\emptyset + L) \equiv_W L$. Consider the Wadge game $W(L', \emptyset + (\emptyset + L))$. Player 2 has a winning strategy in this game which we now describe.

As long as Player 1 remains in the closed set $h(\Sigma^\omega)$ (this means that the word written by Player 1 is a prefix of some infinite word in $h(\Sigma^\omega)$) Player 2 essentially copies the play of player 1 except that Player 2 skips when player 1 writes a letter not in Σ. He continues forever with this strategy if the word written by player 1 is always a prefix of some ω-word of $h(\Sigma^\omega)$. Then after ω steps Player 1 has written an ω-word $h(x)$ for some $x \in \Sigma^\omega$, and Player 2 has written x. So in that case $h(x) \in L'$ iff $x \in L(\mathcal{A})$ iff $x \in \emptyset + (\emptyset + L)$.

But if at some step of the play, Player 1 "goes out of" the closed set $h(\Sigma^\omega)$ because the word he has now written is not a prefix of any ω-word of $h(\Sigma^\omega)$, then Player 1 "enters" in the open set $h(\Sigma^\omega)^- = \mathcal{L} \cup \mathcal{L}'$ and will stay in this set. Two cases may now appear.

First Case. When Player 1 "enters" in the open set $h(\Sigma^\omega)^- = \mathcal{L} \cup \mathcal{L}'$, he actually enters in the open set $\mathcal{L} = V \cdot \Gamma^\omega$ (this means that Player 1 has written an initial segment in V). Then the final word written by Player 1 will surely be inside L'. Player 2 can now write a letter of $\Delta - \Sigma$ in such a way that he is now like a player in charge of the wholeset and he can now writes an ω-word u so that his final ω-word will be inside $\emptyset + L$, and also inside $\emptyset + (\emptyset + L)$. Thus Player 2 wins this play too.

Second Case. When Player 1 "enters" in the open set $h(\Sigma^\omega)^- = \mathcal{L} \cup \mathcal{L}'$, he does not enter in the open set $\mathcal{L} = V \cdot \Gamma^\omega$. Then Player 2, being first like a player in charge of the set $(\emptyset + L)$, can write a letter of $\Delta - \Sigma$ in such a way that he is now like a player in charge of the emptyset and he can now continue, writing an ω-word u. If Player 1 never enters in the open set $\mathcal{L} = V \cdot \Gamma^\omega$ then the final word written by Player 1 will be in $\mathcal{L}'$ and thus surely outside L', and the final word written by Player 2 will be outside the emptyset. So in that case Player 2 wins this play too. If at some step of the play Player 1 enters in the open set $\mathcal{L} = V \cdot \Gamma^\omega$ then his final ω-word will be surely in L'. In that case Player 1, in charge of the set $\emptyset + (\emptyset + L)$, can again write an extra letter and choose to be in charge of the wholeset and he can now write an ω-word v so that his final ω-word will be inside $\emptyset + (\emptyset + L)$. Thus Player 2 wins this play too.

Finally we have proved that $L \leq_W L' \leq_W L$ thus it holds that $L' \equiv_W L$. This ends the proof.   □

End of Proof of Theorem 10

Let $L \subseteq \Sigma^\omega$ be a Borel set accepted by a real time 1-counter Büchi automaton $\mathcal{A}$. If the Wadge degree of L is finite, it is well known that it is Wadge equivalent to a regular ω-language, hence also to an ω-language in the class **r-BCL**$(4)_\omega$. If L is non self dual and its Wadge degree is greater than or equal to ω, then we know from Lemma 19 that there is an ω-language L' accepted by a real time 4-blind counter Büchi automaton such that $L \equiv_W L'$.

It remains to consider the case of self dual Borel sets. The alphabet Σ being finite, a self dual Borel set L is always Wadge equivalent to a Borel set in the form $\Sigma_1 \cdot L_1 \cup \Sigma_2 \cdot L_2$, where (Σ_1, Σ_2) form a partition of Σ, and $L_1, L_2 \subseteq \Sigma^\omega$ are non self dual Borel sets such that $L_1 \equiv_W L_2^-$. Moreover L_1 and L_2 can be taken in the form $L_{(u_1)} = u_1 \cdot \Sigma^\omega \cap L$ and $L_{(u_2)} = u_2 \cdot \Sigma^\omega \cap L$ for some $u_1, u_2 \in \Sigma^\star$, see [5]. So if $L \subseteq \Sigma^\omega$ is a self dual Borel set accepted by a real time 1-counter Büchi automaton then $L \equiv_W \Sigma_1 \cdot L_1 \cup \Sigma_2 \cdot L_2$, where (Σ_1, Σ_2) form a partition of Σ, and $L_1, L_2 \subseteq \Sigma^\omega$ are non self dual Borel sets accepted by real time 1-counter Büchi automata. We have already proved that there is an ω-language L_1' in the class **r-BCL**$(4)_\omega$ such that $L_1' \equiv_W L_1$ and an ω-language L_2' in the class **r-BCL**$(4)_\omega$ such that $L_2'^- \equiv_W L_2$. Thus $L \equiv_W \Sigma_1 \cdot L_1 \cup \Sigma_2 \cdot L_2 \equiv_W \Sigma_1 \cdot L_1' \cup \Sigma_2 \cdot L_2'$ and $\Sigma_1 \cdot L_1' \cup \Sigma_2 \cdot L_2'$ is an ω-language in the class **r-BCL**$(4)_\omega$.

The reverse direction is immediate: if $L \subseteq \Sigma^\omega$ is a Borel set accepted by a 4-blind counter Büchi automaton $\mathcal{A}$, then it is also accepted by a Büchi Turing machine and thus by [13, Theorem 25] there exists a real time 1-counter Büchi automaton $\mathcal{B}$ such that $L(\mathcal{A}) \equiv_W L(\mathcal{B})$.   □

Recall that, for each non-null countable ordinal α, the Σ_α^0-complete sets (respectively, the Π_α^0-complete sets) form a single Wadge degree. Thus we can infer the following result from the above Theorem 10 and from the results of [13, 22].

Corollary 20. *For each non-null recursive ordinal $\alpha < \omega_1^{CK}$ there exist some $\mathbf{\Sigma}_\alpha^0$-complete and some $\mathbf{\Pi}_\alpha^0$-complete ω-languages in the class $\mathbf{r}\text{-}\mathbf{BCL}(4)_\omega$. And the supremum of the set of Borel ranks of ω-languages in the class $\mathbf{r}\text{-}\mathbf{BCL}(4)_\omega$ is the ordinal γ_2^1, which is precisely defined in [22].*

We have only considered Borel sets in the above Theorem 10. However we know that there also exist some non-Borel ω-languages accepted by real time 1-counter Büchi automata, and even some $\mathbf{\Sigma}_1^1$-complete ones, [12].

By Lemma 4.7 of [16] the conclusion of the above Lemma 18 is also true if L is assumed to be an analytic but non-Borel set.

Lemma 21 ([16]). *Let $L \subseteq \Sigma^\omega$ be an analytic but non-Borel set. Then $L \equiv_W \emptyset + L$.*

Next the proof of the above Lemma 19 can be adapted to the case of an analytic but non-Borel set, and we can state the following result.

Theorem 22. *Let $L \subseteq \Sigma^\omega$ be an analytic but non-Borel set accepted by a real time 1-counter Büchi automaton $\mathcal{A}$. Then there is an ω-language L' accepted by a real time 4-blind counter Büchi automaton such that $L \equiv_W L'$.*

Proof. It is very similar to the proof of the above Lemma 19, using Lemma 21 instead of the above Lemma 18.   □

This implies in particular the existence of a $\mathbf{\Sigma}_1^1$-complete, hence non Borel, ω-language accepted by a real-time 4-blind-counter Büchi automaton. But Michał Skrzypczak has recently proved that one blind counter is sufficient.

Theorem 23 (Skrzypczak [34]). *There exists a $\mathbf{\Sigma}_1^1$-complete ω-language accepted by a 1-blind-counter automaton.*

5   High Undecidability of Topological Properties

We prove that it is highly undecidable to determine the topological complexity of a Petri net ω-language. As usual, since there is a finite description of a real time 1-counter Büchi automaton or of a 4-blind-counter Büchi automaton, we can define a Gödel numbering of all 1-counter Büchi automata or of all 4-blind-counter Büchi automata and then speak about the 1-counter Büchi automaton (or 4-blind-counter Büchi automaton) of index z. Recall first the following result, proved in [15], where we denote $\mathcal{A}_z$ the real time 1-counter Büchi automaton of index z reading words over a fixed finite alphabet Σ having at least two letters. We refer the reader to a textbook like [26] for more background about the analytical hierarchy of subsets of the set $\mathbb{N}$ of natural numbers.

Theorem 24. *Let α be a countable ordinal. Then*

1. $\{z \in \mathbb{N} \mid L(\mathcal{A}_z)$ *is in the Borel class* $\mathbf{\Sigma}_\alpha^0\}$ *is* Π_2^1*-hard.*
2. $\{z \in \mathbb{N} \mid L(\mathcal{A}_z)$ *is in the Borel class* $\mathbf{\Pi}_\alpha^0\}$ *is* Π_2^1*-hard.*
3. $\{z \in \mathbb{N} \mid L(\mathcal{A}_z)$ *is a Borel set* $\}$ *is* Π_2^1*-hard.*

Using the previous constructions we can now easily show the following result, where $\mathcal{P}_z$ is the real time 4-blind-counter Büchi automaton of index z.

Theorem 25. *Let $\alpha \geq 2$ be a countable ordinal. Then*

1. $\{z \in \mathbb{N} \mid L(\mathcal{P}_z)$ *is in the Borel class* $\Sigma_\alpha^0\}$ *is* Π_2^1-*hard.*
2. $\{z \in \mathbb{N} \mid L(\mathcal{P}_z)$ *is in the Borel class* $\Pi_\alpha^0\}$ *is* Π_2^1-*hard.*
3. $\{z \in \mathbb{N} \mid L(\mathcal{P}_z)$ *is a Borel set*$\}$ *is* Π_2^1-*hard.*

Proof. It follows from the fact that one can easily get an injective recursive function $g : \mathbb{N} \to \mathbb{N}$ such that $L(\mathcal{P}_{\mathcal{A}_z}) = h(L(\mathcal{A}_z)) \cup \mathcal{L} = L(\mathcal{P}_{g(z)})$ and from the following equivalences which hold for each countable ordinal $\alpha \geq 2$:

1. $L(\mathcal{A}_z)$ is in the Borel class Σ_α^0 (resp., Π_α^0) $\Longleftrightarrow L(\mathcal{P}_{g(z)})$ is in the Borel class Σ_α^0 (resp., Π_α^0).
2. $L(\mathcal{A}_z)$ is a Borel set $\Longleftrightarrow L(\mathcal{P}_{g(z)})$ is a Borel set. $\square$

6 High Undecidability of the Equivalence and the Inclusion Problems

We now add a result obtained from our previous constructions and which is important for verification purposes.

Theorem 26. *The equivalence and the inclusion problems for ω-languages of Petri nets, or even for ω-languages in the class* **r-BCL**$(4)_\omega$*, are* Π_2^1-*complete.*

1. $\{(z, z') \in \mathbb{N} \mid L(\mathcal{P}_z) = L(\mathcal{P}_{z'})\}$ *is* Π_2^1-*complete*
2. $\{(z, z') \in \mathbb{N} \mid L(\mathcal{P}_z) \subseteq L(\mathcal{P}_{z'})\}$ *is* Π_2^1-*complete*

Proof. Firstly, it is easy to see that each of these decision problems is in the class Π_2^1, since the equivalence and the inclusion problems for ω-languages of Turing machines are already in the class Π_2^1, see [2,15]. The completeness part follows from the fact that the equivalence and the inclusion problems for ω-languages accepted by real time 1-counter Büchi automata are Π_2^1-complete [15], and from the fact that there exists an injective recursive function $g : \mathbb{N} \to \mathbb{N}$ such that $\mathcal{P}_{\mathcal{A}_z} = \mathcal{P}_{g(z)}$, and then from the following equivalences:

1. $L(\mathcal{A}_z) = L(\mathcal{A}_{z'}) \Longleftrightarrow L(\mathcal{P}_{g(z)}) = L(\mathcal{P}_{g(z')})$
2. $L(\mathcal{A}_z) \subseteq L(\mathcal{A}_{z'}) \Longleftrightarrow L(\mathcal{P}_{g(z)}) \subseteq L(\mathcal{P}_{g(z')})$. $\square$

7 Concluding Remarks

We have proved that the Wadge hierarchy of Petri nets ω-languages, and even of ω-languages in the class **r-BCL**$(4)_\omega$, is equal to the Wadge hierarchy of effective analytic sets, and that it is highly undecidable to determine the topological complexity of a Petri net ω-language. In some sense our results show that, in contrast with the finite behavior, the infinite behavior of Petri nets is closer to the infinite behavior of Turing machines than to that of finite automata.

It remains open for further study to determine the Borel and Wadge hierarchies of ω-languages accepted by automata with less than four blind counters.

References

1. Carstensen, H.: Infinite behaviour of deterministic Petri nets. In: Chytil, M.P., Koubek, V., Janiga, L. (eds.) MFCS 1988. LNCS, vol. 324, pp. 210–219. Springer, Heidelberg (1988). https://doi.org/10.1007/BFb0017144
2. Castro, J., Cucker, F.: Nondeterministic ω-computations and the analytical hierarchy. J. Math. Logik und Grundlagen d. Math **35**, 333–342 (1989)
3. Cohen, R., Gold, A.: ω-computations on Turing machines. Theor. Comput. Sci. **6**, 1–23 (1978)
4. Duparc, J.: Wadge hierarchy and Veblen hierarchy: part 1: Borel sets of finite rank. J. Symb. Logic **66**(1), 56–86 (2001)
5. Duparc, J.: A hierarchy of deterministic context free ω-languages. Theor. Comput. Sci. **290**(3), 1253–1300 (2003)
6. Duparc, J., Finkel, O., Ressayre, J.P.: Computer science and the fine structure of Borel sets. Theor. Comput. Sci. **257**(1–2), 85–105 (2001)
7. Duparc, J., Finkel, O., Ressayre, J.P.: The Wadge hierarchy of Petri nets ω-languages. In: Special Volume in Honor of Victor Selivanov at the Occasion of his Sixtieth Birthday, Logic, Computation, Hierarchies, Ontos Mathematical Logic, vol. 4, pp. 109–138. De Gruyter, Berlin (2014). http://hal.archives-ouvertes.fr/hal-00743510
8. Engelfriet, J., Hoogeboom, H.J.: X-automata on ω-words. Theor. Comput. Sci. **110**(1), 1–51 (1993)
9. Esparza, J.: Decidability and complexity of Petri net problems—An introduction. In: Reisig, W., Rozenberg, G. (eds.) ACPN 1996. LNCS, vol. 1491, pp. 374–428. Springer, Heidelberg (1998). https://doi.org/10.1007/3-540-65306-6_20
10. Finkel, O.: An effective extension of the Wagner hierarchy to blind counter automata. In: Fribourg, L. (ed.) CSL 2001. LNCS, vol. 2142, pp. 369–383. Springer, Heidelberg (2001). https://doi.org/10.1007/3-540-44802-0_26
11. Finkel, O.: Topological properties of omega context free languages. Theor. Comput. Sci. **262**(1–2), 669–697 (2001)
12. Finkel, O.: Borel hierarchy and omega context free languages. Theor. Comput. Sci. **290**(3), 1385–1405 (2003)
13. Finkel, O.: Borel ranks and Wadge degrees of omega context free languages. Math. Struct. Comput. Sci. **16**(5), 813–840 (2006)
14. Finkel, O.: Wadge degrees of infinitary rational relations. Math. Comput. Sci. **2**(1), 85–102 (2008). https://doi.org/10.1007/s11786-008-0045-7. Special Issue on Intensional Programming and Semantics in Honour of Bill Wadge on the Occasion of his 60th Cycle
15. Finkel, O.: Highly undecidable problems for infinite computations. RAIRO-Theor. Inform. Appl. **43**(2), 339–364 (2009)
16. Finkel, O.: The determinacy of context-free games. J. Symb. Logic **78**(4), 1115–1134 (2013)
17. Finkel, O.: Wadge degrees of ω-languages of Petri nets (2017). Preprint arXiv:1712.07945
18. Finkel, O., Skrzypczak, M.: On the topological complexity of ω-languages of nondeterministic Petri nets. Inf. Process. Lett. **114**(5), 229–233 (2014)
19. Greibach, S.: Remarks on blind and partially blind one way multicounter machines. Theor. Comput. Sci. **7**, 311–324 (1978)
20. Haddad, S.: Decidability and complexity of Petri net problems. In: Diaz, M. (ed.) Petri Nets: Fundamental Models, Verification and Applications, pp. 87–122. Wiley-ISTE, Hoboken (2009)

21. Kechris, A.S.: Classical Descriptive Set Theory. Springer, New York (1995). https://doi.org/10.1007/978-1-4612-4190-4
22. Kechris, A.S., Marker, D., Sami, R.L.: Π_1^1 Borel sets. J. Symb. Logic **54**(3), 915–920 (1989)
23. Landweber, L.: Decision problems for ω-automata. Math. Syst. Theory **3**(4), 376–384 (1969)
24. Thomas, W., Lescow, H.: Logical specifications of infinite computations. In: de Bakker, J.W., de Roever, W.-P., Rozenberg, G. (eds.) REX 1993. LNCS, vol. 803, pp. 583–621. Springer, Heidelberg (1994). https://doi.org/10.1007/3-540-58043-3_29
25. Moschovakis, Y.N.: Descriptive Set Theory. North-Holland Publishing Co., Amsterdam (1980)
26. Odifreddi, P.: Classical Recursion Theory, Volume I. Studies in Logic and the Foundations of Mathematics, vol. 125. North-Holland Publishing Co., Amsterdam (1989)
27. Perrin, D., Pin, J.E.: Infinite Words, Automata, Semigroups, Logic and Games. Pure and Applied Mathematics, vol. 141. Elsevier, Amsterdam (2004)
28. Desel, J., Reisig, W., Rozenberg, G.: Lectures on Concurrency and Petri Nets: Advances in Petri Nets, vol. 3098. Springer, Heidelberg (2004). https://doi.org/10.1007/b98282
29. Selivanov, V.: Fine hierarchy of regular ω-languages. Theor. Comput. Sci. **191**, 37–59 (1998)
30. Selivanov, V.: Wadge degrees of ω-languages of deterministic Turing machines. RAIRO-Theor. Inform. Appl. **37**(1), 67–83 (2003)
31. Selivanov, V.: Fine hierarchies and m-reducibilities in theoretical computer science. Theor. Comput. Sci. **405**(1–2), 116–163 (2008)
32. Selivanov, V.: Wadge reducibility and infinite computations. Math. Comput. Sci. **2**(1), 5–36 (2008). https://doi.org/10.1007/s11786-008-0042-x. Special Issue on Intensional Programming and Semantics in Honour of Bill Wadge on the Occasion of his 60th Cycle
33. Simonnet, P.: Automates et théorie descriptive. Ph.D. thesis, Université Paris VII (1992)
34. Skrzypczak, M.: Büchi VASS recognise Σ_1^1-complete ω-languages. In: Potapov, I., Reynier, P.-A. (eds.) RP 2018. LNCS, vol. 11123, pp. 133–145. Springer, Cham (2018). https://doi.org/10.1007/978-3-030-00250-3_10
35. Staiger, L.: Recursive automata on infinite words. In: Enjalbert, P., Finkel, A., Wagner, K.W. (eds.) STACS 1993. LNCS, vol. 665, pp. 629–639. Springer, Heidelberg (1993). https://doi.org/10.1007/3-540-56503-5_62
36. Staiger, L.: ω-languages. In: Rozenberg, G., Salomaa, A. (eds.) Handbook of Formal Languages, pp. 339–387. Springer, Heidelberg (1997). https://doi.org/10.1007/978-3-642-59126-6_6
37. Thomas, W.: Automata on infinite objects. In: van Leeuwen, J. (ed.) Handbook of Theoretical Computer Science, Volume B. Formal Models and Semantics, pp. 135–191. Elsevier, Amsterdam (1990)
38. Valk, R.: Infinite behaviour of Petri nets. Theor. Comput. Sci. **25**(3), 311–341 (1983)
39. Wadge, W.: Reducibility and determinateness in the Baire space. Ph.D. thesis, University of California, Berkeley (1983)
40. Wagner, K.: On ω-regular sets. Inf. Control **43**(2), 123–177 (1979)

A New Property of Choice-Free Petri Net Systems

Eike Best[1], Raymond Devillers[2(✉)], and Evgeny Erofeev[1]

[1] Department of Computing Science,
Carl von Ossietzky Universität Oldenburg, 26111 Oldenburg, Germany
{eike.best,evgeny.erofeev}@uni-oldenburg.de
[2] Département d'Informatique,
Université Libre de Bruxelles, 1050 Brussels, Belgium
rdevil@ulb.ac.be

Abstract. When a Petri net system of some class is synthesised from a labelled transition system, it may be interesting to derive structural properties of the corresponding reachability graphs and to use them in a pre-synthesis phase in order to quickly reject inadequate transition systems, and provide fruitful error messages. The same is true for simultaneous syntheses problems. This was exploited for the synthesis of choice-free nets for instance, for which several interesting properties have been derived. We exhibit here a new property for this class, and analyse if this gets us closer to a full characterisation of choice-free synthesizable transition systems.

Keywords: Petri nets · Synthesis · Choice-freeness · Structural conditions

1 Introduction: Implementation Issue

Those last years, several works have been dedicated to the (simultaneous) synthesis of choice-free Petri nets[1], among others [1,3,8,9]. Since this means a loss in the expressive power, we may thus wonder why this class of systems could be especially interesting, besides the known structural properties of their behaviours.

The rather general class of weighted place-transition nets and systems is composed of places and transitions, as indicated by its name, together with weighted arcs between them. A place represents a type of resource; it may contain tokens representing the number of resources of that kind presently available; these (consumable) resources are considered interchangeable, at least as far as the control flow is concerned ("black" tokens), but they may carry some information that will be used (in a way not specified) by the transition that will absorb them.

---

[1] Those that would not be familiar with the domain are referred to the Appendix and the Introduction sections.

E. Erofeev—Supported by DFG through grant Be 1267/16-1 ASYST.

R. Janicki et al. (Eds.): PETRI NETS 2020, LNCS 12152, pp. 89–108, 2020.
https://doi.org/10.1007/978-3-030-51831-8_5

A transition needs some number (possibly zero) of tokens of each kind in order to be able to function, and then produces some (other) tokens. As such, those nets allow to model complex intertwined mixtures of sequences, exclusive choices and concurrency, hence exhibit an interesting expressive power.

But they may also serve as static specifications of systems to build. And a (finite) labelled transition system may serve as a behavioural specification, from which a (bounded) synthesised net system of some class may be built and then serves itself as a structural specification. However, here we may encounter some problems. Since those models allow to represent a concurrency feature between transition firings, a natural implementation strategy would be to build a system composed of data structures to model places and tokens, and parallel agents, one for each transition, interacting only through their competition to access their needed resources, one place at a time to obtain a fully distributed realisation. Note that, since memory resources are finite, we need to restrict our attention to bounded net systems and finite transition systems.

The structure of each implemented transition could then be sketched as follows:

repeat
>   check availability of needed resources
>   if some are missing, retry after some time
>   otherwise, collect the needed tokens
>   process the action of the transition
>>   (possibly using the hidden information of the black tokens)
>   produce the output tokens
>>   (possibly with adequate hidden information,
>>   depending on the one of the used tokens.)

There are variants of this schema however; for instance, the production of the output tokens may be performed one by one during the action process, and not all together at the end of the processing.

However there is potentially a big problem with this procedure: between the checking phase and the collection phase, the situation may have changed, due to the parallel action of other transitions, and it may happen that the tokens which were available during the checking phase are no longer there! It may even be possible that this is only discovered in the middle of the collection phase, and that we need to "give back" the tokens that were absorbed during the beginning of the phase. This is illustrated by the system in Fig. 1.

We may partly avoid this problem by absorbing the remainder of the needed tokens whenever the transition discovers they have been produced, but it may still be necessary to give back the absorbed tokens if we observe that it takes too long to get the remaining ones, possibly meaning that we reached a deadlock.

It is possible to avoid this kind of problem by blocking in a critical section all the data structures implementing the places and tokens, in order to check and absorb the needed tokens without being bothered by the other transitions, relaxing the critical section to wait a bit if not all the needed tokens were present, or to start the process and produce phase. However, this seriously reduces the parallelism of the implementation, may lead to starvation problems, and is certainly not distributed in the way we searched for.

We may also slightly alleviate this technique by only blocking together the input places of the transition, by protecting each place separately by a critical section and nesting them while following a predefined ordering in order to avoid deadlocks. However, this still reduces the parallelism of the implementation, may still lead to starvation problems, needs that the transitions first agree on a common ordering of the places, and is again not distributed as expected.

PNS: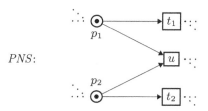

Fig. 1. A Petri net system, where transition u may see that all its needed tokens are present, but when it decides to take them, conflicting transitions t_1 and/or t_2 may have taken them before. If u takes a token from p_1 and then sees that the token in p_2 has disappeared, u must give back the token to p_1.

These problems underpin the difficulties encountered when one tries to realise a distributed hardware implementation of Petri nets, as in [13, 18].

The problem we described disappears, whatever the marking, if the places are not shared, that is if each place has (at most) a single output transition (for instance, in Fig. 1, if transition u is dropped, or t_1 and t_2), which is the definition of choice-free nets (see also Fig. 2 left, Fig. 5 right, and many others in the following).

In that case, it is never necessary to give back some absorbed token(s), and it is even possible to fuse the check and collect phases, leading for each transition to a (parallel) procedure of the kind:

    repeat
        for each input place **do**
            while the needed tokens are not present **do** wait for some time
            grab the needed tokens
        process the action of the transition and produce the output tokens

Of course, it is necessary to protect the data structures representing places, by semaphore-like devices [11] or monitor-like devices [12], in order to ensure that if many producers or the consumer and some producer(s) access the place, the result may be serialised (for instance, if two producers add one token in some common place, the result will be to add two tokens, and not a single one, the last addition wiping out the first one as it can happen in a -badly implemented- parallel system).

There is another class of nets, however, where a distributed implementation may be obtained whatever the initial marking: the join-free nets, where each transition has at most a single input place (called a simple-choice place), as illustrated on the right of Fig. 2. This is in some sense the reverse-dual of the choice-free case, since we have $\forall t : |{}^\bullet t| \leq 1$ instead of $\forall p : |p^\bullet| \leq 1$.

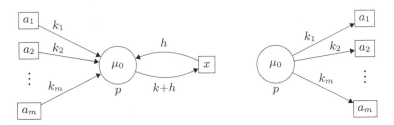

Fig. 2. On the left: a general pure ($h = 0$) or non-pure ($h > 0$) choice-free place p with initial marking μ_0 and unique output x (the inputs of a_i's and other inputs/outputs of x are not constrained); $\{a_1, \ldots, a_m\} = T \setminus \{x\}$, but some k_i's may be null. On the right, a general simple-choice place (the outputs of a_i's and inputs of p are not constrained).

In an implementation, each transition simply has to check the number of tokens in its unique input place and grab the needed ones in one access[2] if present (otherwise, one waits for some time before retrying). If the acquisition succeeded, one then processes the transition action and produces the output tokens, before retrying the whole cycle.

Let us now consider the case of free-choice nets [10], or their extended and weighted version called equal conflict nets [17] (see Appendix A.2 and Fig. 3 left). Here, there possibly are conflicts but they may be solved freely. We may then define an equivalence $t \sim u \iff {}^\bullet t \cap {}^\bullet u \neq \emptyset$ between transitions, and consider its equivalence classes, called clusters, i.e., the sets of transitions sharing their input places. In a distributed implementation, for each cluster, we may introduce a new process to check the availability of the needed resources (it behaves like in a choice-free net since its input places are not shared); when it works, it produces a single token in an intermediate place (with all the hidden information of the absorbed tokens, if any) which is simply shared by the transitions in the cluster, hence may be implemented as in a join-free net. We thus have a mixture of a choice-free net and a join-free one. This is illustrated in Fig. 3.

2 Introduction: Persistence Issue

A well-known [16] feature of CF-systems is the persistence of their reachability graph, meaning that a label, once firable, remains firable forever till fired (see

---

[2] A bit like in a "test and set" instruction, which allow multiprocessors to neatly manage their common memory resource.

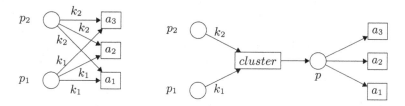

Fig. 3. On the left: a cluster in an equal conflict net, with two places and three transitions. On the right, its transformation for implementation: the cluster behaves as a transition in a choice-free net, and the intermediate place p as a place in a join-free net.

Appendix A.1). Indeed, suppose that in the reachability graph of a choice-free net system there are a state s and two edge labels $a \neq b$ such that $s \xrightarrow{a} s_1$ and $s \xrightarrow{b} s_2$ for two states s_1 and s_2. By the CF property, a and b have no common input place, so that, by the firing rule, firing a cannot disable b and firing b cannot disable a. Hence $s \xrightarrow{a} s_1 \xrightarrow{b} s'_1$ and $s \xrightarrow{b} s_2 \xrightarrow{a} s'_2$ for states s'_1, s'_2. Moreover, $s'_1 = s'_2$ by $\Psi(ab) = \Psi(ba)$ and the fact that the effect of a firing sequence depends only on its Parikh vector. So, the reachability graph is indeed persistent.

In a Petri net, a choice between a and b is indicated structurally by the presence of a place having both a and b as output transitions. In a transition system, on the other hand, a choice may be indicated by a branching state with two or more successor states. These two notions do not coincide however, as illustrated by Fig. 4.

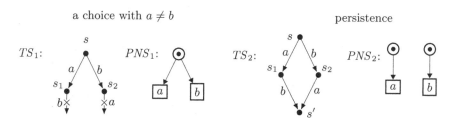

Fig. 4. Different kinds of branchings in an *lts*: choice (l.h.s.) and persistence (r.h.s.).

However, we may have a Petri net system with choice-places which nevertheless yields a persistent reachability graph. This is illustrated by Fig. 5. In the plain Petri net system PNS_3, place p_3 is a choice place between a and b. However, its reachability graph is isomorphic to TS_3, which is persistent (in a trivial way: there is no choice at all). This is due to the fact that, when a is enabled, b is not, and when b is enabled a is not. PNS_3 is not the only Petri net solving TS_3: PNS'_3 is also a solution, and this one is a (non-plain) choice-free net system (it may happen alas, that a persistent transition system is solved by a non-choice-free Petri net system but by no choice-free one: see for instance Figs. 12 and 13 below). There is a difference between these two solutions as to persistence however. If we add an initial token in p_1, a and b are initially enabled, but a disables b and b disables a, so that PNS_3 loses its persistence: its persistence is due to a very specific choice of the initial marking. On the contrary, PNS'_3 remains persistent, whatever its initial marking: its persistence is due to the underlying net, which is choice-free.

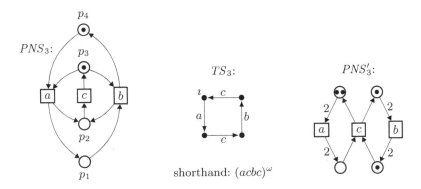

Fig. 5. Two Petri net systems PNS_3 and PNS'_3, and a transition system TS_3 with initial state $\imath$.

This remark may be related to a rather general property.

Proposition 1. *If a pure Petri net presents a choice-place, then there is an initial marking generating a non-persistent reachability graph. This is never the case for a choice-free net.*

Proof. Let p be a choice-place in the considered Petri net and let $w = \max_{t\in p\bullet}\{F(p,t)\}$. Assume we put initially w tokens in p. Initially, all the output transitions of p are enabled, but if $F(p,t) = w$, t is no longer enabled after another output transition of p is fired since the net is assumed to be pure. Hence the claim.

If there is no choice-place, the argument given above is valid and all the reachable markings yield diamonds if they enable more than one transition.   □

If a choice-place presents side-conditions, the previous proposition may be wrong, as shown in Fig. 6: PN_4 is plain but has a choice place p_1 and side conditions between the latter and a, b. If p_1 gets a token, a and b will permanently be enabled, forming (a degenerate kind of) diamonds. However, this does not resist to a splitting of the conflicting transitions into a "check and gather" phase and a "process and produce" phase, as detailed in the previous section and illustrated by the pure Petri net PN'_4 in the same figure. This does not happen with choice-free nets, where the split transforms the net into a pure one, still choice-free (or does not modify the net if we only apply the splitting to conflicting transitions).

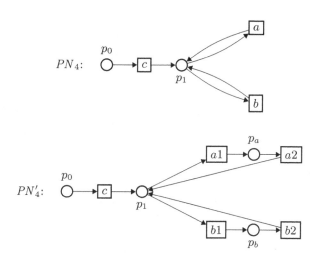

Fig. 6. A choice-place with side-conditions.

3 Simultaneous Choice-Free Synthesis: Preliminary Phase

A generalisation of the synthesis problem, called the simultaneous synthesis problem, has been introduced in [9]. It consists in considering several transition systems $\{TS_1, \ldots, TS_m\}$ simultaneously, and in search for a single Petri net N (of some class) together with m initial markings $\{M_{0,1}, \ldots, M_{0,m}\}$ such that the reachability graph of $(N, M_{0,i})$ is isomorphic to TS_i for each $i \in \{1, \ldots, m\}$. The classic synthesis corresponds to the case $m = 1$, and an example with $m = 2$ is shown in Fig. 7. With a single token on p, PN_5 solves $TS_{5,1}$. Without tokens on p, PN_5 solves $TS_{5,2}$. By contrast, $TS_{5,1}$ and $TS'_{5,2}$ can be CF-solved individually, but not simultaneously.

The simultaneous solvability of m transition systems can be reduced to the solvability of a single one by adding a single artificial initial state with m arcs with fresh labels to the m given initial states $\imath_1, \ldots, \imath_m$. Solving the augmented transition system and dropping the added auxiliary transitions yields a simultaneous solution, with the various initial markings provided by the markings corresponding to the $\imath_i$'s; and conversely, if there is a simultaneous solution, it may be obtained that way.

Such a straightforward approach amounts, however, to consider a large input lts, and we already mentioned that the performance of synthesis procedures is

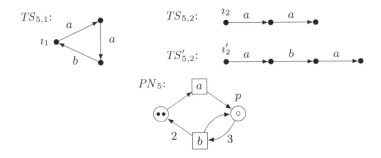

Fig. 7. $TS_{5,1}$ and $TS_{5,2}$ are simultaneously CF-solvable while $TS_{5,1}$ and $TS'_{5,2}$ are not (see below).

very sensitive to the size of the state space, hence this is not very effective. Also, it fails to preserve the choice-freeness of the underlying unmarked Petri net if $m \geq 2$, and cannot, therefore, be applied to CF-solvability.

Fortunately, the simultaneous choice-free synthesis may be related to m separate standard but slightly augmented choice-free syntheses.

Let T be the set of all the labels occurring in the various TS_i's; this will be the set of transitions in any solution to the simultaneous synthesis. Let also $\mathcal{G}_i$ be the set of $(T\text{-})$Parikh vectors of all the small cycles occurring in TS_i, and $\mathcal{G} = \cup_i \mathcal{G}_i$. An augmented synthesis is a synthesis of some TS_i which is compatible with all the members of $\mathcal{G}$, i.e., which reproduces the markings of all places along all the (small) cycles (not only the ones present in TS_i; this only depends on $\mathcal{G}$). We then have:

Theorem 1. *[9] A simultaneous choice-free synthesis problem is solvable if each of its augmented individual synthesis problems is, and the solution may be obtained easily by aggregating all those individual solutions.* □

Like for the classic choice-free synthesis problem (as well as many other targeted synthesis problems [6]), the synthesis procedure may be separated into a pre-synthesis and a proper synthesis. The pre-synthesis allows to quickly check properties of the structure of the reachability graph(s), from the structure of the class of nets aimed for. If this fails, we avoid a lengthy computation of the places of a tentative solution, get a better intuition about the true causes of the failure, hence are able to produce better error messages.

For this first phase, a series of checks arising from the analysis of choice-free systems, has been elaborated in [9], mainly based on the following observation [16]:

Theorem 2. *In the reachability graph of a choice-free system, the Parikh vector of any small cycle (if any) is a minimal semiflow of the underlying net.* □

This links a behavioural property (on small cycles, from some initial marking) and a structural one (on semiflows of the unmarked net). It also explains why, in a simultaneous synthesis problem, we have to consider the small cycles in all the given transition systems together.

These checks may be summarised as follows:

1. General: Each (finite) TS_i must be deterministic, totally reachable and persistent.
2. The small cycles property: All the members of $\mathcal{G}$ must be prime (no common nontrivial divisor) and two different members must be disjoint.
3. The short distance property: For each state s_i of each TS_i, all the shortest paths from $\imath_i$ to s_i have the same Parikh vector, called the distance Δ_{s_i} to s_i, and none of those distances may dominate a member of $\mathcal{G}$.
4. The reduced distance property: For the next checks we need the notion of residue of a T-vector μ by another one ν which builds the T-vector $\mu \overset{\bullet}{-} \nu$ in such a way that $\forall t \in T : \mu \overset{\bullet}{-} \nu(t) = \max(\mu(t) - \nu(t), 0)$; then, for each $i \in \{1, \ldots, m\}$, state s_i in TS_i and member $\Phi \in \mathcal{G}$, there must be a state r_i in TS_i such that $\Delta_{r_i} = \Delta_{s_i} \overset{\bullet}{-} \Phi$.
5. The earliest Parikh cycles property: For any member $\Upsilon \in \mathcal{G}$ and state s in some TS_i with a (small) cycle around s with Parikh vector Υ and member $\Phi \in \mathcal{G}$ disjoint from Υ, there is a cycle with Parikh vector Υ around the (unique) state in TS_i at distance $\Delta_s \overset{\bullet}{-} \Phi$ (which exists by the previous property).[3]

For instance, Condition 2 is not valid in TS_6 (Fig. 8) since the (small) cycle $\imath[aabbcc\rangle\imath$ has Parikh vector $(2, 2, 2)$, which is not prime.

Condition 3 is not valid in TS_7 since there are two short paths $\imath \overset{bac}{\to} s$ and $\imath \overset{dae}{\to} s$ whose Parikh vectors differ, so that the distance from $\imath$ to s is not defined here (also, they dominate $\Psi(bc)$ and $\Psi(de)$, respectively, which are small cycles in TS_7); it is not valid either for the pair $TS_{5,1}$–$TS'_{5,2}$ in Fig. 7, since there is a short path $\imath'_2 \overset{aba}{\to}$ whose Parikh vector is equal to (hence dominates) the Parikh vector of the small cycle $\imath_1 \overset{aab}{\to} \imath_1$ in $TS_{5,1}$.

Condition 4 is not valid in TS_8 since there is a path $\imath[abb\rangle s$ and a small cycle $s[abc\rangle s$: if we reduce abb by abc, we get b but the latter is not enabled at $\imath$.

The importance of Condition 5 is illustrated by the system TS_9 in Fig. 8: it satisfies conditions 1 to 4, but not the earliest cycles property since there is a cycle ba at s_3, a cycle d at s_2 and $\Delta_{s_2} = (1, 0, 1, 0)$, so that there should be a cycle d at distance $\Delta_{s_2} \overset{\bullet}{-} \Psi(ba) = (0, 0, 1, 0)$, i.e. at s_1; hence TS_9 (or $\{TS_9\}$) has no choice-free solution; this is corrected in TS'_9, which has the choice-free solution PN'_9.

---

[3] It is known [1] that in a finite deterministic and persistent transition system, loops are forward Parikh-equivalently transported (in the sense that, if there is a loop $s[\tau\rangle s$ and a directed path $s[\sigma\rangle s'$, then there is a loop $s'[\phi\rangle s'$ with $\Psi(\phi) = \Psi(\tau)$). The earliest Parikh cycles property means they may also be backward transported, up to some extent.

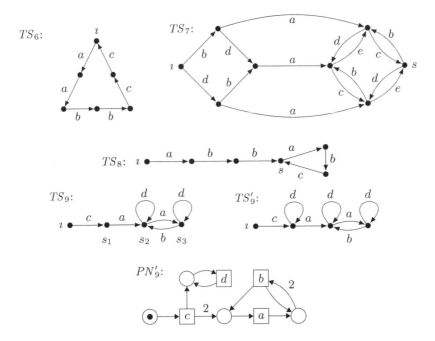

Fig. 8. Illustration of Conditions 2 to 5 for a choice-free pre-synthesis.

4   A New Criterion for the Choice-Free Simultaneous (and Individual) Synthesis

We shall now consider general paths in a transition system (see Appendix A.1) in addition to the usual directed ones. Referring to the general form of a place $p \in {}^\bullet x$ in a choice-free net as illustrated in the left of Fig. 2, we shall denote ${}^{\bullet\bullet}x \setminus \{x\}$ by $A(x)$.

Lemma 1. Co-enabling along a general path: *In a choice-free net N, if $M_1[x\rangle$, x belongs to the support of some (minimal) semiflow Υ, $M_1[\sigma\rangle M_2$ with $\sigma \in (\pm T)^*$ and $\forall a \in A(x) : \Psi(\sigma)(a) \geq \Psi(\sigma)(x) \cdot \Upsilon(a)/\Upsilon(x)$, then $M_2[x\rangle$.*

Proof. We only need to show that $\forall p \in {}^\bullet x : M_2(p) \geq M_1(p)$.
In the general form of a place $p \in {}^\bullet x$ (left of Fig. 2), $k_i = 0$ if $a_i \notin A(x)$.
We know that $M_2(p) = M_1(p) + \sum_{a_i \in A(x)} k_i \cdot \Psi(\sigma)(a_i) - k \cdot \Psi(\sigma)(x)$, and that $k \cdot \Upsilon(x) = \sum_{a_i \in A(x)} k_i \cdot \Upsilon(a_i)$.
Hence, $M_2(p) = M_1(p) + \sum_{a_i \in A(x)} k_i \cdot [\Psi(\sigma)(a_i) - \Psi(\sigma)(x) \cdot \Upsilon(a_i)/\Upsilon(x)]$. The claimed property arises, since no k_i is negative. □

This may be interpreted as follows: if there is a general path σ from some marking M_1 to some marking M_2, and some (possibly fractional, possibly negative) factor $f \in \mathbb{Q}$ such that $\Psi(\sigma) \geq f \cdot \Upsilon$ on $A(x)$, with $f = \Psi(\sigma)(x)/\Upsilon(x)$ (or $\Psi(\sigma)(x) = f \cdot \Upsilon(x)$), then $M_1[x\rangle \Rightarrow M_2[x\rangle$, i.e., we have a kind of co-enabling of x.
If we do not know exactly $A(x)$, we may use any over-approximation:

Corollary 1. *In a choice-free net N, if $M_1[x\rangle$, x belongs to the support of some (minimal) semiflow Υ, $A(x) \subseteq A_x$ for some set $A_x \subseteq T$, $M_1[\sigma\rangle M_2$ with $\sigma \in (\pm T)^*$ and $\forall a \in A_x : \Psi(\sigma)(a) \geq \Psi(\sigma)(x) \cdot \Upsilon(a)/\Upsilon(x)$, then $M_2[x\rangle$.* □

Note that the satisfaction of the constraint on a in this property relies on the sign of $\Psi(\sigma)(x)$, i.e. of f, and of the membership of a to the support of Υ:

- if $\Upsilon(a) = 0$, we need $\Psi(\sigma)(a) \geq 0$, otherwise
- if $\Psi(\sigma)(x) > 0$ (or $f > 0$) and $\Psi(\sigma)(a) \leq 0$, the constraint is not satisfied; if $\Psi(\sigma)(a) > 0$, one needs $\Psi(\sigma)(a) \geq f \cdot \Upsilon(a)$;
- if $\Psi(\sigma)(x) < 0$ (or $f < 0$) and $\Psi(\sigma)(a) \geq 0$, the constraint is satisfied; if $\Psi(\sigma)(a) < 0$, one needs $|\Psi(\sigma)(a)| \leq |f| \cdot \Upsilon(a)$;
- if $\Psi(\sigma)(x) = 0 = f$ and $\Psi(\sigma)(a) \geq 0$, the constraint is satisfied; if $\Psi(\sigma)(a) < 0$, the constraint is not satisfied.

Definition 1. *In a simultaneous choice-free synthesis problem, if $\Upsilon \in \mathcal{G}$ we shall denote by T_Υ the support of Υ, and if x belongs to the support of Υ we also denote $T_x = T_\Upsilon$.*
We shall also denote by T_0 the set of labels in T not belonging to the support of some member of $\mathcal{G}$, i.e., $T_0 = T \setminus \cup_{\Upsilon \in \mathcal{G}} T_\Upsilon$. □

Proposition 2. *In a simultaneous choice-free synthesis problem, if x belongs to the support of some $\Upsilon \in \mathcal{G}$, $s[x\rangle$ and $\neg s'[x\rangle$ in some TS_ℓ, if $\forall a \in T_0 \cup T_x : (\Delta_{s'} - \Delta_s)(a) \geq (\Delta_{s'} - \Delta_s)(x) \cdot \Upsilon(a)/\Upsilon(x)$, then the problem is not solvable.*

Proof. If there is a solution N, from Theorem 2, Υ is a minimal semiflow of N. For any i, if a_i belongs to the support of some $\Upsilon' \in \mathcal{G} \setminus \{\Upsilon\}$, we have that x does not belong to the support of Υ', from the small cycles property above.
From the definition of semiflows, we must have $\sum_j \Upsilon'(a_j) \cdot k_j = k \cdot \Upsilon'(x) = 0$, hence $k_i = 0$. As a consequence we may state that $A(x) \subseteq T_0 \cup T_x$ (we do not have equality since it may happen that $k_i = 0$ while $a_i \notin T_0 \cup T_x$).
Let σ_s be a short path from $\imath_\ell$ to s, and similarly for s'. $(-\sigma_s)(\sigma_{s'})$ is a general path from s to s' in TS_ℓ, and $\Psi((-\sigma_s)(\sigma_{s'})) = \Delta_{s'} - \Delta_s$, from the short distance property above.
The proposition then results from Corollary 1, where we may choose $A_x = T_0 \cup T_x$. □

In this proposition, we used a special g-path from s to s', and we could wonder if choosing another one would lead to another condition. We shall now see that this is not the case.

Lemma 2. *If, in a simultaneous choice-free synthesis problem, conditions 1 to 5 are satisfied, for each g-cycle $s[\sigma\rangle s$ in any TS_ℓ ($\sigma \in (\pm T_\ell)^*$), we have $\Psi(\sigma) = \sum_{\Upsilon \in \mathcal{G}_\ell} z_\Upsilon \cdot \Upsilon$, for some coefficients $z_\Upsilon \in \mathbb{Z}$.*

Proof. First, we may observe that each transition system TS_ℓ is weakly periodic. Indeed, if $s_1 \xrightarrow{\sigma} s_2 \xrightarrow{\sigma} s_3 \xrightarrow{\sigma} s_4 \xrightarrow{\sigma} \ldots$, since TS_ℓ is finite, we must have $s_i = s_j$ for some $i < j$, hence a cycle $s_i[\sigma^{j-i}\rangle s_i$. In a finite, totally reachable, deterministic and persistent system, from Corollary 2 in [8] we have $\Psi(\sigma^{j-i}) = (j-i) \cdot \Psi(\sigma) = \sum_{\Upsilon \in \mathcal{G}_\ell} k_\Upsilon \cdot \Upsilon$ for some coefficients $k_\Upsilon \in \mathbb{N}$. Since, the Υ's are prime and disjoint, $(j-i)$ must divide the $k_\Upsilon \in \mathbb{N}$ and $\Psi(\sigma) = \sum_{\Upsilon \in \mathcal{G}_\ell} k'_\Upsilon \cdot \Upsilon$ for some coefficients $k'_\Upsilon \in \mathbb{N}$. Then, from the properties of distances in [9], we know that $\Delta_{s_2} = \Delta_{s_1}$ and $s_1 = s_2$.

The claimed property then results from Lemma 4 in [7], which states that general cycles are generated by directed ones when the previous properties are satisfied. $\qquad\square$

Corollary 2. *If, in a simultaneous choice-free synthesis problem, conditions 1 to 5 are satisfied, two g-paths σ and σ' go from s to s' in TS_ℓ, and x belongs to the support of some $\Upsilon \in \mathcal{G}$, then $\forall a \in T_0 \cup T_x : \Psi(\sigma)(a) \geq \Psi(\sigma)(x) \cdot \Upsilon(a)/\Upsilon(x) \iff \Psi(\sigma')(a) \geq \Psi(\sigma')(x) \cdot \Upsilon(a)/\Upsilon(x)$.*

Proof. Since $s'[(-\sigma)(\sigma')\rangle s'$ is a g-cycle around s', from Lemma 2, we know that $\Psi(\sigma') = \Psi(\sigma) + \sum_{\Upsilon' \in \mathcal{G}_\ell} z_{\Upsilon'} \cdot \Upsilon'$. In particular, from condition 2, that implies $\Psi(\sigma')(x) = \Psi(\sigma)(x) + z_\Upsilon \cdot \Upsilon(x)$.

From Proposition 2, we only have to consider $a \in T_0 \cup T_x$.

If a belongs to the support of Υ, hence to T_x, $\Psi(\sigma')(a) = \Psi(\sigma)(a) + z_\Upsilon \cdot \Upsilon(a)$. Since $z_\Upsilon \cdot \Upsilon(a) = z_\Upsilon \cdot \Upsilon(x) \cdot \Upsilon(a)/\Upsilon(x)$, we then have the claimed equivalence.

If a belongs to T_0, $\Psi(\sigma')(a) = \Psi(\sigma)(a)$ and $\Upsilon(a) = 0$, which again leads to the claimed equivalence. $\qquad\square$

This corollary immediately leads to the main result of the present paper, a new pre-synthesis condition that may be added to the five ones we already listed for the simultaneous or individual choice-free synthesis (we shall later see that this new condition is not implied by the other ones):

6 Co-enabling: For any $\Upsilon \in \mathcal{G}$ and x in its support, for any TS_i and $s, s' \in S_i$, if $s[x\rangle$ and $\forall a \in T_0 \cup T_x : (\Delta_{s'} - \Delta_s)(a) \geq (\Delta_{s'} - \Delta_s)(x) \cdot \Upsilon(a)/\Upsilon(x)$, then $s'[x\rangle$.

Two special cases may then be distinguished, corresponding to the cases where s' occurs before or after s, i.e., that $s'[\sigma\rangle s$ or $s[\sigma\rangle s'$ for a directed path σ (but we may have both cases simultaneously):

Corollary 3. Forward fractional reduction: *In a choice-free net N, if $M[\sigma' t\rangle$ with $\sigma' \in T^*$ and, for some $f \in \mathbb{Q}_{>0}$ and semiflow Υ, $\Psi(\sigma') \leq f \cdot \Upsilon$ and $\Psi(\sigma')(t) = f \cdot \Upsilon(t)$, then $M[t\rangle$.*

Proof. In this case, we may apply Lemma 1 with $M_2 = M$, $\sigma = -\sigma'$ and $\Psi(\sigma) \leq \mathbf{0}$, hence the claim. $\qquad\square$

From this result, we deduce the following criterion for non-choice-free-solvability.

Corollary 4. Forward fractional reduction in a pre-synthesis: *Let $TS = (S, T, \rightarrow, \imath)$ be a deterministic and persistent labelled transition system, $s \in [\imath\rangle$, $t \in T$, $\sigma \in T^*$, $f \in \mathbb{Q}_{>0}$ and Υ be any T-vector. If $\neg s[t\rangle$, $s[\sigma t\rangle$, $\Psi(\sigma) \leq f \cdot \Upsilon$ and $\Psi(\sigma)(t) = f \cdot \Upsilon(t)$, then there is no choice-free solution of TS with semiflow Υ.*

Applying these results, the two systems TS_{10}, TS_{11} in Fig. 9 do not allow a simultaneous choice-free solution. Indeed, TS_{11} determines a semiflow $(3, 2)$, hence also a fractional semiflow $(3/2, 1)$. With $s = \imath_1$, $\sigma = ab$, $t = b$, $\Upsilon = (3, 2)$ and $f = 1/2$, since we do not have $\imath_1[b\rangle$, we may conclude there is no simultaneous choice-free-solution.

This is also true for the two systems TS_{11}, TS_{12}: applying Corollary 3 with $\imath_3[a\rangle s'$, $s = s'$, $\sigma = ab$, $t = b$, $\Upsilon = (3, 2)$ and $f = 1/2$, since we do not have $s'[b\rangle$, we may conclude there is no simultaneous choice-free-solution.

Since these two pairs of systems satisfy conditions 1 to 5, this also shows the independence of condition 6: this last condition is not implied by the other ones.

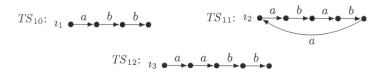

Fig. 9. Three transition systems TS_{10}, TS_{11}, TS_{12}.

There is also the symmetric version of those last two Corollaries.

Corollary 5. Backward fractional reduction: *In a choice-free net N, if $M[t\rangle$ and $M[\sigma'\rangle$ with $\sigma' \in T^*$ and, for some $f \in \mathbb{Q}_{>0}$ and semiflow Υ, $\Psi(\sigma') \leq f \cdot \Upsilon$ and $\Psi(\sigma')(t) = f \cdot \Upsilon(t)$, then $M[\sigma't\rangle$.*

Proof. In this case, we may apply Lemma 1 with $M_1 = M$, $\sigma = \sigma'$ and $\Psi(\sigma) \geq \mathbf{0}$, hence the claim.   □

Corollary 6. Backward fractional reduction in a pre-synthesis: *Let $TS = (S, T, \to, \imath)$ be a deterministic and persistent labelled transition system, $s \in [\imath\rangle$, $t \in T$, $\sigma \in T^*$, $f \in \mathbb{Q}_{>0}$ and Υ be any T-vector. If $s[t\rangle$, $s[\sigma\rangle$, $\Psi(\sigma) \geq f \cdot \Upsilon$, $\Psi(\sigma)(t) = f \cdot \Upsilon(t)$ and $\neg s[\sigma t\rangle$, then there is no choice-free-solution of TS with semiflow Υ.*

Applying these results, we get that the two systems in Fig. 10, are not simultaneously choice-free-solvable either. Indeed, in TS_{13} we have the minimal semiflow $\Upsilon = (3, 2)$; in TS_{14}, we have $\imath_{12}[b\rangle$, $\imath_4[aab\rangle$, $\Psi(aab) = (2, 1)$, so that with $f = 1/2$ and $t = b$, from Corollary 6 we should have $\imath_{12}[aabb\rangle$.

Fig. 10. Two transition systems TS_{13}, TS_{14}.

The two systems in Fig. 11, are not simultaneously choice-free-solvable either. Indeed, in TS_{16} we have the minimal semiflow $\Upsilon = (3,2)$; in TS_{15}, we have $\imath_{13}[b\rangle$, $\imath_{13}[baa\rangle$, $\Psi(baa) = (2,1)$, so that with $f = 1/2$ and $t = b$, from Corollary 6 we should have $\imath_{13}[baab\rangle$.

Fig. 11. Transition systems TS_{15}, TS_{16}.

5   But Still, Our Conditions Are Not Sufficient

Now that we accumulated several structural properties of reachability graphs of bounded choice-free systems, and devised corresponding checks to be performed during the (simultaneous) pre-synthesis, we may wonder if they are sufficient to ascertain the existence of a solution, at least in some interesting subclasses of problems.

Since most of these properties concern the small cycles, they are of no help for acyclic transition systems. Let us then restrict our attention to a single ($m = 1$) reversible system, since then there are small cycles around each state (we mentioned before that cycles are transported Parikh-equivalently along the flow in finite deterministic and persistent systems). We shall also assume that all small cycles have the same Parikh vector, and that there is at least a (non-necessarily choice-free) solution. However, even in these very specific class of systems, there is not always a choice-free solution. We shall give two different counter-examples.

The first one, illustrated in Fig. 12, has already been used in [3]. It has five labels and is remarkable by its small cycles, that all have the Parikh vector $1 = (1,1,1,1,1)$ (so that fractional semiflows do not introduce any additional constraint since at least one of the components must be an integer: $f \cdot \Upsilon(t)$ must count the number of occurrences of t in some path). It has a (non-choice-free) Petri net solution, but not a choice-free one: the numerical construction of the proper synthesis phase (for instance with the use of the APT tool [14]) finds an unsolvable ESSP problem.

The next transition system has four labels, and has a unique small cycle with Parikh vector $(5,3,1,1)$, hence has the form of a simple circle. Here too it has a non-choice-free solution, but no choice-free one (again found by APT). It is illustrated by Fig. 13.

Both counterexamples share the additional condition that they have a Petri net solution. We may then wonder what happens if we still reinforce this condition, in the following way. Let us assume that $m > 1$, that each TS_i is individually choice-free solvable (not only Petri net-solvable), and reversible, that

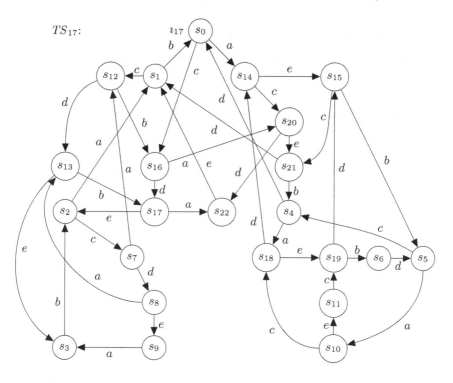

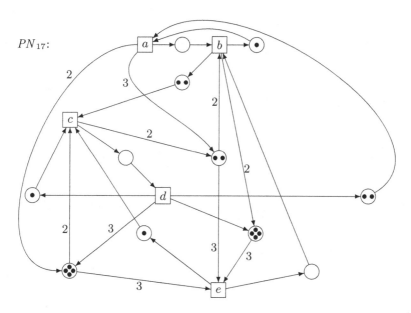

Fig. 12. A reversible *lts* with unitary small cycles, with a possible Petri net solution. It survives all the structural checks mentioned in this paper for the pre-synthesis phase, but does not have a choice-free solution.

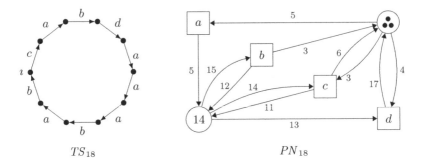

Fig. 13. TS_{18} has the form of a circle, survives all the pre-synthesis checks we mention above and has a Petri net solution PN_{18}, but not a choice-free one.

all the small cycles have the same Parikh vector and that all conditions 1 to 6 are satisfied. Is it still possible that the simultaneous choice-free synthesis fails? Presently, we do not know and we are working on it.

6   Concluding Remarks

We added a new structural criterion to the already long list of available necessary conditions for a labelled transition system to have a bounded choice-free solution. This allows to quickly rule out more inadequate (simultaneous) synthesis problems. However, we are still not in reach of a necessary and sufficient set of criteria fully characterising the state spaces of choice-free systems, in the style of the conditions developed for (bounded or unbounded) marked graphs and T-systems in [2,4,5].

Acknowledgements. We are indebted to the anonymous referees for their detailed and helpful remarks.

A   Basic Definitions

A.1   Labelled Transition Systems

A labelled transition system with initial state, *lts* for short, is a quadruple $TS = (S, T, \rightarrow, \imath)$ where S is a set of states, T is a set of labels, $\rightarrow \subseteq (S \times T \times S)$ is the set of labelled edges, and $\imath \in S$ is an initial state. TS is finite if S and T are finite. Let $(-T)$ (called backward edges) be a disjoint copy of T:

$$(-T) = \{(-a) \mid a \in T\}, \text{ with } T \cap (-T) = \emptyset, \text{ and } (-(-a)) = a \text{ for all } a \in T.$$

A g-path is a sequence $\sigma \in (T \cup (-T))^*$; it is a (directed) path if $\sigma \in T^*$. For $s, s' \in S$, a g-path $\sigma = a_1 \ldots a_m$ leads from s to s' (denoted by $s \xrightarrow{\sigma} s'$) if $\exists r_0, r_1, \ldots, r_m \in S: \ s = r_0 \ \wedge \ r_m = s'$

$$\wedge \ \forall j \in \{1, \ldots, m\}: \ \begin{cases} (r_{j-1}, a_j, r_j) \in \to \text{ if } a_j \in T \\ (r_j, a_j, r_{j-1}) \in \to \text{ if } a_j \in (\!-\!T) \end{cases}$$

The back-arrow notation is extended to sequences in $(T \cup (\!-\!T))^*$ inductively as follows:

$$(\!-\!\varepsilon) = \varepsilon, \quad \text{and} \quad (\!-\!(\sigma a)) = (\!-\!a)(\!-\!\sigma), \quad \text{for } a \in T \cup (\!-\!T) \text{ and } \sigma \in (T \cup (\!-\!T))^*$$

A label a is enabled at $s \in S$, or $s \xrightarrow{a}$ for short, if $\exists s' \in S: s \xrightarrow{a} s'$. A state s' is reachable from state s if $\exists \sigma \in T^*: s \xrightarrow{\sigma} s'$.[4] Sometimes, we shall use the notation $s[t\rangle$ instead of $s \xrightarrow{a}$, and $[s\rangle$ to denote the set of states reachable from s.

For a g-path $\sigma \in (T \cup (\!-\!T))^*$, the Parikh vector of σ is a T-vector $\Psi(\sigma)$, defined inductively as follows:

$$\Psi(\varepsilon) = 0 \ \text{(the null vector)} \quad \text{and} \quad (\Psi(\sigma a))(t) = \begin{cases} (\Psi(\sigma))(t) + 1 \text{ if } t = a \in T \\ (\Psi(\sigma))(t) - 1 \text{ if } (\!-\!t) = a \in (\!-\!T) \\ (\Psi(\sigma))(t) \quad \text{if } t \neq a \neq (\!-\!t) \end{cases}$$

For example, $\Psi(ab(\!-\!a)b(\!-\!c)aa) = (2, 2, -1)$. Two finite sequences are Parikh-equivalent if they have the same Parikh vector, and Parikh-disjoint if no label occurs in both of them. The support of a T-vector, hence in particular of a Parikh vector, is the set of indices for which the vector is non-null.

A g-path $s[\sigma\rangle s'$ is called a g-cycle, or more precisely a g-cycle at (or around) state s, if $s = s'$.[5]

A finite transition system $TS = (S, T, \to, \imath)$ is called

- totally reachable if $\forall s \in S \, \exists \sigma \in T^*: \imath \xrightarrow{\sigma} s$;
- deterministic if, for all states $s, s', s'' \in S$, and for any label $t \in T$, $s \xrightarrow{t} s'$ and $s \xrightarrow{t} s''$ imply $s' = s''$;
- backward deterministic if, for all states $s, s', s'' \in S$, and for any label $t \in T$, $s' \xrightarrow{t} s$ and $s'' \xrightarrow{t} s$ imply $s' = s''$;
- fully forward deterministic if, for all states $s, s', s'' \in S$ and for all sequences $\alpha, \alpha' \in T^*$, $(s \xrightarrow{\alpha} s' \wedge s \xrightarrow{\alpha'} s'' \wedge \Psi(\alpha) = \Psi(\alpha'))$ entails $s' = s''$;
- weakly periodic if for every $s_1 \in S$, label sequence $\sigma \in T^*$, and infinite sequence $s_1 \xrightarrow{\sigma} s_2 \xrightarrow{\sigma} s_3 \xrightarrow{\sigma} s_4 \xrightarrow{\sigma} \ldots$, $\forall i, j \geq 1: s_i = s_j$;
- persistent if for all states $s, s', s'' \in S$, and labels $t \neq u$, if $s \xrightarrow{t} s'$ and $s \xrightarrow{u} s''$, then there is some state $r \in S$ such that $s' \xrightarrow{u} r$ and $s'' \xrightarrow{t} r$.
- live if $\forall t \in T \ \forall s \in [s_0\rangle \ \exists s' \in [s\rangle$ such that $s'[t\rangle$;
- reversible if $\forall s \in [s_0\rangle, s_0 \in [s\rangle$.

Two *lts* with the same label set $TS = (S, T, \to, \imath)$ and $TS' = (S', T, \to', \imath')$ are isomorphic if there is a bijection $\zeta: S \to S'$ with $\zeta(\imath) = \imath'$ and $(r, t, s) \in \to \Leftrightarrow (\zeta(r), t, \zeta(s)) \in \to'$, for all $r, s \in S$ and $t \in T$.

---

[4] Note that enabledness refers only to outgoing edges and reachability refers only to directed paths, rather than to g-paths.

[5] In this paper, whenever we speak of a path or a cycle, a directed path or cycle is meant.

A cycle $s \xrightarrow{\sigma} s$ with $\sigma \in T^*$ is small if $\sigma \neq \varepsilon$ and there is no cycle $s' \xrightarrow{\sigma'} s'$ with $\sigma' \neq \varepsilon$ and $\Psi(\sigma') \lneqq \Psi(\sigma)$ (this should not be confused with a simple cycle, not visiting twice any state). A cycle $s[\sigma\rangle s$ is prime if the greatest common divisor of the entries in its Parikh vector is 1. A (directed) path $s[\sigma\rangle s'$ is short if there is no other path between the same states with a smaller Parikh vector: $s[\sigma'\rangle s' \Rightarrow \neg(\Psi(\sigma') \lneqq \Psi(\sigma))$.

A.2   Petri Nets

A (finite, place-transition, arc-weighted) Petri net is a triple $PN = (P, T, F)$ such that P is a finite set of places, T is a finite set of transitions, with $P \cap T = \emptyset$, F is a flow function $F \colon ((P \times T) \cup (T \times P)) \to \mathbb{N}$.

The predecessors of a node x is the set $^\bullet x = \{y | F(y,x) > 0\}$. Symmetrically its successor set is $x^\bullet = \{y | F(x,y) > 0\}$. The incidence matrix C of a Petri net is the integer place-transition matrix with components $C(p,t) = F(t,p) - F(p,t)$, where p is a place and t is a transition. An elementary property of Petri nets is the state equation which expresses that, if $M[\sigma\rangle M'$, then $M' = M + C \cdot \Psi(\sigma)$. A semiflow of a net N is a T-vector Φ such that $\Phi \gneqq 0$ and $C \cdot \Phi = 0$. A semiflow is minimal if there is no smaller one.

PN is plain if no arc weight exceeds 1; pure if $\forall p \in P \colon (p^\bullet \cap {}^\bullet p) = \emptyset$; choice-free if $\forall p \in P \colon |p^\bullet| \leq 1$; join-free if $\forall t \in T \colon |{}^\bullet t| \leq 1$; fork-attribution if choice-free and, in addition, $\forall t \in T \colon |{}^\bullet t| \leq 1$; a marked graph if PN is plain and $|p^\bullet| = 1$ and $|{}^\bullet p| = 1$ for all places $p \in P$; and a T-system if PN is plain and $|p^\bullet| \leq 1$ and $|{}^\bullet p| \leq 1$ for all places $p \in P$; free-choice if it is plain and $\forall t, t' \in T \colon {}^\bullet t \cap {}^\bullet t' \neq \emptyset \Rightarrow {}^\bullet t = {}^\bullet t'$; equal conflict if $\forall t, t' \in T \colon {}^\bullet t \cap {}^\bullet t' \neq \emptyset \Rightarrow \forall p \in P \colon F(p,t) = F(p,t')$.

A marking is a mapping $M \colon P \to \mathbb{N}$, indicating the number of (black) tokens in each place. A Petri net system is a net provided with an initial marking (P, T, F, M_0); the subclasses defined above for Petri nets extend immediately for Petri net systems. A transition $t \in T$ is enabled by a marking M, denoted by $M \xrightarrow{t}$, if for all places $p \in P$, $M(p) \geq F(p,t)$. If t is enabled at M, then t can occur (or fire) in M, leading to the marking M' defined by $M'(p) = M(p) - F(p,t) + F(t,p)$ and denoted by $M \xrightarrow{t} M'$. The reachability graph of PN is the labelled transition system whose initial state is M_0, whose vertices are the reachable markings, and whose edges are $\{(M, t, M') \mid M \xrightarrow{t} M'\}$. A Petri net system (P, T, F, M_0) is k-bounded for some fixed $k \in \mathbb{N}$, if $\forall M \in [M_0\rangle \forall p \in P \colon M(p) \leq k$; bounded if $\exists k \in \mathbb{N} \colon N$ is k-bounded; and safe if it is 1-bounded. Properties of transition systems extend immediately to Petri net systems through their reachability graphs.

A labelled transition system is PN-solvable if it is isomorphic to the reachability graph of a Petri net system (called the solution); it is CF-solvable if there is a choice-free solution.

A.3   Regions

When linking transition systems and Petri nets, it is useful to introduce regions. A region $(\rho, \mathbb{B}, \mathbb{F})$ is a triple of functions ρ (from states to $\mathbb{N}$), and $\mathbb{B}, \mathbb{F}$ (both from labels to $\mathbb{N}$), satisfying the property that for any states s, s' and label a:

$$(s, a, s') \text{ is an edge of } TS \quad \Rightarrow \quad \rho(s') - \rho(s) = \mathbb{F}(a) - \mathbb{B}(a)$$

since this is the typical behaviour of a place p with token count ρ during the firing of a with backward and forward connections $\mathbb{B}, \mathbb{F}$ to p (anywhere in TS). To solve an $SSP(s_1, s_2)$ (for State Separation Problem, where $s_1 \neq s_2$), we need to find an appropriate region $(\rho, \mathbb{B}, \mathbb{F})$ satisfying $\rho(s_1) \neq \rho(s_2)$, i.e., separating states s_1 and s_2. For an $ESSP(s, a)$ (for Event-State Separation Problem, where e is not enabled at s), we need to find a region $(\rho, \mathbb{B}, \mathbb{F})$ with $\rho(s) < \mathbb{B}(a)$. This can be done by solving suitable systems of linear inequalities which arise from these two requirements, and from the requirement that regions not be too restrictive. If such a system can be solved, an appropriate place has been found, otherwise there does not exist any and TS is unsolvable.

References

1. Best, E., Darondeau, P.: A decomposition theorem for finite persistent transition systems. Acta Informatica **46**, 237–254 (2009). https://doi.org/10.1007/s00236-009-0095-6
2. Best, E., Devillers, R.: Characterisation of the state spaces of live and bounded marked graph Petri nets. In: Dediu, A.-H., Martín-Vide, C., Sierra-Rodríguez, J.-L., Truthe, B. (eds.) LATA 2014. LNCS, vol. 8370, pp. 161–172. Springer, Cham (2014). https://doi.org/10.1007/978-3-319-04921-2_13
3. Best, E., Devillers, R.: Synthesis of live and bounded persistent systems. Fundamenta Informaticae **140**(1), 39–59 (2015)
4. Best, E., Devillers, R.: State space axioms for T-systems. Acta Informatica **52**(2–3), 133–152 (2015). https://doi.org/10.1007/s00236-015-0219-0. In: Lüttgen, G., Corradini, F. (eds.) Special Volume on the Occasion of Walter Vogler's 60th Birthday
5. Best, E., Devillers, R.: Characterisation of the state spaces of marked graph Petri nets. Inf. Comput. **253**, 399–410 (2017)
6. Best, E., Devillers, R., Erofeev, E., Wimmel, H.: Target-Oriented Petri Net Synthesis (2019, submitted paper)
7. Best, E., Devillers, R., Schlachter, U.: A graph-theoretical characterisation of state separation. In: Steffen, B., Baier, C., van den Brand, M., Eder, J., Hinchey, M., Margaria, T. (eds.) SOFSEM 2017. LNCS, vol. 10139, pp. 163–175. Springer, Cham (2017). https://doi.org/10.1007/978-3-319-51963-0_13
8. Best, E., Devillers, R., Schlachter, U.: Bounded choice-free Petri net synthesis: algorithmic issues. Acta Informatica **55**(7), 575–611 (2017). https://doi.org/10.1007/s00236-017-0310-9
9. Best, E., Devillers, R., Schlachter, U., Wimmel, H.: Simultaneous Petri net synthesis. Sci. Ann. Comput. Sci. **28**(2), 199–236 (2018)
10. Desel, J., Esparza, J.: Free Choice Petri Nets. Cambridge Tracts in Theoretical Computer Science. Cambridge University Press, Cambridge (1995)
11. Dijkstra, E.W.: The structure of "THE"-multiprogramming system. Commun. ACM **11**(5), 341–346 (1968)
12. Hoare, C.A.R.: Monitors: an operating system structuring concept. Commun. ACM **17**(10), 549–557 (1974)
13. Patil, S.: Circuit Implementation of Petri Nets. Computation Structures Group: Memo 73, Project MAC MIT (1972)

14. Schlachter, U., et al.: APT: Analysis of Petri Nets and Transition Systems. https:// github.com/CvO-Theory/apt
15. Wimmel, H.: Presynthesis of bounded choice-free or fork-attribution nets. Inf. Comput. (2019). https://doi.org/10.1016/j.ic.2019.104482
16. Teruel, E., Colom, J.M., Silva Suárez, M.: Choice-free Petri nets: a model for deterministic concurrent systems with bulk services and arrivals. IEEE Trans. Systems Man Cybern. Part A **27**(1), 73–83 (1997)
17. Teruel, E., Silva, M.: Structure theory of equal conflict systems. Theor. Comput. Sci **153**(1 & 2), 271–300 (1996)
18. Yakovlev, A., Gomes, L., Lavagno, L. (eds.): Hardware Design and Petri Nets. Springer, Boston (2000). https://doi.org/10.1007/978-1-4757-3143-9. XII+332 p.

On-the-Fly Synthesis for Strictly Alternating Games

Shyam Lal Karra, Kim Guldstrand Larsen, Marco Muñiz, and Jiří Srba[✉]

Aalborg University, Aalborg, Denmark
{shyamlal,kgl,muniz,srba}@cs.aau.dk

Abstract. We study two-player zero-sum infinite reachability games with strictly alternating moves of the players allowing us to model a race between the two opponents. We develop an algorithm for deciding the winner of the game and suggest a notion of alternating simulation in order to speed up the computation of the winning strategy. The theory is applied to Petri net games, where the strictly alternating games are in general undecidable. We consider soft bounds on Petri net places in order to achieve decidability and implement the algorithms in our prototype tool. Finally, we compare the performance of our approach with an algorithm proposed in the seminal work by Liu and Smolka for calculating the minimum fixed points on dependency graphs. The results show that using alternating simulation almost always improves the performance in time and space and with exponential gain in some examples. Moreover, we show that there are Petri net games where our algorithm with alternating simulation terminates, whereas the algorithm without the alternating simulation loops for any possible search order.

1 Introduction

An embedded controller often has to interact continuously with an external environment and make decisions in order for the overall system to evolve in a safe manner. Such systems may be seen as (alternating) games, where two players—the controller and the environment—race against each other in order to achieve their individual objectives. The environment and controller alternate in making moves: the environment makes a move and gives the turn to the controller who can correct the behaviour of the system and give the control back to the environment and so on. We consider zero-sum turn-based games where the objective of the controller is to reach a set of goal states, while the objective of the environment is to avoid these states. Winning such a game requires to synthesize a strategy for the moves of the controller, so that any run under this strategy leads to a goal state no matter the moves of the environment. We talk about *synthesizing a winning controller strategy*.

Consider the game in Fig. 1. The game has six configurations $\{s_0, \ldots, s_5\}$ and the solid edges indicate controller moves, while the dashed edges are environmental moves. The game starts at s_0 and the players alternate in taking turns,

© Springer Nature Switzerland AG 2020
R. Janicki et al. (Eds.): PETRI NETS 2020, LNCS 12152, pp. 109–128, 2020.
https://doi.org/10.1007/978-3-030-51831-8_6

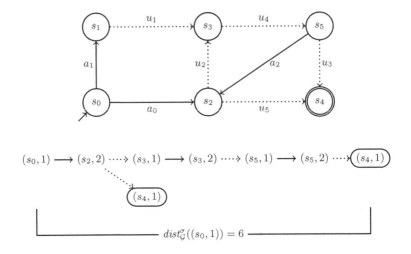

Fig. 1. A game graph and the distance of its winning strategy σ where $\sigma(s_0, 1) = a_0$, $\sigma(s_1, 1) = \sigma(s_2, 1) = \sigma(s_3, 1) = \sigma(s_4, 1) = \sigma(s_5, 1) = \epsilon_1$

assuming that the controller has to make the first move. It has three choices, i.e., move to s_1 or s_2 (shown by solid arrows) or stay in s_0 without moving. Assume that the controller chooses to move to s_2 and gives the control to environment. The environment now has the choice to go to s_3 or s_4 (which is a goal config- uration that the environment tries to avoid). Observe that there can be states like s_5 from which both the environment and controller can make a move, based on whose turn it is. Also the game may have states like s_0 or s_3 where one of the players does not have any explicit move in which case they play an empty move and give the control to the other. The controller can play the empty move also in the situation where other controllable moves are enabled, whereas the environment is allowed to make the empty move only if no other environmental moves are possible (in order to guarantee progress in the game). The goal of the controller is to reach s_4 and it has a winning strategy to achieve this goal as shown below the game graph (here the second component in the pair denotes the player who has the turn).

Petri nets [9] is a standard formalism that models the behaviour of a con- current system. Petri net games represent the reactive interaction between a controller and an environment by distinguishing controllable transitions from uncontrollable ones. In order to account for both the adversarial behaviour of the environment and concurrency, one can model reactive systems as alternat- ing games (as mentioned before) played on these nets where each transition is designated to either environment or controller and the goal of the controller is to eventually reach a particular marking of the net.

Contribution: We define the notion of alternating simulation and prove that if a configuration c' simulates c then c' is a winning configuration whenever c is a winning configuration. We then provide an on-the-fly algorithm which uses this alternating simulation relation and we prove that the algorithm is partially correct. The termination of the algorithm is in general not guaranteed. However, for finite game graphs we have total correctness. We apply this algorithm to games played on Petri nets and prove that these games are in general undecidable and therefore we consider a subclass of these games where the places in nets have a bounded capacity, resulting in a game over a finite graph. As an important contribution, we propose an efficiently computable (linear in the size of the net) alternating simulation relation for Petri nets. We also show an example where this specific simulation relation allows for a termination on an infinite Petri net game, while the adaptation of Liu-Smolka algorithm from [2] does not terminate for all possible search strategies that explore the game graph. Finally, we demonstrate the practical usability of our approach on three case studies.

Related Work: The notion of *dependency graphs* and their use to compute fixed points was originally proposed in [7]. In [2,3] an adaptation of these fixed point computations has been extended to two player games with reachability objectives, however, without the requirement on the alternation of the player moves as in our work. Two player coverability games on Petri nets with strictly alternating semantics were presented in [10] and [1], but these games are restricted in the sense that they assume that the moves in the game are monotonic with respect to a partial order which makes the coverability problem decidable for the particular subclass (B-Petri games). Our games are more general and hence the coverability problem for our games played on Petri nets, unlike the games in [10], are undecidable. Our work introduces an on-the-fly algorithm closely related to the work in [3] which is an adaptation of the classical Liu Smolka algorithm [7], where the set of losing configurations is stored, which along with the use of our alternating simulation makes our algorithm more efficient with respect to both space and time consumption. Our main novelty is the introduction of alternating simulation as a part of our algorithm in order to speed up the computation of the fixed point and we present its syntactic approximation for the class of Petri net games with encouraging experimental results.

2   Alternating Games

We shall now introduce the notion of an alternating game where the players strictly alternate in their moves such that the first player tries to enforce some given reachability objective and the other player's aim is to prevent this from happening.

Definition 1. *A* Game graph *is a tuple* $\mathcal{G} = (S, Act_1, Act_2, \longrightarrow_1, \longrightarrow_2, Goal)$ *where:*

- *S is the set of states,*
- *Act_i is a finite set of Player-i actions where $i \in \{1, 2\}$ and $Act_1 \cap Act_2 = \emptyset$,*
- *$\longrightarrow_i \subseteq S \times Act_i \times S$, $i \in \{1, 2\}$ is the Player-i edge relation such that the relations $\longrightarrow_i$ are deterministic i.e. if $(s, \alpha, s') \in \longrightarrow_i$ and $(s, \alpha, s'') \in \longrightarrow_i$ then $s' = s''$, and*
- *$Goal \subseteq S$ is a set of goal states.*

If $(s, a, s') \in \longrightarrow_i$ then we write $s \xrightarrow{a}_i s'$ where $i \in \{1, 2\}$. Also, we write $s \longrightarrow_i$ if there exists an $s' \in S$ and $a \in Act_i$ such that $s \xrightarrow{a}_i s'$; otherwise we write $s \not\longrightarrow_i$. In the rest of this paper, we restrict ourselves to game graphs that are *finitely branching*, meaning that for each action a (of either player) there are only finitely many a-successors.

Alternating Semantics for Game Graphs. Given a game graph $\mathcal{G} = (S, Act_1, Act_2, \longrightarrow_1, \longrightarrow_2, Goal)$, the set $C = (S \times \{1, 2\})$ is the set of configurations of the game graph and $C_1 = (S \times \{1\}), C_2 = (S \times \{2\})$ represent the set of configurations of *Player-1* and *Player-2* respectively, $\Longrightarrow \subseteq (C_1 \times (Act_1 \cup \{\epsilon_1\}) \times C_2) \cup (C_2 \times (Act_2 \cup \{\epsilon_2\}) \times C_1)$ is the set of transitions of the game such that

- if $(s, \alpha, s') \in \longrightarrow_i$ then $((s, i), \alpha, (s', 3 - i)) \in \Longrightarrow$
- for all configurations $(s, 1) \in C_1$ we have $((s, 1), \epsilon_1, (s, 2)) \in \Longrightarrow$
- for all configurations $(s, 2) \in C_2$ where $s \not\longrightarrow_2$ we have $((s, 2), \epsilon_2, (s, 1)) \in \Longrightarrow$

Apart from the actions available in Act_1 and Act_2, the controller and the environment have two empty actions ϵ_1 and ϵ_2, respectively. Given a state $s \in S$, when it is its turn, the controller or the environment can choose to take any action in Act_1 or Act_2 enabled at s respectively. While, the controller can always play ϵ_1 at a given state s, the environment can play ϵ_2 only when there are no actions in Act_2 enabled at s so that the environment can not delay or deadlock the game.

We write $(s, i) \xrightarrow{\alpha}_i (s', 3 - i)$ to denote $((s, i), \alpha, (s', 3 - i)) \in \Longrightarrow$ and $(s, i) \xrightarrow{\alpha}_i$ if there exists $s' \in S$ such that $(s, i) \xrightarrow{\alpha} (s', 3 - i)$. A configuration $c_g = (s_g, i)$ is called *goal configuration* ifs$_g \in Goal$. By abuse of notation, for configuration c we use $c \in Goal$ to denote that c is a goal configuration. A run in the game $\mathcal{G}$ starting at a configuration c_0 is an infinite sequence of configurations and actions $c_0, \alpha_0, c_1, \alpha_1, c_2, \alpha_2 \ldots$ where for every $j \in \mathbb{N}^0$, $c_j \xrightarrow{\alpha_j} c_{j+1}$. The set $Runs(c_0, \mathcal{G})$ denotes the set of all runs in the game $\mathcal{G}$ starting from configuration c_0. A run $c_0, \alpha_0, c_1, \alpha_1, c_2, \alpha_2 \ldots$ is *winning* if there exists j such that c_j is a goal configuration. The set of all winning runs starting from c_0 is denoted by $WinRuns(c_0, \mathcal{G})$. A *strategy* σ for *Player-1* where $\sigma : C_1 \to Act_1 \cup \{\epsilon_1\}$ is a function such that if $\sigma(c) = \alpha$ then $c \xrightarrow{\alpha}$. For a given strategy σ, and a configuration $c_0 = (s_0, i_0)$, we define $OutcomeRuns_{\mathcal{G}}^\sigma(c_0) = \{c_0, \alpha_0, c_1, \alpha_1 \ldots \in Runs(c_0, \mathcal{G}) \mid$ for every $k \in \mathbb{N}^0, \alpha_{2k+i_0-1} = \sigma(c_{2k+i_0-1})\}$. A strategy σ is *winning* at a configuration c if $OutcomeRuns_{\mathcal{G}}^\sigma(c) \subseteq WinRuns(c, \mathcal{G})$. A configuration c is a *winning configuration* if there is a winning strategy at c. A winning strategy for our running example is given in the caption of Fig. 1.

Given a configuration $c = (s, i)$ and one of its *winning* strategies σ, we define the quantity $dist_{\mathcal{G}}^\sigma(c) := \max(\{n \mid c_0, \alpha_0 \ldots c_n \ldots \in OutcomeRuns_{\mathcal{G}}^\sigma(c)$ with $c = c_0$ and $c_n \in Goal$ while $c_i \notin Goal$ for all $i < n\})$. The quantity represents the *distance* to the goal state, meaning that during any play by the given strategy there is a guarantee that the goal state is reached within that many steps. Due to our assumption on finite branching of our game graph, we get that distance is well defined.

Lemma 1. *Let σ be a winning strategy for a configuration c. The distance function $dist_{\mathcal{G}}^{\sigma}(c)$ is well defined.*

Proof. Should $dist_{\mathcal{G}}^{\sigma}(c)$ not be well defined, meaning that the maximum does not exist, then necessarily the set $T = \{c_0, \alpha_0 \ldots c_n \mid c_0, \alpha_0 \ldots c_n \ldots \in OutcomeRuns_{\mathcal{G}}^{\sigma}(c)$ where $c = c_0$ and $c_n \in Goal$ with $c_i \notin Goal$ for all $i < n\}$ induces an infinite tree. By definition the game graph $\mathcal{G}$ is a deterministic transition system with finite number of actions and hence the branching factor of $\mathcal{G}$ and consequently of T is finite. By Köning's Lemma the infinite tree T must contain an infinite path without any goal configuration, contradicting the fact that σ is a winning strategy.   □

Consider again the game graph shown in Fig. 1 where solid arrows indicate the transitions of *Player-1* while the dotted arrows indicate those of *Player-2*. A possible tree of all runs under a (winning) strategy σ is depicted in the figure and its distance is 6. We can also notice that the distance strictly decreases once *Player-1* performs a move according to the winning strategy σ, as well as by any possible move of *Player-2*. We can now observe a simple fact about alternating games.

Lemma 2. *If $(s, 2)$ is a winning configuration in the game $\mathcal{G}$, then $(s, 1)$ is also a winning configuration.*

Proof. Since $(s_0, 1) \overset{\epsilon_1}{\Longrightarrow} (s_0, 2)$ and there is a winning strategy from $(s_0, 2)$, we can conclude $(s_0, 1)$ is also a winning configuration.   □

We shall be interested in finding an efficient algorithm for deciding the following problem.

Definition 2 (Reachability Control Problem). *For a given game $\mathcal{G}$ and a configuration (s, i), the reachability control problem is to decide if (s, i) is a winning configuration.*

3   Alternating Simulation and On-the-Fly Algorithm

We shall now present the notion of alternating simulation that will be used as the main component of our on-the-fly algorithm for determining the winner in the alternating game. Let $\mathcal{G} = (S, Act_1, Act_2, \longrightarrow_1, \longrightarrow_2, Goal)$ be a game graph and let us adopt the notation used in the previous section.

Definition 3. *A reflexive binary relation $\preceq \subseteq C_1 \times C_1 \cup C_2 \times C_2$ is an alternating simulation relation iff whenever $(s_1, i) \preceq (s_2, i)$ then*

- if $s_1 \in Goal$ then $s_2 \in Goal$,
- if $(s_1, 1) \overset{a}{\Longrightarrow}_1 (s'_1, 2)$ then $(s_2, 1) \overset{a}{\Longrightarrow}_1 (s'_2, 2)$ such that $(s'_1, 2) \preceq (s'_2, 2)$, and
- if $(s_2, 2) \overset{u}{\Longrightarrow}_2 (s'_2, 1)$ then $(s_1, 2) \overset{u}{\Longrightarrow}_2 (s'_1, 1)$ such that $(s'_1, 1) \preceq (s'_2, 1)$.

An important property of alternating simulation is that it preserves winning strategies as stated in the following theorem.

Theorem 1. Let (s, i) be a winning configuration and $(s, i) \preceq (s', i)$ then (s', i) is also a winning configuration.

Proof. By induction on k we shall prove the following claim: if $(s, i) \preceq (s', i)$ and (s, i) is a winning configuration with $dist^\sigma_\mathcal{G}((s, i)) \leq k$ then (s', i) is also a winning configuration.

Base case ($k = 0$): Necessarily (s, i) is a goal state and by the definition of alternating simulation (s', i) is also a goal configuration and hence the claim trivially holds.

Induction step ($k > 0$): Case $i = 1$. Let σ be a winning strategy for $(s, 1)$ such that $(s, 1) \overset{\sigma((s,1))}{\Longrightarrow}_1 (s_1, 2)$. We define a winning strategy σ' for $(s', 1)$ by $\sigma'((s', 1)) = \sigma((s, 1))$. By the property of alternating simulation we get that $(s', 1) \overset{\sigma((s,1))}{\Longrightarrow}_1 (s'_1, 2)$ such that $(s_1, 2) \preceq (s'_1, 2)$ and $dist^\sigma_\mathcal{G}((s_1, 2)) < k$. Hence we can apply induction hypothesis and claim that $(s'_1, 2)$ has a winning strategy which implies $(s', 1)$ is a winning configuration as well. Case $i = 2$. If $(s', 2) \overset{u}{\Longrightarrow}_2 (s'_1, 1)$ for some $u \in Act_2$ then by the property of alternating simulation also $(s, 2) \overset{u}{\Longrightarrow}_2 (s_1, 1)$ such that $(s_1, 1) \preceq (s'_1, 1)$ and $dist^\sigma_\mathcal{G}((s_1, 2)) < k$. By the induction hypothesis $(s'_1, 1)$ is a winning configuration and so is $(s', 2)$. $\square$

As a direct corollary of this result, we also know that if $(s, i) \preceq (s', i)$ and (s', i) is not a winning configuration then (s, i) cannot be a winning configuration either. Before we present our on-the-fly algorithm, we need to settle some notation. Given a configuration $c = (s, i)$ we define

- $Succ(c) = \{c' \mid c \Longrightarrow c'\}$,
- $MaxSucc(c) \subseteq Succ(c)$ is a set of states such that for all $c' \in Succ(c)$ there exists a $c'' \in MaxSucc(c)$ such that $c' \preceq c''$, and
- $MinSucc(c) \subseteq Succ(c)$ is a set of states such that for all $c' \in Succ(c)$ there exists a $c'' \in MinSucc(c)$ such that $c'' \preceq c'$.

Remark 1. There can be many candidate sets of states that satisfy the definition of $MaxSucc(c)$ (and $MinSucc(c)$). In the rest of the paper, we assume that there is a way to fix one among these candidates and the theory proposed here works for any such candidate.

Given a game graph $\mathcal{G}$, we define a subgraph of $\mathcal{G}$ denoted by $\mathcal{G}'$, where for all *Player-1* sucessors we only keep the maximum ones and for every *Player-2* sucessors we preserve only the minimum ones. In the following lemma we show that in order to solve the reachability analysis problem on $\mathcal{G}$, it is sufficient to solve it on $\mathcal{G}'$.

Definition 4. *Given a game graph* $\mathcal{G} = (S, Act_1, Act_2, \longrightarrow_1, \longrightarrow_2, Goal)$, *we define a pruned game graph* $\mathcal{G}' = (S, Act_1, Act_2, \longrightarrow_1', \longrightarrow_2', Goal)$ *such that*
$\longrightarrow_1' = \{(s_1, 1, s_2) \mid (s_1, 1, s_2) \in \longrightarrow_1 \text{ where } (s_2, 2) \in MaxSucc(s_1, 1)\}$ *and*
$\longrightarrow_2' = \{(s_2, 2, s_1) \mid (s_2, 2, s_1) \in \longrightarrow_2 \text{ where } (s_1, 1) \in MinSucc(s_2, 2)\}$.

Lemma 3. *Given a game graph* $\mathcal{G}$, *a configuration* (s, i) *is winning in* $\mathcal{G}$ *iff* (s, i) *is winning in* $\mathcal{G}'$.

Proof. We prove the case for $i = 1$ (the argument for $i = 2$ is similar).

"$\Longrightarrow$": Let $(s, 1)$ be winning in $\mathcal{G}$ under a winning strategy σ. We define a new strategy σ' in $\mathcal{G}'$ such that for every $(s_1, 1)$ where $\sigma((s_1, 1)) = (s_1', 2)$ we define $\sigma'((s_1, 1)) = (s_1'', 2)$ for some $(s_1'', 2) \in MaxSucc(s_1, 1)$ such that $(s_1', 2) \preceq (s_1'', 2)$. By induction on $dist_{\mathcal{G}}^{\sigma}((s, 1))$ we prove that if $(s, 1)$ is winning in $\mathcal{G}$ then it is winning also in $\mathcal{G}'$. Base case $(dist_{\mathcal{G}}^{\sigma}((s, 1)) = 0)$ clearly holds as $(s, 1)$ is a goal configuration. Let $dist_{\mathcal{G}}^{\sigma}((s, 1)) = k$ where $k > 0$ and let $\sigma((s, 1)) = (s', 2)$ and $\sigma'((s, 1)) = (s'', 2)$. Clearly, $dist_{\mathcal{G}}^{\sigma}((s', 2)) < k$ and by induction hypothesis $(s', 2)$ is winning also in $\mathcal{G}'$. Because $(s', 2) \preceq (s'', 2)$ we get by Theorem 1 that $(s'', 2)$ is also winning and $dist_{\mathcal{G}}^{\sigma}((s'', 2)) < k$. Since $s \xrightarrow{\sigma'(s,1)}_1' s''$ we get that σ' is a winning strategy for $(s, 1)$ in $\mathcal{G}'$.

"$\Longleftarrow$": Let $(s, 1)$ be winning in $\mathcal{G}'$ by strategy σ'. We show that $(s, 1)$ is also winning in $\mathcal{G}$ under the same strategy σ'. We prove this fact by induction on $dist_{\mathcal{G}'}^{\sigma'}((s, 1))$. If $dist_{\mathcal{G}'}^{\sigma'}((s, 1)) = 0$ then $(s, 1)$ is a goal configuration and the claim follows. Let $dist_{\mathcal{G}'}^{\sigma'}((s, 1)) = k$ where $k > 0$ and let $\sigma'(s, 1) = (s'', 2)$. Clearly $dist_{\mathcal{G}'}^{\sigma'}((s'', 2)) < k$. So by induction hypothesis, $(s'', 2)$ is a winning configuration in $\mathcal{G}$. Since $s \xrightarrow{\sigma'(s,1)}_1 s''$ we get that σ' is a winning strategy for $(s, 1)$ in $\mathcal{G}$. $\square$

We can now present an algorithm for the reachability problem in alternating games which takes some alternating simulation relation as a parameter, along with the game graph and the initial configuration. In particular, if we initialize $\preceq$ to the identity relation, the algorithm results essentially (modulo the additional *Lose* set construction) as an adaption of Liu-Smolka fixed point computation as given in [3]. Our aim is though to employ some more interesting alternating simulation that is fast to compute (preferably in syntax-driven manner as we demonstrate in the next section for the case of Petri net games). Our algorithm uses this alternating simulation at multiple instances which improves the efficiency of deciding the winning and losing status of configurations without having to explore unnecessary state space.

The algorithm uses the following data structures:

- W is the set of edges (in the alternating semantics of game graph) that are *waiting* to be processed,
- *Disc* is the set of already *discovered* configurations,
- *Win* and *Lose* are the sets of currently known *winning* resp. *losing* configurations, and
- D is a *dependency* function that to each configuration assigns the set of edges to be reinserted to the waiting set W whenever the configuration is moved to the set *Win* or *Lose* by the help functions AddToWin resp. AddToLose.

Algorithm 1. Game graph algorithm

Input Game graph $\mathcal{G} = (S, Act_1, Act_2, \longrightarrow_1, \longrightarrow_2, G)$, alternating simulation $\preceq$ and the initial configuration (s_0, i_0)

Output true if (s_0, i_0) is a winning configuration, false otherwise

1: $W = \emptyset$; $Disc := \{(s_0, i_0)\}$; $Lose := \emptyset$; $Win := \emptyset$; $D((s_0, i_0)) = \emptyset$;

2: If $s_0 \in G$ then ADDTOWIN$((s_0, i_0))$;

3: ADDSUCCESSORS(W, s_0, i_0);

4: **while** (

5:     $\nexists(s_0', i_0).(s_0', i_0) \preceq (s_0, i_0) \wedge (s_0', i_0) \in Win$ **and**

6:     $\nexists(s_0', i_0).(s_0, i_0) \preceq (s_0', i_0) \wedge (s_0', i_0) \in Lose$ **and**

7:     $W \neq \emptyset$

8:     ) **do**

9:     Pick $(s_1, i, s_2) \in W$; $W := W \setminus \{(s_1, i, s_2)\}$;

10:     **if** $(s_1, i) \in Win \cup Lose$ **then** Continue at line 4;

11:     **if** (

12:         $\exists s_1'. (s_1, i) \preceq (s_1', i) \wedge (s_1', i) \in Lose$ **or**

13:         $i = 1 \wedge \forall s_3.s_1 \Longrightarrow_1 s_3$ implies $\exists s_3'.(s_3, 2) \preceq (s_3', 2) \wedge (s_3', 2) \in Lose$ **or**

14:         $i = 2 \wedge \exists s_3, s_3'.(s_3, 1) \preceq (s_3', 1) \wedge s_1 \Longrightarrow_2 s_3 \wedge (s_3', 1) \in Lose$ **or**

15:         $i = 2 \wedge \exists s_1'. (s_1, 1) \preceq (s_1', 1) \wedge (s_1', 1) \in Lose$ **or**

16:         $s_1 \not\longrightarrow_1 \wedge s_1 \not\longrightarrow_2$

17:     ) **then**

18:         ADDTOLOSE(W, s_1, i); Continue at line 4;

19:     **end if**

20:     **if** (

21:         $\exists s_1'. (s_1', i) \preceq (s_1, i) \wedge (s_1', i) \in Win$ **or**

22:         $i = 1 \wedge \exists s_3, s_3'.(s_1, 1) \Longrightarrow_1 (s_3, 2) \wedge (s_3', 2) \preceq (s_3, 2) \wedge (s_3', 2) \in Win$ **or**

23:         $i = 2 \wedge \forall s_3.(s_1, 2) \Longrightarrow_2 (s_3, 1)$ implies $\exists s_3'.(s_3', 1) \preceq (s_3, 1) \wedge (s_3', 1) \in Win$

24:     ) **then**

25:         ADDTOWIN(W, s_1, i); Continue at line 4;

26:     **end if**

27:     **if** $(s_2, 3 - i) \notin Win \cup Lose$ **then**

28:         **if** $(s_2, 3 - i) \in Disc$ **then**

29:             $D((s_2, 3 - i)) := D((s_2, 3 - i)) \cup \{(s_1, i, s_2)\}$;

30:         **else**

31:             $Disc := Disc \cup \{(s_2, 3 - i)\}$;

32:             $D((s_2, 3 - i)) := \{(s_1, i, s_2)\}$;

33:             **if** $s_2 \in G$ **then**

34:                 ADDTOWIN$(W, s_2, 3 - i)$;

35:             **else**

36:                 ADDSUCCESSORS$(W, s_2, 3 - i)$;

37:             **end if**

38:         **end if**

39:     **end if**

40: **end while**

41: **if** $\exists s_0'. s_0' \preceq s_0 \wedge (s_0', i_0) \in Win$ **then return** true

42: **else return** false

43: **end if**

Algorithm 2. Helper procedures for Algorithm 1

1: **procedure** ADDSUCCESSORS(W, s, i)
2:     **if** ($i = 1$) **then**
3:         $L := \{(s, i, s') \mid$ such that $(s', 3 - i) \in MaxSucc(s,i)\}$;
4:     **end if**
5:     **if** ($i = 2$) **then**
6:         $L := \{(s, i, s') \mid$ such that $(s', 3 - i) \in MinSucc(s,i)\}$;
7:     **end if**
8:     $W := W \cup L$
9: **end procedure**
10: **procedure** ADDTOWIN(W, s, i)
11:     $Win := Win \cup \{(s, i)\}$;
12:     $W := W \cup D(s, i) \setminus \{(s'', 3 - i, s) \mid (s'', 3 - i) \in (Win \cup Lose)\}$;
13: **end procedure**
14: **procedure** ADDTOLOSE(W, s, i)
15:     $Lose := Lose \cup \{(s, i)\}$;
16:     $W := W \cup D(s, i) \setminus \{(s'', 3 - i, s) \mid (s'', 3 - i) \in (Win \cup Lose)\}$;
17: **end procedure**

As long as the waiting set W is nonempty and no conclusion about the initial configuration can be drawn, we remove an edge from the waiting set and check whether the source configuration of the edge can be added to the losing or winning set. After this, the target configuration of the edge is explored. If it is already discovered, we only update the dependencies. Otherwise, we also check if it is a goal configuration (and call AddToWin if this is the case), or we add the outgoing edges from the configuration to the waiting set.

In order to argue about the correctness of the algorithm, we introduce a number of loop invariants for the while loop in Algorithm 1 in order to argue that if the algorithm terminates then $(s_0, turn)$ is winning if and only if $(s_0, turn) \in Win$.

Lemma 4. *Loop Invariant 1 for Algorithm 1: If $(s, i) \in Win$ and $(s, i) \preceq (s', i)$ then (s', i) is a winning configuration.*

Proof Sketch. Initially, $Win = \emptyset$ and the invariant holds trivially before the loop is entered. Let us assume that the invariant holds before we execute the body of the while loop and we want to argue that after the body is executed the invariant still holds. During the current iteration, a new element can be added to the set Win at lines 25 or 34. If line 25 is executed then at least one of the conditions at lines 21, 22, 23 must hold. We argue in each of these cases, along with the cases in which line 34 is executed, that the invariant continues to hold after executing the call to the function AddToWin. □

Lemma 5. *Loop Invariant 2 for Algorithm 1: If $(s, i) \in Lose$ and $s' \preceq s$ then (s', i) is not a winning configuration.*

Proof Sketch. Initially $Lose = \emptyset$ and loop invariant 2 holds trivially before the while loop at line 4 is entered. Similarly to the previous lemma, we argue that in all the cases where the function AddToLose is called, the invariant continues to hold after the call as well.   □

The next invariant is essential for the correctness proof.

Lemma 6. *Loop Invariant 3 for Algorithm 1: During the execution of the algorithm for any $(s, j) \in Disc \setminus (Win \cup Lose)$ invariantly holds*

(a) if $j = 1$ then for every $(s', 2) \in MaxSucc(s, 1)$

> **I.** $(s, 1, s') \in W$ *or*
> **II.** $(s, 1, s') \in D(s', 2)$ *and* $(s', 2) \in Disc \setminus Win$ *or*
> **III.** $(s', 2) \in Lose$

(b) if $j = 2$ then for every $(s', 1) \in MinSucc(s, 2)$

> **I.** $(s, 2, s') \in W$ *or*
> **II.** $(s, 2, s') \in D(s', 1)$ *and* $(s', 1) \in Disc \setminus Lose$ *or*
> **III.** $(s', 1) \in Win.$

This third invariant claims that for any discovered but yet undetermined configuration (s, j) and for any of its outgoing edges (maximum ones in case it is *Player-1* move and minimum ones for *Player-2* moves), the edge is either on the waiting list, or the target configuration is determined as winning or losing, or otherwise the edge is in the dependency set of the target configuration, so that in case the target configuration is later on determined as winning or losing, the edge gets reinserted to the waiting set and the information can possibly propagate to the parent configuration of the edge. This invariant is crucial in the proof of partial correctness of the algorithm which is given below.

Theorem 2. *If the algorithm terminates and returns true then the configuration (s_0, i) is winning and if the algorithm upon termination returns false then the configuration (s_0, i) is losing.*

Proof. Upon termination, if the algorithm returns true (this can happen only at the Line 41), then it means there exists a $s'_0 \in S$ such that $(s'_0, i) \in Win$ and $(s'_0, i) \preceq (s_0, i)$. From Lemma 4, we can deduce that (s_0, i) is winning. On the other hand if the algorithm returns false, then there are two cases.

Case 1: There exists a $s'_0 \in S$ such that $(s_0, i) \preceq (s'_0, i)$ and $(s'_0, i) \in Lose$. From Lemma 5, we can deduce that (s_0, i) does not have a winning strategy.

Case 2: $W = \emptyset$ and there exists no $s'_0 \in S$ such that $(s'_0, i) \in Win$ and $(s'_0, i) \preceq (s_0, i)$ and there exists no $s'_0 \in S$ such that $(s'_0, i) \in Lose$ and $(s_0, i) \preceq (s'_0, i)$. We prove that (s_0, i) is not winning for *Player-1* by contradiction. Let (s_0, i) be a winning configuration. Let σ be a *Player-1* winning strategy. Let $L = Disc \setminus (Win \cup Lose)$. Now we make the following two observations.

From Lemma 6, we can deduce that for all $(s, 1) \in L$, if $(s, 1) \Longrightarrow_1 (s_1, 2)$ then there exists an s'_1 such that $(s_1, 2) \preceq (s'_1, 2)$ and $(s'_1, 2) \in L$ or $(s'_1, 2) \in Lose$.

Also, we can deduce that the case that for all $(s_2, 1) \in Succ(s, 2)$ we have $(s_2, 1) \in Win$ is not possible as follows. Among all $(s_2, 1) \in Succ(s, 2)$ consider the last $(s_2, 1)$ entered the set Win. During this process in the call to function AddToWin, the edge $(s, 2, s_2)$ is added to the waiting list. Since by the end of the algorithm, this edge is processed and it would have resulted in adding $(s, 2)$ to Win because of the condition at line 23. But this is a contradiction as $(s, 2) \in L$ and $L \cap Win = \emptyset$. Hence for all $(s, 2) \in L$, there exists an s_2 such that $(s, 2) \Longrightarrow_2 (s_2, 1)$ and $(s_2, 1) \in L$.

Given these two observations, and the definition of $\preceq$, one can show that *Player-2* can play the game such that for the resulting run $\rho \in OutcomeRuns_{\mathcal{G}}^{\sigma}(s_0)$ where $\rho = \langle (s_0, i), a_0, (s_1, 3 - i), a_1 \ldots \rangle$, there exists a $\rho' = \langle (s_0', i), b_0, (s_1', 3 - i), b_1 \ldots \rangle \in Runs(s_0, \mathcal{G})$ such that $s_0' = s_0$, for all $k \in \mathbb{N}$, $(s_k, 2 - ((i + k)\%2)) \preceq (s_k', 2 - ((i + k)\%2))$ and $(s_k', 2 - ((i + k)\%2)) \in L$. In other words the configurations in the run ρ are all (one by one) simulated by the corresponding configurations in the run ρ' and the configurations from the run ρ' moreover all belong to the set L. Since σ is a winning strategy, there must exist an index $j \in \{0, 1, 2, \ldots\}$ such that $s_j \in G$. Since the set of goal states are upward closed, it also means $s_j' \in G$. But $L \cap \{(s, 1), (s, 2) \mid s \in G\} = \emptyset$ because the line 34 of the algorithm adds a configuration in $\{(s, 1), (s, 2) \mid s \in G\}$ to Win when it enters the set $Disc$ for the first time. Hence our assumption that (s_0, i) has a winning strategy is contradicted. □

The algorithm does not necessarily terminate on general game graphs, however, termination is guaranteed on finite game graphs.

Theorem 3. *Algorithm 1 terminates on finite game graphs.*

Proof. In the following, we shall prove that any edge $e = (s_1, i, s_2)$ can be added to W at most twice and since there are only finitely many edges and every iteration of the while loop removes an edge from W (at line 9), W eventually becomes empty and the algorithm terminates on finite game graphs.

During the execution of while loop, an edge can only be added W through the call to functions AddSuccessors at line 36, AddToWin at lines 25, 34, or AddToLose at line 18. We shall show that these three functions can add an edge e atmost twice to waiting list W.

Let $e = (s_1, i, s_2)$ be an edge added to W in iteration k through a call to AddToWin at line 34. This implies that, during iteration k, the condition in line 27 is true. Hence $(s_2, 3 - i) \notin Win \cup Lose$ before iteration k is executed and after line 34 is executed, $(s_2, 3 - i) \in Win \cup Lose$ (hence the condition in line 27 is now false). So the call to function AddToWin at line 34 can not add e to W after iteration k.

Let $e = (s_1, i, s_2)$ be an edge added to W in iteration k through a call to AddToWin at line 25. This implies that, during iteration k, the condition in line 10 is not true. Hence $(s_2, 3 - i) \notin Win \cup Lose$ before iteration k is executed and after line 25 is executed, $(s_2, 3 - i) \in Win \cup Lose$ (hence the condition in line 10 is false). So the call to function AddToWin at line 25 can not add e to W after iteration k.

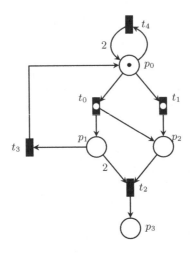

Fig. 2. A Petri net game

Similar to previous cases, we can argue that the call to AddToLose at line 18 can add e to W atmost once. Also it is easy to observe that once $(s_2, 3 - i)$ is added to set $Win \cup Lose$ by any of the lines 25, 34, 18, the other two can not add e to W. So, all together, all these three lines can add e to W atmost once.

Now consider case when $e = (s_1, i, s_2)$ is added to W in iteration k through a call to AddSuccessors at line 36. This implies, during iteration k, the condition in the line 28 is not true. Hence $(s_1, i) \notin Disc$ before the iteration k is executed and after the line 31 is executed $(s_1, i) \in Disc$ by the end of iteration k and the condition in the line 28 is true . So the call to function AddSuccessors at line 36 can add e to W atmost once. In total, edge e can enter W atmost twice. □

4   Application to Petri Games

We are now ready to instantiate our general framework to the case of Petri net games where the transitions are partitioned between the controller and environment transitions and we moreover consider a soft bound for the number of tokens in places (truncating the number of tokens that exceed the bound).

Definition 5 (Petri Net Games). *A Petri Net Game* $\mathcal{N} = (P, T_1, T_2, W, B, \varphi)$ *is a tuple where*

- *P is a finite set of places,*
- *T_1 and T_2 are finite sets of transitions such that $T_1 \cap T_2 = \emptyset$,*
- *$W : (P \times (T_1 \cup T_2)) \cup ((T_1 \cup T_2) \times P) \to \mathbb{N}^0$ is the weight function,*
- *$B : P \to \mathbb{N}^0 \cup \{\infty\}$ is a function which assigns a (possible infinite) soft bound to every $p \in P$, and*
- *φ is a formula in* coverability logic *given by the grammar $\varphi ::= \varphi_1 \wedge \varphi_2 \mid p \geq c$ where $p \in P$ and $c \in \mathbb{N}^0$.*

A *marking* is a function $M : P \to \mathbb{N}^0$ such that $M(p) \leq B(p)$ for all $p \in P$. The set $\mathcal{M}(\mathcal{N})$ is the set of all markings of $\mathcal{N}$.

In Fig. 2 we show an example of a Petri game where $P = \{p_0, \ldots, p_3\}$ are the places, there are three controller transitions $T_1 = \{t_2, t_3, t_4\}$ and two environment transitions (indicated with a circle inside) $T_2 = \{t_0, t_1\}$. The function W assigns each arc a weight of 1, except the arcs from p_1 to t_2 and t_4 to p_0 that have weight 2. The bound function B assigns every place ∞ and $\varphi ::= p_3 \geq 1$ requires that in the goal marking the place p_3 must have at least one token. The initial marking of the net is $\langle 1, 0, 0, 0 \rangle$ where the place p_0 has one token and the places p_1 to p_3 (in this order) have no tokens.

For the rest of this section, let $\mathcal{N} = (P, T_1, T_2, W, B, \varphi)$ be a fixed Petri net game. The satisfaction relation for a coverability formula φ in marking M is defined as expected:

- $M \models p \geq n$ iff $M(p) \geq n$ where $n \in \mathbb{N}^0$, and
- $M \models \varphi_1 \wedge \varphi_2$ iff $M \models \varphi_1$ and $M \models \varphi_2$.

For a given formula φ, we now define the set of goal markings $\mathcal{M}_\varphi = \{M \in \mathcal{M}(\mathcal{N}) \mid M \models \varphi\}$.

Firing a transition $t \in T_1 \cup T_2$ from a marking M results in marking M' if for all $p \in P$ we have $M(p) \geq W((p, t))$ and

$$M'(p) = \begin{cases} M(p) - W((p, t)) + W((t, p)) & \text{if } M(p) - W((p, t)) + W((t, p)) \leq B(p) \\ B(p) & \text{otherwise.} \end{cases}$$

We denote this by $M \xrightarrow{t} M'$ and note that this is a standard Petri net firing transition in case the soft bound for each place is infinity, otherwise if the number of tokens in a place p of the resulting marking exceeds the integer bound $B(p)$ then it is truncated to $B(p)$.

Petri net game $\mathcal{N}$ induces a game graph $(\mathcal{M}(\mathcal{N}), T_1, T_2, \longrightarrow_1, \longrightarrow_2, \mathcal{M}_\varphi)$ where $\longrightarrow_i = \{(M, t, M') \mid M \xrightarrow{t} M' \text{ and } t \in T_i\}$ for $i \in \{1, 2\}$.

For example, the (infinite) game graph induced by Petri game from Fig. 2 is shown in Fig. 3a. Next we show that reachability control problem for game graphs induced by Petri games is in general undecidable.

Theorem 4. *The reachability control problem for Petri games is undecidable.*

Proof. In order to show that reachability control problem for Petri games is undecidable, we reduce the undecidable problem of reachability for two counter Minsky machine [6] to the reachability control problem.

A Minsky Counter Machine with two non-negative counters c_1 and c_2 is a sequence of labelled instructions of the form:

$$1 : instr_1, 2 : instr_2, \ldots n : \text{HALT} \tag{1}$$

where for all $i \in \{1, 2, \ldots, n-1\}$ each $instr_i$ is of the form

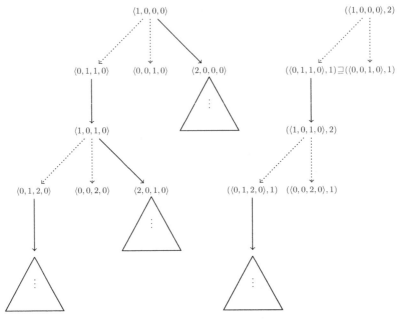

(a) Game graph for PN from Figure 2   (b) Alternating semantics for Figure 3a

Fig. 3. Game graph and its alternating semantics

- $instr_i$: $c_r = c_r + 1$; goto j or
- $instr_i$: if $c_r = 0$ then goto j else $c_r = c_r - 1$ goto k

for $j, k \in \{1, 2, \ldots, n\}$ and $r \in \{1, 2\}$.

A *computation* of a Minsky Counter Machine is a sequence of configurations (i, c_1, c_2) where $i \in \{1, 2 \ldots n\}$ is the label of instruction to be performed and c_1, c_2 are values of the counters. The starting configuration is $(1, 0, 0)$. The computation sequence is deterministic and determined by the instructions in the obvious way. The *halting problem* of a Minsky Counter Machine as to decide whether the computation of the machine ever halts (reaches the instruction with label n).

Given a two counter machine, we construct a Petri game $\mathcal{N} = \langle P, T_1, T_2, W, B, \varphi \rangle$ where $P = \{p_1, \ldots, p_n, p_{c_1}, p_{c_2}\} \cup \{p_{i:c_r=0}, p_{i:c_r\neq0} \mid i \in \{1 \ldots n\}\}$ is the set of places containing place for each of n instructions, a place for each counter and some helper places. The set of transitions is $T_1 = \{t_{i:1}, t_{i:2}, t_{i:3}, t_{i:4} \mid i$ is a decrement instruction$\} \cup \{t_i \mid i$ is an increment instruction$\}$ and $T_2 = \{t_{i:cheat} \mid i$ is a decrement transition$\}$. Figures 4a, 4b define the gadgets that are added for each increment and decrement instruction, respectively. The function B binds every place to 1, except for p_{c_1}, p_{c_2} each of which have a bound ∞. The formula to be satisfied by the goal marking is $\varphi = p_n \geq 1$, stating that a token is put to the place p_n. The initial marking contains just one token in the place p_1 and all other places are empty. We now show that the counter machine halts iff the controller has a winning strategy in the constructed game.

"$\Longrightarrow$": Assume that the Minsky machine halts. The controller's strategy is to faithfully simulate the counter machine, meaning that if the controller is in place p_i for a decrement instruction labelled with i, then it selects the transition $t_{i:1}$ in case the counter c_r is not empty and transition $t_{i:2}$ in case the counter is empty. For increment instructions there is only one choice for the controller. As the controller is playing faithfully, the transition $t_{i:cheat}$ is never enabled and hence the place p_n is eventually reached and the controller has a winning strategy.

"$\Longleftarrow$": Assume that the Minsky machine loops. We want to argue that there is no winning controller's strategy. For the contradiction assume that the controller can win the game also in this case. Clearly, playing faithfully as described in the previous direction will never place a token into the place p_n. Hence at some point the controller must cheat when executing the decrement instruction (no cheating is possible in increment instruction). There are two cheating scenarios. Either the place p_{c_r} is empty and the controller fires $t_{i:1}$ which leads to a deadlock marking and clearly cannot reach the goal marking that marks the place p_n. The other cheating move is to play $t_{i:2}$ in case the place p_{c_r} is not empty. However, now in the next round the environment has the transition $t_{i:cheat}$ enabled and can deadlock the net by firing it. In any case, it is impossible to mark the place p_n, which contradicts the existence of a winning strategy for the controller.   □

This means that running Algorithm 1 specialized to the case of Petri nets does not necessarity terminates on general unbounded Petri nets (nevertheless, if it terminates even for unbounded nets then it still provides a correct answer). In case a Petri net game is bounded, e.g. for each $p \in P$ we have $B(p) \neq \infty$, we can guarantee also termination of the algorithm.

Theorem 5. *The reachability control problem for bounded Petri net games is decidable.*

Proof. Theorem 2 guarantees the correctness of Algorithm 1 (where we assume that the alternating simulation relation is the identity relation). As the presence of soft bounds for each place makes the state space finite, we can apply Theorem 3 to conclude that Algorithm 1 terminates. Hence the total correctness of the algorithm is established.   □

Clearly the empty alternating simulation that was used in the proof of the above theorem guarantees correctness of the algorithm, however, it is not necessarily the most efficient one. Hence we shall now define a syntax-based and linear time computable alternating simulation for Petri net games. For a transition $t \in T_1 \cup T_2$, let $^\bullet t = \{p \in P \mid W(p, t) > 0\}$. We partition the places P in *equality* places P_e and *remaining* places P_r such that $P_e = \bigcup_{t \in T_2} {}^\bullet t$ and $P_r = P \setminus P_e$. Recall that a Petri net game $\mathcal{N}$ induces a game graph with configurations $C_i = \mathcal{M}(\mathcal{N}) \times \{i\}$ for $i \in \{1, 2\}$.

Definition 6 (Alternating Simulation for Petri Net Games). *We define a relation $\sqsubseteq \subseteq C_1 \times C_1 \cup C_2 \times C_2$ on configurations such that $(M, i) \sqsubseteq (M', i)$ iff for all $p \in P_e$ we have $M(p) = M'(p)$ and for all $p \in P_r$ we have $M(p) \leq M'(p)$.*

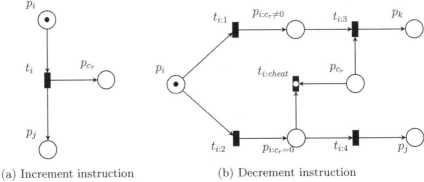

(a) Increment instruction   (b) Decrement instruction

Fig. 4. Petri game gadgets for wwo counter machine instructions

Theorem 6. *The relation $\sqsubseteq$ is an alternating simulation and it is computable in linear time.*

Proof. In order to prove that $\sqsubseteq$ is alternating simulation, let us assume that $(M_1, i) \sqsubseteq (M_1', i)$ and we need to prove the three conditions of the alternating simulation relation from Definition 3.

- If (M_1, i) is a goal configuration, then by definition since the set of goal markings are upward closed, (M_1', i) is also a goal configuration.
- If $(M_1, 1) \xRightarrow{t} (M_2, 2)$ where $t \in T_1$ then by definition of transition firing for every $p \in P$ holds that $M_1(p) \geq W(p, t)$. Since $(M_1, 1) \sqsubseteq (M_1', 1)$ then for every $p \in P$ holds $M_1'(p) \geq M_1(p)$ implying that $M_1'(p) \geq W(p, t)$. Hence t can be fired at M_1' and let $M_1' \xrightarrow{t} M_2'$. It is easy to verify that $(M_2, 2) \sqsubseteq (M_2', 2)$.
- The third condition claiming that if $(M_1', 2) \xRightarrow{t} (M_2', 1)$ where $t \in T_2$ then $(M_1, 2) \xRightarrow{t} (M_2, 1)$ such that $(M_2, 1) \sqsubseteq (M_2', 1)$ follows straightforwardly as for all input places to the environment transition t we assume an equal number of tokens in both M_1 and M_1'.

We can conclude with the fact that $\sqsubseteq$ is an alternating simulation. The relation can be clearly decided in linear time as it requires only comparison of number of tokens for each place in the markings. □

Remark 2. In Fig. 3b, at the initial configuration $c_0 = (\langle 1, 0, 0, 0 \rangle, 2)$ *Player-2* has two transitions enabled. The two successor configurations to c_0 are $c_1 = (\langle 0, 1, 1, 0 \rangle, 1)$ and $c_2 = (\langle 0, 0, 1, 0 \rangle, 1)$ which are reached by firing the transitions t_0 and t_1 respectively. Since $c_1 \sqsupseteq c_2$, Algorithm 1 explores c_2 and ignores the successor configuration c_1. Hence our algorithm terminates on this game and correctly declares it as winning for *Player-2*.

We now point out the following facts about Algorithm 1 and how it is different from the classical Liu and Smolka's algorithm [7]. First of all, Liu and Smolka

Table 1. Fire alarm system

Fire alarm		Winning	Time (sec.)		States		Reduction in %	
Sensors	Channels		LS	ALT	LS	ALT	Time	States
2	2	True	1.5	0.2	116	19	87.5	83.6
3	2	True	30	0.7	434	56	97.6	87.1
4	2	False	351.7	1.1	934	60	99.7	93.6
5	2	False	2249.8	1.1	1663	63	99.9	96.2
6	2	False	8427.1	1.3	3121	64	99.9	98.0
7	2	False	T.O.	1.4	–	66	–	–
2	3	True	20.2	0.5	385	25	97.5	93.5
2	4	True	622.7	1.0	1233	31	99.8	97.5
2	5	True	10706.7	2.3	3564	37	99.9	98.9
2	6	True	T.O.	3.4	–	43	–	–

do not consider any alternating simulation and they do not use the set *Lose* in order to ensure early termination of the algorithm. As a result, Liu and Smolka's algorithm does not terminate (for any search strategy) on the Petri net game from Fig. 2, whereas our algorithm always terminates (irrelevant of the search strategy) and provides a conclusive answer. This fact demonstrates that not only our approach is more efficient but it also terminates on a larger class of Petri net games than the previously studied approaches.

5   Implementation and Experimental Evaluation

We implemented both Algorithm 1 (referred to as ALT and using the alternating simulation on Petri nets given in Definition 6) and the classical Liu and Smolka algorithm (referred to as LS) in the prototype tool sAsEt [8]. We present three use cases where each of the experiments is run 20 times and the average of the time/space requirements is presented in the summary tables. The columns in the tables show the scaling, a Boolean value indicating whether the initial configuration has a winning strategy or not, the time requirements, number of explored states (the size of the set *Disc*) and a relative reduction of the running time and state space in percentage. The experiments are executed on AMD Opteron 6376 processor running at 2.3 GHz with 10 GB memory limit.

5.1   Fire Alarm Use Case

The German company SeCa GmbH produces among other products fire alarm systems. In [4] the formal verification of a wireless communication protocol for a fire alarm system is presented. The fire alarm system consists of a central unit and a number of sensors that communicate using a wireless Time Division Multiple

Table 2. Student teacher use case

Model	Winning	Time (sec.)		States		Reduction in %	
		LS	ALT	LS	ALT	Time	States
$s_1 a_6 d_2 w_2$	False	14.4	12.0	278	213	16.7	23.4
$s_1 a_6 d_2 w_3$	False	220.9	101.0	734	527	54.3	28.2
$s_1 a_6 d_2 w_4$	False	2107.7	620.1	1546	1027	70.6	33.6
$s_1 a_6 d_2 w_5$	False	9968.1	2322.3	2642	1672	76.7	36.7
$s_1 a_6 d_2 w_{10}$	True	7159.3	2710.0	2214	1744	62.1	21.2
$s_1 a_6 d_2 w_{11}$	True	8242.0	2662.7	2226	1756	67.7	21.1
$s_1 a_6 d_2 w_{12}$	True	7687.7	2867.2	2238	1768	62.7	21.0
$s_1 a_6 d_2 w_{13}$	True	7819.7	2831.4	2250	1780	63.8	20.9
$s_1 a_6 d_2 w_{14}$	True	7674.1	2960.3	2262	1792	61.4	20.8
$s_1 a_6 d_2 w_{15}$	True	8113.8	2903.7	2274	1804	64.2	20.7
$s_1 a_7 d_2 w_2$	False	35.3	25.8	382	286	26.9	25.1
$s_1 a_7 d_2 w_3$	False	744.7	284.0	1114	783	61.9	29.7
$s_1 a_7 d_2 w_4$	False	9400.1	2637.2	2604	1674	71.9	35.7
$s_1 a_7 d_2 w_5$	False	T.O.	10477.3	T.O.	2961	–	–
$s_4 a_6 d_1 w_2$	True	40.6	28.9	290	250	28.8	13.8
$s_4 a_6 d_1 w_3$	True	60.7	45.3	326	286	25.4	12.3
$s_4 a_6 d_1 w_4$	True	90.4	62.0	362	322	31.4	11.0
$s_4 a_6 d_1 w_5$	True	122.9	93.9	398	358	23.6	10.1
$s_4 a_6 d_1 w_6$	True	172.1	122.6	434	394	28.8	9.2
$s_4 a_6 d_1 w_7$	True	197.7	163.6	470	430	17.2	8.5
$s_4 a_6 d_1 w_8$	True	263.3	190.6	506	466	27.6	7.9
$s_2 a_4 d_1 w_2$	True	1.7	1.9	90	82	−11.8	8.9
$s_2 a_4 d_1 w_3$	True	2.9	3.1	110	102	−6.9	7.3
$s_2 a_4 d_1 w_4$	True	4.3	4.7	130	122	−9.3	6.2
$s_2 a_4 d_1 w_5$	True	6.3	6.4	150	142	−1.6	5.3
$s_2 a_4 d_1 w_6$	True	9.5	8.3	170	162	12.6	4.7
$s_2 a_4 d_1 w_7$	True	11.9	11.7	190	182	1.7	4.2
$s_2 a_4 d_1 w_8$	True	16.4	15.7	210	202	4.3	3.8

Access protocol. We model a simplified version of the fire alarm system from [4] as a Petri net game and use our tool to guarantee a reliable message passing even under interference. Table 1 shows that as the number of sensors and channels increases, we achieve an exponential speed up using ALT algorithm compared to LS. The reason is that the alternating simulation significantly prunes out the state space that is necessary to explore.

Table 3. Cat and mice use case

Model	Winning	Time (sec.)		States		Reduction in %	
		LS	ALT	LS	ALT	Time	States
m211	False	159.2	109.7	525	405	31.1	22.9
m222	False	562.1	298.7	798	579	46.9	27.4
m322	False	5354.2	1914.4	1674	1096	64.2	34.5
m332	False	9491.7	4057.6	2228	1419	57.3	36.3
m333	False	T.O.	8174.5	T.O.	1869	–	–
m2000	False	127.7	50.2	474	341	60.7	28.1
m2100	False	404.6	157.5	762	490	61.1	35.7
m2200	False	776.4	148.3	921	490	80.9	46.8

5.2 Student-Teacher Scheduling Use Case

In [5] the student semester scheduling problem is modelled as a workflow net. We extend this Petri net into a game by introducing a teacher that can observe the student behaviour (work on assignments) and there is a race between the two players. In Table 2 we see the experimental results. Model instances are of the form $s_i a_j d_k w_l$ where i is the number of students, j is the number of assignments, k is the number of deliverables in each assignment and l is the number of weeks in a semester. We can observe a general trend that with a higher number of assignments ($j = 6$ and $j = 7$) and two deliverables per assignment ($d = 2$), the alternating simulation reduces a significant amount of the state space leading to considerable speed up. As the number of assignment and deliverables gets lower, there is less and less to save when using the ALT algorithm, ending in a situation where only a few percents of the state space can be reduced at the bottom of the table. This leaves us only with computational overhead for the additional checks related to alternating simulation in the ATL algorithm, however, the overhead is quite acceptable (typically less than 20% of the overall running time).

5.3 Cat and Mouse Use Case

This Petri net game consists of an arena of grid shape with number of mice starting at the first row. The mice try to reach the goal place at the other end of the arena without being caught while doing so by the cats (that can move only on one row of the arena). In Table 3 we consider arena of size 2×3 and 2×4 and the model instances show the initial mice distribution in the top row of the grid. The table contains only instances where the mice do not have a winning strategy and again shows an improvement both in the running time as well as the number of explored states. On the positive instances there was no significant difference between ALT and LS algorithms as the winning strategy for mice were relatively fast discovered by both approaches.

6   Conclusion

We presented an on-the-fly algorithm for solving strictly alternating games and introduced the notion of alternating simulation that allows us to significantly speed up the strategy synthesis process. We formally proved the soundness of the method and instantiated it to the case of Petri net games where the problems are in general undecidable and hence the algorithm is not guaranteed to terminate. However, when using alternating simulation there are examples where it terminates for any search strategy, even though classical approaches like Liu and Smolka dependency graphs do not terminate at all. Finally, we demonstrated on examples of Petri net games with soft bounds on places (where the synthesis problem is decidable) the practical applicability of our approach. In future work, we can consider an extension of our framework to other types of formalism like concurrent automata or time Petri nets.

Acknowledgments. The work was supported by the ERC Advanced Grant LASSO and DFF project QASNET.

References

1. Abdulla, P.A., Bouajjani, A., d'Orso, J.: Deciding monotonic games. In: Baaz, M., Makowsky, J.A. (eds.) CSL 2003. LNCS, vol. 2803, pp. 1–14. Springer, Heidelberg (2003). https://doi.org/10.1007/978-3-540-45220-1_1
2. Cassez, F., David, A., Fleury, E., Larsen, K.G., Lime, D.: Efficient on-the-fly algorithms for the analysis of timed games. In: Abadi, M., de Alfaro, L. (eds.) CONCUR 2005. LNCS, vol. 3653, pp. 66–80. Springer, Heidelberg (2005). https://doi.org/10.1007/11539452_9
3. Dalsgaard, A.E., et al.: Extended dependency graphs and efficient distributed fixed-point computation. In: van der Aalst, W., Best, E. (eds.) PETRI NETS 2017. LNCS, vol. 10258, pp. 139–158. Springer, Cham (2017). https://doi.org/10.1007/978-3-319-57861-3_10
4. Feo-Arenis, S., Westphal, B., Dietsch, D., Muñiz, M., Andisha, S., Podelski, A.: Ready for testing: ensuring conformance to industrial standards through formal verification. Formal Aspects Comput. **28**(3), 499–527 (2016). https://doi.org/10.1007/s00165-016-0365-3
5. Juhásova, A., Kazlov, I., Juhás, G., Molnár, L.: How to model curricula and learn-flows by petri nets - a survey. In: 2016 International Conference on Emerging eLearning Technologies and Applications (ICETA), pp. 147–152, November 2016
6. Minsky, M.L.: Computation: finite and infinite machines. Am. Math. Mon. **75** (1968)
7. Liu, X., Smolka, S.A.: Simple linear-time algorithms for minimal fixed points. In: Larsen, K.G., Skyum, S., Winskel, G. (eds.) ICALP 1998. LNCS, vol. 1443, pp. 53–66. Springer, Heidelberg (1998). https://doi.org/10.1007/BFb0055040
8. Muñiz, M.: Model checking for time division multiple access systems. Ph.D. thesis, Freiburg University (2015). https://doi.org/10.6094/UNIFR/10161
9. Petri, C.A.: Kommunikation mit Automaten. Ph.D. thesis, Universität Hamburg (1962)
10. Raskin, J., Samuelides, M., Begin, L.V.: Games for counting abstractions. Electr. Notes Theor. Comput. Sci. **128**(6), 69–85 (2005). https://doi.org/10.1016/j.entcs.2005.04.005

Semantics

Interleaving vs True Concurrency: Some Instructive Security Examples

Roberto Gorrieri(✉)

Dipartimento di Informatica—Scienza e Ingegneria, Università di Bologna,
Mura A. Zamboni, 7, 40127 Bologna, Italy
roberto.gorrieri@unibo.it

Abstract. Information flow security properties were defined some years ago in terms of suitable equivalence checking problems. These definitions were provided by using sequential models of computations (e.g., labeled transition systems [17,26]), and interleaving behavioral equivalences (e.g., bisimulation equivalence [27]). More recently, the distributed model of Petri nets has been used to study non-interference in [1,5,6], but also in these papers an interleaving semantics was used. By exploiting a simple process algebra, called CFM [18] and equipped with a Petri net semantics, we provide some examples showing that team equivalence, a truly-concurrent behavioral equivalence proposed in [19,20], is much more suitable to define information flow security properties. The distributed non-interference property we propose, called DNI, is very easily checkable on CFM processes, as it is compositional, so that it does not suffer from the state-space explosion problem. Moreover, DNI is characterized syntactically on CFM by means of a type system.

1 Introduction

The title of this paper reminds [8], where a strong argument is presented in support of the use of truly concurrent semantics: some of them, differently from interleaving semantics, can be a congruence for *action refinement* (see, e.g., the tutorial [15]). The purpose of this note is similar: to present one further argument in support of true concurrency, by showing by means of examples that, by adopting an interleaving semantics to describe distributed systems, some real information flows cannot be detected.

Information flow security properties were defined some years ago (see, e.g., the surveys [10,32]) in terms of suitable equivalence checking problems. Usually, a distributed system is described by means of a term of some process algebra (typically CCS [17,27] or CSP [23]), whose operational semantics is often given in terms of labeled transition systems (LTSs, for short) [26] and whose observational semantics is usually defined by means of some *interleaving* behavioral equivalence such as trace semantics or bisimulation semantics (see, e.g., the textbook [17] for an overview of behavioral equivalences on LTSs). More recently, the distributed model of Petri nets [18,31] was used to study non-interference in [1,5,6], but

© Springer Nature Switzerland AG 2020
R. Janicki et al. (Eds.): PETRI NETS 2020, LNCS 12152, pp. 131–152, 2020.
https://doi.org/10.1007/978-3-030-51831-8_7

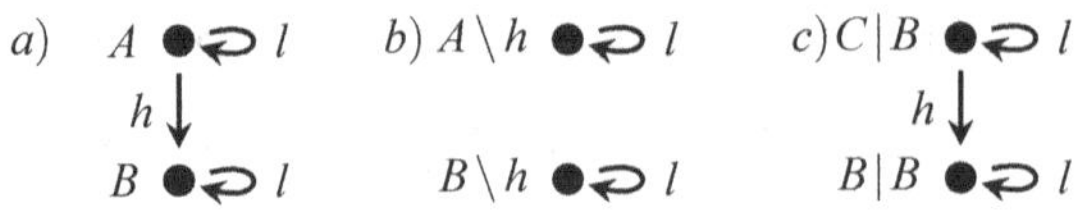

$a)$ A ●⊋ l $b)$ $A \setminus h$ ●⊋ l $c)$ $C | B$ ●⊋ l

$\quad\quad h \downarrow$ $\quad\quad\quad\quad\quad\quad\quad\quad\quad\quad\quad h \downarrow$

$\quad B$ ●⊋ l $\quad\quad\quad B \setminus h$ ●⊋ l $\quad\quad B | B$ ●⊋ l

Fig. 1. An example of a secure process with never-ending behavior

also in these papers the security properties are based on interleaving behavioral semantics and the distributed model is used only to show that, under some conditions, these (interleaving) information flows can be characterized by the presence (or absence) of certain causal or conflict structures in the net.

The thesis of this paper is that, for security analysis, it is necessary to describe the behavior of distributed systems by means of a distributed model of computation, such as Petri nets, but also to observe the distributed model by means of some truly-concurrent behavioral semantics. There is a wide range of possible truly-concurrent equivalences (see, e.g., [13,14] for a partial overview) and it may be not obvious to understand which is more suitable. Our intuition is that, in order to capture all the possible information flows, it is necessary that the observational semantics is very concrete, observing not only the partial order of events that have occurred, as in *fully-concurrent bisimilarity* [4], but also the structure of the distributed state, as in *team equivalence* [19], *state-sensitive fully-concurrent bisimilarity* [22] and *structure-preserving bisimilarity* [14].

Our aim is to analyze systems that can perform two kinds of actions: *high-level* actions, representing the interaction of the system with high-level users, and *low-level* actions, representing the interaction with low-level users. We want to verify whether the interplay between the high user and the high part of the system can affect the view of the system as observed by a low user. We assume that the low user knows the structure of the system, and we check if, in spite of this, (s)he is not able to infer the behavior of the high user by observing the low view of the execution of the system. Hence, we assume that the set of actions is partitioned into two subsets: the set H of high-level actions and the set L of low-level actions.

To explain our point of view, we use the process algebra CFM [18,20], extending finite-state CCS [27] with a top-level operator of asynchronous (i.e., without communication capabilities) parallelism. Hence, the systems we are going to define, onto which high-level users and low-level users interact, are simply a collection of non-interacting, independent, sequential processes. Consider the CFM sequential process

$$A \doteq l.A + h.B \quad\quad B \doteq l.B$$

where A and B are constants, each equipped with a defining equation, and where l is a low action and h is a high one. The LTS semantics for A is in Fig. 1(a). Intuitively, this system is secure because the execution of the high-level action h does not add any information to a low-level observer: what such a low observer can see is just a sequence of the low action l in any case. The most restrictive non-

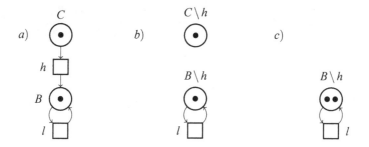

Fig. 2. The Petri net for the process $C \mid B$ in (a), of $(C \mid B) \backslash h$ in (b) and of $(B \mid B) \backslash h$ in (c)

interference property discussed in [6,10], called SBNDC, requires that whenever the system performs a high-level action, the states of the system *before* and *after* executing that high-level action are indistinguishable for a low-level observer. In our example, this means that A is SBNDC if $A \backslash h$ is bisimilar to $B \backslash h$, where $A \backslash h$ denotes the process A where the transition h is pruned. By observing the LTSs in Fig. 1(b), we conclude that A is SBNDC.

However, the LTS in Fig. 1(a) is isomorphic to the LTS in 1(c), which is the semantics of the parallel process $C \mid B$, where $C \doteq h.B$. Therefore, we can conclude that also $C \mid B$ enjoys the SBNDC non-interference property. Unfortunately, this parallel system is not secure: if a low observer can realize that two occurrences of actions l have been performed in parallel (which is a possible behavior of $B \mid B$, but this behavior is not represented in the LTS semantics),

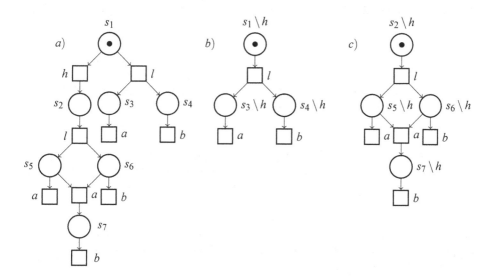

Fig. 3. An example of an insecure process

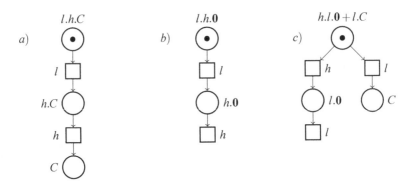

Fig. 4. By observing the state, information flows can be detectable

then (s)he is sure that the high-level action h has been performed. This trivial example show that SBNDC, based on interleaving bisimulation equivalence, is unable to detect a real information flow from high to low.

However, if we use a Petri net semantics for the process algebra CFM, as described in Sect. 3, we have that the net semantics for $C \mid B$, outlined in Fig. 2(a), is such that $(C \mid B) \backslash h$ (whose net semantics is outlined in Fig. 2(b)) is not "truly-concurrent" bisimilar to $(B \mid B) \backslash h$ (whose net semantics is in Fig. 2(c)). In fact, it is now visible that the two tokens on place $B \backslash h$ in the net in (c) can perform transition l at the same time, and such a behavior is impossible for the net in (b). This example calls for a truly-concurrent semantics which is at least as discriminating as *step bisimulation* equivalence [18,28], which is able to observe the parallel execution of actions.

However, it is not difficult to elaborate on the example above in order to show that, actually, a more discriminating truly-concurrent semantics is necessary. E.g., the net in Fig. 3(a) is such that the two low-observable subnets *before* and *after* the execution of h (depicted in Fig. 3(b) and 3(c), resp.) are step bisimilar, even if they generate different partial orders: if an observer realizes that b is causally dependent on a, then (s)he infers that h has been performed. (To be precise, this example is not expressible in our simple CFM process algebra, but it can in the more generous process algebra FNM [18].) One of the most accurate (and popular) truly-concurrent behavioral equivalences for Petri nets is *fully-concurrent bisimilarity* [4] (fc-bisimilarity, for short), whose intuition originates from *history-preserving bisimilarity* [13]. So, it seems that we can simply change the SBNDC definition by replacing interleaving bisimilarity (on the LTS semantics) with fc-bisimilarity (on the Petri net semantics). However, fc-bisimilarity observes only the partial order of the performed events, while it abstracts from the size of the distributed state, because it may relate markings composed of a completely different number of tokens. The following example shows that also the size of the marking is an important feature, that cannot be ignored, as fc-bisimilarity does.

Consider the net in Fig. 4(a), which is the semantics of the CFM process $l.h.C$, with $C \doteq \mathbf{0}$. Since h is not causing any low action, we conclude that this net is secure. However, the very similar net in (b), which is the semantics of $l.h.\mathbf{0}$, is not secure: if the low observer realizes that the token disappears in the end, then (s)he can infer that h has been performed. This simple observation is about the observability of the size of the distributed system: if the execution of a high-level action modifies the current number of tokens (i.e., the number of currently active sequential subprocesses), then its execution has an observable effect that can be recognized by a low observer. Similarly, the net in (c), which is the semantics of $h.l.\mathbf{0} + l.C$ with $C \doteq \mathbf{0}$, is such that, before and after h, the same partial order of event can be performed, but if the token disappears in the end, then the low observer knows that h has been performed. Therefore, it is necessary to use a truly-concurrent behavioral equivalence slightly finer than fully-concurrent bisimilarity, able to observe also the structure of the distributed state, such as *state-sensitive fc-bisimilarity* [22] and *structure-preserving bisimilarity* [14], which can be characterized in a very simple and effective way on CFM as *team equivalence* [19,20,22]. So, we are now ready to propose the property DNI (acronym of Distributed Non-Interference) as follows: a CFM process p is DNI if for each reachable p', p'' and for each $h \in H$ such that $p' \xrightarrow{h} p''$, we have $p' \backslash H \sim^{\oplus} p'' \backslash H$, where $\sim^{\oplus}$ denotes team equivalence.

The DNI check can be done in a simple way. Given a process $p = p_1 \,|\, p_2 \,|\, \dots \,|\, p_n$, where each p_i is sequential, then we prove that p is DNI if and only if each p_i is DNI. Hence, instead of inspecting the state-space of p, we can simply inspect the state-spaces for $p_1, p_2, \dots p_n$. If the state-space of each p_i is composed of 10 states, then the state-space of p is composed of 10^n states, so that a direct check of DNI on the state space of p is impossible for large values of n. However, the distributed analysis on each p_i can be done in linear time w.r.t. n, hence also for very large values of n. Moreover, a structural characterization of DNI can be provided, as well as a typing system: we prove that a CFM process p is DNI iff p (or a slight variant of it) is typed. The typing system is based on a finite, sound and complete, axiomatization of team equivalence [20].

The paper is organized as follows. Section 2 introduces the simple Petri net model we are using, namely *finite-state machines*, together with the definitions of *bisimulation on places* and *team equivalence*. Section 3 defines the syntax of the process algebra CFM, its net semantics and recalls the finite, sound and complete, axiomatization of team equivalence from [20]. Section 4 introduces the distributed non-interference property DNI on CFM processes, proving that it can be really checked in a distributed manner, and also describes the typing system for DNI. Finally, Sect. 5 comments on related work, outlines some possible future research and lists some additional reasons to prefer truly-concurrent semantics over interleaving semantics.

2   Finite-State Machines and Team Equivalence

By finite-state machine (FSM, for short) we mean a simple type of finite Petri net [18,31] whose transitions have singleton pre-set and singleton, or empty, post-set.

The name originates from the fact that an unmarked net of this kind is essentially isomorphic to a nondeterministic finite automaton [24] (NFA, for short), usually called a finite-state machine as well. However, semantically, our FSMs are richer than NFAs because, as their initial marking may be not a singleton, these nets can also exhibit concurrent behavior, while NFAs are strictly sequential.

Definition 1 (Multiset). *Let* $\mathbb{N}$ *be the set of natural numbers. Given a finite set* S, *a multiset over* S *is a function* $m : S \rightarrow \mathbb{N}$. *Its support set* $dom(m)$ *is* $\{s \in S \mid m(s) \neq 0\}$. *The set of all multisets over* S *is* $\mathcal{M}(S)$, *ranged over by* m. *We write* $s \in m$ *if* $m(s) > 0$. *The multiplicity of* s *in* m *is the number* $m(s)$. *The size of* m, *denoted by* $|m|$, *is the number* $\sum_{s \in S} m(s)$, *i.e., the total number of its elements. A multiset* m *such that* $dom(m) = \emptyset$ *is called* empty *and is denoted by* θ. *We write* $m \subseteq m'$ *if* $m(s) \leq m'(s)$ *for all* $s \in S$.

Multiset union $\_ \oplus \_$ *is defined as follows:* $(m \oplus m')(s) = m(s) + m'(s)$; *it is commutative, associative and has* θ *as neutral element. Multiset difference* $\_ \ominus \_$ *is defined as follows:* $(m_1 \ominus m_2)(s) = max\{m_1(s) - m_2(s), 0\}$. *The scalar product of a number* j *with* m *is the multiset* $j \cdot m$ *defined as* $(j \cdot m)(s) = j \cdot (m(s))$. *By* s_i *we also denote the multiset with* s_i *as its only element. Hence, a multiset* m *over* $S = \{s_1, \ldots, s_n\}$ *can be represented as* $k_1 \cdot s_1 \oplus k_2 \cdot s_2 \oplus \ldots \oplus k_n \cdot s_n$, *where* $k_j = m(s_j) \geq 0$ *for* $j = 1, \ldots, n$. $\square$

Definition 2 (Finite-state machine). *A* labeled *finite-state machine (FSM, for short) is a tuple* $N = (S, A, T)$, *where*

- S *is the finite set of* places, *ranged over by* s *(possibly indexed),*
- A *is the finite set of* labels, *ranged over by* ℓ *(possibly indexed), and*
- $T \subseteq S \times A \times (S \cup \{\theta\})$ *is the finite set of* transitions, *ranged over by* t.

Given a transition $t = (s, \ell, m)$, *we use the notation* $^\bullet t$ *to denote its* pre-set s *(which is a single place) of tokens to be consumed;* $l(t)$ *for its label* ℓ, *and* $t^\bullet$ *to denote its* post-set m *(which is a place or the empty multiset* θ*) of tokens to be produced. Hence, transition* t *can be also represented as* $^\bullet t \xrightarrow{l(t)} t^\bullet$. $\square$

Definition 3 (Marking, FSM net system). *A multiset over* S *is called a* marking. *Given a marking* m *and a place* s, *we say that the place* s *contains* $m(s)$ *tokens, graphically represented by* $m(s)$ *bullets inside place* s. *An FSM net system* $N(m_0)$ *is a tuple* (S, A, T, m_0), *where* (S, A, T) *is an FSM and* m_0 *is a marking over* S, *called the* initial marking. *We also say that* $N(m_0)$ *is a* marked *net. An FSM net system* $N(m_0) = (S, A, T, m_0)$ *is* sequential *if* m_0 *is a singleton, i.e.,* $|m_0| = 1$; *while it is* concurrent *if* m_0 *is arbitrary.* $\square$

Definition 4 (Firing sequence, reachable markings). *Given an FSM* $N = (S, A, T)$, *a transition* t *is* enabled *at marking* m, *denoted by* $m[t\rangle$, *if* $^\bullet t \subseteq m$. *The execution (or firing) of* t *enabled at* m *produces the marking* $m' = (m \ominus {}^\bullet t) \oplus t^\bullet$. *This is written usually as* $m[t\rangle m'$, *but also as* $m \xrightarrow{l(t)} m'$. *A firing sequence starting at* m *is defined inductively as*

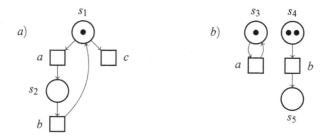

Fig. 5. A sequential finite-state machine in (a), and a concurrent finite-state machine in (b)

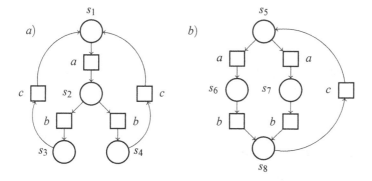

Fig. 6. Two bisimilar FSMs

- $m[\varepsilon\rangle m$ is a firing sequence (where ε denotes an empty sequence of transitions) and
- if $m[\sigma\rangle m'$ is a firing sequence and $m'[t\rangle m''$, then $m[\sigma t\rangle m''$ is a firing sequence.

If $\sigma = t_1 \ldots t_n$ (for $n \geq 0$) and $m[\sigma\rangle m'$ is a firing sequence, then there exist $m_1, \ldots, m_{n+1}$ such that $m = m_1[t_1\rangle m_2[t_2\rangle \ldots m_n[t_n\rangle m_{n+1} = m'$, and $\sigma = t_1 \ldots t_n$ is called a transition sequence starting at m and ending at m'. The set of reachable markings from m is $reach(m) = \{m' \mid \exists\sigma.m[\sigma\rangle m'\}$.   □

Definition 5. An FSM system $N(m_0) = (S, A, T, m_0)$ is dynamically reduced if $\forall s \in S \; \exists m \in reach(m_0).m(s) \geq 1$ and $\forall t \in T \exists m, m' \in reach(m_0)$ such that $m[t\rangle m'$.   □

Example 1. By using the usual drawing convention for Petri nets, Fig. 5 shows in (a) a sequential FSM, which performs a, possibly empty, sequence of a's and b's, until it performs one c and then stops *successfully* (the token disappears in the end). Note that a sequential FSM is such that any reachable marking is a singleton or empty. Hence, a sequential FSM is a *safe* (or 1-bounded) net: each place in any reachable marking can hold one token at most. In (b), a concurrent

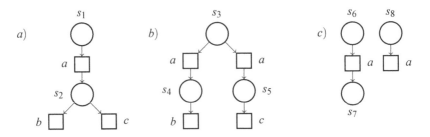

Fig. 7. Some (non-)-bisimilar FSMs

FSM is depicted: it can perform a forever, interleaved with two occurrences of b, only: the two tokens in s_4 will eventually reach s_5, which is a place representing unsuccessful termination (deadlock). Note that a concurrent FSM is a k-*bounded* net, where k is the size of the initial marking: each place in any reachable marking can hold k tokens at most. Hence, the set $reach(m)$ is finite for any m. As a final comment, we mention that for each FSM $N = (S, A, T)$ and each place $s \in S$, the set $reach(s)$ is a subset of $S \cup \{\theta\}$.   □

2.1   Bisimulation on Places

We provide the definition of (strong) *bisimulation on places* for unmarked FSMs, originally introduced in [19]. Because of the shape of FSMs transitions, in the definition below the post-sets m_1 and m_2 can be either the empty marking θ or a single place.

Definition 6 (Bisimulation on places). *Let $N = (S, A, T)$ be an FSM. A bisimulation is a relation $R \subseteq S \times S$ such that if $(s_1, s_2) \in R$ then for all $\ell \in A$*

- *$\forall m_1$ such that $s_1 \xrightarrow{\ell} m_1$, $\exists m_2$ such that $s_2 \xrightarrow{\ell} m_2$ and either $m_1 = \theta = m_2$ or $(m_1, m_2) \in R$,*
- *$\forall m_2$ such that $s_2 \xrightarrow{\ell} m_2$, $\exists m_1$ such that $s_1 \xrightarrow{\ell} m_1$ and either $m_1 = \theta = m_2$ or $(m_1, m_2) \in R$.*

Two places s and s' are bisimilar *(or bisimulation equivalent), denoted $s \sim s'$, if there exists a bisimulation R such that $(s, s') \in R$.*   □

Example 2. Consider the nets in Fig. 6. It is not difficult to realize that relation $R = \{(s_1, s_5), (s_2, s_6), (s_2, s_7), (s_3, s_8), (s_4, s_8)\}$ is a bisimulation on places.   □

Example 3. Consider the nets in Fig. 7. It is not difficult to realize that $s_1 \nsim s_3$. In fact, s_1 may reach s_2 by performing a; s_3 can reply to this move in two different ways by reaching either s_4 or s_5; however, while s_2 offers both b and c, s_4 may perform only b and s_5 only c; hence, s_4 and s_5 are not bisimilar to s_2 and so also s_1 is not bisimilar to s_3. This example shows that bisimulation

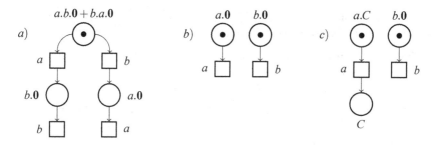

Fig. 8. Three net systems: $a.b.0 + b.a.0$, $a.0 \,|\, b.0$ and $a.C \,|\, b.0$ (with $C \doteq 0$)

equivalence is sensitive to the timing of choice. Furthermore, also s_6 and s_8 are not bisimilar. In fact, s_6 can reach s_7 by performing a, while s_8 can reply by reaching the empty marking, but $\theta \nsim s_7$. This example shows that bisimulation equivalence is sensitive to the kind of termination of a process: even if s_7 is stuck, it is not equivalent to θ because the latter is the marking of a properly terminated process, while s_7 denotes a deadlock situation. □

Remark 1 (**Complexity**). If m is the number of net transitions and n of places, checking whether two places of an FSM are bisimilar can be done in $O(m \log (n+1))$ time, by adapting the algorithm in [30] for ordinary bisimulation on LTSs. Indeed, this very same algorithm can compute $\sim$ by starting from an initial partition of $S \cup \{\theta\}$ composed of two subsets: S and $\{\theta\}$. Bisimulation on places enjoys the same properties of bisimulation on LTSs, i.e., it is coinductive and with a fixed-point characterization. The largest bisimulation is an equivalence and can be used to minimize the net [19]. □

2.2   Team Equivalence

Definition 7 (**Additive closure**). *Given an FSM net $N = (S, A, T)$ and a place relation $R \subseteq S \times S$, we define a marking relation $R^{\oplus} \subseteq \mathcal{M}(S) \times \mathcal{M}(S)$, called the additive closure of R, as the least relation induced by the following axiom and rule.*

$$\frac{}{(\theta, \theta) \in R^{\oplus}} \qquad \frac{(m_1, m_2) \in R \quad (m_1', m_2') \in R^{\oplus}}{(m_1 \oplus m_1', m_2 \oplus m_2') \in R^{\oplus}}$$

□

Note that, by definition, two markings are related by $R^{\oplus}$ only if they have the same size; in fact, the axiom states that the empty marking is related to itself, while the rule, assuming by induction that m_1 and m_2 have the same size, ensures that $s_1 \oplus m_1$ and $s_2 \oplus m_2$ have the same size. An alternative way to define that two markings m_1 and m_2 are related by $R^{\oplus}$ is to state that m_1 can be represented as $s_1 \oplus s_2 \oplus \ldots \oplus s_k$, m_2 can be represented as $s_1' \oplus s_2' \oplus \ldots \oplus s_k'$

and $(s_i, s_i') \in R$ for $i = 1, \ldots, k$. Note also that if R is an equivalence relation, then $R^{\oplus}$ is also an equivalence relation.

Definition 8 (Team equivalence). *Given a finite-state machine* $N = (S, A, T)$ *and the bisimulation equivalence* $\sim \subseteq S \times S$, *we define team equivalence* $\sim^{\oplus} \subseteq \mathcal{M}(S) \times \mathcal{M}(S)$, *as the additive closure of* $\sim$. □

Example 4. Continuing Example 2 about Fig. 6, we have, e.g., that the marking $s_1 \oplus 2 \cdot s_2 \oplus s_3 \oplus s_4$ is team equivalent to any marking with one token in s_5, two tokens distributed over s_6 and s_7, and two tokens in s_8; e.g., $s_5 \oplus s_6 \oplus s_7 \oplus 2 \cdot s_8$ or $s_5 \oplus 2 \cdot s_7 \oplus 2 \cdot s_8$. Note that $m_1 = s_1 \oplus 2 \cdot s_2$ has the same size of $m_2 = 2 \cdot s_5 \oplus s_6$, but the two are not team equivalent; in fact, we can first match s_1 with one instance of s_5; then, one instance of s_2 with s_6; but, now, we are unable to find a match for the second instance of s_2, because the only element left in m_2 is s_5, and s_2 is not bisimilar to s_5. □

Example 5. Team equivalence is a truly concurrent equivalence. The FSM in Fig. 8(a) denotes the net for the sequential CFM process term $a.b.\mathbf{0} + b.a.\mathbf{0}$, which can perform the two actions a and b in either order. On the contrary, the concurrent FSM in (b) denotes the net for the parallel CFM process term $a.\mathbf{0}|b.\mathbf{0}$. Note that place $a.b.\mathbf{0} + b.a.\mathbf{0}$ is not team equivalent to the marking $a.\mathbf{0} \oplus b.\mathbf{0}$, because the two markings have different size. Nonetheless, $a.b.\mathbf{0} + b.a.\mathbf{0}$ and $a.\mathbf{0} \oplus b.\mathbf{0}$ are interleaving bisimilar. □

Example 6. If two markings m_1 and m_2 are interleaving bisimilar and have the same size, then they may be not team equivalent. For instance, consider Fig. 8(c), which denotes the net for the CFM process term $a.C \mid b.\mathbf{0}$, where C is a constant with empty body, i.e., $C \doteq \mathbf{0}$. Markings $a.\mathbf{0} \oplus b.\mathbf{0}$ and $a.C \oplus b.\mathbf{0}$ have the same size, they are interleaving bisimilar (actually, they are even fully concurrent bisimilar [4]), but they are not team equivalent. In fact, if $a.C \oplus b.\mathbf{0} \xrightarrow{a} C \oplus b.\mathbf{0}$, then $a.\mathbf{0} \oplus b.\mathbf{0}$ may try to respond with $a.\mathbf{0} \oplus b.\mathbf{0} \xrightarrow{a} b.\mathbf{0}$, but $C \oplus b.\mathbf{0}$ and $b.\mathbf{0}$ are not team bisimilar markings, as they have different size. □

Remark 2 **(Complexity).** Once the place relation $\sim$ has been computed once and for all for the given net in $O(m \cdot \log (n + 1))$ time, the algorithm in [19] checks whether two markings m_1 and m_2 are team equivalent in $O(k^2)$ time, where k is the size of the markings. In fact, if $\sim$ is implemented as an adjacency matrix, then the complexity of checking if two markings m_1 and m_2 (represented as a list of places with multiplicities) are related by $\sim^{\oplus}$ is $O(k^2)$, because the problem is essentially that of finding for each element s_1 of m_1 a matching, $\sim$-related element s_2 of m_2. □

3   CFM: Syntax, Semantics, Axiomatization

3.1   Syntax

Let Act be a finite set of actions, ranged over by μ, partitioned into two subsets L and H of low-level actions and high-level ones, respectively. Let $\mathscr{C}$ be a finite

Table 1. Structural operational LTS semantics for CFM

(Pref)	$\dfrac{}{\mu.p \xrightarrow{\mu} p}$	(Cons)	$\dfrac{p \xrightarrow{\mu} p'}{C \xrightarrow{\mu} p'} \quad C \doteq p$
(Sum$_1$)	$\dfrac{p \xrightarrow{\mu} p'}{p+q \xrightarrow{\mu} p'}$	(Sum$_2$)	$\dfrac{q \xrightarrow{\mu} q'}{p+q \xrightarrow{\mu} q'}$
(Par$_1$)	$\dfrac{p \xrightarrow{\mu} p'}{p \mid q \xrightarrow{\mu} p' \mid q}$	(Par$_2$)	$\dfrac{q \xrightarrow{\mu} q'}{p \mid q \xrightarrow{\mu} p \mid q'}$

set of constants, disjoint from Act, ranged over by $A, B, C, \ldots$. The CFM *terms* (where CFM is the acronym of *Concurrent Finite-state Machines*) are generated from actions and constants by the following abstract syntax (with three syntactic categories):

$$
\begin{aligned}
s &::= \mathbf{0} \mid \mu.q \mid s+s & \text{\textit{guarded processes}} \\
q &::= s \mid C & \text{\textit{sequential processes}} \\
p &::= q \mid p \mid p & \text{\textit{parallel processes}}
\end{aligned}
$$

where $\mathbf{0}$ is the empty process, $\mu.q$ is a process where action μ prefixes the residual q ($\mu.-$ is the *action prefixing* operator), $s_1 + s_2$ denotes the alternative composition of s_1 and s_2 ($-+-$ is the *choice* operator), $p_1 \mid p_2$ denotes the asynchronous parallel composition of p_1 and p_2 and C is a constant. A constant C may be equipped with a definition, but this must be a guarded process, i.e., $C \doteq s$. A term p is a CFM *process* if each constant in $Const(p)$ (the set of constants used by p; see [18] for details) is equipped with a defining equation (in category s). The set of CFM processes is denoted by $\mathscr{P}_{CFM}$, the set of its sequential processes, i.e., those in syntactic category q, by $\mathscr{P}_{CFM}^{seq}$ and the set of its guarded processes, i.e., those in syntactic category s, by $\mathscr{P}_{CFM}^{grd}$. By $sort(p) \subseteq Act$ we denote the set of all the actions occurring in p.

3.2 Semantics

The interleaving LTS semantics for CFM is given by the structural operational rules in Table 1. Note that each state of the LTS is a CFM process. As an example, the LTS for $C \mid B$, with $C \doteq h.B$ and $B \doteq l.B$, is described in Fig. 1(c). It is possible to prove that, for any p, the set of states reachable from p is finite [18].

The Petri net semantics for CFM, originally in [18], is such that the set S_{CFM} of pla-ces is the set of the sequential CFM processes, without $\mathbf{0}$, i.e., $S_{CFM} = \mathscr{P}_{CFM}^{seq} \setminus \{\mathbf{0}\}$. The decomposition function $dec : \mathscr{P}_{CFM} \to \mathcal{M}(S_{CFM})$, mapping process terms to markings, is defined in Table 2. An easy induction proves that for any $p \in \mathscr{P}_{CFM}$, $dec(p)$ is a finite multiset of sequential processes. Note that, if $C \doteq \mathbf{0}$, then $\theta = dec(\mathbf{0}) \neq dec(C) = \{C\}$. Note also that $\theta = dec(\mathbf{0}) \neq dec(\mathbf{0} + \mathbf{0}) = \{\mathbf{0} + \mathbf{0}\}$, which is a deadlock place.

Table 2. Decomposition function

$$dec(\mathbf{0}) = \theta \qquad dec(\mu.p) = \{\mu.p\}$$
$$dec(p + p') = \{p + p'\} \qquad dec(C) = \{C\}$$
$$dec(p \,|\, p') = dec(p) \oplus dec(p')$$

Table 3. Denotational net semantics

$\llbracket \mathbf{0} \rrbracket_I = (\emptyset, \emptyset, \emptyset, \theta)$

$\llbracket \mu.p \rrbracket_I = (S, A, T, \{\mu.p\})$ given $\llbracket p \rrbracket_I = (S', A', T', dec(p))$ and where

$\qquad S = \{\mu.p\} \cup S', \ A = \{\mu\} \cup A', \ T = \{(\{\mu.p\}, \mu, dec(p))\} \cup T'$

$\llbracket p_1 + p_2 \rrbracket_I = (S, A, T, \{p_1 + p_2\})$ given $\llbracket p_i \rrbracket_I = (S_i, A_i, T_i, dec(p_i))$ for $i = 1, 2$, and where

$\qquad S = \{p_1 + p_2\} \cup S_1' \cup S_2'$, with, for $i = 1, 2$,

$$S_i' = \begin{cases} S_i & \exists t \in T_i \text{ such that } t^\bullet(p_i) > 0 \\ S_i \setminus \{p_i\} & \text{otherwise} \end{cases}$$

$\qquad A = A_1 \cup A_2, \ T = T' \cup T_1' \cup T_2'$, with, for $i = 1, 2$,

$$T_i' = \begin{cases} T_i & \exists t \in T_i \,.\, t^\bullet(p_i) > 0 \\ T_i \setminus \{t \in T_i \mid {}^\bullet t(p_i) > 0\} & \text{otherwise} \end{cases}$$

$\qquad T' = \{(\{p_1 + p_2\}, \mu, m) \mid (\{p_i\}, \mu, m) \in T_i, i = 1, 2\}$

$\llbracket C \rrbracket_I = (\{C\}, \emptyset, \emptyset, \{C\})$ if $C \in I$

$\llbracket C \rrbracket_I = (S, A, T, \{C\})$ if $C \notin I$, given $C \doteq p$ and $\llbracket p \rrbracket_{I \cup \{C\}} = (S', A', T', dec(p))$

$\qquad A = A', S = \{C\} \cup S''$, where

$$S'' = \begin{cases} S' & \exists t \in T' \,.\, t^\bullet(p) > 0 \\ S' \setminus \{p\} & \text{otherwise} \end{cases}$$

$\qquad T = \{(\{C\}, \mu, m) \mid (\{p\}, \mu, m) \in T'\} \cup T''$ where

$$T'' = \begin{cases} T' & \exists t \in T' \,.\, t^\bullet(p) > 0 \\ T' \setminus \{t \in T' \mid {}^\bullet t(p) > 0\} & \text{otherwise} \end{cases}$$

$\llbracket p_1 \,|\, p_2 \rrbracket_I = (S, A, T, m_0)$ given $\llbracket p_i \rrbracket_I = (S_i, A_i, T_i, m_i)$ for $i = 1, 2$, and where

$\qquad S = S_1 \cup S_2, \ A = A_1 \cup A_2, \ T = T_1 \cup T_2, \ m_0 = m_1 \oplus m_2$

Now we provide a construction of the net system $\llbracket p \rrbracket_\emptyset$ associated with process p, which is compositional and denotational in style. The details of the construction are outlined in Table 3. The mapping is parametrized by a set of constants that have already been found while scanning p; such a set is initially empty and it is used to avoid looping on recursive constants. The definition is syntax driven and also the places of the constructed net are syntactic objects, i.e., CFM sequential process terms. For instance, the net system $\llbracket a.\mathbf{0} \rrbracket_\emptyset$ is a net composed of one single marked place, namely process $a.\mathbf{0}$, and one single transition $(\{a.\mathbf{0}\}, a, \theta)$. A bit of care is needed in the rule for choice: in order to include only strictly necessary places and transitions, the initial place p_1 (or p_2) of the subnet $\llbracket p_1 \rrbracket_I$ (or $\llbracket p_2 \rrbracket_I$) is to be kept in the net for $p_1 + p_2$ only if there exists a transition reaching place p_1 (or p_2) in $\llbracket p_1 \rrbracket_I$ (or $\llbracket p_2 \rrbracket_I$), otherwise p_1 (or p_2) can be safely removed in the new net. Similarly, for the rule for constants.

We now list some properties of the semantics, whose proofs are in [18], which state that CFM really represents the class of FSMs.

Theorem 1 (Only Concurrent FSMs). *For each CFM process p, $[\![p]\!]_\emptyset$ is a concurrent finite-state machine.*   □

Definition 9 (Translating Concurrent FSMs into CFM Process Terms). *Let $N(m_0) = (S, A, T, m_0)$—with $S = \{s_1, \ldots, s_n\}$, $A \subseteq Act$, $T = \{t_1, \ldots, t_k\}$, and $l(t_j) = \mu_j$—be a concurrent finite-state machine. Function $\mathscr{T}_{CFM}(-)$, from concurrent finite-state machines to CFM processes, is defined as*

$$\mathscr{T}_{CFM}(N(m_0)) = \underbrace{C_1 | \cdots | C_1}_{m_0(s_1)} | \cdots | \underbrace{C_n | \cdots | C_n}_{m_0(s_n)}$$

where each C_i is equipped with a defining equation $C_i \doteq c_i^1 + \cdots + c_i^k$ (with $C_i \doteq 0$ if $k = 0$), and each summand c_i^j, for $j = 1, \ldots, k$, is equal to

- 0, *if* $s_i \notin {}^\bullet t_j$;
- $\mu_j.0$, *if* ${}^\bullet t_j = \{s_i\}$ *and* $t_j^\bullet = \emptyset$;
- $\mu_j.C_h$, *if* ${}^\bullet t_j = \{s_i\}$ *and* $t_j^\bullet = \{s_h\}$.   □

Theorem 2 (All Concurrent FSMs). *Let $N(m_0) = (S, A, T, m_0)$ be a dynamically reduced, concurrent finite-state machine such that $A \subseteq Act$, and let $p = \mathscr{T}_{CFM}(N(m_0))$. Then, $[\![p]\!]_\emptyset$ is isomorphic to $N(m_0)$.*   □

Therefore, thanks to these results (proved in [18]), we can conclude that the CFM process algebra truly represents the class of FSMs. Hence, we can transfer the definition of team equivalence from FSMs to CFM process terms in a simple way.

Definition 10. *Two CFM processes p and q are team equivalent, denoted $p \sim^\oplus q$, if, by taking the (union of the) nets $[\![p]\!]_\emptyset$ and $[\![q]\!]_\emptyset$, we have that $dec(p) \sim^\oplus dec(q)$.*   □

Of course, for sequential CFM processes, team equivalence $\sim^\oplus$ coincides with bisimilarity on places $\sim$.

Finally, as we are going to use an auxiliary restriction operator over CFM terms of the form $p\backslash H$, we define its net semantics as follows.

Definition 11 (Semantics of the auxiliary restriction operator). *Given a CFM process p, whose net semantics is $[\![p]\!]_\emptyset = (S, A, T, dec(p))$, we define the net associated to $p\backslash H$ as the net $[\![p\backslash H]\!]_\emptyset = (S', A', T', m)$ where*

- $S' = \{s\backslash H \mid s \in S\}$, *i.e., each place is decorated by the restriction operator;*
- $A' = A\backslash H$, *i.e.,* $\{\mu \mid \mu \in A, \mu \notin H\}$;
- $T' = \{({}^\bullet t\backslash H, l(t), t^\bullet\backslash H) \mid t \in T, l(t) \notin H\}$;
- $m = dec(p)\backslash H$, *where the restriction operator is applied element-wise to the places, if any, of the marking $dec(p)$.*   □

As an example, the net for $C | B$, with $C \doteq h.B$ and $B \doteq l.B$, is outlined in Fig. 2(a), while, assuming that H is composed of a single action h, the net for $(C | B)\backslash h$ is in Fig. 2(b).

3.3   Axiomatization

In this section we recall the sound and complete, finite axiomatization of team equivalence over CFM outlined in [20]. For simplicity's sake, the syntactic definition of *open* CFM (i.e., CFM with variables) is given with only one syntactic category, but each ground instantiation of an axiom must respect the syntactic definition of CFM given (by means of three syntactic categories) in Sect. 3.1; this means that we can write the axiom $x + (y + z) = (x + y) + z$, but it is invalid to instantiate it to $C + (a.\mathbf{0} + b.\mathbf{0}) = (C + a.\mathbf{0}) + b.\mathbf{0}$ because these are not legal CFM processes (the constant C cannot be used as a summand).

The set of axioms are outlined in Table 4. We call E the set of axioms {**A1**, **A2**, **A3**, **A4**, **R1**, **R2**, **R3**, **P1**, **P2**, **P3**}. By the notation $E \vdash p = q$ we mean that there exists an equational deduction proof of the equality $p = q$, by using the axioms in E. Besides the usual equational deduction rules of reflexivity, symmetry, transitivity, substitutivity and instantiation (see, e.g., [17]), in order to deal with constants we need also the following recursion congruence rule:

$$\frac{p = q \ \wedge \ A \doteq p\{A/x\} \ \wedge \ B \doteq q\{B/x\}}{A = B}$$

where $p\{A/x\}$ denotes the open term p where all occurrences of the variable x are replaced by A. The axioms **A1**–**A4** are the usual axioms for choice where, however, **A3**–**A4** have the side condition $x \neq \mathbf{0}$; hence, it is not possible to prove $E \vdash \mathbf{0} + \mathbf{0} = \mathbf{0}$, as expected, because these two terms have a completely different semantics. The conditional axioms **R1**–**R3** are about process constants. Note that **R2** requires that p is not (equal to) $\mathbf{0}$ (condition $p \neq \mathbf{0}$). Note also that these conditional axioms are actually a finite collection of axioms, one for each constant definition: since the set $\mathscr{C}$ of process constants is finite, the instances of **R1**–**R3** are finitely many. Finally, we have axioms **P1**–**P3** for parallel composition.

Theorem 3 [20] **(Sound and Complete).** *For every* $p, q \in \mathscr{P}_{CFM}$, $E \vdash p = q$ *if and only if* $p \sim^{\oplus} q$.   □

Table 4. Axioms for team equivalence

A1	Associativity	$x + (y + z) = (x + y) + z$	
A2	Commutativity	$x + y = y + x$	
A3	Identity	$x + \mathbf{0} = x$	if $x \neq \mathbf{0}$
A4	Idempotence	$x + x = x$	if $x \neq \mathbf{0}$
R1	Stuck	if $C \doteq \mathbf{0}$, then $C = \mathbf{0} + \mathbf{0}$	
R2	Unfolding	if $C \doteq p \ \wedge \ p \neq \mathbf{0}$, then $C = p$	
R3	Folding	if $C \doteq p\{C/x\} \ \wedge \ q = p\{q/x\}$, then $C = q$	
P1	Associativity	$x \mid (y \mid z) = (x \mid y) \mid z$	
P2	Commutativity	$x \mid y = y \mid x$	
P3	Identity	$x \mid \mathbf{0} = x$	

4 DNI: Distributed Non-interference

4.1 Definition and Compositional Verification

Definition 12 (Distributed Non-Interference (DNI)). *A CFM process p enjoys the distributed non-interference property (DNI, for short) if for each p', p'', reachable from p, and for each $h \in H$, such that $p' \xrightarrow{h} p''$, we have that $p' \backslash H \sim^{\oplus} p'' \backslash H$ holds.*   □

This intuitive and simple definition is somehow hybrid, because, on the one hand, it refers to reachable states p' and p'' in the LTS semantics, while, on the other hand, it requires that $p' \backslash H \sim^{\oplus} p'' \backslash H$, a condition that can be checked on the Petri net semantics. We can reformulate this definition in such a way that it refers only to the Petri net semantics, a reformulation that will be very useful in proving the following theorem.

Definition 13 (DNI on Petri nets). *Given a CFM process p and the FSMs $[\![p]\!]_\emptyset$ and $[\![p\backslash H]\!]_\emptyset$, we say that p satisfies DNI if for each m', m'', reachable from $dec(p)$ in $[\![p]\!]_\emptyset$, and for each $h \in H$, such that $m' \xrightarrow{h} m''$, we have that the two markings $m' \backslash H$ and $m'' \backslash H$ of $[\![p\backslash H]\!]_\emptyset$ are team equivalent.*   □

Theorem 4. *A process p is not DNI iff there exists $p_i \in dec(p)$ such that p_i is not DNI.*

Proof. *If p is not DNI, then there exist m, m' reachable from $dec(p)$, and $h \in H$, such that $m' \xrightarrow{h} m''$, but the two markings $m' \backslash H$ and $m'' \backslash H$ of $[\![p\backslash H]\!]_\emptyset$ are not team equivalent. Because of the shape of FSM transitions, $m' \xrightarrow{h} m''$ is a move that must be due to a transition $s \xrightarrow{h} m$, so that $m' = s \oplus \overline{m}$ and $m'' = m \oplus \overline{m}$. Therefore, $m' \backslash H = s \backslash H \oplus \overline{m} \backslash H$ and $m'' \backslash H = m \backslash H \oplus \overline{m} \backslash H$. If $m' \backslash H$ is not team equivalent to $m'' \backslash H$, then necessarily $s \backslash H \nsim m \backslash H$. Since m is reachable from $dec(p)$, because of the shape of net transitions, there exists $p_i \in dec(p)$ such that s is reachable from p_i. Summing up, if p is not DNI, we have found a $p_i \in dec(p)$ which is not DNI, because p_i can reach s, transition $s \xrightarrow{h} m$ is firable and $s \backslash H \nsim m \backslash H$. The reverse implication is obvious.*   □

Corollary 1. *A CFM process p is DNI if and only if each $p_i \in dom(dec(p))$ is DNI.*

Proof. *The thesis is just the contranominal of Theorem 4.*   □

Hence, in order to check whether p is DNI, we first compute $dec(p)$ to single out its sequential components; then, we consider only the elements of $dom(dec(p))$, because it is necessary to check each sequential component only once. For instance, if $p = (q_1 \mid q_2) \mid (q_1 \mid q_2)$, then, assuming q_1 and q_2 sequential, $dec(p) = 2 \cdot q_1 \oplus 2 \cdot q_2$, so that $dom(dec(p)) = \{q_1, q_2\}$, and so we have simply to check whether q_1 and q_2 are DNI.

Corollary 2. *If $p \sim^\oplus q$ and p is DNI, then also q is DNI.*

Proof. By Corollary 1, p is DNI if and only if each $p_i \in dom(dec(p))$ is DNI. Since $p \sim^\oplus q$, there exists a $\sim$-relating bijection between $dec(p)$ and $dec(q)$. Therefore, the thesis is implied by the following obvious fact: given two sequential CFM processes p_i, q_j such that $p_i \sim q_j$, if p_i is DNI, then q_j is DNI.   □

4.2   Efficient Verification Based on a Structural Characterization

A very efficient DNI verification can be done with the following algorithm. Given the CFM process p, first compute the nets $[\![p]\!]_\emptyset = (S, A, T, dec(p))$ and $[\![p \backslash H]\!]_\emptyset$. Then, compute bisimilarity $\sim$ on the places of the net $[\![p \backslash H]\!]_\emptyset$. Finally, for each $t \in T$ such that $l(t) \in H$, check whether ${}^\bullet t \backslash H$ and $t^\bullet \backslash H$ are bisimilar: if this is the case for all the high-transitions of $[\![p]\!]_\emptyset$, then p is DNI; on the contrary, if for some t the check fails (e.g., because $t^\bullet = \theta$), then p is not DNI. The correctness of this polynomial algorithm follows by the fact that the net $[\![p]\!]_\emptyset$ is dynamically reduced. Its complexity is essentially related to the problem of computing $\sim$ (cf. Remark 1) for $[\![p \backslash H]\!]_\emptyset$ and to give it a suitable adjacency matrix representation in order to check easily, for each high-transition $t \in T$, whether the two relevant places ${}^\bullet t \backslash H$ and $t^\bullet \backslash H$ are related by $\sim$.

4.3   Typing System

In this section we provide a syntactic characterization of DNI by means of a typing proof system, which exploits the axiomatization of team equivalence. Let us first define an auxiliary operator $r(-)$, which takes in input a CFM process p and returns a CFM process p' obtained from p by pruning its high-level actions, so that $sort(p') \subseteq L$. Its definition is outlined in Table 5, where $r(h.p) = \mathbf{0} + \mathbf{0}$ because the pruning of $h.p$ is to be considered as a deadlock place. For instance, consider $C \doteq h.l.C + l.C$; then $r(C) = C'$, where $C' \doteq \mathbf{0} + \mathbf{0} + l.C'$. Similarly, if $D \doteq l.h.D$, then $r(D) = D'$ where $D' \doteq l.(\mathbf{0} + \mathbf{0})$. It is a trivial observation that the net semantics of $r(p)$ is isomorphic to the (reachable part of the) net semantics of $p \backslash H$ for any CFM process p. Moreover, we also define the set of initial actions of p, denoted by $In(p)$, whose definition is outlined in Table 6.

Table 5. Restriction function

$r(\mathbf{0}) = \mathbf{0}$	$r(l.p) = l.r(p)$	$r(h.p) = \mathbf{0} + \mathbf{0}$
$r(p + p') = r(p) + r(p')$	$r(p \,|\, p') = r(p) \,|\, r(p')$	$r(C) = C'$ where $C' \doteq r(p)$ if $C \doteq p$

Then we define a typing system on CFM processes such that, if a process p is typed, then p satisfies DNI and, moreover, if p is DNI, then there exists a process p', obtained by possibly reordering its summands via axioms $\mathbf{A_1} - \mathbf{A_4}$, which is typed. The typing system is outlined in Table 7, where we are using a set I of already scanned constants. A process p is typed if $(p, \emptyset) : dni$ is derivable

Table 6. Initial function

$In(0) = \emptyset$	$In(\mu.p) = \{\mu\}$	$In(p + p') = In(p) \cup In(p')$
$In(p \,\vert\, p') = In(p) \cup In(p')$	$In(C) = In(p)$ if $C \doteq p$	

Table 7. Typing proof system

	$(p,I) : dni, (q,I) : dni$	$(p,I) : dni, (q,I) : dni, In(p+q) \subseteq L$
$(0,I) : dni$	$(p \,\vert\, q, I) : dni$	$(p+q,I) : dni$
$(p,I) : dni$	$E \vdash p = 0+0$	$C \notin I, C \doteq p, (p, I \cup \{C\}) : dni$
$(l.p,I) : dni$	$(h.p,I) : dni$	$(C,I) : dni$
$C \in I$	$\emptyset \neq sort(p) \subseteq H, (p,I) : dni$	$(p,I) : dni, p \neq 0, (q,I) : dni, E \vdash r(p) = r(q)$
$(C,I) : dni$	$(h.p,I) : dni$	$(h.p+q,I) : dni$

by the rules; this is often simply denoted by $p : dni$. The need for the argument I is clear in the two rules for the constant C: if C has been already scanned (i.e. $C \in I$), then C is typed; otherwise it is necessary to check that its body p is typed, using a set enriched by C (condition $(p, I \cup \{C\}) : dni$).

We are implicitly assuming that the formation rules in this table respect the syntax of CFM; this means that, for instance, the third rule requires also that p and q are actually guarded processes because this is the case in $p + q$. Note that in this rule we are requiring that $In(p + q)$ is a subset of L, so that, as no high-level action is executable by p and q as their first action, no DNI check is required at this level of the syntax.

The interesting cases are the three rules about action prefixing and the rule about the choice operator when a summand starts with a high-level action. The first rule states that if p is typed, then also $l.p$ is typed, for each $l \in L$. The second rule states that if p is a deadlock place (condition $E \vdash p = 0 + 0$), then $h.p$ is typed, for each $h \in H$; note that p cannot be 0, because $h.0$ is not secure. The third rule states that if p is a typed term that can perform at least one action, but only high-level actions (condition $\emptyset \neq sort(p) \subseteq H$), then $h.p$ is typed. The rule about the choice operator states that if we prefix a generic typed process p with a high-level action h, then it is necessary that an additional typed summand q is present such that $p \backslash H$ and $q \backslash H$ are team equivalent; this semantic condition is expressed syntactically by requiring that $E \vdash r(p) = r(q)$, thanks to Theorem 3; note that $p \neq 0$, because $h.0 + 0$ is not secure. It is interesting to observe that this rule covers also the case when many summands start with high-level actions. For instance, $h.l.l.0 + (h.l.(h.l.0 + l.0) + l.l.0)$ is typed because $l.l.0$ and $h.l.(h.l.0 + l.0) + l.l.0$ are typed and $E \vdash r(l.l.0) = r(h.l.(h.l.0 + l.0) + l.l.0)$. This strategy is intuitively correct (i.e., it respects DNI) because, by checking that the subterm $h.l.(h.l.\mathbf{0} + l.\mathbf{0}) + l.l.\mathbf{0}$ is typed/DNI, we can safely ignore the other summand $h.l.l.\mathbf{0}$, as it does not contribute any initial low-visible behavior.

Now we want to prove that the typing system characterizes DNI correctly. To get convinced of this result, consider again $C \doteq h.l.C + l.C$. This process is DNI because, if $C \xrightarrow{h} l.C$, then $C\backslash H \sim (l.C)\backslash H$, which is equivalent to say that $r(C) \sim r(l.C)$. As a matter of fact, $(C, \emptyset) : dni$ holds, because $(h.l.C + l.C, \{C\}) : dni$ holds, because, in turn, $(l.C, \{C\}) : dni$ and $E \vdash r(l.C) = r(l.C)$. On the contrary, $D \doteq l.h.D$ is not DNI because if $h.D \xrightarrow{h} D$, then $h.D\backslash H \nsim D\backslash H$ as $h.D\backslash H$ is stuck, while $D\backslash H$ can perform l. As a matter of fact, D is not typed: to get $(D, \emptyset) : dni$, we need $(l.h.D, \{D\}) : dni$, which would require $(h.D, \{D\}) : dni$, which is false, as no rule for high-level prefixing is applicable.

Proposition 1. *For each CFM process p, if $p : dni$, then p satisfies DNI.*

Proof. By induction on the proof of $(p, \emptyset) : dni$.   □

Note that the reverse implication is not alway true, because of the ordering of summands. For instance, $l.\mathbf{0} + h.l.\mathbf{0}$, which is clearly DNI, is not typed. However, $\{\mathbf{A_1} - \mathbf{A_4}\} \vdash l.\mathbf{0} + h.l.\mathbf{0} = h.l.\mathbf{0} + l.\mathbf{0}$, where the process $h.l.\mathbf{0} + l.\mathbf{0}$ is typed.

Proposition 2. *For each CFM process p, if p is DNI, then there exists p' such that $\{\mathbf{A_1} - \mathbf{A_4}\} \vdash p = p'$ and $p' : dni$.*

Proof. By induction on the structure of p. By considering an additional parameter I of already scanned constants, the base cases are $(\mathbf{0}, I)$ and (C, I) with $C \in I$. Both cases are trivial as these two terms cannot do anything. The only non-trivial inductive case is about summation. Assume that $p_1 + p_2$ is DNI. If $In(p_1 + p_2) \subseteq L$, then both p_1 and p_2 are DNI; by induction, there exist $p'_i : dni$ such that $\{\mathbf{A_1} - \mathbf{A_4}\} \vdash p_i = p'_i$ for $i = 1, 2$; hence, $\{\mathbf{A_1} - \mathbf{A_4}\} \vdash p_1 + p_2 = p'_1 + p'_2$ and also $p'_1 + p'_2 : dni$, as required. On the contrary, if there exists $h \in In(p_1 + p_2)$, then there exist q_1 and q_2 such that $\{\mathbf{A_1} - \mathbf{A_4}\} \vdash p_1 + p_2 = h.q_1 + q_2$. Since also $h.q_1 + q_2$ is DNI by Corollary 2, it is necessary that $q_1\backslash H \sim q_2\backslash H$, which is equivalent to $r(q_1) \sim r(q_2)$, in turn equivalent to stating that $E \vdash r(q_1) = r(q_2)$. Moreover, the DNI property has to be satisfied by q_1 and q_2. Hence, by induction, there exist q'_1, q'_2 such that $\{\mathbf{A_1} - \mathbf{A_4}\} \vdash q_i = q'_i$ and $q'_i : dni$, for $i = 1, 2$. By transitivity and substitutivity, we have $\{\mathbf{A_1} - \mathbf{A_4}\} \vdash p_1 + p_2 = h.q'_1 + q'_2$. Moreover, since $E \vdash r(q_1) = r(q_2)$, we also have that $E \vdash r(q'_1) = r(q'_2)$, and so, by the proof system, $h.q'_1 + q'_2 : dni$, as required.   □

5   Conclusion

Related Literature. The non-interference problem in a distributed model of computation was first addressed in [5,6]. There, the Petri net class of *unlabeled elementary net systems* (i.e., 1-safe, contact-free nets) was used to describe some information flow security properties, notably *BNDC* (Bisimulation Non-Deducibility on Composition) and *SBNDC* (Strong BNDC), based on interleaving bisimilarity. These two properties do coincide on unlabeled elementary net systems, but actually SBNDC is stronger on *labeled* FSMs; for instance, the CFM process $l.h.l.0 + l.0 + l.l.0$ is BNDC [10], while it is not SBNDC; this explains why we have chosen SBNDC as our starting point towards the formulation of DNI.

In [6] it is shown that BNDC can be characterized as a structural property of the net concerning two special classes of places: *causal places*, i.e., places for which there are an incoming high transition and an outgoing low transition; and *conflict places*, i.e. places for which there are both low and high outgoing transitions. The main theorem in [6] states that if places of these types are not present or cannot be reached from the initial marking, then the net is BNDC. An algorithm following this definition was implemented in [11], but it suffers from the high complexity of the reachability problem.

Starting from [6], Baldan and Carraro in [1] provide a *causal characterization* of BNDC on safe Petri nets (with injective labeling), in terms of the unfolding semantics, resulting in an algorithm much better than [11]. Nonetheless, the BNDC property is based on an interleaving semantics and the true-concurrency semantics is used only to provide efficient algorithms to check the possible presence of interferences.

Another paper studying non-interference over a distributed model is [7]. Bérard et al. study a form of non-interference similar to SNNI [10] for *High-level Message Sequence Charts* (HMSC), a scenario language for the description of distributed systems, based on composition of partial orders. The model allows for unbounded parallelism and the observable semantics they use is interleaving and linear-time (i.e., language-based). They prove that non-interference is undecidable in general, while it becomes decidable for regular behaviors, or for weaker variants based on observing local behavior only. Also in this case, however, the truly-concurrent semantics based on partial orders is used mainly for algorithmic purpose; in fact, the authors shows that their decidable properties are PSPACE-complete, with procedures that never compute the interleaving semantics of the original HMSC.

The use of causal semantics in security analysis was advocated also in [12], where Fröschle suggests that for modeling, verification, decidability and complexity reasons (i.e., mainly for algorithmic convenience), this kind of equivalences should be used.

On the contrary, Baldan et al. in [2] define security policies, similar to non-interference, where causality is used as a first-class concept. So, their notion of non-interference is more restrictive than those based on interleaving semantics. However, their approach is linear-time, while non-interference is usually defined on a branching-time semantics, i.e., on bisimulation. Moreover, it seems overly restrictive; for instance, the CFM process $h.l.0 + l.0$, which is DNI, would be considered insecure in their approach.

Future Research. Our proposal of DNI is customized for CFM and for the simple subclass of Petri nets called *finite-state machines*. However, we think that for finite Place/Transition Petri nets [31] (and its corresponding process algebra FNM [18]), the non-interference property DNI can be formulated as follows:

Given a finite P/T Petri net N, a marking m is DNI if, for each m', m'' reachable from m, and $\forall h \in H$ such that $m' \xrightarrow{h} m''$, we have that $m' \backslash H \sim_{sfc} m'' \backslash H$, where $\sim_{sfc}$ denotes state-sensitive fc-bisimulation equivalence [22].

We may wonder if such a property is decidable. We conjecture that for 1-safe nets, DNI is decidable because *history-preserving bisimilarity* is decidable [33] on this class of nets and $\sim_{sfc}$ is very similar to it. About unbounded P/T nets, we conjecture that DNI is decidable for BPP nets (i.e., nets with singleton preset transitions) because $\sim_{sfc}$ is characterizable as a decidable team-like equivalence [22]. However, for general P/T nets, the problem is less clear. On the one hand, $\sim_{sfc}$ (as any other bisimulation-based behavioral equivalence) is undecidable for *labeled* finite P/T nets with at least two *unbounded* places [9,25]; so, we would expect that DNI is undecidable on this class of nets. On the other hand, for unbounded *partially observed* finite P/T nets (i.e., net with unobservable high transitions and *injective labeling* on low transitions), Best et al. proved in [3] that SBNDC is decidable; so, this positive result gives some hope that it might be possible to prove decidability of DNI, at least on this restricted class of unbounded finite nets.

We plan to extend this approach to a setting with silent transitions as well as to intransitive non-interference [2, 3, 7, 16].

Why True Concurrency? To conclude this paper, we want to recapitulate some further reasons for preferring truly-concurrent semantics over interleaving semantics. A first observation was that interleaving semantics are not a congruence for the action refinement operator [8,13,15], nor for (some forms of) multi-party synchronization (see Chapter 6 of [17]), while this property holds for some truly concurrent semantics. One further (and stronger) observation is about expressiveness. In [18] six process algebras are compared w.r.t. an interleaving semantics and w.r.t. a concrete truly-concurrent semantics based on Petri nets. The resulting two hierachies are quite different; in particular, the hierarchy based on interleaving semantics equates process algebras that have very different expressive power in terms of classes of solvable problems in distributed computing. As it may happen that two Turing-complete process algebras can solve different classes of problems, classic Turing (or sequential) computability, based on computable functions on natural numbers, needs to be extended to deal with computable objects in a distributed model such as Petri nets. This observation calls for the study of a generalization of Turing computability, we call *distributed computability* [21].

Acknowledgments. The anonymous referees are thanked for their comments.

References

1. Baldan, P., Carraro, A.: A causal view on non-intereference. Fundam. Infor. **140**(1), 1–38 (2015)
2. Baldan, P., Beggiato, A., Lluch Lafuente, A.: Many-to-many information flow policies. In: Jacquet, J.-M., Massink, M. (eds.) COORDINATION 2017. LNCS, vol. 10319, pp. 159–177. Springer, Cham (2017). https://doi.org/10.1007/978-3-319-59746-1_9
3. Best, E., Darondeau, Ph., Gorrieri, R.: On the decidability of non-interference over unbounded Petri nets. In: Proceedings of the 8th International Workshop on Security Issues in Concurrency (SecCo 2010), EPTCS, vol. 51, pp. 16–33 (2010)
4. Best, E., Devillers, R., Kiehn, A., Pomello, L.: Concurrent bisimulations in Petri nets. Acta Informatica **28**(3), 231–264 (1991). https://doi.org/10.1007/BF01178506
5. Busi, N., Gorrieri, R.: A survey on non-interference with Petri nets. In: Desel, J., Reisig, W., Rozenberg, G. (eds.) ACPN 2003. LNCS, vol. 3098, pp. 328–344. Springer, Heidelberg (2004). https://doi.org/10.1007/978-3-540-27755-2_8
6. Busi, N., Gorrieri, R.: Structural non-interference in elementary and trace nets. Math. Struct. Comput. Sci. **19**(6), 1065–1090 (2009)
7. Bérard, B., Hélouët, L., Mullins, J.: Non-interference in partial order models. In: Proceedings of the ACSD 2015, pp. 80–89. IEEE Computer Society (2015)
8. Castellano, L., De Michelis, G., Pomello, L.: Concurrency versus interleaving: an instructive example. Bull. EATCS **31**, 12–14 (1987)
9. Esparza, J.: Decidability and complexity of petri net problems—an introduction. In: Reisig, W., Rozenberg, G. (eds.) ACPN 1996. LNCS, vol. 1491, pp. 374–428. Springer, Heidelberg (1998). https://doi.org/10.1007/3-540-65306-6_20
10. Focardi, R., Gorrieri, R.: Classification of security properties. In: Focardi, R., Gorrieri, R. (eds.) FOSAD 2000. LNCS, vol. 2171, pp. 331–396. Springer, Heidelberg (2001). https://doi.org/10.1007/3-540-45608-2_6
11. Frau, S., Gorrieri, R., Ferigato, C.: Petri net security checker: structural non-interference at work. In: Degano, P., Guttman, J., Martinelli, F. (eds.) FAST 2008. LNCS, vol. 5491, pp. 210–225. Springer, Heidelberg (2009). https://doi.org/10.1007/978-3-642-01465-9_14
12. Fröschle, S.: Causality, behavioural equivalences, and the security of cyberphysical systems. In: [30], pp. 83–98 (2015)
13. van Glabbeek, R., Goltz, U.: Equivalence notions for concurrent systems and refinement of actions. In: Kreczmar, A., Mirkowska, G. (eds.) MFCS 1989. LNCS, vol. 379, pp. 237–248. Springer, Heidelberg (1989). https://doi.org/10.1007/3-540-51486-4_71
14. van Glabbeek, R.J.: Structure preserving bisimilarity: supporting an operational Petri net semantics of CCSP. In: [30], pp. 99–130. Springer (2015)
15. Gorrieri, R., Rensink, A.: Action Refinement. In: Handbook of Process Algebra, pp. 1047–1147. North-Holland (2001)
16. Gorrieri, R., Vernali, M.: On intransitive non-interference in some models of concurrency. In: Aldini, A., Gorrieri, R. (eds.) FOSAD 2011. LNCS, vol. 6858, pp. 125–151. Springer, Heidelberg (2011). https://doi.org/10.1007/978-3-642-23082-0_5
17. Gorrieri, R., Versari, C.: Introduction to Concurrency Theory: Transition Systems and CCS. EATCS Texts in Theoretical Computer Science. Springer, Cham (2015). https://doi.org/10.1007/978-3-319-21491-7

18. Gorrieri, R.: Process Algebras for Petri Nets: The Alphabetization of Distributed Systems. EATCS Monographs in Theoretical Computer Science. Springer, Cham (2017). https://doi.org/10.1007/978-3-319-55559-1

19. Gorrieri, R.: Verification of finite-state machines: a distributed approach. J. Log. Algebraic Methods Program. **96**, 65–80 (2018)

20. Gorrieri, R.: Axiomatizing team equivalence for finite-state machines. In: Alvim, M.S., Chatzikokolakis, K., Olarte, C., Valencia, F. (eds.) The Art of Modelling Computational Systems: A Journey from Logic and Concurrency to Security and Privacy. LNCS, vol. 11760, pp. 14–32. Springer, Cham (2019). https://doi.org/10.1007/978-3-030-31175-9_2

21. Gorrieri, R.: Toward distributed computability theory. In: Reisig, W., Rozenberg, G. (eds.) Carl Adam Petri: Ideas, Personality, Impact, pp. 141–146. Springer, Cham (2019). https://doi.org/10.1007/978-3-319-96154-5_18

22. Gorrieri, R.: A study on team bisimulations for BPP nets (extended abstract). In: Janicki, R., et al. (eds.) Petri Nets 2020. LNCS, vol. 12152, pp. xx–yy. Springer, Cham (2020)

23. Hoare, C.A.R.: Communicating Sequential Processes. Prentice-Hall International Series in Computer Science (1985)

24. Hopcroft, J.E., Motwani, R., Ullman, J.D.: Introduction to Automata Theory, Languages and Computation, 2nd edn. Addison-Wesley, Boston (2001)

25. Jančar, P.: Undecidability of bisimilarity for Petri nets and some related problems. Theor. Comput. Sci. **148**(2), 281–301 (1995)

26. Keller, R.: Formal verification of parallel programs. Commun. ACM **19**(7), 561–572 (1976)

27. Milner, R.: Communication and Concurrency. Prentice-Hall, Upper Saddle River (1989)

28. Nielsen, M., Thiagarajan, P.S.: Degrees of non-determinism and concurrency: a Petri net view. In: Joseph, M., Shyamasundar, R. (eds.) FSTTCS 1984. LNCS, vol. 181, pp. 89–117. Springer, Heidelberg (1984). https://doi.org/10.1007/3-540-13883-8_66

29. Meyer, R., Platzer, A., Wehrheim, H. (eds.): Correct System Design-Symposium in Honor of Ernst-Rüdiger Olderog on the Occasion of His 60th Birthday. LNCS, vol. 9360. Springer, Cham (2015). https://doi.org/10.1007/978-3-319-23506-6

30. Paige, R., Tarjan, R.E.: Three partition refinement algorithms. SIAM J. Comput. **16**(6), 973–989 (1987)

31. Peterson, J.L.: Petri Net Theory and the Modeling of Systems. Prentice-Hall, Upper Saddle River (1981)

32. Ryan, P.Y.A.: Mathematical models of computer security. In: Focardi, R., Gorrieri, R. (eds.) FOSAD 2000. LNCS, vol. 2171, pp. 1–62. Springer, Heidelberg (2001). https://doi.org/10.1007/3-540-45608-2_1

33. Vogler, W.: Deciding history preserving bisimilarity. In: Albert, J.L., Monien, B., Artalejo, M.R. (eds.) ICALP 1991. LNCS, vol. 510, pp. 495–505. Springer, Heidelberg (1991). https://doi.org/10.1007/3-540-54233-7_158

A Study on Team Bisimulations for BPP Nets

Roberto Gorrieri[(✉)]

Dipartimento di Informatica—Scienza e Ingegneria, Università di Bologna,
Mura A. Zamboni, 7, 40127 Bologna, Italy
roberto.gorrieri@unibo.it

Abstract. BPP nets, a subclass of finite P/T nets, were equipped in [13] with an efficiently decidable, truly concurrent, behavioral equivalence, called *team bisi-milarity*. This equivalence is a very intuitive extension of classic bisimulation equivalence (over labeled transition systems) to BPP nets and it is checked in a distributed manner, without building a global model of the overall behavior of the marked BPP net. This paper has three goals. First, we provide BPP nets with various causality-based equivalences, notably a novel one, called *causal-net* bisimilarity, and (a version of) *fully-concurrent* bisimilarity [3]. Then, we define a variant equivalence, *h-team* bisimilarity, coarser than team bisimilarity. Then, we complete the study by comparing them with the causality-based semantics we have introduced: the main results are that team bisimilarity coincides with causal-net bisimilarity, while h-team bisimilarity with fully-concurrent bisimilarity.

1 Introduction

A BPP net is a simple type of finite Place/Transition Petri net [18] whose transitions have singleton pre-set. Nonetheless, as a transition can produce more tokens than the only one consumed, there can be infinitely many reachable markings of a BPP net. BPP is the acronym of *Basic Parallel Processes* [4], a simple CCS [11,15] subcalculus (without the restriction operator) whose processes cannot communicate. In [12] a variant of BPP, which requires guarded sum and guarded recursion, is actually shown to represent all and only the BPP nets, up to net isomorphism, and this explains the name of this class of nets.

In a recent paper [13], we proposed a novel behavioral equivalence for BPP nets, based on a suitable generalization of the concept of bisimulation [15], originally defined over labeled transition systems (LTSs, for short). A *team bisimulation* R over the places of an *unmarked* BPP net is a relation such that if two places s_1 and s_2 are related by R, then if s_1 performs a and reaches the marking m_1, then s_2 may perform a reaching a marking m_2 such that m_1 and m_2 are element-wise, bijectively related by R (and vice versa if s_2 moves first). *Team bisimilarity* is the largest team bisimulation over the places of the *unmarked* BPP net, and then such a relation is lifted to markings by *additive closure*: if place s_1 is team bisimilar to place s_2 and the marking m_1 is team bisimilar to

© Springer Nature Switzerland AG 2020
R. Janicki et al. (Eds.): PETRI NETS 2020, LNCS 12152, pp. 153–175, 2020.
https://doi.org/10.1007/978-3-030-51831-8_8

m_2 (the base case relates the empty marking to itself), then also $s_1 \oplus m_1$ is team bisimilar to $s_2 \oplus m_2$, where $\_ \oplus \_$ is the operator of multiset union. Note that to check if two markings are team bisimilar we need not to construct an LTS, such as the *interleaving marking graph*, describing the global behavior of the whole system, but only to find a *bijective*, team bisimilarity-preserving match among the elements of the two markings. In other words, two distributed systems, each composed of a *team* of sequential, non-cooperating processes (i.e., the tokens in the BPP net), are equivalent if it is possible to match each sequential component of the first system with one team-bisimilar, sequential component of the other system, as in any sports where two competing (distributed) teams have the same number of (sequential) players.

The complexity of checking whether two markings of equal size are team bisimilar is very low. First, by adapting the optimal algorithm for standard bisimulation equivalence over LTSs [19], team bisimulation equivalence over places can be computed in $O(m \cdot p^2 \cdot \log(n+1))$ time, where m is the number of net transitions, p is the size of the largest post-set (i.e., p is the least natural such that $|t^\bullet| \leq p$ for all t) and n is the number of places. Then, checking whether two markings of size k are team bisimilar can be done in $O(k^2)$ time. Of course, we proved that team bisimilar markings respect the global behavior; in particular, we proved that team bisimilarity implies interleaving bisimilarity and that team bisimilarity coincides with *strong place bisimilarity* [1].

In this paper, we complete the comparison between team bisimilarity on markings and the causal semantics of BPP nets. In particular, we propose a novel coinductive equivalence, called *causal-net* bisimulation equivalence, inspired by [9], which is essentially a bisimulation semantics over the causal nets [2,17] of the BPP net under scrutiny. We prove that team bisimilarity on markings coincides with causal-net bisimilarity, hence proving that our distributed semantics is coherent with the expected causal semantics of BPP nets. Moreover, we adapt the definition of *fully-concurrent* bisimulation (fc-bisimulation, for short) in [3], in order to be better suited for our aims. Fc-bisimilarity was inspired by previous notions of equivalence on other models of concurrency, in particular, by *history-preserving bisimulation* (hpb, for short) [8]. Moreover, we define also a slight strengthening of fc-bisimulation, called *state-sensitive* fc-bisimulation, which requires additionally that, for each pair of related processes, the current markings have the same size. We also prove that causal-net bisimilarity coincides with state-sensitive fc-bisimilarity. These behavioral causal semantics have been provided for BPP nets, but they can be easily adapted for general P/T nets.

The other main goal of this paper is to show that fc-bisimilarity (hence also hpb) can be characterized for BPP nets in a team-style, by means of *h-team* bisimulation equivalence. (The prefix *h-* is used to remind that h-team bisimilarity is connected to hpb.) The essential difference between a team bisimulation and an h-team bisimulation is that the former is a relation on the set of places only, while the latter is a relation on the set composed of the places *and* the empty marking θ.

The paper is organized as follows. Section 2 introduces the basic definitions about BPP nets and recalls interleaving bisimilarity. Section 3 discusses the causal semantics of BPP nets. First, the novel causal-net bisimulation is introduced, then (state-sensitive) fully-concurrent bisimilarity, as an improvement of the original one [3], which better suits our aims. Section 4 recalls the main definitions and results about team bisimilarity from [13]; in this section we also prove a novel result: causal-net bisimilarity coincides with team bisimilarity for BPP nets. Section 5 defines h-team bisimulation equivalence and studies its properties; in particular, we prove that h-team bisimilarity coincides with fc-bisimilarity. Finally, Sect. 6 discusses related literature.

2 Basic Definitions

Definition 1 (Multiset). *Let $\mathbb{N}$ be the set of natural numbers. Given a finite set S, a* multiset *over S is a function $m : S \to \mathbb{N}$. Its* support set $dom(m)$ *is $\{s \in S \mid m(s) \neq 0\}$. The set $\mathcal{M}(S)$ of all multisets over S is ranged over by m. We write $s \in m$ if $m(s) > 0$. The* multiplicity *of s in m is the number $m(s)$. The* size *of m, denoted by $|m|$, is the number $\sum_{s \in S} m(s)$, i.e., the total number of its elements. A multiset m such that $dom(m) = \emptyset$ is called* empty *and is denoted by θ. We write $m \subseteq m'$ if $m(s) \leq m'(s)$ for all $s \in S$.*

Multiset union $\_ \oplus \_$ is defined as follows: $(m \oplus m')(s) = m(s) + m'(s)$; it is commutative, associative and has θ as neutral element. Multiset difference $\_ \ominus \_$ *is defined as follows: $(m_1 \ominus m_2)(s) = max\{m_1(s) - m_2(s), 0\}$. The* scalar product *of a number j with m is the multiset $j \cdot m$ defined as $(j \cdot m)(s) = j \cdot (m(s))$. By s_i we also denote the multiset with s_i as its only element. Hence, a multiset m over $S = \{s_1, \ldots, s_n\}$ can be represented as $k_1 \cdot s_1 \oplus k_2 \cdot s_2 \oplus \ldots \oplus k_n \cdot s_n$, where $k_j = m(s_j) \geq 0$ for $j = 1, \ldots, n$.* ☐

Definition 2 (BPP net). *A labeled BPP net is a tuple $N = (S, A, T)$, where*

- *S is the finite set of* places, *ranged over by s (possibly indexed),*
- *A is the finite set of* labels, *ranged over by ℓ (possibly indexed), and*
- *$T \subseteq S \times A \times \mathcal{M}(S)$ is the finite set of* transitions, *ranged over by t (possibly indexed).*

Given a transition $t = (s, \ell, m)$, we use the notation:

- *$^\bullet t$ to denote its* pre-set *s (which is a single place) of tokens to be consumed;*
- *$l(t)$ for its* label *ℓ, and*
- *$t^\bullet$ to denote its* post-set *m (which is a multiset) of tokens to be produced.*

Hence, transition t can be also represented as $^\bullet t \xrightarrow{l(t)} t^\bullet$. We also define pre-sets and post-sets for places as follows: $^\bullet s = \{t \in T \mid s \in t^\bullet\}$ and $s^\bullet = \{t \in T \mid s \in {}^\bullet t\}$. Note that the pre-set (post-set) of a place is a set. ☐

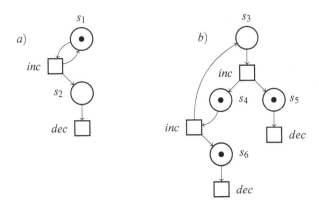

Fig. 1. The net representing a semi-counter in (a), and a variant in (b)

Definition 3 (Marking, BPP net system). *A multiset over S is called a marking. Given a marking m and a place s, we say that the place s contains m(s) tokens, graphically represented by m(s) bullets inside place s. A BPP net system N(m_0) is a tuple (S, A, T, m_0), where (S, A, T) is a BPP net and m_0 is a marking over S, called the* initial marking. *We also say that N(m_0) is a* marked net. □

Definition 4 (Firing sequence). *A transition t is* enabled *at m, denoted m[t⟩, if •t ⊆ m. The* firing *of t enabled at m produces the marking m' = (m ⊖ •t) ⊕ t•, written m[t⟩m'. A* firing sequence *starting at m is defined inductively as follows:*

- *m[ε⟩m is a firing sequence (where ε denotes an empty sequence of transitions) and*
- *if m[σ⟩m' is a firing sequence and m'[t⟩m'', then m[σt⟩m'' is a firing sequence.*

If σ = $t_1 \ldots t_n$ (for n ≥ 0) and m[σ⟩m' is a firing sequence, then there exist $m_1, \ldots, m_{n+1}$ such that m = $m_1[t_1⟩m_2[t_2⟩\ldots m_n[t_n⟩m_{n+1}$ = m', and σ = $t_1 \ldots t_n$ is called a transition sequence *starting at m and ending at m'. The set of* reachable markings *from m is [m⟩ = {m' | ∃σ.m[σ⟩m'}.*

Note that the reachable markings can be countably infinite. A BPP net system N(m_0) = (S, A, T, m_0) is safe *if each marking m reachable from the initial marking m_0 is a set, i.e., ∀m ∈ [m_0⟩, m(s) ≤ 1 for all s ∈ S. The set of* reachable places *from s is reach(s) = $\bigcup_{m∈[s⟩}$ dom(m).*

Note that reach(s) is always a finite set, even if [s⟩ is infinite. □

Example 1. Figure 1(a) shows the simplest BPP net representing a semi-counter, i.e., a counter unable to test for zero. Note that the number represented by this semi-counter is the number of tokens which are present in s_2, i.e., in the place ready to perform *dec*; hence, Fig. 1(a) represents a semi-counter holding 0; note also that the number of tokens which can be accumulated in s_2 is unbounded. Indeed, the set of reachable markings for a BPP net can be countably infinite. In (b), a variant semi-counter is outlined, which holds number 2 (i.e., two tokens are ready to perform *dec*). □

Definition 5 (Interleaving Bisimulation). *Let* $N = (S, A, T)$ *be a BPP net. An* interleaving bisimulation *is a relation* $R \subseteq \mathscr{M}(S) \times \mathscr{M}(S)$ *such that if* $(m_1, m_2) \in R$ *then*

- $\forall t_1$ *such that* $m_1[t_1\rangle m_1'$, $\exists t_2$ *such that* $m_2[t_2\rangle m_2'$ *with* $l(t_1) = l(t_2)$ *and* $(m_1', m_2') \in R$,
- $\forall t_2$ *such that* $m_2[t_2\rangle m_2'$, $\exists t_1$ *such that* $m_1[t_1\rangle m_1'$ *with* $l(t_1) = l(t_2)$ *and* $(m_1', m_2') \in R$.

Two markings m_1 *and* m_2 *are* interleaving bisimilar, *denoted by* $m_1 \sim_{int} m_2$, *if there exists an interleaving bisimulation* R *such that* $(m_1, m_2) \in R$. □

Interleaving bisimilarity $\sim_{int}$, which is defined as the union of all the interleaving bisimulations, is the largest interleaving bisimulation and also an equivalence relation.

Remark 1 (**Interleaving bisimulation between two nets**). The definition above covers also the case of an interleaving bisimulation between two BPP nets, say, $N_1 = (S_1, A, T_1)$ and $N_2 = (S_2, A, T_2)$ with $S_1 \cap S_2 = \emptyset$, because we may consider just one single BPP net $N = (S_1 \cup S_2, A, T_1 \cup T_2)$: An interleaving bisimulation $R \subseteq \mathscr{M}(S_1) \times \mathscr{M}(S_2)$ is also an interleaving bisimulation on $\mathscr{M}(S_1 \cup S_2) \times \mathscr{M}(S_1 \cup S_2)$. Similar considerations hold for all the bisimulation-like definitions we propose in the following. □

Remark 2 (**Comparing two marked nets**). The definition above of interleaving bisimulation is defined over an *unmarked* BPP net, i.e., a net without the specification of an initial marking m_0. Of course, if one desires to compare two marked nets, then it is enough to find an interleaving bisimulation (over the union of the two nets, as discussed in the previous remark), containing the pair composed of the respective initial markings. This approach is followed for all the other bisimulation-like definitions we propose. □

Example 2. Continuing Example 1 about Fig. 1, it is easy to realize that relation $R = \{(s_1 \oplus k \cdot s_2, s_3 \oplus k_1 \cdot s_5 \oplus k_2 \cdot s_6) \mid k = k_1 + k_2 \text{ and } k, k_1, k_2 \geq 0\} \cup \{(s_1 \oplus k \cdot s_2, s_4 \oplus k_1 \cdot s_5 \oplus k_2 \cdot s_6) \mid k = k_1 + k_2 \text{ and } k, k_1, k_2 \geq 0\}$ is an interleaving bisimulation. □

3   Causality-Based Semantics

We start with the most concrete equivalence definable over BPP nets: isomorphism equivalence.

Definition 6 (Isomorphism). *Given two BPP nets $N_1 = (S_1, A, T_1)$ and $N_2 = (S_2, A, T_2)$, we say that N_1 and N_2 are isomorphic via f if there exists a type-preserving bijection $f : S_1 \cup T_1 \to S_2 \cup T_2$ (i.e., a bijection such that $f(S_1) = S_2$ and $f(T_1) = T_2$), satisfying the following condition:*

$$\forall t \in T_1, \; if\, t = (\,^\bullet t, \ell, t^\bullet), \; then \; f(t) = (f(\,^\bullet t), \ell, f(t^\bullet)),$$

where f is homomorphically extended to markings (i.e., f is applied element-wise to each component of the marking: $f(\theta) = \theta$ and $f(m_1 \oplus m_2) = f(m_1) \oplus f(m_2)$.)

  Two BPP net systems $N_1(m_1)$ and $N_2(m_2)$ are rooted isomorphic if the isomorphism f ensures, additionally, that $f(m_1) = m_2$.   □

  In order to define our approach to causality-based semantics for BPP nets, we need some auxiliary definitions, adapting those in, e.g., [3,9,10].

Definition 7 (Acyclic net). *A BPP net $N = (S, A, T)$ is acyclic if there exists no sequence $x_1 x_2 \ldots x_n$ such that $n \geq 3$, $x_i \in S \cup T$ for $i = 1, \ldots, n$, $x_1 = x_n$, $x_1 \in S$ and $x_i \in\, ^\bullet x_{i+1}$ for $i = 1, \ldots, n-1$, i.e., the arcs of the net do not form any cycle.*   □

Definition 8 (Causal net). *A BPP causal net is a marked BPP net $\mathsf{N}(m_0) = (\mathsf{S}, \mathsf{A}, \mathsf{T}, m_0)$ satisfying the following conditions:*

1. *N is acyclic;*
2. *$\forall s \in \mathsf{S} \; |\,^\bullet s| \leq 1 \wedge |s^\bullet| \leq 1$ (i.e., the places are not branched);*
3. *$\forall s \in \mathsf{S} \; m_0(s) = \begin{cases} 1 & if \; ^\bullet s = \emptyset \\ 0 & otherwise; \end{cases}$*
4. *$\forall t \in \mathsf{T} \; t^\bullet(s) \leq 1$ for all $s \in \mathsf{S}$ (i.e., all the arcs have weight 1).*

We denote by $Min(\mathsf{N})$ the set m_0, and by $Max(\mathsf{N})$ the set $\{s \in \mathsf{S} \; | \; s^\bullet = \emptyset\}$.   □

  Note that a BPP causal net, being a BPP net, is finite; since it is acyclic, it represents a finite computation. Note also that any reachable marking of a BPP causal net is a set, i.e., this net is *safe*; in fact, the initial marking is a set and, assuming by induction that a reachable marking m is a set and enables t, i.e., $m[t\rangle m'$, then also $m' = (m \ominus\, ^\bullet t) \oplus t^\bullet$ is a set, because the net is acyclic and because of the condition on the shape of the post-set of t (weights can only be 1).

Definition 9 (Partial orders of events from a causal net). *From a BPP causal net $\mathsf{N}(m_0) = (\mathsf{S}, \mathsf{A}, \mathsf{T}, m_0)$, we can extract the partial order of its events $\mathsf{E}_\mathsf{N} = (\mathsf{T}, \preceq)$, where $t_1 \preceq t_2$ iff there exists a sequence $x_1 x_2 x_3 \ldots x_n$ such that $n \geq 3$, $x_i \in \mathsf{S} \cup \mathsf{T}$ for $i = 1, \ldots, n$, $t_1 = x_1, t_2 = x_n$, and $x_i \in\, ^\bullet x_{i+1}$ for $i = 1, \ldots, n-1$; in other words, $t_1 \preceq t_2$ if there is a path from t_1 to t_2.*

  Two partial orders $(T_1, \preceq_1)$ and $(T_2, \preceq_2)$ are isomorphic if there is a label-preserving, order-preserving bijection $g : T_1 \to T_2$, i.e., a bijection such that $l_1(t) = l_2(g(t))$ and $t \preceq_1 t'$ if and only if $g(t) \preceq_2 g(t')$.

  We also say that g is an event isomorphism between the causal nets N_1 and N_2 if it is an isomorphism between their associated partial orders of events $\mathsf{E}_{\mathsf{N}_1}$ and $\mathsf{E}_{\mathsf{N}_2}$.   □

Remark 3. As the initial marking of a causal net is fixed by its shape (according to item 3 of Definition 8), in the following, in order to make the notation lighter, we often omit the indication of the initial marking, so that the causal net $N(m_0)$ is denoted by N. □

Definition 10 (Moves of a causal net). *Given two BPP causal nets* $N = (S, A, T, m_0)$ *and* $N' = (S', A, T', m_0)$, *we say that* N *moves in one step to* N' *through* t, *denoted by* $N[t\rangle N'$, *if* $^\bullet t \in Max(N)$, $T' = T \cup \{t\}$ *and* $S' = S \cup t^\bullet$; *in other words,* N' *extends* N *by one event* t. □

Definition 11 (Folding and Process). *A* folding *from a BPP causal net* $N = (S, A, T, m_0)$ *into a BPP net system* $N(m_0) = (S, A, T, m_0)$ *is a function* $\rho : S \cup T \to S \cup T$, *which is type-preserving, i.e., such that* $\rho(S) \subseteq S$ *and* $\rho(T) \subseteq T$, *satisfying the following:*

- $A = A$ *and* $l(t) = l(\rho(t))$ *for all* $t \in T$;
- $\rho(m_0) = m_0$, *i.e.,* $m_0(s) = |\rho^{-1}(s) \cap m_0|$;
- $\forall t \in T, \rho(^\bullet t) = {}^\bullet \rho(t)$, *i.e.,* $\rho(^\bullet t)(s) = |\rho^{-1}(s) \cap {}^\bullet t|$ *for all* $s \in S$;
- $\forall t \in T, \rho(t^\bullet) = \rho(t)^\bullet$, *i.e.,* $\rho(t^\bullet)(s) = |\rho^{-1}(s) \cap t^\bullet|$ *for all* $s \in S$.

A pair (N, ρ), *where* N *is a BPP causal net and* ρ *a folding from* N *to a BPP net system* $N(m_0)$, *is a* process *of* $N(m_0)$. □

Definition 12 (Isomorphic processes). *Given a BPP net system* $N(m_0)$, *two of its processes* (N_1, ρ_1) *and* (N_2, ρ_2) *are* isomorphic via f *if* N_1 *and* N_2 *are* rooted isomorphic *via bijection* f *(see Definition 6) and* $\rho_1 = \rho_2 \circ f$. □

Definition 13 (Moves of a process). *Let* $N(m_0) = (S, A, T, m_0)$ *be a BPP system and let* (N_i, ρ_i), *for* $i = 1, 2$, *be two processes of* $N(m_0)$. *We say that* (N_1, ρ_1) *moves in one step to* (N_2, ρ_2) *through* t, *denoted by* $(N_1, \rho_1) \xrightarrow{t} (N_2, \rho_2)$, *if* $N_1[t\rangle N_2$ *and* $\rho_1 \subseteq \rho_2$. □

3.1 Causal-Net Bisimulation

We would like to define a bisimulation-based equivalence which is coarser than the branching-time semantics of *isomorphism of (nondeterministic) occurrence nets* (or unfoldings) [5,7,16] and finer than the linear-time semantics of *isomorphism of causal nets* [2,17]. The proposed novel behavioral equivalence is the following *causal-net bisimulation*, inspired by [9].

Definition 14 (Causal-net bisimulation). *Let* $N = (S, A, T)$ *be a BPP net. A* causal-net bisimulation *is a relation* R, *composed of triples of the form* (ρ_1, N, ρ_2), *where, for* $i = 1, 2$, (N, ρ_i) *is a process of* $N(m_{0_i})$ *for some* m_{0_i}, *such that if* $(\rho_1, N, \rho_2) \in R$ *then*

i) $\forall t_1$ *such that* $\rho_1(Max(N))[t_1\rangle m_1$, $\exists t_2, m_2, t, N', \rho_1', \rho_2'$ *such that*
   1. $\rho_2(Max(N))[t_2\rangle m_2$,
   2. $(N, \rho_1) \xrightarrow{t} (N', \rho_1')$, $\rho_1'(t) = t_1$ *and* $\rho_1'(Max(N')) = m_1$,

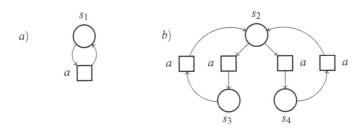

Fig. 2. Two cn-bisimilar BPP nets

3. $(N, \rho_2) \xrightarrow{t} (N', \rho_2')$, $\rho_2'(t) = t_2$ and $\rho_2'(Max(N')) = m_2$; and finally,
4. $(\rho_1', N', \rho_2') \in R$;

ii) symmetrically, if $\rho_2(Max(N))$ moves first.

Two markings m_1 and m_2 of N are cn-bisimilar (or cn-bisimulation equivalent), denoted by $m_1 \sim_{cn} m_2$, if there exists a causal-net bisimulation R containing a triple $(\rho_1^0, N^0, \rho_2^0)$, where N^0 contains no transitions and $\rho_i^0(Min(N^0)) = \rho_i^0(Max(N^0)) = m_i$ for $i = 1, 2$. ☐

Let us denote by $\sim_R^{cn} = \{(m_1, m_2) \mid m_1$ is cn-bisimilar to m_2 thanks to $R\}$. Of course, cn-bisimilarity $\sim_{cn}$ can be seen as $\bigcup\{ \sim_R^{cn} \mid R$ is a causal-net bisimulation $\} = \sim_{\mathscr{R}}^{cn}$, where $\mathscr{R} = \bigcup\{R \mid R$ is a causal-net bisimulation$\}$ is the largest causal-net bisimulation by item 4 of the following proposition.

Proposition 1. *For each BPP net $N = (S, A, T)$, the following hold:*

1. *the identity relation $\mathscr{I} = \{(\rho, N, \rho) \mid \exists m \in \mathscr{M}(S).(N, \rho)$ is a process of $N(m)\}$ is a causal-net bisimulation;*
2. *the inverse relation $R^{-1} = \{(\rho_2, N, \rho_1) \mid (\rho_1, N, \rho_2) \in R\}$ of a causal-net bisimulation R is a causal-net bisimulation;*
3. *the relational composition, up to net isomorphism, $R_1 \circ R_2 = \{(\rho_1, N, \rho_3) \mid \exists \rho_2. (\rho_1, N, \rho_2) \in R_1 \wedge (\bar{\rho}_2, \bar{N}, \bar{\rho}_3) \in R_2 \wedge (N, \rho_2)$ and $(\bar{N}, \bar{\rho}_2)$ are isomorphic processes via $f \wedge \rho_3 = \bar{\rho}_3 \circ f\}$ of two causal-net bisimulations R_1 and R_2 is a causal-net bisimulation;*
4. *the union $\bigcup_{i \in I} R_i$ of causal-net bisimulations R_i is a causal-net bisimulation.*

Proof. Trivial for 1, 2 and 4. For case 3, assume that $(\rho_1, N, \rho_3) \in R_1 \circ R_2$ and that $\rho_1(Max(N))[t_1\rangle m_1$. Since R_1 is a causal-net bisimulation and $(\rho_1, N, \rho_2) \in R_1$, we have that $\exists t_2, m_2, t, N', \rho_1', \rho_2'$ such that

1. $\rho_2(Max(N))[t_2\rangle m_2$,
2. $(N, \rho_1) \xrightarrow{t} (N', \rho_1')$, $\rho_1'(t) = t_1$ and $\rho_1'(Max(N')) = m_1$,
3. $(N, \rho_2) \xrightarrow{t} (N', \rho_2')$, $\rho_2'(t) = t_2$ and $\rho_2'(Max(N')) = m_2$; and finally,
4. $(\rho_1', N', \rho_2') \in R_1$;

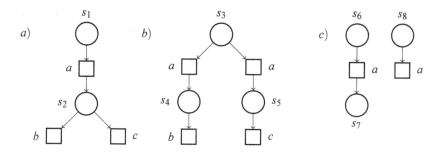

Fig. 3. Some non-cn-bisimilar BPP nets

Since (N, ρ_2) and $(\overline{N}, \overline{\rho}_2)$ are isomorphic via f, it follows that $\rho_2(Max(N)) = \overline{\rho}_2(Max(\overline{N}))$, so that the move $\overline{\rho}_2(Max(\overline{N}))[t_2\rangle m_2$ is derivable, too. As $(\overline{\rho}_2, \overline{N}, \overline{\rho}_3) \in R_2$ and R_2 is a causal-net bisimulation, for $\overline{\rho}_2(Max(\overline{N}))[t_2\rangle m_2$, $\exists t_3, m_3, \overline{t}, \overline{N}', \overline{\rho}_2', \overline{\rho}_3'$ such that

1. $\overline{\rho}_3(Max(\overline{N}))[t_3\rangle m_3$,
2. $(\overline{N}, \overline{\rho}_2) \xrightarrow{\overline{t}} (\overline{N}', \overline{\rho}_2')$, $\overline{\rho}_2'(\overline{t}) = t_2$ and $\overline{\rho}_2'(Max(\overline{N}')) = m_2$,
3. $(\overline{N}, \overline{\rho}_3) \xrightarrow{\overline{t}} (\overline{N}', \overline{\rho}_3')$, $\overline{\rho}_3'(\overline{t}) = t_3$ and $\overline{\rho}_3'(Max(\overline{N}')) = m_3$; and finally,
4. $(\overline{\rho}_2', \overline{N}', \overline{\rho}_3') \in R_2$.

Note that (N', ρ_2') and $(\overline{N}', \overline{\rho}_2')$ are isomorphic via f', where f' extends f in the obvious way (notably, by mapping transition t to $\overline{t}$). As $\rho_3 = \overline{\rho}_3 \circ f$, it follows that (N, ρ_3) and $(\overline{N}, \overline{\rho}_3)$ are isomorphic via f. Therefore, $\rho_3(Max(N)) = \overline{\rho}_3(Max(\overline{N}))$, so that the move $\rho_3(Max(N))[t_3\rangle m_3$ is derivable, too. Since (N, ρ_3) and $(\overline{N}, \overline{\rho}_3)$ are isomorphic via f, transition $(\overline{N}, \overline{\rho}_3) \xrightarrow{\overline{t}} (\overline{N}', \overline{\rho}_3')$, can be matched by $(N, \rho_3) \xrightarrow{t} (N', \rho_3')$, where $\rho_3' = \overline{\rho}_3' \circ f'$, so that (N', ρ_3') and $(\overline{N}', \overline{\rho}_3')$ are isomorphic via f'. Hence, if $(\rho_1, N, \rho_3) \in R_1 \circ R_2$ and $\rho_1(Max(N))[t_1\rangle m_1$, then $\exists t_3, m_3, t, N', \rho_1', \rho_3'$ such that

1. $\rho_3(Max(N))[t_3\rangle m_3$,
2. $(N, \rho_1) \xrightarrow{t} (N', \rho_1')$, $\rho_1'(t) = t_1$ and $\rho_1'(Max(N')) = m_1$,
3. $(N, \rho_3) \xrightarrow{t} (N', \rho_3')$, $\rho_3'(t) = t_3$ and $\rho_3'(Max(N')) = m_3$; and finally,
4. $(\rho_1', N', \rho_3') \in R_1 \circ R_2$.

The symmetric case when (N, ρ_3) moves first is analogous, hence omitted. Therefore, $R_1 \circ R_2$ is a causal-net bisimulation, indeed.   □

Proposition 2. *For each BPP net* $N = (S, A, T)$, *relation* $\sim_{cn} \subseteq \mathcal{M}(S) \times \mathcal{M}(S)$ *is an equivalence relation.*

Proof. Standard, by exploiting Proposition 1.   □

Example 3. Consider the nets in Fig. 1. Clearly the net in a) with initial marking s_1 and the net in b) with initial marking s_3 are not isomorphic; however, we can prove that they have isomorphic unfoldings [5,7,16]; moreover, $s_1 \sim_{cn} s_3$, even if the required causal-net bisimulation contains infinitely many triples.   □

Example 4. Consider the nets in Fig. 2. Of course, the initial markings s_1 and s_2 do not generate isomorphic unfoldings; however, $s_1 \sim_{cn} s_2$.   □

Example 5. Look at Fig. 3. Of course, $s_1 \not\sim_{cn} s_3$, even if they generate the same causal nets. In fact, $s_1 \xrightarrow{a} s_2$ might be matched by s_3 either with $s_3 \xrightarrow{a} s_4$ or with $s_3 \xrightarrow{a} s_5$, so that it is necessary that $s_2 \sim_{cn} s_4$ or $s_2 \sim_{cn} s_5$; but this is impossible, because only s_2 can perform both b and c. Moreover, $s_6 \not\sim_{cn} s_8$ as they generate different causal nets.   □

3.2   (State-Sensitive) Fully-Concurrent Bisimulation

Behavioral equivalences for distributed systems, usually, observe only the events. Hence, causal-net bisimulation, which also observes the structure of the distributed state, may be considered too concrete an equivalence. We disagree with this view, as the structure of the distributed state is not less observable than the events this distributed system can perform. Among the equivalences not observing the state, the most prominent is *fully-concurrent bisimulation* (fc-bisimulation, for short) [3]. As we think that the definition in [3] is not very practical (as it assumes implicitly a universal quantification over the infinite set of all the possible extensions of the current process), we prefer to offer here an equivalent definition, by considering a universal quantification over the finite set of the net transitions only. We define also a novel, slightly stronger version, called *state-sensitive* fc-bisimulation equivalence, that we prove to coincide with cn-bisimilarity.

Definition 15 (Fully-concurrent bisimulation). *Let* $N = (S, A, T)$ *be a BPP net. An fc-bisimulation is a relation* R, *composed of triples of the form* $((N_1, \rho_1), g, (N_2, \rho_2))$, *where, for* $i = 1, 2$, (N_i, ρ_i) *is a process of* $N(m_{0_i})$ *for some* m_{0_i} *and* g *is an event isomorphism between* N_1 *and* N_2, *such that if* $((N_1, \rho_1), g, (N_2, \rho_2)) \in R$ *then*

i) $\forall t_1$ *such that* $\rho_1(Max(N_1))[t_1\rangle m_1$, $\exists t_2, m_2, t_1', t_2', N_1', N_2', \rho_1', \rho_2', g'$ *such that*

    1. $\rho_2(Max(N_2))[t_2\rangle m_2$;

    2. $(N_1, \rho_1) \xrightarrow{t_1'} (N_1', \rho_1')$, $\rho_1'(t_1') = t_1$ *and* $\rho_1'(Max(N_1')) = m_1$,

    3. $(N_2, \rho_2) \xrightarrow{t_2'} (N_2', \rho_2')$, $\rho_2'(t_2') = t_2$ *and* $\rho_2'(Max(N_2')) = m_2$;

    4. $g' = g \cup \{(t_1', t_2')\}$, *and finally,*

    5. $((N_1', \rho_1'), g', (N_2', \rho_2')) \in R$;

ii) symmetrically, if $\rho_2(Max(N_2))$ *moves first.*

Two markings m_1 *and* m_2 *of* N *are fc-bisimilar, denoted by* $m_1 \sim_{fc} m_2$, *if there exists an fc-bisimulation* R *containing a triple* $((N_1^0, \rho_1^0), g_0, (N_2^0, \rho_2^0))$, *where* N_i^0 *contains no transitions,* g_0 *is empty and* $\rho_i^0(Min(N_i^0)) = \rho_i^0(Max(N_i^0)) = m_i$ *for* $i = 1, 2$.   □

Let us denote by $\sim_R^{fc} = \{(m_1, m_2) \mid m_1$ is fc-bisimilar to m_2 thanks to $R\}$. Of course, $\sim_{fc} = \bigcup \{\sim_R^{fc} \mid R$ is a fully-concurrent bisimulation$\} = \sim_{\mathscr{R}}^{fc}$, where relation

$$\mathscr{R} = \bigcup \{R \mid R \text{ is a fully-concurrent bisimulation}\}$$

is the largest fully-concurrent bisimulation. Similarly to what done in Proposition 1, we can prove that (i) the identity relation $\mathscr{I} = \{((\mathsf{N}, \rho), id, (\mathsf{N}, \rho)) \mid \exists m.(\mathsf{N}, \rho)$ is a process of $N(m)$ and id is the identity event isomorphism on $\mathsf{N}\}$ is an fc-bisimulation; that (ii) the inverse relation R^{-1} of an fc-bisimulation R is an fc-bisimulation; that (iii) the composition $R_1 \circ R_2 = \{((\mathsf{N}_1, \rho_1), g, (\mathsf{N}_3, \rho_3)) \mid \exists \mathsf{N}_2, \rho_2, g_1, g_2.((\mathsf{N}_1, \rho_1), g_1, (\mathsf{N}_2, \rho_2)) \in R_1 \wedge ((\overline{\mathsf{N}}_2, \overline{\rho}_2), g_2, (\overline{\mathsf{N}}_3, \overline{\rho}_3)) \in R_2 \wedge (\mathsf{N}_2, \rho_2)$ and $(\overline{\mathsf{N}}_2, \overline{\rho}_2)$ are isomorphic processes via $f_2 \wedge (\mathsf{N}_3, \rho_3)$ and $(\overline{\mathsf{N}}_3, \overline{\rho}_3)$ are isomorphic processes via $f_3 \wedge g = f_3^{-1} \circ (g_2 \circ (f_2 \circ g_1))\}$ of two fc-bisimulations R_1 and R_2 is an fc-bisimulation; and finally, that (iv) the union $\bigcup_{i \in I} R_i$ of a family of fc-bisimulations R_i is an fc-bisimulation.

Proposition 3. *For each BPP net* $N = (S, A, T)$, *relation* $\sim_{fc} \subseteq \mathscr{M}(S) \times \mathscr{M}(S)$ *is an equivalence relation.* $\qquad \square$

Example 6. In Example 5 about Fig. 3 we argued that $s_6 \nsim_{cn} s_8$; however, $s_6 \sim_{fc} s_8$, because, even if they do not generate the same causal net, still they generate isomorphic partial orders of events. On the contrary, $s_1 \nsim_{fc} s_3$ because, even if they generate the same causal nets, the two markings have a different branching structure. Note that the deadlock place s_7 and the empty marking θ are fc-bisimilar. $\qquad \square$

Definition 16 (State-sensitive fully-concurrent bisimulation). *An fc-bisimulation* R *is state-sensitive if for each triple* $((\mathsf{N}_1, \rho_1), g, (\mathsf{N}_2, \rho_2)) \in R$, *the maximal markings have equal size, i.e.,* $|\rho_1(Max(\mathsf{N}_1))| = |\rho_2(Max(\mathsf{N}_2))|$. *Two markings* m_1 *and* m_2 *of* N *are sfc-bisimilar, denoted by* $m_1 \sim_{sfc} m_2$, *if there exists a state-sensitive fc-bisimulation* R *containing a triple* $((\mathsf{N}_1^0, \rho_1^0), g_0, (\mathsf{N}_2^0, \rho_2^0))$, *where* N_i^0 *contains no transitions,* g_0 *is empty and* $\rho_i^0(Min(\mathsf{N}_i^0)) = \rho_i^0(Max(\mathsf{N}_i^0)) = m_i$ *for* $i = 1, 2$. $\qquad \square$

Of course, also the above definition is defined coinductively; as we can prove an analogous of Proposition 1, it follows that $\sim_{sfc}$ is an equivalence relation, too.

Theorem 1 (cn-bisimilarity and sfc-bisimilarity coincide). *For each BPP net* $N = (S, A, T)$, $m_1 \sim_{cn} m_2$ *if and only if* $m_1 \sim_{sfc} m_2$.

Proof $\Rightarrow$). If $m_1 \sim_{cn} m_2$, then there exists a causal-net bisimulation R such that it contains a triple $(\rho_1^0, \mathsf{N}^0, \rho_2^0)$, where N^0 contains no transitions and $\rho_i^0(Min(\mathsf{N}^0)) = \rho_i^0(Max(\mathsf{N}^0)) = m_i$ for $i = 1, 2$. Relation $\mathscr{R} = \{((\mathsf{N}, \rho_1), id, (\mathsf{N}, \rho_2)) \mid (\rho_1, \mathsf{N}, \rho_2) \in R\}$, where id is the identity event isomorphism on N, is a state-sensitive fc-bisimulation. Since $\mathscr{R}$ contains the triple $((\mathsf{N}^0, \rho_1^0), id, (\mathsf{N}^0, \rho_2^0))$, it follows that $m_1 \sim_{sfc} m_2$.

$\Leftarrow$) (Sketch). If $m_1 \sim_{sfc} m_2$, then there exists a state-sensitive fc-bisimulation $\mathscr{R}$ containing a triple $((N_1^0, \rho_1^0), g_0, (N_2^0, \rho_2^0))$, where N_i^0 contains no transitions, g_0 is empty and $\rho_i^0(Min(N_i^0)) = \rho_i^0(Max(N_i^0)) = m_i$ for $i = 1, 2$, with $|m_1| = |m_2|$. Hence, N_1^0 and N_2^0 are isomorphic, where the isomorphism function f_0 is a suitably chosen bijection from $Min(N_1^0)$ to $Min(N_2^0)$.[1]

We build the candidate causal-net bisimulation R inductively, by first including the triple $(\rho_1^0, N_1^0, \rho_2^0 \circ f_0)$; hence, if R is a causal-net bisimulation, then $m_1 \sim_{cn} m_2$.

Since $((N_1^0, \rho_1^0), g_0, (N_2^0, \rho_2^0)) \in \mathscr{R}$ and $\mathscr{R}$ is a state-sensitive fully-concurrent bisimulation, if $\rho_1^0(N_1^0)[t_1\rangle m_1'$, then $\exists t_2, m_2', t_1', t_2', N_1, N_2, \rho_1, \rho_2, g$ such that

1. $\rho_2^0(Max(N_2^0))[t_2\rangle m_2'$;
2. $(N_1^0, \rho_1^0) \xrightarrow{t_1'} (N_1, \rho_1)$, $\rho_1(t_1') = t_1$ and $\rho_1(Max(N_1)) = m_1'$,
3. $(N_2^0, \rho_2^0) \xrightarrow{t_2'} (N_2, \rho_2)$, $\rho_2(t_2') = t_2$ and $\rho_2(Max(N_2)) = m_2'$;
4. $g = g_0 \cup \{(t_1', t_2')\}$, and finally,
5. $((N_1, \rho_1), g, (N_2, \rho_2)) \in \mathscr{R}$, with $|\rho_1(Max(N_1))| = |\rho_2(Max(N_2))|$.

It is necessary that the isomorphism f_0 has been chosen in such a way that $f_0({}^\bullet t_1') = {}^\bullet t_2'$. As $|\rho_1^0(Max(N_1^0))| = |\rho_2^0(Max(N_2^0))|$ and $|\rho_1(Max(N_1))| = |\rho_2(Max(N_2))|$, it is necessary that t_1 and t_2 have the same post-set size; hence, N_1 and N_2 are isomorphic and the bijection f_0 can be extended to bijection f with the pair $\{(t_1', t_2')\}$ and also with a suitably chosen bijection between the post-sets of these two transitions. Hence, we include into R also the triple $(\rho_1, N_1, \rho_2 \circ f)$. Symmetrically, if $\rho_2^0(N_2^0)$ moves first.

By iterating this procedure, we add (possibly unboundedly many) triples to R. It is an easy observation to realize that R is a causal-net bisimulation. $\square$

Remark 4. For general P/T nets, $\sim_{cn}$ is finer than $\sim_{sfc}$. E.g., consider the nets $N = (\{s_1, s_2, s_3, s_4\}, \{a\}, \{(s_1 \oplus s_2, a, s_3 \oplus s_4)\})$ and $N' = (\{s_1', s_2', s_3'\}, \{a\}, \{(s_1', a, s_3')\})$. Of course, $s_1 \oplus s_2 \sim_{sfc} s_1' \oplus s_2'$, but $s_1 \oplus s_2 \nsim_{cn} s_1' \oplus s_2'$. $\square$

3.3   Deadlock-Free BPP Nets and Fully-Concurrent Bisimilarity

We first define a cleaning-up operation on a BPP net N, yielding a net $d(N)$ where all the deadlock places of N are removed. Then, we show that two markings m_1 and m_2 of N are fc-bisimilar if and only if the markings $d(m_1)$ and $d(m_2)$, obtained by removing all the deadlock places in m_1 and m_2 respectively, are state-senstive fc-bisimilar in $d(N)$.

Definition 17 (Deadlock-free BPP net). *For each BPP net $N = (S, A, T)$, we define its associated* deadlock-free *net $d(N)$ as the tuple $(d(S), A, d(T))$ where*

---

[1] The actual choice of f_0 (among the $k!$ different bijections, where $k = |m_1| = |m_2|$) will be driven by the bisimulation game that follows; in the light of Corollary 2, it would map team bisimilar places.

- $d(S) = S \backslash \{s \in S \mid s \nrightarrow\}$, where $s \nrightarrow$ if and only if $\nexists m, a.s \xrightarrow{a} m$;
- $d(T) = \{d(t) \mid t \in T\}$, where $d(t) = (\bullet t, l(t), d(t^\bullet))$ and $d(m) \in \mathcal{M}(d(S))$ is the marking obtained from $m \in \mathcal{M}(S)$ by removing all its deadlock places.

A BPP net $N = (S, A, T)$ is deadlock-free if all of its places are not a deadlock, i.e., $d(S) = S$ and so $d(T) = T$.   □

Formally, given a marking $m \in \mathcal{M}(S)$, we define $d(m)$ as the marking

$$d(m)(s) = \begin{cases} m(s) & \text{if } s \in d(S) \\ 0 & \text{otherwise.} \end{cases}$$

For instance, let us consider the FSM in Fig. 3(c). Then, $d(2 \cdot s_6 \oplus 3 \cdot s_7) = 2 \cdot s_6$, or $d(s_7) = \theta$. Of course, $d(m)$ is a multiset on $d(S)$.

Example 7. Let us consider the BPP net $N = (S, A, T)$, where $S = \{s_1, s_2\}$, $A = \{a\}$ and $T = \{t_1, t_2\}$, where $t_1 = (s_1, a, s_2)$ and $t_2 = (s_1, a, \theta)$. Then, its associated deadlock-free net is $d(N) = (\{s_1\}, A, \{t_2\})$. Note that $d(t_1) = t_2$ and $d(t_2) = t_2$.   □

Proposition 4 (Fc-bisimilarity and sfc-bisimilarity coincide on deadlock-free nets). *For each deadlock-free BPP net $N = (S, A, T)$, $m_1 \sim_{fc} m_2$ if and only if $m_1 \sim_{sfc} m_2$.*

Proof $\Leftarrow$). Of course, a state-sensitive fc- bisimulation is also an fc-bisimulation.
$\Rightarrow$). If there are no deadlock places, an fc-bisimulation must be state sensitive. In fact, if two related markings have a different size, then, since no place is a deadlock and the BPP net transitions have singleton pre-set, they would originate different partial orders of events.   □

Proposition 5. *Given a BPP net $N = (S, A, T)$ and its associated deadlock-free net $d(N) = (d(S), A, d(T))$, two markings m_1 and m_2 of N are fc-bisimilar if and only if $d(m_1)$ and $d(m_2)$ in $d(N)$ are sfc-bisimilar.*

Proof $\Rightarrow$). If $m_1 \sim_{fc} m_2$, then there exists an fc-bisimulation $\mathcal{R}$ on N containing a triple $((\mathsf{N}_1^0, \rho_1^0), g_0, (\mathsf{N}_2^0, \rho_2^0))$, where N_i^0 contains no transitions, g_0 is empty and $\rho_i^0(Min(\mathsf{N}_i^0)) = \rho_i^0(Max(\mathsf{N}_i^0)) = m_i$ for $i = 1, 2$.
    Relation $R = \{((d(\mathsf{N}_1), d(\rho_1)), \hat{g}, (d(\mathsf{N}_2), d(\rho_2))) \mid ((\mathsf{N}_1, \rho_1), g, (\mathsf{N}_2, \rho_2)) \in \mathcal{R}$, such that $d(\rho_i)$ is the restriction of ρ_i on the places of $d(\mathsf{N}_i)$, for $i = 1, 2$, and $\hat{g}$ is such that $g(t_1) = t_2$ iff $\hat{g}(d(t_1)) = d(t_2)\}$ is an fc-bisimulation on $d(N)$. By Proposition 4, R is actually a state-sensitive fully-concurrent bisimulation on $d(N)$. Note that R contains the triple $((d(\mathsf{N}_1^0), d(\rho_1^0)), g_0, (d(\mathsf{N}_2^0), d(\rho_2^0)))$ such that $d(\rho_i^0)(Min(d(\mathsf{N}_i^0))) = d(\rho_i^0)(Max(d(\mathsf{N}_i^0))) = d(m_i)$ for $i = 1, 2$, and so $d(m_1) \sim_{sfc} d(m_2)$.
    $\Leftarrow$). If $d(m_1) \sim_{sfc} d(m_2)$, then there exists an sfc-bisimulation R on $d(N)$ containing a triple $((\overline{\mathsf{N}}_1^0, \overline{\rho}_1^0), g_0, (\overline{\mathsf{N}}_2^0, \overline{\rho}_2^0))$, where $\overline{\mathsf{N}}_i^0$ contains no transitions, g_0 is empty and $\overline{\rho}_i^0(Min(\overline{\mathsf{N}}_i^0)) = \overline{\rho}_i^0(Max(\overline{\mathsf{N}}_i^0)) = d(m_i)$ for $i = 1, 2$.

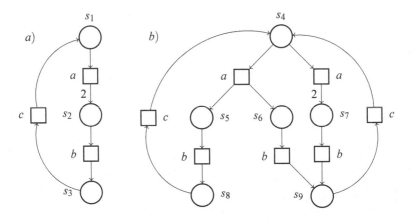

Fig. 4. Two team bisimilar BPP nets

Relation $\mathscr{R} = \{((\mathsf{N}_1, \rho_1), g, (\mathsf{N}_2, \rho_2)) \mid (\mathsf{N}_i, \rho_i)$ is a process of $N(m_{0_i})$ for some m_{0_i}, for $i = 1, 2$, $((d(\mathsf{N}_1), d(\rho_1)), \hat{g}, (d(\mathsf{N}_2), d(\rho_2))) \in R$, such that $d(\rho_i)$ is the restriction of ρ_i on the places of $d(\mathsf{N}_i)$, for $i = 1, 2$, and g is such that $g(t_1) = t_2$ iff $\hat{g}(d(t_1)) = d(t_2)\}$ is an fc-bisimulation on N. Note that $\mathscr{R}$ contains the triple $((\mathsf{N}_1^0, \rho_1^0), g_0, (\mathsf{N}_2^0, \rho_2^0))$ such that, for $i = 1, 2$, $d(\mathsf{N}_i^0) = \overline{\mathsf{N}}_i^0$, $d(\rho_i^0) = \overline{\rho}_i^0$, $\rho_i^0(Min(\mathsf{N}_i^0)) = \rho_i^0(Max(\mathsf{N}_i^0)) = m_i$ and so $m_1 \sim_{fc} m_2$. $\qquad\square$

4   Team Bisimulation Equivalence

In this section, we recall the main definitions and results about team bisimulation equivalence, outlined in [13]. We also include one novel, main result: causal-net bisimilarity coincides with team bisimilarity.

4.1   Additive Closure and Its Properties

Definition 18 (Additive closure). *Given a BPP net $N = (S, A, T)$ and a place relation $R \subseteq S \times S$, we define a marking relation $R^\oplus \subseteq \mathscr{M}(S) \times \mathscr{M}(S)$, called the* additive closure *of R, as the least relation induced by the following axiom and rule.*

$$\frac{}{(\theta, \theta) \in R^\oplus} \qquad \frac{(s_1, s_2) \in R \quad (m_1, m_2) \in R^\oplus}{(s_1 \oplus m_1, s_2 \oplus m_2) \in R^\oplus}$$

$\qquad\square$

Note that, by definition, two markings are related by $R^\oplus$ only if they have the same size; in fact, the axiom states that the empty marking is related to itself, while the rule, assuming by induction that m_1 and m_2 have the same size, ensures that $s_1 \oplus m_1$ and $s_2 \oplus m_2$ have the same size. An alternative way to define that two markings m_1 and m_2 are related by $R^\oplus$ is to state that m_1 can be represented as $s_1 \oplus s_2 \oplus \ldots \oplus s_k$, m_2 can be represented as $s'_1 \oplus s'_2 \oplus \ldots \oplus s'_k$ and $(s_i, s'_i) \in R$ for $i = 1, \ldots, k$.

It is possible to prove that if R is an equivalence relation, then its additive closure $R^\oplus$ is also an equivalence relation. Moreover, if $R_1 \subseteq R_2$, then $R_1^\oplus \subseteq R_2^\oplus$, i.e., the additive closure is monotonic.

4.2 Team Bisimulation on Places

Definition 19 (Team bisimulation). *Let* $N = (S, A, T)$ *be a BPP net. A team bisimulation is a place relation* $R \subseteq S \times S$ *such that if* $(s_1, s_2) \in R$ *then for all* $\ell \in A$

- $\forall m_1$ *such that* $s_1 \xrightarrow{\ell} m_1$, $\exists m_2$ *such that* $s_2 \xrightarrow{\ell} m_2$ *and* $(m_1, m_2) \in R^\oplus$,
- $\forall m_2$ *such that* $s_2 \xrightarrow{\ell} m_2$, $\exists m_1$ *such that* $s_1 \xrightarrow{\ell} m_1$ *and* $(m_1, m_2) \in R^\oplus$.

Two places s *and* s' *are* team *bisimilar (or* team *bisimulation equivalent), denoted* $s \sim s'$, *if there exists a team bisimulation* R *such that* $(s, s') \in R$. □

Example 8. Continuing Example 1 about Fig. 1, it is easy to see that relation $R = \{(s_1, s_3), (s_1, s_4), (s_2, s_5), (s_2, s_6)\}$ is a team bisimulation. In fact, (s_1, s_3) is a team bisimulation pair because, to transition $s_1 \xrightarrow{inc} s_1 \oplus s_2$, s_3 can respond with $s_3 \xrightarrow{inc} s_4 \oplus s_5$, and $(s_1 \oplus s_2, s_4 \oplus s_5) \in R^\oplus$; symmetrically, if s_3 moves first. Also (s_1, s_4) is a team bisimulation pair because, to transition $s_1 \xrightarrow{inc} s_1 \oplus s_2$, s_4 can respond with $s_4 \xrightarrow{inc} s_3 \oplus s_6$, and $(s_1 \oplus s_2, s_3 \oplus s_6) \in R^\oplus$; symmetrically, if s_4 moves first. Also (s_2, s_5) is a team bisimulation pair: to transition $s_2 \xrightarrow{dec} \theta$, s_5 responds with $s_5 \xrightarrow{dec} \theta$, and $(\theta, \theta) \in R^\oplus$. Similarly for the pair (s_2, s_6). Hence, relation R is a team bisimulation, indeed.

The team bisimulation R above is a very simple, finite relation, proving that s_1 and s_3 are team bisimulation equivalent. In Example 2, in order to show that s_1 and s_3 are interleaving bisimilar, we had to introduce a complex relation, with infinitely many pairs. In Example 3 we argued that $s_1 \sim_{cn} s_3$, even if we did not provide any causal-net bisimulation (which would be composed of infinitely many triples). □

Example 9. Consider the nets in Fig. 2. Of course, $s_1 \sim s_2$ because the finite relation $R = \{(s_1, s_2), (s_1, s_3), (s_1, s_4)\}$ is a team bisimulation. Actually, all the places are pairwise team bisimilar. In Example 4 we argued that $s_1 \sim_{cn} s_2$, but the justifying causal-net bisimulation would contain infinitely many triples. □

Example 10. Consider the nets in Fig. 4. It is easy to see that $R = \{(s_1, s_4), (s_2, s_5), (s_2, s_6), (s_2, s_7), (s_3, s_8), (s_3, s_9)\}$ is a team bisimulation. This example shows that team bisimulation is compatible with duplication of behavior and fusion of places. □

It is not difficult to prove [13] that (i) the identity relation $\mathscr{I}_S = \{(s,s) \mid s \in S\}$ is a team bisimulation; that (ii) the inverse relation $R^{-1} = \{(s',s) \mid (s,s') \in R\}$ of a team bisimulation R is a team bisimulation; that (iii) the relational composition $R_1 \circ R_2 = \{(s,s'') \mid \exists s'.(s,s') \in R_1 \wedge (s',s'') \in R_2\}$ of two team bisimulations R_1 and R_2 is a team bisimulation; and, finally, that (iv) the union $\bigcup_{i \in I} R_i$ of team bisimulations R_i is a team bisimulation. Remember that $s \sim s'$ if there exists a team bisimulation containing the pair (s,s'). This means that $\sim$ is the union of all team bisimulations, i.e.,

$$\sim = \bigcup\{R \subseteq S \times S \mid R \text{ is a team bisimulation}\}.$$

Hence $\sim$ is also a team bisimulation, the largest such relation. Moreover, by the property listed above, relation $\sim \subseteq S \times S$ is an equivalence relation.

Remark 5 (**Complexity 1**). It is well-known that the optimal algorithm for computing bisimilarity over a finite-state LTS with n states and m transitions has $O(m \cdot \log n)$ time complexity [19]; this very same partition refinement algorithm can be easily adapted also for team bisimilarity over BPP nets; it is enough to start from an initial partition composed of two blocks: S and $\{\theta\}$, and to consider the little additional cost due to the fact that the reached markings are to be related by the additive closure of the current partition; this extra cost is related to the size of the post-set of the net transitions; if p is the size of the largest post-set of the net transitions, then the time complexity is $O(m \cdot p^2 \cdot \log (n+1))$, where m is the number of the net transitions and n is the number of the net places. $\square$

4.3   Team Bisimilarity over Markings

Starting from team bisimulation equivalence $\sim$, which has been computed over the places of an *unmarked* BPP net N, we can lift it over *the markings* of N in a distributed way: m_1 is team bisimulation equivalent to m_2 if these two markings are related by the additive closure of $\sim$, i.e., if $(m_1, m_2) \in \sim^{\oplus}$, usually denoted by $m_1 \sim^{\oplus} m_2$.

Of course, if $m_1 \sim^{\oplus} m_2$, then $|m_1| = |m_2|$. Moreover, for any BPP net $N = (S, A, T)$, relation $\sim^{\oplus} \subseteq \mathscr{M}(S) \times \mathscr{M}(S)$ is an equivalence relation.

Remark 6 (**Complexity 2**). Once $\sim$ has been computed once and for all for the given net (in $O(m \cdot p^2 \cdot \log (n+1))$ time), the algorithm in [13] checks whether two markings m_1 and m_2 are team bisimulation equivalent in $O(k^2)$ time, where k is the size of the markings. In fact, if $\sim$ is implemented as an adjacency matrix, then the complexity of checking if two markings m_1 and m_2 (represented as an array of places with multiplicities) are related by $\sim^{\oplus}$ is $O(k^2)$, because the problem is essentially that of finding for each element s_1 of m_1 a matching, $\sim$-related element s_2 of m_2. Moreover, if we want to check whether other two markings of the same net are team bisimilar, we can reuse the already computed $\sim$ relation, so that the time complexity is again quadratic. $\square$

Example 11. Continuing Example 8 about the semi-counters, the marking $s_1 \oplus 2 \cdot s_2$ is team bisimilar to the following markings of the net in (b): $s_3 \oplus 2 \cdot s_5$, or $s_3 \oplus s_5 \oplus s_6$, or $s_3 \oplus 2 \cdot s_6$, or $s_4 \oplus 2 \cdot s_5$, or $s_4 \oplus s_5 \oplus s_6$, or $s_4 \oplus 2 \cdot s_6$.   □

Of course, two markings m_1 and m_2 are *not* team bisimilar if there is no bijective, team-bisimilar-preserving mapping between them; this is the case when m_1 and m_2 have different size, or if the algorithm in [13] ends with b holding *false*, i.e., by singling out a place s'_i in (the residual of) m_1 which has no matching team bisimilar place in (the residual of) m_2.

The following theorem provides a characterization of team bisimulation equivalence $\sim^\oplus$ as a suitable bisimulation-like relation over markings. It is interesting to observe that this characterization gives a dynamic interpretation of team bisimulation equivalence, while Definition 18 gives a structural definition of team bisimulation equivalence $\sim^\oplus$ as the additive closure of $\sim$. The proof is outlined in [13].

Theorem 2. *Let $N = (S, A, T)$ be a BPP net. Two markings m_1 and m_2 are team bisimulation equivalent, $m_1 \sim^\oplus m_2$, if and only if $|m_1| = |m_2|$ and*

- $\forall t_1$ *such that* $m_1[t_1\rangle m'_1$, $\exists t_2$ *such that* ${}^\bullet t_1 \sim {}^\bullet t_2$, $l(t_1) = l(t_2)$, $t_1^\bullet \sim^\oplus t_2^\bullet$, $m_2[t_2\rangle m'_2$ *and* $m'_1 \sim^\oplus m'_2$, *and symmetrically,*
- $\forall t_2$ *such that* $m_2[t_2\rangle m'_2$, $\exists t_1$ *such that* ${}^\bullet t_1 \sim {}^\bullet t_2$, $l(t_1) = l(t_2)$, $t_1^\bullet \sim^\oplus t_2^\bullet$, $m_1[t_1\rangle m'_1$ *and* $m'_1 \sim^\oplus m'_2$.   □

By the theorem above, it is clear that $\sim^\oplus$ is an interleaving bisimulation.

Corollary 1 (Team bisimilarity is finer than interleaving bisimilarity). *Let $N = (S, A, T)$ be a BPP net. If $m_1 \sim^\oplus m_2$, then $m_1 \sim_{int} m_2$.*   □

4.4   Team Bisimilarity and Causal-Net Bisimilarity Coincide

Theorem 3 (Team bisimilarity implies cn-bisimilarity). *Let $N = (S, A, T)$ be a BPP net. If $m_1 \sim^\oplus m_2$, then $m_1 \sim_{cn} m_2$.*

Proof. Let $R = \{(\rho_1, \mathsf{N}, \rho_2) \mid (\mathsf{N}, \rho_1)$ is a process of $N(m_1)$ and (N, ρ_2) is a process of $N(m_2)$ such that $\rho_1(s) \sim \rho_2(s)$, for all $s \in Max(\mathsf{N})\}$. We want to prove that R is a causal-net bisimulation. First, observe that, any triple of the form $(\rho_1^0, \mathsf{N}^0, \rho_2^0)$, where N^0 is a BPP causal net with no transitions, $\rho_i^0(Max(\mathsf{N}^0)) = m_i$ and $\rho_1^0(s) \sim \rho_2^0(s)$, for all $s \in Max(\mathsf{N}^0)$, belongs to R and its existence is justified by the hypothesis $m_1 \sim^\oplus m_2$. Note also that if the relation R is a causal-net bisimulation, then this triple ensures that $m_1 \sim_{cn} m_2$. Now assume $(\rho_1, \mathsf{N}, \rho_2) \in R$. In order to be a causal-net bisimulation triple, it is necessary that

i) $\forall t_1$ such that $\rho_1(Max(\mathsf{N}))[t_1\rangle m'_1$, $\exists t_2, m'_2, t, \mathsf{N}', \rho'_1, \rho'_2$ such that
   1. $\rho_2(Max(\mathsf{N}))[t_2\rangle m'_2$,
   2. $(\mathsf{N}, \rho_1) \xrightarrow{t} (\mathsf{N}', \rho'_1)$, $\rho'_1(t) = t_1$ and $\rho'_1(Max(\mathsf{N}')) = m'_1$,
   3. $(\mathsf{N}, \rho_2) \xrightarrow{t} (\mathsf{N}', \rho'_2)$, $\rho'_2(t) = t_2$ and $\rho'_2(Max(\mathsf{N}')) = m'_2$; and finally,

4. $(\rho_1', \mathsf{N}', \rho_2') \in R$;

$ii)$ symmetrically, if $\rho_2(Max(\mathsf{N}))$ moves first.

Let t_1 be any transition such that $\rho_1(Max(\mathsf{N}))[t_1\rangle m_1'$ and let $s_1 = {}^\bullet t_1$. Since by hypothesis we have that $\rho_1(s) \sim \rho_2(s)$, for all $s \in Max(\mathsf{N})$, if $s_1 = \rho_1(s')$, then there exists $s_2 = \rho_2(s')$ such that $s_1 \sim s_2$. Hence, there exists t_2 such that $s_1 = {}^\bullet t_1 \sim {}^\bullet t_2 = s_2$, $l(t_1) = l(t_2)$, $t_1^\bullet \sim^\oplus t_2^\bullet$, so that, by Theorem 2, $\rho_2(Max(\mathsf{N}))[t_2\rangle m_2'$ and $m_1' \sim^\oplus m_2'$. Therefore, it is really possible to extend the causal net N to the causal net N' through a suitable transition t such that ${}^\bullet t = s'$, as required above, and to extend ρ_1 and ρ_2 to ρ_1' and ρ_2', respectively, in such a way that $\rho_1'(s) \sim \rho_2'(s)$, for all $s \in t^\bullet$ because $t_1^\bullet \sim^\oplus t_2^\bullet$.

Summing up, for the move t_1, we have that $(\rho_1', \mathsf{N}', \rho_2') \in R$ because $\rho_1'(s) \sim \rho_2'(s)$, for all $s \in Max(\mathsf{N}')$, as required. Symmetrically, if $\rho_2(Max(\mathsf{N}))$ moves first. □

Theorem 4 (Cn-bisimilarity implies team bisimilarity). *Let* $N = (S, A, T)$ *be a BPP net. If* $m_1 \sim_{cn} m_2$ *then* $m_1 \sim^\oplus m_2$.

Proof. If $m_1 \sim_{cn} m_2$, then there exists a causal-net bisimulation R containing a triple $(\rho_1^0, \mathsf{N}^0, \rho_2^0)$, where N^0 is a BPP causal net which has no transitions and $\rho_i^0(Max(\mathsf{N}^0)) = m_i$ for $i = 1, 2$.

Let us consider $\mathscr{R} = \{(\rho_1(s), \rho_2(s)) \mid (\rho_1, \mathsf{N}, \rho_2) \in R \wedge s \in Max(\mathsf{N})\}$. If we prove that $\mathscr{R}$ is a team bisimulation, then, since $(\rho_1^0(s), \rho_2^0(s)) \in \mathscr{R}$ for each $s \in Max(\mathsf{N}^0)$, it follows that $(m_1, m_2) \in \mathscr{R}^\oplus$. As $\mathscr{R} \subseteq \sim$, we also have that $m_1 \sim^\oplus m_2$.

Let us consider a pair $(s_1, s_2) \in \mathscr{R}$. Hence, there exist a triple $(\rho_1, \mathsf{N}, \rho_2) \in R$ and a place $s \in Max(\mathsf{N})$ such that $s_1 = \rho_1(s)$ and $s_2 = \rho_2(s)$. If s_1 moves, e.g., $t_1 = s_1 \xrightarrow{\ell} m_1'$, then $\rho_1(Max(\mathsf{N}))[t_1\rangle \overline{m}_1$, where $\overline{m}_1 = \rho_1(Max(\mathsf{N})) \ominus s_1 \oplus m_1'$. Since R is a causal-net bisimulation, $\exists t_2, \overline{m}_2, t, \mathsf{N}', \rho_1', \rho_2'$ such that

1. $\rho_2(Max(\mathsf{N}))[t_2\rangle \overline{m}_2$,
2. $(\mathsf{N}, \rho_1) \xrightarrow{t} (\mathsf{N}', \rho_1')$, $\rho_1'(t) = t_1$ and $\rho_1'(Max(\mathsf{N}')) = \overline{m}_1$,
3. $(\mathsf{N}, \rho_2) \xrightarrow{t} (\mathsf{N}', \rho_2')$, $\rho_2'(t) = t_2$ and $\rho_2'(Max(\mathsf{N}')) = \overline{m}_2$; and finally,
4. $(\rho_1', \mathsf{N}', \rho_2') \in R$;

Note that t is such that ${}^\bullet t = s$, and so ${}^\bullet t_2 = s_2$. This means that $\overline{m}_2 = \rho_2(Max(\mathsf{N})) \ominus s_2 \oplus m_2'$, where $m_2' = t_2^\bullet$; in other words, $t_2 = s_2 \xrightarrow{\ell} m_2'$. Note also that ρ_1' extends ρ_1 by mapping t to t_1 and, similarly, ρ_2' extends ρ_2 by mapping t to t_2; in this way, $\rho_1'(t^\bullet) = t_1^\bullet$ and $\rho_2'(t^\bullet) = t_2^\bullet$. Since $(\rho_1', \mathsf{N}', \rho_2') \in R$, it follows that the set $\{(\rho_1'(s'), \rho_2'(s')) \mid s' \in t^\bullet)\}$ is a subset of $\mathscr{R}$, so that $(m_1', m_2') \in \mathscr{R}^\oplus$.

Summing up, for $(s_1, s_2) \in \mathscr{R}$, if $s_1 \xrightarrow{\ell} m_1'$, then $s_2 \xrightarrow{\ell} m_2'$ such that $(m_1', m_2') \in \mathscr{R}^\oplus$; symmetrically, if s_2 moves first. Therefore, $\mathscr{R}$ is a team bisimulation. □

Corollary 2 (Team bisimilarity and cn-bisimilarity coincide). *Let* $N = (S, A, T)$ *be a BPP net. Then,* $m_1 \sim_{cn} m_2$ *if and only if* $m_1 \sim^\oplus m_2$.

Proof. By Theorems 3 and 4, we get the thesis.   □

Corollary 3 (Team bisimilarity and sfc-bisimilarity coincide). *Let $N = (S, A, T)$ be a BPP net. Then, $m_1 \sim_{sfc} m_2$ if and only if $m_1 \sim^{\oplus} m_2$.*

Proof. By Corollary 2 and Theorem 1, we get the thesis.   □

Therefore, our characterization of cn-bisimilarity and sfc-bisimilarity, which are, in our opinion, the intuitively correct (strong) causal semantics for BPP nets, is quite appealing because it is based on the very simple technical definition of team bisimulation on the places of the unmarked net, and, moreover, offers a very efficient algorithm to check if two markings are cn-bisimilar (see Remarks 5 and 6).

5   H-Team Bisimulation

We provide the definition of *h-bisimulation on places* for unmarked BPP nets, adapting the definition of team bisimulation on places (cf. Definition 19). In this definition, the empty marking θ is considered as an additional place, so that the relation is defined not on S, rather on $S \cup \{\theta\}$; therefore, the symbols p_1 and p_2 that occur in the definition below can only denote either the empty marking θ or a single place.

Definition 20 (H-team bisimulation). *Let $N = (S, A, T)$ be a BPP net. An h-team bisimulation is a place relation $R \subseteq (S \cup \{\theta\}) \times (S \cup \{\theta\})$ such that if $(p_1, p_2) \in R$ then for all $\ell \in A$*

- *$\forall m_1$ such that $p_1 \xrightarrow{\ell} m_1$, $\exists m_2$ such that $p_2 \xrightarrow{\ell} m_2$ and $(m_1, m_2) \in R^{\oplus}$,*
- *$\forall m_2$ such that $p_2 \xrightarrow{\ell} m_2$, $\exists m_1$ such that $p_1 \xrightarrow{\ell} m_1$ and $(m_1, m_2) \in R^{\oplus}$.*

p_1 and p_2 are h-team bisimilar (or h-team bisimulation equivalent), denoted $p_1 \sim_h p_2$, if there exists an h-team bisimulation R such that $(p_1, p_2) \in R$.   □

Since a team bisimulation is also an h-team bisimulation, we have that team bisimilarity $\sim$ implies h-team bisimilarity $\sim_h$. This implication is strict as illustrated in the following example.

Example 12. Consider the nets in Fig. 3. It is not difficult to realize that s_6 and s_8 are h-team bisimilar because $R = \{(s_6, s_8), (s_7, \theta)\}$ is an h-team bisimulation. In fact, s_6 can reach s_7 by performing a, and s_8 can reply by reaching the empty marking θ, and $(s_7, \theta) \in R$. In Example 6 we argued that $s_6 \sim_{fc} s_8$ and in fact we will prove that h-team bisimilarity coincide with fc-bisimilarity.   □

Remark 7 **(Additive closure properties).** Note that the additive closure of an h-team bisimulation R does not ensure that if two markings are related by $R^{\oplus}$, then they must have the same size. For instance, considering the above relation $R = \{(s_6, s_8), (s_7, \theta)\}$, we have that, e.g., $s_6 \oplus s_7 R^{\oplus} s_8$, because θ is the identity for multiset union. However, the other properties of the additive closure described in Sect. 4.1 hold also for these more general place relations.   □

It is not difficult to prove that, for any BPP net $N = (S, A, T)$, the following hold:

1. The identity relation $\mathscr{I}_S = \{(s, s) \mid s \in S\}$ is an h-team bisimulation;
2. the inverse relation $R^{-1} = \{(p', p) \mid (p, p') \in R\}$ of an h-team bisimulation R is an h-team bisimulation;
3. the relational composition $R_1 \circ R_2 = \{(p, p'') \mid \exists p'.(p, p') \in R_1 \wedge (p', p'') \in R_2\}$ of two h-team bisimulations R_1 and R_2 is an h-team bisimulation;
4. the union $\bigcup_{i \in I} R_i$ of h-team bisimulations R_i is an h-team bisimulation.

Relation $\sim_h$ is the union of all h-team bisimulations, i.e.,

$$\sim_h = \bigcup \{R \subseteq (S \cup \{\theta\}) \times (S \cup \{\theta\}) \mid R \text{ is a h-team bisimulation}\}.$$

Hence, $\sim_h$ is also an h-team bisimulation, the largest such relation. Moreover, by the observations above, relation $\sim_h \subseteq (S \cup \{\theta\}) \times (S \cup \{\theta\})$ is an equivalence relation.

Starting from h-team bisimulation equivalence $\sim_h$, which has been computed over the places (and the empty marking) of an *unmarked* BPP net N, we can lift it over *the markings* of N in a distributed way: m_1 is h-team bisimulation equivalent to m_2 if these two markings are related by the additive closure of $\sim_h$, i.e., if $(m_1, m_2) \in \sim_h^{\oplus}$, usually denoted by $m_1 \sim_h^{\oplus} m_2$. Since $\sim_h$ is an equivalence relation, then also relation $\sim_h^{\oplus} \subseteq \mathscr{M}(S) \times \mathscr{M}(S)$ is an equivalence relation.

Remark 8 (**Complexity 3**). Computing $\sim_h$ is not more difficult than computing $\sim$. The partition refinement algorithm in [19] can be adapted also in this case. It is enough to consider the empty marking θ as an additional, special place which is h-team bisimilar to each deadlock place. Hence, the initial partition considers two sets: one composed of all the deadlock places and θ, the other one with all the non-deadlock places. Therefore, the time complexity is also in this case $O(m \cdot p^2 \cdot \log(n + 1))$, where m is the number of the net transitions, n is the number of the net places and p the size of the largest post-set of the net transitions.

Once $\sim_h$ has been computed once and for all for the given net, the complexity of checking whether two markings m_1 and m_2 are h-team bisimulation equivalent, according to the algorithm in [13], is $O(k^2)$, where k is the size of the largest marking, since the problem is essentially that of finding for each element s_1 (not $\sim_h$-related to θ) of m_1 a matching, $\sim_h$-related element s_2 of m_2 (and then checking that all the remaining elements of m_1 and m_2 are $\sim_h$-related to θ). $\square$

5.1 H-Team Bisimilarity and Fully-Concurrent Bisimilarity Coincide

In this section, we first show that h-team bisimilarity over a BPP net N coincides with team-bisimilarity over its associated deadlock-free net $d(N)$. A consequence of this result is that h-team bisimilarity coincides with fc-bisimilarity on BPP nets.

Proposition 6. *Given a BPP net $N = (S, A, T)$ and its associated deadlock-free net $d(N) = (d(S), A, d(T))$, two markings m_1 and m_2 of N are h-team bisimilar if and only if $d(m_1)$ and $d(m_2)$ in $d(N)$ are team bisimilar.*

Proof $\Rightarrow$). If $m_1 \sim_h^\oplus m_2$, then there exists an h-team bisimulation R_1 on N such that $m_1 R_1^\oplus m_2$. If we take relation $R_2 = \{(s_1, s_2) \mid s_1, s_2 \in d(S) \wedge (s_1, s_2) \in R_1\}$, then it is easy to see that R_2 is a team bisimulation on $d(N)$, so that $d(m_1) R_2^\oplus d(m_2)$, hence $d(m_1) \sim^\oplus d(m_2)$.

$\Leftarrow$). If $d(m_1) \sim^\oplus d(m_2)$, then there exists a team bisimulation R_2 on $d(N)$ such that $d(m_1) R_2^\oplus d(m_2)$. Now, take relation $R_1 = R_2 \cup (S' \cup \{\theta\}) \times (S' \cup \{\theta\})$, where the set S' is $\{s \in S \mid s \nrightarrow\}$. It is easy to observe that R_1 is an h-team bisimulation on N, so that $m_1 R_1^\oplus m_2$, hence $m_1 \sim_h^\oplus m_2$. □

Theorem 5 (Fully concurrent bisimilarity and h-team bisimilarity coincide). *Given a BPP net $N = (S, A, T)$, $m_1 \sim_{fc} m_2$ if and only if $m_1 \sim_h^\oplus m_2$.*

Proof. By Proposition 5, $m_1 \sim_{fc} m_2$ in N if and only if $d(m_1) \sim_{sfc} d(m_2)$ in the associated deadlock-free net $d(N)$. By Corollary 3, $d(m_1) \sim_{sfc} d(m_2)$ iff $d(m_1) \sim^\oplus d(m_2)$ in $d(N)$. By Proposition 6, $d(m_1) \sim^\oplus d(m_2)$ in $d(N)$ if and only if $m_1 \sim_h^\oplus m_2$ in N. □

6 Conclusion

Team bisimilarity is the most natural, intuitive and simple extension of LTS bisimilarity to BPP nets; it also has a very low complexity, actually lower than any other equivalence for BPP nets. Moreover, it coincides with causal-net bisimilarity and state-sensitive fully-concurrent bisimilarity, hence it corresponds to the intuitively correct bisimulation-based causal semantics for BPP nets. Moreover, it coincides also with *structure-preserving bisimilarity*, because our causal-net bisimilarity is rather similar to its process-oriented characterization in [9]. From a technical point of view, team bisimulation seems a sort of *egg of Columbus*: a simple (actually, a bit surprising in its simplicity) solution for a presumedly hard problem. This paper is not only an addition to [13], where team bisimilarity was originally introduced, but also an extension to a team-style characterization of fully-concurrent bisimilarity, namely h-team bisimilarity.

We think that state-sensitive fc-bisimilarity (hence, also team bisimilarity) is more accurate than fc-bisimilarity (hence, h-team bisimilarity) because it is *resource-aware*, i.e., it is sensitive to the number of resources that are present in the net. This more concrete equivalence is justified in, e.g., the area of information flow security [14].

Our complexity results for fc-bisimilarity in terms of the equivalent h-team bisimilarity (cf. Remark 8), seem comparable with those in [6], where, by using an event structure [20] semantics, Fröschle et al. show that history-preserving bisimilarity (hpb, for short) is decidable for the BPP process algebra with guarded summation in $O(n^3 \cdot \log n)$ time, where n is the *size of the involved BPP terms*.

However, this value n is strictly larger than the size of the corresponding BPP net. In fact, in [6] the size of a BPP term p is defined as "the total number of occurrences of symbols (including parentheses)", where p is defined by means of a concrete syntax. E.g., $p = (a.0) \mid (a.0)$ has size 11, while the net semantics for p generates one place and one transition (and 2 tokens). For a comparison of team bisimilarity with other equivalences for BPP, we refer you to [13].

In [13] we presented a modal logic characterization of $\sim^{\oplus}$ and also a finite axiomatization for the process algebra BPP (with guarded sum and guarded recursion). As a future work, we plan to extend these results to $\sim_h^{\oplus}$, hence equipping fc-bisimilarity (and hpb) with a logic characterization and an axiomatic one for the process algebra BPP.

Acknowledgments. The anonymous referees are thanked for their comments.

References

1. Autant, C., Belmesk, Z., Schnoebelen, P.: Strong bisimilarity on nets revisited. In: Aarts, E.H.L., van Leeuwen, J., Rem, M. (eds.) PARLE 1991. LNCS, vol. 506, pp. 295–312. Springer, Heidelberg (1991). https://doi.org/10.1007/3-540-54152-7_71

2. Best, E., Devillers, R.: Sequential and concurrent behavior in Petri net theory. Theor. Comput. Sci. **55**(1), 87–136 (1987)

3. Best, E., Devillers, R., Kiehn, A., Pomello, L.: Concurrent bisimulations in Petri nets. Acta Informatica **28**(3), 231–264 (1991). https://doi.org/10.1007/BF01178506

4. Christensen, S.: Decidability and decomposition in process algebra. Ph.D. Thesis, University of Edinburgh (1993)

5. Engelfriet, J.: Branching processes of Petri nets. Acta Informatica **28**(6), 575–591 (1991). https://doi.org/10.1007/BF01463946

6. Fröschle, S., Jančar, P., Lasota, S., Sawa, Z.: Non-interleaving bisimulation equivalences on basic parallel processes. Inf. Comput. **208**(1), 42–62 (2010)

7. van Glabbeek, R., Vaandrager, F.: Petri net models for algebraic theories of concurrency. In: de Bakker, J.W., Nijman, A.J., Treleaven, P.C. (eds.) PARLE 1987. LNCS, vol. 259, pp. 224–242. Springer, Heidelberg (1987). https://doi.org/10.1007/3-540-17945-3_13

8. van Glabbeek, R., Goltz, U.: Equivalence notions for concurrent systems and refinement of actions. In: Kreczmar, A., Mirkowska, G. (eds.) MFCS 1989. LNCS, vol. 379, pp. 237–248. Springer, Heidelberg (1989). https://doi.org/10.1007/3-540-51486-4_71

9. Glabbeek, R.J.: Structure preserving bisimilarity, supporting an operational petri net semantics of CCSP. In: Meyer, R., Platzer, A., Wehrheim, H. (eds.) Correct System Design. LNCS, vol. 9360, pp. 99–130. Springer, Cham (2015). https://doi.org/10.1007/978-3-319-23506-6_9

10. Goltz, U., Reisig, W.: The non-sequential behaviour of Petri nets. Inf. Control **57**(2–3), 125–147 (1983)

11. Gorrieri, R., Versari, C.: Introduction to Concurrency Theory: Transition Systems and CCS. EATCS Texts in Theoretical Computer Science. Springer, Cham (2015). https://doi.org/10.1007/978-3-319-21491-7

12. Gorrieri, R.: Process Algebras for Petri Nets: The Alphabetization of Distributed Systems. EATCS Monographs in Theoretical Computer Science. Springer, Cham (2017). https://doi.org/10.1007/978-3-319-55559-1
13. Gorrieri, R.: Team bisimilarity, and its associated modal logic, for BPP nets. Acta Informatica. to appear. https://doi.org/10.1007/s00236-020-00377-4
14. Gorrieri, R.: Interleaving vs true concurrency: some instructive security examples. In: Janicki, R., et al. (eds.) Petri Nets 2020. LNCS, vol. 12152, pp. xx–yy. Springer, Cham (2020)
15. Milner, R.: Communication and Concurrency. Prentice-Hall, Upper Saddle River (1989)
16. Nielsen, M., Plotkin, G.D., Winskel, G.: Petri nets, event structures and domains, part I. Theor. Comput. Sci. **13**(1), 85–108 (1981)
17. Olderog, E.R.: Nets, Terms and Formulas. Cambridge Tracts in Theoretical Computer Science, vol. 23. Cambridge University Press, Cambridge (1991)
18. Peterson, J.L.: Petri Net Theory and the Modeling of Systems. Prentice-Hall, Upper Saddle River (1981)
19. Paige, R., Tarjan, R.E.: Three partition refinement algorithms. SIAM J. Comput. **16**(6), 973–989 (1987)
20. Winskel, G.: Event structures. In: Brauer, W., Reisig, W., Rozenberg, G. (eds.) ACPN 1986. LNCS, vol. 255, pp. 325–392. Springer, Heidelberg (1987). https://doi.org/10.1007/3-540-17906-2_31

Circular Traffic Queues and Petri's Cycloids

Rüdiger Valk$^{(\boxtimes)}$

Department of Informatics, University of Hamburg, Hamburg, Germany
valk@informatik.uni-hamburg.de

Abstract. Two sorts of circular traffic systems are defined and their minimal length of recurrent transition sequences is computed. The result is used for finding cycloids that have an isomorphic reachability graph. Cycloids are particular Petri nets for modelling processes of actions or events, belonging to the fundaments of Petri's general systems theory. They have very different interpretations, ranging from Einstein's relativity theory to elementary information processing gates. The cycloid representation of circular traffic systems allows to identify basic synchronisation mechanisms and leads to a structure theory of such systems.

Keywords: Cycloids · Analysis and synthesis of Petri nets · Structure of Petri nets · Circular traffic queues

1 Introduction

Queues of traffic items like cars, trains, aircrafts, production goods, computer tasks or electronic particles are widespread instances of dynamical systems. Therefore they have been frequently used to illustrate the modelling by Petri nets. Petri himself usually introduced the concept of coordination and synchronization by the regimen or organization rule for people carrying buckets to extinguish a fire [4] or by cars driving in line on a road with varying distances. As Petri always followed the principle of discrete modelling, resulting in finite structures, he defined cycloids by folding such structures with respect to space and time. Cyloids define a subclass of partial cyclic orders and hence generalize the well known token ring structure (a total cyclic order) that is at the core of many solutions to the distributed mutual exclusion problem. This also includes virtual token rings that have been employed in group communication middleware (e.g. the Spread system). Hence, we conjecture that cycloids could more generally play a role as coordination models in new middleware architectures. Another potential application of cycloids might be their use as parallel data flow models for iterative computations. Such data flow models can be directly implemented in hardware (e.g. as systolic arrays) or in a combination of hardware and software, for example on many-core architectures like graphics processing units (GPUs) that are attracting a lot of interest for stream processing and machine learning

© Springer Nature Switzerland AG 2020
R. Janicki et al. (Eds.): PETRI NETS 2020, LNCS 12152, pp. 176–195, 2020.
https://doi.org/10.1007/978-3-030-51831-8_9

applications. While Petri introduced cycloids in [4], a more formal definition has been given in [2,7] and [9]. As a consequence of this formalization some properties could be derived, leading for instance to a synthesis procedure based on observable parameters of the cycloid net, among which is the length of a minimal cycle. However, the concrete structure of a circular traffic queue given by a definite number of c traffic items and g gap instances is not known. In this article two different such systems are defined as transition systems $tq\text{-}1(c,g)$ and $tq\text{-}2(c,g)$. For finding the behaviour equivalent cycloids the synthesis theorem from [7] and [9] is used. It requires the number of cycloid transitions, which is computed by the length of recurrent transition sequences of the circular traffic queues in Sect. 2. Also the minimal length cyc of cycles is needed in this theorem. To find it out a particular sequence of c transitions, called *release message chain*, is introduced, which is the most important synchronisation mechanism of the model. The release message chain is extended to a *release message cycle*. The length rm of the latter varies for different models and contributes to the determination of cyc. After having found these cycloids their behaviour equivalence is established by proving the isomorphism between their transition systems. Here also the release message cycle plays an important role. As an intermediate step *regular cycloid systems* are introduced. The obtained cycloids are related to unfoldings of coloured net models of circular traffic queues.

We recall some standard notations for set theoretical relations. If $R \subseteq A \times B$ is a relation and $U \subseteq A$ then $R[U] := \{b \,|\, \exists u \in U : (u,b) \in R\}$ is the *image* of U and $R[a]$ stands for $R[\{a\}]$. R^{-1} is the *inverse relation* and R^+ is the *transitive closure* of R if $A = B$. Also, if $R \subseteq A \times A$ is an equivalence relation then $[a]_R$ is the *equivalence class* of the quotient A/R containing a. Furthermore $\mathbb{N}_+$, $\mathbb{Z}$ and $\mathbb{R}$ denote the sets of positive integer, integer and real numbers, respectively. For integers: $a|b$ if a is a factor of b. The *modulo*-function is used in the form $a \bmod b = a - b \cdot \lfloor \frac{a}{b} \rfloor$, which also holds for negative integers $a \in \mathbb{Z}$. In particular, $-a \bmod b = b - a$ for $0 < a \le b$. Due to naming conventions (e.g. the function ind in Definition 1), the range of the *modulo*-function is supposed to be $\{1, \cdots, b\}$ instead of $\{0, \cdots, b-1\}$ in some cases. In these cases however, b can be seen to be equivalent to 0.

The author is grateful to U. Fenske, O. Kummer, M.O. Stehr and the anonymous referees for numerous corrections and improvements.

2   Circular Traffic Queues

We define a model of circular traffic queues with two sorts of traffic items. Traffic items $a \in C$, going from left to right, exchange their position with traffic items $u \in G$ moving in the opposite direction. See the example after the next definition. By a transition $\langle\!\langle t_i, a_j \rangle\!\rangle$ traffic item a_j is moved from the queue position i.

Definition 1. *A circular traffic queue $tq(c,g)$ is defined by two positive integers c and g. Implicitly with these integers we consider two finite and disjoint sets of traffic items $C = \{a_1, \cdots, a_c\}$ and $G = \{u_1, \cdots, u_g\}$ with cardinalities c and g,* *respectively. A state is a bijective index function* $ind : \{1, \cdots, n\} \to C \cup G$, *hence* $c + g = n$. *The labelled transition system* $LTS(c, g) = (States, T, tr, ind_0)$ *of* $tq(c, g)$ *is defined by a set States of states, a set of transitions* $T = \{\langle\!\langle t_i, a_j \rangle\!\rangle | 1 \le i \le n, 1 \le j \le c\}$, *a transition relation* tr *and a regular initial state* ind_0. *The regular initial state is given by* $ind_0(i) = a_i$ *for* $1 \le i \le c$ *and* $ind_0(i) = g_{i-c}$ *for* $c < i \le n$. *The transition relation* $tr \subseteq States \times T \times States$ *is defined by*

$$(ind_1, \langle\!\langle t_i, a_j \rangle\!\rangle, ind_2) \in tr \Leftrightarrow$$
$$ind_1((i + 1) \bmod n) = ind_2(i) \in G \wedge ind_2((i + 1) \bmod n) = ind_1(i) = a_j \wedge$$
$$ind_2(m) = ind_1(m) \ \text{for all} \ m \notin \{i, (i + 1) \bmod n\}.$$

This is written as $ind_1 \xrightarrow{\langle\!\langle t_i, a_j \rangle\!\rangle} ind_2$ *or* $ind_1 \to ind_2$. *A transition sequence* $ind_0 \to ind_1 \to \cdots \to ind_0$ *of minimal length, leading from the initial state* ind_0 *back to* ind_0 *is called a recurrent sequence. As usual* $ind_1 \xrightarrow{*} ind_2$ *denotes the reflexive and transitive closure of* tr. *We restrict the set of states to the states reachable from the initial state:* $States := \mathcal{R}(LTS(c, g), ind_0) := \{ind | ind_0 \xrightarrow{*} ind\}$.

A more intuitive notation would be to consider a state as a word of length n over the alphabet $C \cup G$ with distinct letters only, and the rewrite rule $au \to ua$ with $a \in C, u \in G$ when inside the word and $u \cdots a \to a \cdots u$ at the borders. An example of two such transitions from $tq(3, 4)$ with $C = \{a, b, c\}, G = \{u, v, w, x\}$ and transitions $\langle\!\langle t_3, b \rangle\!\rangle, \langle\!\langle t_7, c \rangle\!\rangle$ is $u\,a\,b\,v\,w\,x\,c \to u\,a\,v\,b\,w\,x\,c \to c\,a\,v\,b\,w\,x\,u$. When defining the elements of G to be indistinguishable, they can be interpreted as gaps interchanging with the traffic items from C.

Definition 2. *A circular traffic queue with gaps* $tq\text{-}g(c, g)$ *is defined as in Definition 1, with the difference that* $|G| = 1$ *and the index function* ind *is not bijective in general, but only on the co-image* $ind^{-1}(C)$. *In addition we require that there is at least one gap:* $ind^{-1}(G) \ne \emptyset$. *As the number* g *from Definition 1 is not longer needed, we use it here to define the number of gaps:* $g := n - c \ge 1$.

With $G = \{\times\}$ for $tq\text{-}g(3, 4)$ the example above modifies to $\times a\,b \times \times \times c \to \times a \times b \times \times c \to c\,a \times b \times \times \times$. While the regular initial state is natural in the sense that the traffic items start without gaps in between, in a different context (as defined for cycloids by Petri) it is useful that the gaps are equally distributed. If for instance, as in the example after the following definition of a *standard initial state*, the numbers c and g are even, the queue in its initial state is composed of two equal subsystems with the parameters $\frac{c}{2}$ and $\frac{g}{2}$. An analogous situation holds for larger divisors.

Definition 3. *A standard initial state* ind_0 *of a circular traffic queue is defined by the form* $ind_0(1)ind_0(2) \cdots ind_0(n) = a_1 w_1 a_2 w_2 \cdots a_c w_c$ *with* $a_j \in C, w_j \in G^{r_j}$ *and* $r_j = |\{ x \in \mathbb{N} \mid j - 1 < \frac{c}{g} \cdot x \le j\}|$ *for* $1 \le j \le c$. G^{r_j} *is the set of words of length* r_j *over* G.

To be a consistent definition it should be verified that all the r_j sum up together to $\sum_{j=1}^{c} r_j = g$. This follows from the observation that the intervals in the definition of r_j are disjoint and all together define a set $\{x \in \mathbb{N} \mid 0 < \frac{c}{g} \cdot x \leq c\} = \{x \in \mathbb{N} \mid 0 < x \leq g\}$ of cardinality g. To give an example, we consider the case $c = 4, g = 6$. We obtain $(r_1, r_2, r_3, r_4) = (1, 2, 1, 2)$ and $a_1 \times a_2 \times \times a_3 \times a_4 \times \times$ for the resulting standard initial state.

Theorem 4. *Let be* $\Delta = gcd(c, g)$ *the greatest common divisor of* c *and* g. *The length of each recurrent sequence (Definition 1) of* $tq(c, g)$ *is* $\Xi(c, g) := \frac{g}{\Delta} \cdot (c + g) \cdot c$. *For a circular traffic queue with gaps* $tq\text{-}g(c, g)$ *this reduces to* $\Gamma(c, g) := c \cdot (c + g)$.

Proof. In both cases we start with a regular initial state $ind_0 = a_1 a_2 \cdots a_c u_1 u_2 \cdots u_g$ with $a_i \in C$ and $u_h \in G$. In the second case all u_h equal $\times$. To reach the initial state ind_0 for the first time, each traffic item $a_j \in C$ has to make $n = c + g$ steps. Hence in total, we have $\Gamma(c, g) := c \cdot (c + g)$. Since the general model $tq(c, g)$ is symmetric with respect to G and C, as well as the result $\Xi(c, g) := \frac{1}{\Delta} c(c + g) g$ to be proved, it is sufficient to consider the case $g \geq c$. Furthermore we assume $g > 1$ since for $g = 1$ we also have $c = 1$ and the problem reduces to the case of $\Gamma(c, g)$. The recurrent sequence to be constructed is split into several pieces. We start by shifting all traffic items to the end of the queue requiring g steps for each and $g \cdot c$ steps in total: (I) $ind_0 = a_1 a_2 \cdots a_c u_1 u_2 \cdots u_g \xrightarrow{g \cdot c - times} u_1 u_2 \cdots u_g a_1 a_2 \cdots a_c$. Then we move the traffic items back to their initial position, needing $c \cdot c$ steps: (II) $u_1 u_2 \cdots u_g a_1 a_2 \cdots a_c \xrightarrow{c \cdot c} a_1 a_2 \cdots a_c u_{c+1} \cdots u_g u_1 \cdots u_c$. Ignoring the individual names of the elements $u_i \in G$ we have the situation described in the case of $c \cdot g + c \cdot c = c \cdot (c + g)$ steps. But here in the first case the initial state is not yet reached since the items u_h are not necessary in their initial order. Therefore the $c \cdot (c + g)$ steps have to be repeated in a number of different levels. In the following, the step from level k to $k + 1$ is shown. The letter $u_{i_j^k}$ denotes the item $u_{i_j} \in G$ of level k in position $ind^{-1}(u_{i_j}) = j$.

(III) $a_1 a_2 \cdots a_c u_{i_1^k} u_{i_2^k} \cdots u_{i_g^k} \xrightarrow{c(c+g)} a_1 a_2 \cdots a_c u_{i_1^{k+1}} u_{i_2^{k+1}} \cdots u_{i_g^{k+1}}$ with $u_{i_1^{k+1}} = u_{(i_1^k + c) \bmod g}$. We now consider the sequence $u_{i_1^1} u_{i_1^2} \cdots u_{i_1^k} \cdots$ of items from G for all levels k in equation (III): $u_{i_1^1} u_{i_1^2} \cdots u_{i_1^k} \cdots$ starting with $u_{i_1^1} = u_1$ from ind_0. The initial state is reached when $u_{i_1^k} = u_{i_1^1} = u_1$ for the first time. By induction, beginning with equations (I) and (II): $u_{i_1^1} = u_1, u_{i_1^2} = u_{1+c}$ and by the induction step from equation (III): $u_{i_1^{k+1}} = u_{(i_1^k + c) \bmod g}$ we conclude $u_{i_1^k} = u_{(k \cdot c + 1) \bmod g}$. Hence to determine the value of k for a recurrent sequence we have to find the smallest nontrivial solution in the following equation: (IV) $(k \cdot c + 1) \bmod g = 1$. Next we prove $k = \frac{g}{x}$ to be a solution, where $x \in \mathbb{N}$. Applying the formula $(a + b) \bmod g = (a \bmod g + b \bmod g) \bmod g$ this is done by $(\frac{g}{x} \cdot c + 1) \bmod g = ([(\frac{c}{x} \cdot g) \bmod g] + [1 \bmod g]) \bmod g = (0 + 1) \bmod g = 1$ since $g > 1$. In this calculation $\frac{c}{x}$ has to be an integer, hence $x \mid c$. The same holds for $k = \frac{g}{x}$ and also $x \mid g$. For a minimal non-trivial solution we obtain $x = gcd(c, g) = \Delta$ and $k = \frac{g}{\Delta}$. Recall that k is the number times the step from equation (III) has to be repeated to obtain a recurrent transition sequence. This gives the result $\Xi(c,g) = k \cdot c \cdot (c+g) = \frac{c \cdot (c+g) \cdot g}{\Delta}$. All different recurrent transition sequences are obtained by permutations and cannot be shorter.   □

The proof has shown that a sequence has to be repeated $\frac{g}{\Delta}$ times to reach a state where all traffic items from C are in their initial order. To obtain an adequate labelling of the transitions we add a counter $k, 0 \le k < r$ to represent the repetitive behaviour. The counter is implemented as an exponent of the names of traffic items a_j^k and transitions t_v^k. Each time the traffic item a_j starts a new round in its (initial) position $j \in \{1, \cdots, c\}$ (with respect to the regular initial state) the counter is increased. With the values of i and k in the following definition, for each traffic item a_j a number $p = n \cdot r$ of process states is reached. Later we will restrict to the cases $r = 1$ (no repetition) and $r = \frac{g}{\Delta}$.

Definition 5. *Let be* $r \in \mathbb{N}_+$ *and* $C^r := \{a_j^k | a_j \in C, 0 \le k < r\}$. *A (r-repetitive) labelled transition system* $LTS_p(c,g) = (States, T, tr, ind_0)$ *with* $p = r \cdot n$ *and* $c, g, n = c + g \in \mathbb{N}_+$ *is defined by a set States of states as in Definition 1 with* C *replaced by* C^r, $T = \{\langle\langle t_v^k, a_j \rangle\rangle | 1 \le v \le p, 0 \le k < r, 1 \le j \le c\}$ *as a set of transitions, The transition relation* $tr \subseteq States \times T \times States$ *is defined by*

$$(ind_1, \langle\langle t_v^k, a_j \rangle\rangle, ind_2) \in tr \Leftrightarrow \exists u \in G \; \exists i \in \{1, \cdots, n\} : v = k \cdot n + i \land$$
$$ind_1(i) = a_j^k \in C^r \land ind_1((i+1) \bmod n) = u \land$$
$$ind_2(i) = u \land [ind_2((i+1) \bmod n) = a_j^k \; \textbf{if} \; i \neq j \; \textbf{else} \; a_j^{(k+1) \bmod r}] \land$$
$$ind_2(m) = ind_1(m) \; \text{for all} \; m \notin \{i, (i+1) \bmod n\}.$$

In particular, we consider the special cases $tq\text{-}2(c,g) := LTS_p(c,g)$ *with* $p = \frac{g}{\Delta} \cdot n$ *and* $tq\text{-}1(c,g) := LTS_n(c,g)$ *with* $G = \{\times\}$. *In the latter cases we have* $r = 1$ *and the labelling of the transitions can be simplified to* $\langle\langle t_i, a_j \rangle\rangle$.

For preparing the modelling of alternative formalisms, in particular of Petri nets in Sect. 4, we give a specification of circular traffic queues by their properties. To be free from a more sequential specification we prefer a message-oriented formulation. This is similar to Petri's notion of a *permit signal* in [5].

Definition 6. *A circular traffic queue* $tq\text{-}1(c,g)$ *respectively* $tq\text{-}2(c,g)$ *has the following properties. Next,* $u \in G$ *denotes a gap in the case of* $tq\text{-}1(c,g)$ *and a traffic item in the case of* $tq\text{-}2(c,g)$.

a) *Each traffic item* $a \in C$ *and* $u \in G$ *is in exactly in one of* $n = c + g$ *positions.*
b) *Each traffic item* $a \in C$ *can make a step from position* $i \in \{1, \cdots, n\}$ *to position* $(i + 1) \bmod n$, *if it has received a message from* $u \in G$ *in position* $(i + 1) \bmod n$. *After this step the* $u \in G$ *is in position* i.
c) *The length of recurrent transition sequences is* $\Gamma(c,g) = c \cdot n$ *and* $\Xi(c,g) := \frac{g}{\Delta} \cdot c \cdot n$, *respectively.*
d) $tq\text{-}2(c,g)$ *is isomorphic to* $\frac{g}{\Delta}$ *copies of transition systems of type* $tq\text{-}1(c,g)$ *with* $|G| = g$. *After* $c \cdot n$ *steps each transition sequence enters the next of these copies.*

This specification is denoted as a definition, but requires some justification. The items a), b) and c) follow from the definitions of a circular traffic queue. The supplement concerning $tq\text{-}2(c,g)$ follows from the proof of Theorem 4.

3   Petri Space and Cycloids

In this section (Petri-) nets and cycloids are defined. Some results are cited from [7] and [9], whereas concepts like regular cycloids and the use of matrix algebra are new. We define Petri nets in the form of condition/event-nets or safe T-nets, while coloured nets are used in Sect. 4.3 without formal definition.

Definition 7. *As usual, a net $\mathcal{N} = (S, T, F)$ is defined by non-empty, disjoint sets S of places and T of transitions, connected by a flow relation $F \subseteq (S \times T) \cup (T \times S)$. $X := S \cup T$. $\mathcal{N} \simeq \mathcal{N}'$ denote isomorphic nets. A transition $t \in T$ is active or enabled in a marking $M \subseteq S$ if $^{\bullet}t \subseteq M \wedge t^{\bullet} \cap M = \emptyset$ and in this case $M \xrightarrow{t} M'$ if $M' = M \backslash ^{\bullet}t \cup t^{\bullet}$, where $^{\bullet}x := F^{-1}[x]$, $x^{\bullet} := F[x]$ denote the input and output elements of an element $x \in S \cup T$, respectively. $\xrightarrow{*}$ is the reflexive and transitive closure of $\rightarrow$. A net together with an initial marking $M_0 \subseteq S$ is called a net-system $(\mathcal{N}, M_0)$ with its reachability set $\mathcal{R}(\mathcal{N}, M_0) := \{M | M_0 \xrightarrow{*} M\}$. The reachability graph $\mathcal{RG}(\mathcal{N}, M_0) = (\mathcal{R}(\mathcal{N}, M_0), \rightarrow, M_0)$ is defined by the reachability set as the set of nodes, the relation $\rightarrow$ as its set of arrows and the initial marking as distinguished node.*

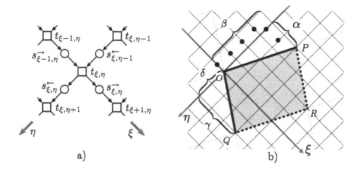

Fig. 1. a) Petri space, b) Fundamental parallelogram of $\mathcal{C}(\alpha, \beta, \gamma, \delta) = \mathcal{C}(2, 4, 3, 2)$ with regular initial marking.

Definition 8. *A Petri space is defined by the net $\mathcal{PS}_1 := (S_1, T_1, F_1)$ where $S_1 = S_1^{\rightarrow} \cup S_1^{\leftarrow}$, $S_1^{\rightarrow} = \{s_{\xi,\eta}^{\rightarrow} | \xi, \eta \in \mathbb{Z}\}$, $S_1^{\leftarrow} = \{s_{\xi,\eta}^{\leftarrow} | \xi, \eta \in \mathbb{Z}\}$, $S_1^{\rightarrow} \cap S_1^{\leftarrow} = \emptyset$, $T_1 = \{t_{\xi,\eta} | \xi, \eta \in \mathbb{Z}\}$, $F_1 = \{(t_{\xi,\eta}, s_{\xi,\eta}^{\rightarrow}) | \xi, \eta \in \mathbb{Z}\} \cup \{(s_{\xi,\eta}^{\rightarrow}, t_{\xi+1,\eta}) | \xi, \eta \in \mathbb{Z}\} \cup \{(t_{\xi,\eta}, s_{\xi,\eta}^{\leftarrow}) | \xi, \eta \in \mathbb{Z}\} \cup \{(s_{\xi,\eta}^{\leftarrow}, t_{\xi,\eta+1}) | \xi, \eta \in \mathbb{Z}\}$ (cutout in Fig. 1a). $S_1^{\rightarrow}$ is the set of forward places and $S_1^{\leftarrow}$ the set of backward places.*
$^{\bullet}t_{\xi,\eta} := s_{\xi-1,\eta}^{\rightarrow}$ *is the forward input place of $t_{\xi,\eta}$ and in the same way*
$^{\bullet}t_{\xi,\eta} := s_{\xi,\eta-1}^{\leftarrow}$, $t_{\xi,\eta}^{\bullet} := s_{\xi,\eta}^{\rightarrow}$ *and $t_{\xi,\eta}^{\bullet} := s_{\xi,\eta}^{\leftarrow}$ (see Fig. 1a).*

By a twofold folding with respect to time and space we obtain the cyclic structure of a cycloid. See [7,9] for motivation and Fig. 2 a) for an example of a cycloid.

Definition 9 ([7,9]).   *A cycloid is a net* $\mathcal{C}(\alpha,\beta,\gamma,\delta) = (S,T,F)$, *defined by parameters* $\alpha,\beta,\gamma,\delta \in \mathbb{N}_+$, *by a quotient of the Petri space* $\mathcal{PS}_1 := (S_1, T_1, F_1)$ *with respect to the equivalence relation* $\equiv \subseteq X_1 \times X_1$ *with* $\equiv[S_1^{\rightarrow}] \subseteq S_1^{\rightarrow}, \equiv[S_1^{\leftarrow}] \subseteq S_1^{\leftarrow}, \equiv[T_1] \subseteq T_1$, $x_{\xi,\eta} \equiv x_{\xi+m\alpha+n\gamma,\, \eta-m\beta+n\delta}$ *for all* $\xi,\eta,m,n \in \mathbb{Z}$, $X = X_1/_\equiv$, $[\![x]\!]_\equiv \, F \, [\![y]\!]_\equiv \Leftrightarrow \exists x' \in [\![x]\!]_\equiv \exists y' \in [\![y]\!]_\equiv : x' F_1 y'$ *for all* $x, y \in X_1$.

The matrix $\mathbf{A} = \begin{pmatrix} \alpha & \gamma \\ -\beta & \delta \end{pmatrix}$ *is called the matrix of the cycloid. Petri denoted the number* $|T|$ *of transitions as the area* A *of the cycloid and determined in [4] its value to* $|T| = A = \alpha\delta + \beta\gamma$ *which equals the determinant* $A = \det(\mathbf{A})$. *Cycloids are safe T-nets with* $|{}^\bullet s| = |s^\bullet| = 1$ *for all places* $s \in S$. *The embedding of a cycloid in the Petri space is called* fundamental parallelogram *(see Fig. 1 b), but ignore the tokens for the moment). If the cycloid is represented as a net* $\mathcal{N}$ *without explicitly giving the parameters* $\alpha,\beta,\gamma,\delta$, *we call it a cycloid in net form* $\mathcal{C}(\mathcal{N})$.

For proving the equivalence of two points in the Petri space the following procedure is useful.

Theorem 10.   *Two points* $x_1, x_2 \in X_1$ *are equivalent* $x_1 \equiv x_2$ *if and only if the parameter vector* $\pi(v) = \pi(x_2 - x_1)$ *has integer values, where* $\pi(v) = \frac{1}{A} \cdot \mathbf{B} \cdot v$ *with area* A *and* $\mathbf{B} = \begin{pmatrix} \delta & -\gamma \\ \beta & \alpha \end{pmatrix}$.

Proof. For $x_1 := (\xi_1, \eta_1), x_2 := (\xi_2, \eta_2), v := x_2 - x_1$ from Definition 9 we obtain in vector form: $x_1 \equiv x_2 \Leftrightarrow \exists m, n \in \mathbb{Z} : \begin{pmatrix} \xi_2 \\ \eta_2 \end{pmatrix} = \begin{pmatrix} \xi_1 + m\alpha + n\gamma \\ \eta_1 - m\beta + n\delta \end{pmatrix} \Leftrightarrow \exists m, n \in$

$\mathbb{Z} : v = \begin{pmatrix} \xi_2 - \xi_1 \\ \eta_2 - \eta_1 \end{pmatrix} = \begin{pmatrix} m\alpha + n\gamma \\ -m\beta + n\delta \end{pmatrix} = \begin{pmatrix} \alpha & \gamma \\ -\beta & \delta \end{pmatrix} \begin{pmatrix} m \\ n \end{pmatrix} = \mathbf{A} \begin{pmatrix} m \\ n \end{pmatrix} \Leftrightarrow \begin{pmatrix} m \\ n \end{pmatrix} =$

$\mathbf{A}^{-1} v \in \mathbb{Z} \times \mathbb{Z}$. It is well-known that $\mathbf{A}^{-1} = \frac{1}{\det(\mathbf{A})}\mathbf{B}$ if $\det(\mathbf{A}) > 0$ (see any book on linear algebra). The condition $\det(\mathbf{A}) = A = \alpha\delta + \beta\gamma > 0$ is satisfied by the definition of a cycloid.   □

Theorem 11 ([7,9]).   *The following cycloids are isomorphic to* $\mathcal{C}(\alpha,\beta,\gamma,\delta)$:

a) $\mathcal{C}(\beta,\alpha,\delta,\gamma)$,   (*The* dual cycloid *of* $\mathcal{C}(\alpha,\beta,\gamma,\delta)$.)
b) $\mathcal{C}(\alpha,\beta,\gamma - q \cdot \alpha, \delta + q \cdot \beta)$ *if* $q \in \mathbb{N}_+$ *and* $\gamma > q \cdot \alpha$,
c) $\mathcal{C}(\alpha,\beta,\gamma + q \cdot \alpha, \delta - q \cdot \beta)$ *if* $q \in \mathbb{N}_+$ *and* $\delta > q \cdot \beta$.

Proof. Part a) has been proved in [7,9] as well as b) and c) for the special case of $q = 1$. The current form is derived by iterating the result.   □

Definition 12.   *For a cycloid* $\mathcal{C}(\alpha,\beta,\gamma,\delta)$ *we define a cycloid-system* $\mathcal{C}(\alpha,\beta,\gamma,\delta,M_0)$ *or* $\mathcal{C}(\mathcal{N}, M_0)$ *by adding the standard initial marking:*

$$M_0 = \{s_{\xi,\eta}^{\rightarrow} \in S_1^{\rightarrow} \mid \beta\xi + \alpha\eta \leq 0 \,\wedge\, \beta(\xi+1) + \alpha\eta > 0\}/_\equiv \;\cup$$
$$\{s_{\xi,\eta}^{\leftarrow} \in S_1^{\leftarrow} \mid \beta\xi + \alpha\eta \leq 0 \,\wedge\, \beta\xi + \alpha(\eta+1) > 0\}/_\equiv$$

The motivation of this definition is given in [7] and [9]. See Fig. 2 a) for an example of a cycloid with standard initial marking. As in the case of circular traffic queues we define a regular initial marking for cycloids, but not necessarily within the fundamental parallelogram. It is characterized by the absence of gaps between the traffic items whereby only a single transition on the top of the queue is enabled. In the example of $\mathcal{C}(2,4,3,2)$ from Fig. 1 b) this is the transition $t_{0,1-\beta} = t_{0,-3}$, which is enabled by the tokens in $s^{\rightarrow}_{-1,-3}$ and $s^{\leftarrow}_{0,-4}$. After occurring the $\beta - 1 = 3$ transitions $t_{0,-2}, t_{0,-1}$ and $t_{0,0}$ can occur in some order. They model the steps of the traffic items following the item at the head of the queue. In this example the regular initial marking is $\{s^{\rightarrow}_{-1,0}, s^{\rightarrow}_{-1,-1}, s^{\rightarrow}_{-1,-2}, s^{\rightarrow}_{-1,-3},\} \cup \{s^{\leftarrow}_{0,-4}, s^{\leftarrow}_{1,-4}\}$, as shown in Fig. 1 b). Equivalent places within the fundamental parallelogram can be computed. For instance, the regular initial marking of $\mathcal{C}(4,3,3,3)$ is represented by highlighted places in Fig. 2 a).

Definition 13. *For a cycloid $\mathcal{C}(\alpha, \beta, \gamma, \delta)$ a regular initial marking is defined by a number of β forward places $\{s^{\rightarrow}_{-1,i}|\ 0 \geq i \geq 1-\beta\}$ and a number of α backward places $\{s^{\leftarrow}_{i,-\beta}|\ 0 \leq i \leq \alpha - 1\}$.*

Theorem 14 ([7,9]). *The length of a minimal cycle of a cycloid $\mathcal{C}(\alpha, \beta, \gamma, \delta)$ is*

$$cyc(\alpha, \beta, \gamma, \delta) = cyc = \gamma + \delta + \begin{cases} \lfloor \frac{\delta}{\beta} \rfloor (\alpha - \beta) & \text{if } \alpha \leq \beta \\ -\lfloor \frac{\gamma}{\alpha} \rfloor (\alpha - \beta) & \text{if } \alpha > \beta \end{cases}$$

As proved in [7,9] by the *Synthesis-Theorem* for cycloids we can compute the parameters α, β, γ and δ of a cycloid from its net presentation using the system parameters τ_0, τ_a, A and cyc. τ_0 is the number of transitions having at least one marked input place, τ_a is the number of active transitions, both w.r.t. the standard initial marking M_0. A is (as before) the number of all transitions and cyc is the minimal length of transition cycles. The following procedures do not necessarily give a unique result, but for $\alpha \neq \beta$ the resulting cycloids are isomorphic.

Theorem 15 (Synthesis-Theorem, [7,9]). *Cycloid systems with identical system parameters τ_0, τ_a, A and cyc are called σ-equivalent. Given a cycloid system $\mathcal{C}(\alpha, \beta, \gamma, \delta, M_0)$ in its net representation (S, T, F, M_0) where the parameters τ_0, τ_a, A and cyc are known (but the parameters $\alpha, \beta, \gamma, \delta$ are not). Then a σ-equivalent cycloid $\mathcal{C}(\alpha', \beta', \gamma', \delta')$ can be computed by $\alpha' = \tau_0$, $\beta' = \tau_a$ and for γ', δ' by some positive integer solution of the following formulas using these settings of α' and β':*

a) *case $\alpha' > \beta'$: $\gamma' \bmod \alpha' = \frac{\alpha' \cdot cyc - A}{\alpha' - \beta'}$ and $\delta' = \frac{1}{\alpha'}(A - \beta' \cdot \gamma')$,*

b) *case $\alpha' < \beta'$: $\delta' \bmod \beta' = \frac{\beta' \cdot cyc - A}{\beta' - \alpha'}$ and $\gamma' = \frac{1}{\beta'}(A - \alpha' \cdot \delta')$,*

c) *case $\alpha' = \beta'$: $\gamma' = \lceil \frac{cyc}{2} \rceil$ and $\delta' = \lfloor \frac{cyc}{2} \rfloor$.*

These equations may result in different cycloid parameters, however the cycloids are isomorphic in the cases a) and b) as in Theorem 11.

Circular traffic queues are composed by a number of c sequential and inter-acting processes of equal length. In the formalism of cycloids this corresponds to a number of β disjoint processes of equal length p. Cycloids with such a property are called regular.

Definition 16. *A cycloid* $\mathcal{C}(\alpha, \beta, \gamma, \delta)$ *with area A is called* regular *if for each* $\eta \in \{0, \cdots, 1 - \beta\}$ *the set* $\{t_{\xi,\eta} | 1 \leq \xi \leq p\}$ *with* $p = \frac{A}{\beta}$ *of transitions forms an elementary cycle*[1] *and all these sets are disjoint.* $p \in \mathbb{N}_+$ *is called the process length of the regular cycloid. A regular cycloid together with its regular initial marking M_0 is called a* regular cycloid system $\mathcal{C}(\alpha, \beta, \gamma, \delta, M_0)$.

Theorem 17. *A cycloid* $\mathcal{C}(\alpha, \beta, \gamma, \delta)$ *is regular if and only if* $\beta | \delta$.

Proof. We first prove that starting in any point $t_{\xi,\eta}$ of the fundamental para-llelogram of the cycloid and proceeding in direction of the ξ-axis we will return to $t_{\xi,\eta}$ after passing $p = \frac{A}{\beta}$ transitions. By Theorem 10, to decide $t_{\xi+p,\eta} \equiv t_{\xi,\eta}$ it is sufficient to check whether

$$\pi\left(\begin{pmatrix} \xi + p \\ \eta \end{pmatrix} - \begin{pmatrix} \xi \\ \eta \end{pmatrix}\right) = \pi\left(\begin{pmatrix} p \\ 0 \end{pmatrix}\right) = \frac{1}{A}\begin{pmatrix} \delta & -\gamma \\ \beta & \alpha \end{pmatrix}\begin{pmatrix} p \\ 0 \end{pmatrix} = \begin{pmatrix} \frac{\delta \cdot p}{A} \\ \frac{\beta \cdot p}{A} \end{pmatrix} \in \mathbb{Z} \times \mathbb{Z}.$$

$\frac{A}{\beta}$ is the smallest value for p to obtain an integer value in the second com-ponent of the last vector. Therefore an equivalent point is not reached before passing p transitions and the cycle is elementary. By the first component, to fulfill $\frac{\delta \cdot p}{A} = \frac{\delta \cdot A}{A \cdot \beta} = \frac{\delta}{\beta} \in \mathbb{Z}$ it is necessary and sufficient that $\beta | \delta$. It follows that there is a number of β such elementary cycles of length $\frac{A}{\beta}$ covering the entire set of A transitions. Therefore no pair of these cycles can have a common transition.   □

A regular cycloid can be seen as a system of β disjoint sequential and cooper-ating processes. To exploit this structure we define specific coordinates, called *regular coordinates*. The process of a traffic item a_1 starts with transition $t_{0,0}$ which is denoted $[t_1, a_1]$, having the input place $[s_0, a_1]$. The next transitions are $[t_2, a_1]$ up to $[t_p, a_1]$ and then returning to $[t_1, a_1]$. The other processes for a_2 to a_c (with $\beta = c$) are denoted in the same way (see Fig. 2 b). As the process of a_j starts in position j of the queue, its initial token is in $[s_{j-1}, a_j]$.

Definition 18. *Given a regular cycloid* $\mathcal{C}(\alpha, \beta, \gamma, \delta)$, *regular coordinates are defined as follows: transitions of process* $j \in \{1, \cdots, \beta\}$ *each with length p are denoted by* $\{[t_1, a_j], \cdots, [t_p, a_j]\}$. *For each transition we define* $[t_i, a_j]^{\overset{\rightarrow}{\bullet}} := [s_i, a_j]$ *and* $[t_i, a_j]^{\overset{\leftarrow}{\bullet}} := [s'_i, a_j]$, *where we also use* $s_0 := s_p$ *and* $s'_0 := s'_p$. *Furthermore for* $1 \leq j \leq \beta$ *let* $[s_i, a_j]^{\bullet} := [t_{(i+1) \bmod p}, a_j]$. *Regular coordinates are related to standard coordinates of the Petri space by defining the following initial condition* $[t_1, a_j] := t_{1-j,1-j}$ *for* $1 \leq j \leq \beta$.

---

[1] An elementary cycle is a cycle where all nodes are different.

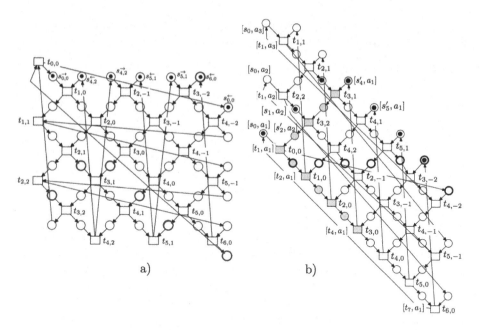

Fig. 2. Cycloid $\mathcal{C}(4,3,3,3)$ in a) and with regular coordinates in b).

In Fig. 2 b) we obtain for the last formula in Definition 18: $[t_1, a_3] := t_{-2,-2} \equiv t_{1,1}$. While the output place $[s'_i, a_j]$ in regular coordinates takes its name from the input transition, it remains to determine its output transition according to the corresponding standard coordinates.

Lemma 19. *The injective mapping* stand *from regular to standard coordinates is given by* $stand([t_i, a_j]) = t_{i-j,1-j}$ *for* $1 \le i \le p$ *and* $1 \le j \le \beta$ *(modulo equivalent transitions). The output transition is* $[s'_i, a_1]^\bullet = [t_{(i+\alpha+\beta-1) \bmod p}, a_\beta]$ *while for* $1 < j \le \beta$ *we have* $[s'_i, a_j]^\bullet = [t_{(i-1) \bmod p}, a_{(j-1) \bmod c}]$. *If* $\beta = \gamma = \delta$ *the two cases coincide.*

Proof. For a given j by Definition 18 we have $[t_1, a_j] := t_{1-j,1-j}$. Adding a value $i - 1$ to the index of t_1 we obtain the index of t_i, hence $stand([t_i, a_j]) := t_{1-j+(i-1),1-j}$. To prove $[s'_i, a_j]^\bullet = [t_{(i-1) \bmod p}, a_{j-1}]$ for $1 < j \le \beta$ we first compute the corresponding standard coordinate $stand([t_i, a_j]) = t_{i-j,1-j}$. To obtain the output transition of $[s'_i, a_j]$ we go to the next transition in η-direction $t_{i-j,1-j+1} = t_{i-j,2-j} = stand([t_{i-1}, a_{j-1}])$, where $\bmod\, p$ and $\bmod\, c$ are omitted. To make the proof for $[s'_i, a_1]$ we start with $[t_i, a_1]$ and compute again $stand([t_i, a_1]) = t_{i-1,0}$. The next transition in η-direction is $t_{i-1,1}$. Using Theorem 10 we prove $t_{i-1,1} \equiv t_{i-1+\alpha,1-\beta} : \pi(\begin{pmatrix} i-1 \\ 1 \end{pmatrix} - \begin{pmatrix} i-1+\alpha \\ 1-\beta \end{pmatrix}) = \pi(\begin{pmatrix} -\alpha \\ \beta \end{pmatrix}) = \frac{1}{A} \begin{pmatrix} \delta & -\gamma \\ \beta & \alpha \end{pmatrix} \begin{pmatrix} -\alpha \\ \beta \end{pmatrix} = \frac{1}{A} \begin{pmatrix} -A \\ 0 \end{pmatrix} \in \mathbb{Z} \times \mathbb{Z}$. Going back to the corresponding regular coordinates the desired result is obtained: $stand([t_{i+\alpha+\beta-1}, a_\beta)]) =$

$t_{i+\alpha+\beta-1-\beta,1-\beta} = t_{i+\alpha-1,1-\beta}$ where $mod\, p$ and $mod\, c$ are omitted again. If $\beta = \gamma = \delta$ the two cases coincide as $p = \alpha + \beta$ in this case.   □

Corollary 20. *The regular initial marking of a regular cycloid system* $\mathcal{C}(\alpha, \beta, \gamma, \delta, M_0)$ *with process length* p *is* $M_0 = \{[s_i, a_{i+1}]| 0 \le i \le \beta - 1\} \cup \{[s'_i, a_1]|p - \alpha + 1 \le i \le p\}$. *For later reference, we note* $\overset{\bullet\leftarrow}{}[t_\beta, a_\beta] = [s'_{p-\alpha+1}, a_1]$.

Proof. Since the mapping *stand* is defined on transitions, from the first place of M_0 in Corollary 20 $[s_0, a_{(0+1)\, mod\, \beta}]$ we go to its output transition $[t_1, a_1]$ and apply $stand([t_1, a_1]) = t_{0,0}$. Going back in ξ-direction we obtain $s^{\rightarrow}_{-1,0}$, which is the first element in Definition 13. Doing the same with $[s_{\beta-1}, a_{\beta\, mod\, \beta}]$ we come via $stand([t_\beta, a_\beta]) = t_{\beta-\beta,1-\beta} = t_{0,1-\beta}$ to $s^{\rightarrow}_{-1,1-\beta}$. By this we obtain the entire forward places from Definition 13. Since the mapping *stand* is injective we can conclude also in the inverse direction. To prove the second part of the union recall that the last traffic item a_β is active (enabled). Therefore also the backward input place $[s'_i, a_1]$ of $[t_\beta, a_\beta]$ must be marked. Using Lemma 19 the value of i must satisfy $[s'_i, a_1]^\bullet = [t_{(i+\alpha+\beta-1)\, mod\, p}, a_\beta] = [t_\beta, a_\beta]$, hence $(i+\alpha+\beta-1)\, mod\, p = \beta$ and $i = (1 - \alpha)\, mod\, p = p - \alpha + 1$. This holds since $\beta|\delta \Rightarrow \beta \le \delta$ and therefore $p = \frac{A}{\beta} = \alpha\frac{\delta}{\beta} + \gamma > \alpha$. The marked place in question is therefore $[s'_i, a_1] = [s'_{p-\alpha+1}, a_1]$. To determine the other elements of $\{[s'_i, a_1]|p - \alpha + 1 \le i \le p\}$ recall that the last traffic item a_β should be able to make α steps before any other transition has to occur. Therefore also the places $[s'_{p-\alpha+2}, a_1]$ to $[s'_{p-\alpha+\alpha}, a_1]$ must be marked in the regular initial marking.   □

In the regular cycloid system $\mathcal{C}(4, 3, 3, 3, M_0)$ in Fig. 2 b) we obtain $\overset{\bullet\leftarrow}{}[t_3, a_3] = [s'_{p-\alpha+1}, a_1] = [s'_4, a_1]$. The given regular initial marking is $\{[s_0, a_1], [s_1, a_2], [s_2, a_3], [s'_4, a_1], [s'_5, a_1], [s'_6, a_1], [s'_7, a_1]\}$. The standard initial marking is given by bold circles. It is useful in some cases to express the minimal cycle length cyc of a regular cycloid by its process length p. While this can be perfectly done for the case $\alpha \le \beta$ in the complementary case only partial results are achieved. Compared with general cycloids the lack of symmetry of regular cycloids becomes apparent by these results. However, they cover all the cases required in Sect. 4. In Theorem 23 we consider $\mathcal{C}(\alpha, \beta, \beta, \beta)$ (in the form $\alpha = g, \beta = c$) with $p = \alpha+\beta$ and $cyc = min\{\alpha + \beta, 2 \cdot \beta\}$, which is covered by the cases a) and c) in Lemma 21. Afterwards in Theorem 24 we have $\mathcal{C}(\alpha, \beta, \frac{1}{A}\alpha\beta, \frac{1}{A}\alpha\beta)$ with $p = \frac{1}{A}\alpha(\alpha + \beta)$ and $cyc = min\{p, \frac{\beta}{\alpha} \cdot p\}$, which is covered by the cases a) and b).

Lemma 21. *Let be* $\mathcal{C}(\alpha, \beta, \gamma, \delta)$ *a regular cycloid with process length* p *and minimal cycle length* cyc. *Then a)* $cyc = p$ *if* $\alpha \le \beta$, *b) If* $\alpha > \beta$ *and* $\alpha|\gamma$ *then* $cyc = \frac{\beta}{\alpha} \cdot p$ *and c)* $cyc = 2 \cdot \beta$ *if* $\alpha > \beta = \gamma = \delta$.

Proof. a) By Theorem 14 and $\delta = m \cdot \beta$ for some $m \in \mathbb{Z}$ we obtain $cyc = \gamma + \delta + \lfloor\frac{\delta}{\beta}\rfloor(\alpha - \beta) = \gamma + m \cdot \beta + \lfloor\frac{m\cdot\beta}{\beta}\rfloor(\alpha - \beta) = \gamma + m \cdot \alpha$. This term equals $p = \frac{A}{\beta} = \frac{1}{\beta}(\alpha\delta + \beta\gamma) = \frac{1}{\beta}(\alpha \cdot m \cdot \beta + \beta\gamma) = \alpha \cdot m + \gamma$.

b) If $\alpha > \beta$ case a) applies to the dual cycloid $\mathcal{C}(\beta, \alpha, \delta, \gamma)$ (Theorem 11) which is regular since $\alpha|\gamma$. Hence $cyc = p'$ where $p' = \frac{A}{\alpha} = \frac{\beta\cdot p}{\alpha}$ is the process length of the dual cycloid which is isomorphic to $\mathcal{C}(\alpha, \beta, \gamma, \delta)$ by Theorem 11.

c) If $\alpha > \beta = \gamma = \delta$ then $cyc = \gamma + \delta - \lfloor \frac{\gamma}{\alpha} \rfloor (\alpha - \beta) = \beta + \beta - 0 \cdot (\alpha - \beta)$. If $\alpha = \beta = \gamma = \delta$ then $cyc = \gamma + \delta + \lfloor \frac{\delta}{\beta} \rfloor (\alpha - \beta) = \beta + \beta + 1 \cdot 0$. $\qquad\square$

The regular cycloid $\mathcal{C}(4,3,3,6)$ does not satisfy any of the conditions of Lemma 21. The parameters in question are $cyc = 9$, $p = 11$ and $\frac{\beta}{\alpha} \cdot p = \frac{3}{4} \cdot 11$.

4   Net Representations of Traffic Queues

From the specification in Definition 6 we deduce different types of cycloids using Theorem 15. This is, however, not a complete proof since it was not shown, that the specifications are correct and complete, but has the great advantage that structural properties of these cycloids are found. Afterwards formal proofs are added, showing that the cycloids are indeed behavioural equivalent to circular traffic queues.

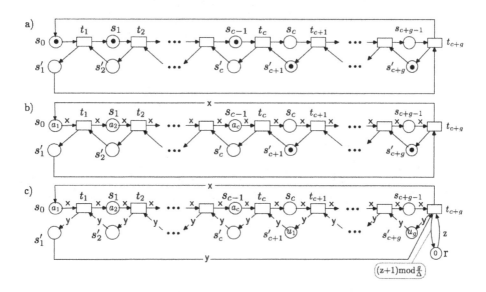

Fig. 3. Nets of circular traffic queues a) $\mathcal{N}_{basic}(c,g)$ b) $\mathcal{N}_{coul}(c,g)$ and c) $\mathcal{N}_{sym}(c,g)$.

4.1   Cycloids from the Circular Traffic Queue Specification

Theorem 22. *From the specifications of a circular traffic queue with gaps $tq\text{-}1(c,g)$ (Definition 6 a, b) using the Synthesis-Theorem 15 the cycloid $\mathcal{C}_0(g,c) := \mathcal{C}(g,c,1,1)$ can be deduced, which is isomorphic to the net $\mathcal{N}_{basic}(c,g)$ of Fig. 3 a).*

Proof. From Petri's papers [3,4] or [5] it follows definitely that the first two parameters are $\beta = c$ and $\alpha = g$. To follow a more formal approach we determine the parameters τ_0, τ_a, A and cyc and then apply Theorem 15 to determine the parameters α, β, γ and δ of the cycloid $\mathcal{C}(g, c, 1, 1)$. Consider first the case $g \geq c$ and the state $a_1 \times {}^{k_1} a_2 \times {}^{k_2} \cdots a_c \times {}^{k_c}$. Due to the assumption $g \geq c$ it follows $k_j \geq 1$ $(1 \leq j \leq c)$. This means that all traffic items a_j are able to move right, i.e they are enabled and therefore $\tau_a = \beta = c$. τ_0 is the number of transitions having initially at least one marked input place. From all g positions of the queue which are empty a release message was sent to the left. Therefore there is a number of $\tau_0 = \alpha = g$ such transitions, among these the c active transitions mentioned before. Given a fixed traffic item a and a fixed position in the queue, in a recurrent sequence the traffic item enters the position exactly once. Hence by Definition 6 a) there are $A = g + c$ transitions, since in this case all $a_j \in C$ are using the same transitions. In a step the traffic item a gives a release message to enable the access to the position it is leaving. This results in a minimal cycle of length 2. With the parameters $\tau_a = \beta = c, \tau_0 = \alpha = g, A = g + c$ and $cyc = 2$ obtained, we compute in the case $\alpha > \beta$ with Theorem 15: $\gamma \mod \alpha = \frac{\alpha \cdot cyc - A}{\alpha - \beta} = \frac{g \cdot 2 - (g+c)}{g - c} = \frac{g - c}{g - c} = 1$. With a solution $\gamma = 1$ of this equation we obtain $\delta = \frac{1}{\alpha}(A - \beta \cdot \gamma) = \frac{1}{g}(c + g - c \cdot 1) = 1$. For the case $\alpha = \beta$ of Theorem 15 we obtain $\gamma = \delta = \lceil \frac{cyc}{2} \rceil = 1$.

Next we prove formally that this cycloid $\mathcal{C}(g, c, 1, 1)$ is isomorphic to the basic tq-net $\mathcal{N}_{basic}(c, g)$ from Fig. 3 a). Starting in the origin $(0,0)$ of the fundamental parallelogram of $\mathcal{C}(g, c, 1, 1)$ using Theorem 10 we compute the smallest point $(\xi, 0)$ on the ξ-axis equivalent to $(0, 0)$: $\begin{pmatrix} m \\ n \end{pmatrix} = \frac{1}{A} \begin{pmatrix} \delta & -\gamma \\ \beta & \alpha \end{pmatrix} v =$

$\frac{1}{g+c} \begin{pmatrix} 1 & -1 \\ c & g \end{pmatrix} \begin{pmatrix} \xi \\ 0 \end{pmatrix} = \frac{1}{g+c} \begin{pmatrix} \xi \\ \xi \cdot c \end{pmatrix}$. The smallest positive integer for m is $\xi = g + c$.

Since $A = g + c$ all transitions have their position on the ξ-axis from $(0,0)$ to $(g + c - 1, 0)$ and form a cycle. This cycle is isomorphic to the cycle $t_1 \cdots t_{c+g}$ in the net of Fig. 3 a). It remains to prove that $t_{i,0}$ $(1 \leq i \leq c + g)$ is connected to $t_{i-1,0}$ via a place isomorphic to s'_i forming a cycle of length 2. As $t_{i,0}^{\bullet} = s_{i,0}^{\leftarrow}$ and $(s_{i,0}^{\leftarrow})^{\bullet} = t_{i,1}$ (see Fig. 4 a) we have to prove: $t_{i-1,0} \equiv t_{i,1}$. By Theorem 10 using $v = (i, 1) - (i - 1, 0)$: $\begin{pmatrix} m \\ n \end{pmatrix} = \frac{1}{A} \begin{pmatrix} \delta & -\gamma \\ \beta & \alpha \end{pmatrix} v =$

$\frac{1}{g+c} \begin{pmatrix} 1 & -1 \\ c & g \end{pmatrix} \begin{pmatrix} 1 \\ 1 \end{pmatrix} = \frac{1}{g+c} \begin{pmatrix} 0 \\ c + g \end{pmatrix} = \begin{pmatrix} 0 \\ 1 \end{pmatrix} \in \mathbb{Z} \times \mathbb{Z}$. For the case $g < c$ we observe that the net from Fig. 3 a) is symmetric in the following sense. Interpreting the places $s'_1, s'_2, \cdots s'_{c+g}$ to be the slots of the traffic items instead of $s_1, s_2, \cdots s_{c+g}$ we obtain an isomorphic system. By the construction in the first part of this proof we obtain the cycloid $\mathcal{C}(c, g, 1, 1)$ which is isomorphic to $\mathcal{C}(g, c, 1, 1)$ by Theorem 11 a). Therefore the theorem holds also in this case.   □

As an example, in Fig. 4 b) the cycloid $\mathcal{C}(4, 2, 1, 1)$ is shown. To illustrate the preceding proof by dashed lines the following equivalent transitions are given: $t_{3,0} \equiv t_{2,-1}, t_{4,0} \equiv t_{3,-1}$ and $t_{5,0} \equiv t_{4,-1}$. The transitions $t_{0,0}, \cdots, t_{5,0}$ on the

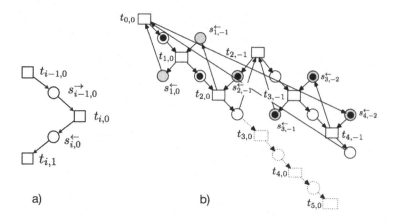

Fig. 4. Cycloid $C(4, 2, 1, 1)$ with regular initial marking and minimal cycle in a).

ξ-axis are instances of the transitions $t_{0,0}, \cdots, t_{g+c-1,0}$ in the proof. The places $s_{1,0}^{\leftarrow}, s_{1,-1}^{\leftarrow}, s_{2,-1}^{\leftarrow}, s_{3,-1}^{\leftarrow}, s_{3,-2}^{\leftarrow}$ and $s_{4,-2}^{\leftarrow}$ correspond to the *complementary places* $s_1', s_2', \cdots, s_{c+g}'$ of the net from Fig. 3 a).

As motivated in the introduction, cyloids are of particular interest and therefore in the focus of this paper. To distinguish them from the preceding model in Theorem 22 we assume $c > 1$ to obtain secure cycloids. To be secure is probably the most important property of cycloids with $\gamma, \delta > 1$. This has been intensively discussed by Petri in [3,4] and [5]. For the definition of safe and secure see [4] or [9]. Starting with the specification of circular traffic queues in Definition 6 in Theorems 23 and 24 we will derive cycloid models, that are behavioural equivalent to the models $tq\text{-}1(c, g)$ and $tq\text{-}2(c, g)$, respectively. Again, as in Theorem 22, the Synthesis-Theorem 15 will have a prominent role. Therefore, the minimal length cyc of cycles has to be computed. Candidates are the length of the processes $p = c + g$ and of the *release message cycle* $rm = 2 \cdot c$. The latter starts with the *release message chain* which is the sequence of release messages from traffic item a_c to a_{c-1}, from a_{c-1} to a_{c-2} and so on. This can be seen as the synchronization principle of the system. Finally we will obtain $cyc = min\{p, rm\}$. After this derivation, in Theorem 25 it will be proved by their transition systems, that the cycloid systems $(C_1(g, c), M_0)$ and $(C_2(g, c), M_0)$ are in fact behaviour equivalent to $tq\text{-}1(c, g)$ and $tq\text{-}2(c, g)$, respectively.

Theorem 23. *Within the class of regular cycloids $C(\alpha, \beta, \gamma, \delta)$ with $\gamma, \delta \geq 2$, from the specifications of a circular traffic queue with gaps $tq\text{-}1(c, g)$ with $c > 1$ (Definition 6) using the Synthesis-Theorem 15 the cycloid $C_1(g, c) := C(g, c, c, c)$ can be deduced. It has process length $p = g + c$ and minimal cycle length $cyc = min\{g + c, 2 \cdot c\}$ (in accordance with Lemma 21 a) and c)).*

Proof. For the determination of α and β we argue as in Theorem 22, i.e. $\alpha = g$ and $\beta = c$. If the total number of transitions is A and there are c traffic items with

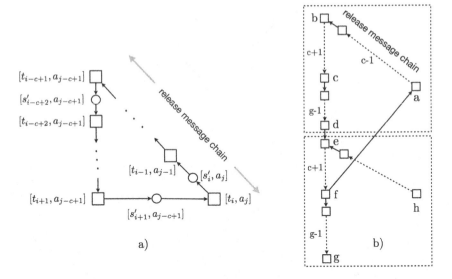

Fig. 5. Release message cycle a) in $\mathcal{C}(g, c, c, c)$ and b) in $\mathcal{C}(g, c, \frac{g \cdot c}{\Delta}, \frac{g \cdot c}{\Delta})$ with $\frac{g}{\Delta} = 2$.

the same process length, this process length is $p = \frac{A}{c}$. Petri made the assertion [5] that a cycloid with $\gamma, \delta \geq 2$ is secure, if and only if every pair of successive arcs lies on a basic cycle. A basic circuit is a cycle with exactly one place marked. This has been proved (also for the more general case of T-systems) by Stehr [6] and it is therefore called the Lemma of Petri/Stehr. Due to this lemma all the marked process cycles are disjoint. As there is a number c of disjoint communicating processes of equal length p the net to be constructed is a regular cycloid (Definition 16) and we can use the naming of Definition 18. The missing places of this net are obtained by considering condition b) of the specification in Definition 6 as follows. Transition $\langle\!\langle t_i, a_j \rangle\!\rangle$ models the step of traffic item a_j in position i. To be enabled for this step, it received a release message from $a_{(j+1) \bmod c}$ and when occurring is sending such a message to $a_{(j-1) \bmod c}$. This is modelled by new places $^\bullet[t_i, a_j] =: [s'_{i+1}, a_{j+1}]$ and $[t_i, a_j]^\bullet =: [s'_i, a_j]$ (Figs. 2 b and 6 c). As a result, we obtain additional $c \cdot p$ places, i.e. in total $|S| = 2 \cdot |T|$. To find a minimum length cycle, we next consider a sequence of transitions, which we call the *release message chain*, rm-chain for short. The rm-chain starts in some transition $[t_i, a_j]$ and continues via place $[s'_i, a_j]$ and transition $[t_{i-1}, a_{j-1}]$ down to $[t_{i-c+1}, a_{j-c+1}]$, i.e. the process of all traffic items are passed (see Fig. 5 a). Again, first and second index is computed modulo n and c, respectively. In the following the rm-chain is closed to a *release message cycle*. We continue within the process cycle a_{j-c+1} a number of c transitions until $[t_{i-c+1+c}, a_{j-c+1}] = [t_{i+1}, a_{j-c+1}]$ and $[t_{i+1}, a_{j-c+1}]^\bullet = [s'_{i+1}, a_{j-c+1}]$. For the final step by Lemma 19 there are two cases to compute $[s'_{i+1}, a_{j-c+1}]^\bullet = [t_i, a_j]$ (Fig. 5 a). If $a_{j-c+1} = a_1$ then $[s'_{i+1}, a_{j-c+1}]^\bullet = [t_{i+1+g+c-1}, a_c] = [t_i, a_j]$ since $(i+1+g+c-1) \bmod (g+$

$c) = i$ and $j - c + 1 = 1 \Rightarrow c = j$. If $a_{j-c+1} \neq a_1$ then $[s'_{i+1}, a_{j-c+1}]^\bullet = [t_i, a_{j-c+1-1}] = [t_i, a_j]$ since $(j - c) \bmod c = j$.

The length of the rm-cycle is $rm = 2 \cdot c$ transitions and there is no shorter cycle connecting the process cycles. These process cycles have the length of $p = \frac{A}{c} = \frac{c \cdot (c+g)}{c} = c + g$ transitions and are also candidates for an overall minimal cycle. Therefore the overall minimal cycle of the cycloid is $cyc = min\{2 \cdot c, c + g\}$. For a regular cycloid we construct the following recurrent transition sequence, starting with $[t_c, a_c]$ in the standard initial marking: $[t_c, a_c], [t_{c-1}, a_{c-1}], \cdots, [t_1, a_1], \cdots\cdots, [t_{c+n-1}, a_c], [t_{c+n}, a_{c-1}], \cdots][t_n, a_1]$ of length $c \cdot n$. By [9] the cycloid is strongly connected and therefore has a T-invariant of the form $(x, x, \cdots, x), x \in \mathbb{Z}$ (see reference [1]). This means that all recurrent sequences contain each transition exactly once, as it is in the constructed one and we have $A = \Gamma(c, g) = c \cdot (g + c)$ (by Definition 6 c, Theorem 4). Together with the value of cyc and the parameters $\alpha = g$, $\beta = c$ we can apply Theorem 15 to obtain $\gamma = \delta = c$ in all cases (see [8] for details).   □

In Fig. 2 b) the cycloid $\mathcal{C}(4, 3, 3, 3)$ is represented as a regular cycloid. Some regular coordinates, like $[s_0, a_1]$ and $[t_1, a_1]$, are added. One of the rm-cycles is highlighted. It has the length $rm = 2 \cdot c = 2 \cdot \beta = 6$. The step in the proof where Lemma 19 is applied becomes here $[s'_4, a_1]^\bullet = [t_3, a_3]$. Also in the next theorem the length rm of the release message cycle is important to determine cyc.

Theorem 24. *Within the class of regular cycloids $\mathcal{C}(\alpha, \beta, \gamma, \delta)$ with $\gamma, \delta \geq 2$, from the specifications of a circular traffic queue with gaps $tq\text{-}2(c, g)$ with $c > 1$ (Definition 6) using the Synthesis-Theorem 15 the cycloid $\mathcal{C}_2(g, c) := \mathcal{C}(g, c, \frac{g \cdot c}{\Delta}, \frac{g \cdot c}{\Delta})$ can be deduced. It has process length $p = \frac{g}{\Delta}(g + c)$ and minimal cycle length $cyc = min\{p, \frac{c}{g} \cdot p\}$ (in accordance with Lemma 21 a) and b)).*

Proof. The proof is the same as for Theorem 23 until the computation of the length of the rm-chain. Due to the increased length of processes the rm-chain cannot be closed as in this case. By the specification of Definition 6 d) the transition system of $tq\text{-}2(c, g)$ is a composition of $\frac{g}{\Delta}$ copies of the transition system of $tq\text{-}1(c, g)$. As informally discussed in Sect. 5 this holds also for the corresponding cycloids and is schematically represented in Fig. 5 b) for the case $\frac{g}{\Delta} = 2$. At full arcs of this graph one place is omitted. At dotted arcs with label λ a number of λ transitions is supposed (including beginning and end of the arc). For the upper copy the rm-chain from transition a via transitions b and c to d cannot be closed, but is continued to the lower copy. To this end a number $g - 1$ of additional transitions are to be passed. From e via f the rm-chain is closed to complete a rm-cycle. In the general case the structure of the upper copy is repeated a number of $\frac{g}{\Delta}$ times. Hence, the overall length of the rm-cycle is the number of transitions on a path from transition a to transition f by repeating the path from b to d a number of $\frac{g}{\Delta} - 1$ times: $rm = (c - 1) + (\frac{g}{\Delta} - 1) \cdot (c + 1 + g - 1) + c + 1 = \frac{g}{\Delta} \cdot (g + c) + (c - g) = p + (c - g)$. If $c \geq g$ the length of the rm-cycle is not smaller than the process-length p, which is the minimal cycle in this case, hence $cyc = p = \frac{g}{\Delta} \cdot (g + c)$. The model $tq\text{-}2(c, g)$ is symmetric with respect to c and g. The same proof can be made with c and g interchanged and $cyc = \frac{c}{\Delta} \cdot (g + c)$ if $c \leq g$. Using these values of cyc, and the parameters $\alpha = g$, $\beta = c$ and $A = \Xi(c, g) = \frac{g}{\Delta} \cdot (g + c) \cdot c$ (by the analogous argument as in the proof of Theorem 23) we apply Theorem 15 with $g = \alpha > \beta = c$. The resulting cycloid is $\mathcal{C}(g, c, g, c \cdot (\frac{g+c}{\Delta} - 1))$. By Theorem 11 c) this cycloid is equivalent to $\mathcal{C}(\alpha, \beta, \gamma + q \cdot \alpha, \delta - q \cdot \beta)$. With $q = \frac{c}{\Delta} - 1$ we obtain $\gamma' = \gamma + q \cdot \alpha = g + (\frac{c}{\Delta} - 1) \cdot g = \frac{g \cdot c}{\Delta}$ and $\delta' = \delta - q \cdot \beta = c \cdot (\frac{g+c}{\Delta} - 1) - (\frac{c}{\Delta} - 1) \cdot c = \frac{g \cdot c}{\Delta}$. Case $\alpha < \beta$ is similar. See [8] for details of the computation. For $\alpha = \beta$ we obtain $\gamma = \delta = \lfloor \frac{cyc}{2} \rfloor = \frac{g \cdot g}{\Delta}$. $\qquad\square$

4.2   Isomorphisms

The formal synthesis of cycloids from a less formal specification in Sect. 4.1 is completed in this section by formal analysis. We will prove that the obtained cycloids are behaviour equivalent to the circular traffic queues. This is done using operational semantics, i.e. by comparing the labelled transition systems.

Theorem 25. *a) The reachability graph of the cycloid system $(\mathcal{C}_1(g, c), M_0) :=$ $\mathcal{C}(g, c, c, c, M_0)$ is isomorphic to the labelled transition system $LTS_p(c, g)$ with $p = n = c + g$, where M_0 and ind_0 are the regular initial marking and state, respectively. The same holds for the standard initial marking.*
b) The reachability graph of the cycloid system $(\mathcal{C}_2(g, c), M_0) := \mathcal{C}(g, c, \frac{g \cdot c}{\Delta}, \frac{g \cdot c}{\Delta}, M_0)$ is isomorphic to the labelled transition system $LTS_p(c, g)$ with $p = \frac{g}{\Delta} \cdot n = \frac{g}{\Delta} \cdot (c + g)$, where M_0 and ind_0 are the regular initial marking and state, respectively. The same holds for the standard initial marking.

Proof. The reachability graph $\mathcal{RG}(\mathcal{N}, M_0) = (\mathcal{R}(\mathcal{N}, M_0), \rightarrow)$ of the cycloid net can be seen as a labelled transition system $LTS' = (States', T', tr', ind_0')$. Each marking contains the same number $n = g + c$ of marked places as this holds in the initial marking of the T-net. Therefore we also consider such a marking as an ordered set by an index function $ind' : \{1, \cdots, n\} \rightarrow S$, where S is the set of places. The labelled transition systems $LTS := LTS_p(c, g)$ and LTS' are isomorphic if there are bijective mappings φ and ψ. φ gives for each state $ind \in States$ a corresponding state $ind' = \varphi(ind) \in States'$ and ψ gives for each transition $t \in T$ a corresponding transition $t' = \psi(t) \in T'$. The following condition is required: $(ind_1, t, ind_2) \in tr \Leftrightarrow (\varphi(ind_1), \psi(t), \varphi(ind_2)) \in tr'$. The cycloids of the theorem are regular by Theorem 17 and we can use its regular coordinates (Definition 18, Lemma 19). In the remaining proof, all indices containing j are understood *modulo c* whereas indices containing i are understood *modulo n* in a) and *modulo p* in part b). We start with part a) of the theorem:

1. Definition of φ and ψ. We define $\varphi(ind) := \{ind'(i) | 1 \leq i \leq n\}$ with
$$ind'(i) := [\hat{s}_i, next(i)] \text{ and } \hat{s}_i = \begin{cases} s_{i-1} \text{ if } ind(i) \in C \\ s_i' \quad \text{if } ind(i) = u \in G \end{cases}.$$ The function *next* is defined by: $next(i) = a_j \in C$ in a position including or next to i (*mod n*). Hence, if $ind(i) = a_j \in C$ then $next(i) = a_j$ and $ind'(i) = [s_{i-1}, a_j]$. If $ind(i) \in G$ then $next(i) = a_j \in C$ is the first $a_j \in C$ after position i (*mod n*) and $ind'(i) = [s_i', a_j]$. The mapping φ is bijective since the image of a state is an

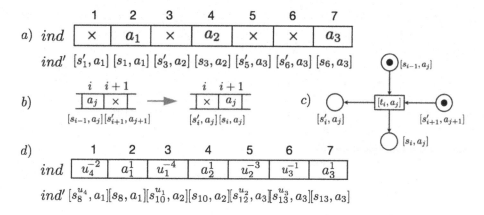

Fig. 6. Illustrating the bijections φ and ψ in the proof of Theorem 25.

encoding of this state. In Fig. 6 a) the case for a state of size $n = 7$ is shown. The mapping ψ is obviously injective since $\psi(\langle\!\langle t_i, a_j \rangle\!\rangle) = [t_i, a_j]$ and also surjective as $LTS_1(c, g)$ has a number of $\Gamma(c, g) = (c + g) \cdot c$ transitions (Theorem 4) as well as $(\mathcal{C}_1(g, c), M_0) := \mathcal{C}(g, c, c, c, M_0)$.

2. $(s_1, t, s_2) \in tr \Rightarrow (\varphi(s_1), \psi(t), \varphi(s_2)) \in tr'$. The occurrence of a transition $\langle\!\langle t_i, a_j \rangle\!\rangle$ of LTS implies that two positions are involved. $ind(i)ind(i + 1) = a_j\times$ is changed to $ind(i)ind(i + 1) = \times a_j$. By transformation with the mapping φ we obtain that $ind'(i)ind'(i + 1) = [s_{i-1}, a_j][s'_{i+1}, a_{j+1}]$ is changed to $ind'(i)ind'(i + 1) = [s'_i, a_j][s_i, a_j]$ (Fig. 6 b). This is just the result of the occurrence of the transition $\psi(\langle\!\langle t_i, a_j \rangle\!\rangle) = [t_i, a_j]$ of LTS' (Fig. 6 c).

3. $(s_1, t, s_2) \notin tr \Rightarrow (\varphi(s_1), \psi(t), \varphi(s_2)) \notin tr'$. A move of a traffic item a_j at position i by transition $\langle\!\langle t_i, a_j \rangle\!\rangle$ is impossible if and only if there is a sequence of traffic items $a_{j+1}, \cdots, a_{j+r}$ ($r \geq 1$ and all $mod\, c$) at positions $i+1, \cdots, i+r$ (all $mod\, n$) followed by a gap $\times \in G$ at position $(i + r + 1)\, mod\, n$. By the map φ in the cycloid the corresponding transitions $[t_{i+s}, a_{j+s}], (0 \leq s \leq r)$ have an empty input place $[s'_{i+s+1}, a_{j+s+1}]$, with the exception of $[t_{i+r}, a_{j+r}]$. For this transition also $[s'_{i+r+1}, a_{j+r+1}]$ is marked since a_{j+r} can make a move to exchange with the gap $\times$. This sequence is part of a release message chain (Fig. 5). Since the cycloid is live and safe [9] the chain is part of a cycle with exactly one token, which implies that $[s'_{i+1}, a_{j+1}]$ is unmarked. Hence, transition $\psi(\langle\!\langle t_i, a_j \rangle\!\rangle) = [t_i, a_j]$ cannot occur.

4. Initial state. Applying the mapping φ to Definition 1 of a regular initial state we obtain $ind'(i) = [s_{i-1}, a_i]$ for $1 \leq i \leq c$ and $ind'(i) = [s'_i, a_1]$ for $c < i \leq n$. Hence, for $1 \leq i \leq c$ traffic item a_i is in position i and for $c < i \leq n$ there is a gap and the next traffic item $modulo\, n$ is a_1. To prove the same for the standard initial state (Definition 3) we have to show that this meets Definition 12 of a standard initial marking of the cycloid. If we substitute j by $-\eta$ the bound $1 \leq j \leq c$ becomes $-1 \leq \eta \leq -c$. Then we substitute g, c and x in $j - 1 < \frac{c}{g} \cdot x \leq j$ by the cycloid parameters α, β and ξ, respectively. As a result we obtain $-\eta - 1 < \frac{\beta}{\alpha} \cdot \xi \leq -\eta$ which is equivalent to $\beta\xi + \alpha\eta \leq 0 \wedge \beta\xi + \alpha(\eta + 1) > 0$ from the definition of the backward places $s^{\leftarrow}_{\xi,\eta} \in S^{\leftarrow}_1$ of M_0 (Definition 12). In this definition the bound $-1 \leq \eta \leq -c$ is not needed since it is replaced by the quotient $/_{\equiv}$. The standard initial state is completely determined by the distribution of the gaps by the definition of the r_i. The same holds for the cycloid. In fact, the definition of the forward places $s^{\rightarrow}_{\xi,\eta} \in S^{\rightarrow}_1$ is unambiguously deducible from the definition of the backward places $s^{\leftarrow}_{\xi,\eta} \in S^{\leftarrow}_1$. For a more formal proof of this assertion consider the definition of a m-path in [9].

Figure 6 d) is the analogon to a) in the proof of part b) with respect to $\mathcal{C}(4,3,12,12)$ with 84 transitions. In principle, the proof is similar to part a), but is omitted due to space limitations and can be found in [8].   □

4.3   Representations of Circular Queues by T- and Coloured Nets

In Fig. 3 a) Petri's queue of cars is represented as net $\mathcal{N}_{basic}(c,g)$ with a regular initial marking. The positions of a number of c cars are represented by black tokens in the places $s_0, \cdots, s_{c-1}$ followed by g tokens in the complementary places $s'_{c+1}, \cdots, s'_{c+g}$ representing the gaps. By the complementary places the net is safe and the cars cannot pass each other. They cannot be distinguished which is different in the net $\mathcal{N}_{coul}(c,g)$ from part b) of Fig. 3, where the cars have identifiers $a_1, \cdots, a_c$. To handle such individual tokens a coloured net is used containing a variable x. It has been proved in [8] that applying a light modification of the standard procedure of unfolding the net into a T-net with black tokens only, a net is obtained that is isomorphic to the cycloid $\mathcal{C}_1(g,c) = \mathcal{C}(g,c,c,c)$. As a result, this cycloid is a folding of the Petri space and an unfolding of the coloured net $\mathcal{N}_{coul}(c,g)$. As shown, the net has the behaviour of a *circular traffic queue with gaps* tq-1(c,g). In the coloured net $\mathcal{N}_{sym}(c,g)$ from part c) of Fig. 3 also the places s'_k have individual tokens representing traffic items moving to the left hand side. A counter place named r has the effect that the initial marking is not reached before a number of $\frac{g}{\Delta}$ iterations of the loop. In [8] it is proved that the unfolding of this coloured net into a T-net is isomorphic to $\mathcal{C}_2(g,c) = \mathcal{C}(g,c,\frac{g \cdot c}{\Delta},\frac{g \cdot c}{\Delta})$.

5   Composition of Cyloids and Summary

In this paper the theory of cycloids is extended by new formal methods and new results concerning circular traffic queues. The use of matrix algebra leads to a more mathematical and easier handling of the cycloid equivalence relation. The formalism of regular cycloids is introduced having a process structure like circular traffic queues, but miss some of the clear mathematical properties of general cycloids. The proof of isomorphism of circular traffic queues and special cycloids is facilitated by the use of these regular cycloids as a link. The concept of release message chain and cycle is shown to be useful. It appears to be closely connected to the notion of minimal cycles which was so important in earlier publications on cycloids. Using Theorem 10 it is straight forward to deduce (see [8]), that the equivalence relation $\equiv$ becomes finer if the cycloid parameters are integer multiples. In particular $\equiv_1 \subseteq \equiv_2$ for the equivalence relations of $\mathcal{C}_1(\alpha, \beta, n \cdot \gamma, n \cdot \delta)$ and $\mathcal{C}_2(\alpha, \beta, \gamma, \delta)$ with $n \in \mathbb{N}_+$, respectively. For instance, the cycloid $\mathcal{C}(\alpha, \beta, 2 \cdot \gamma, 2 \cdot \delta)$ has a fundamental parallelogram consisting of two copies of the fundamental parallelogram of $\mathcal{C}(\alpha, \beta, \gamma, \delta)$, pasted together at the line segment $\overline{QR}$ (Fig. 1 b) of the first and the line segment $\overline{OP}$ of the second. Iterating this construction n times the so-called n-fold temporal iteration of $\mathcal{C}(\alpha, \beta, \gamma, \delta)$ is obtained and denoted by $\mathcal{C}(\alpha, \beta, \gamma, \delta)^{[n]}$. As some kind of a formal summary, this notion allows to characterize the two most important cycloids of this article by iterations of the basic cycloid $\mathcal{C}(\alpha, \beta, 1, 1)$. Table 1 contains the two models of circular traffic queues $tq\text{-}1(c, g)$ and $tq\text{-}2(c, g)$, their modelling by cycloids $\mathcal{C}_0(g, c)$, $\mathcal{C}_1(g, c)$ and $\mathcal{C}_2(g, c)$, the corresponding values of minimal cycles, the numbers of transitions and their representations as iterations of $\mathcal{C}_0(g, c)$.

Table 1. Summary of some results.

Model	Cycloid	Minmal cycle cyc	No of transitions	Iteration
$tq\text{-}1(c,g)$	$\mathcal{C}_0(g,c) = \mathcal{C}(g,c,1,1)$	2	$1 \cdot (g+c)$	$\mathcal{C}_0(g,c)^{[1]}$
$tq\text{-}1(c,g)$	$\mathcal{C}_1(g,c) = \mathcal{C}(g,c,c,c)$	$\begin{cases} 2 \cdot c & \text{if } g \geq c \\ g + c & \text{if } c > g \end{cases}$	$c \cdot (g+c)$	$\mathcal{C}_0(g,c)^{[c]}$
$tq\text{-}2(c,g)$	$\mathcal{C}_2(g,c) = \mathcal{C}(g,c,\frac{g \cdot c}{\Delta},\frac{g \cdot c}{\Delta})$	$\begin{cases} \frac{c}{\Delta}(g+c) & \text{if } g \geq c \\ \frac{g}{\Delta}(g+c) & \text{if } c > g \end{cases}$	$\frac{g \cdot c}{\Delta} \cdot (g+c)$	$\mathcal{C}_0(g,c)^{[\frac{g \cdot c}{\Delta}]}$

References

1. Desel, J., Esparza, J.: Free Choice Petri Nets. Cambridge Tracts in Theoretical Computer Science. Cambridge University Press, Cambridge (1995)
2. Kummer, O., Stehr, M.-O.: Petri's axioms of concurrency - a selection of recent results. In: Azéma, P., Balbo, G. (eds.) ICATPN 1997. LNCS, vol. 1248, pp. 195–214. Springer, Heidelberg (1997). https://doi.org/10.1007/3-540-63139-9_37
3. Petri, C.A.: On technical safety and security. Petri Net Newsl. **33**, 25–30 (1989). https://www.informatik.uni-augsburg.de/pnnl/
4. Petri, C.A.: Nets, time and space. Theor. Comput. Sci. **153**, 3–48 (1996)
5. Petri, C.A., Yuan, C.Y.: On technical safety and security (continued). Petri Net Newsl. **35**, 8–15 (1990). https://www.informatik.uni-augsburg.de/pnnl/
6. Stehr, M.O.: Characterizing security in synchronization graphs. Petri Net Newsl. **56**, 17–26 (1999). https://www.informatik.uni-augsburg.de/pnnl/
7. Valk, R.: On the structure of cycloids introduced by Carl Adam Petri. In: Khomenko, V., Roux, O.H. (eds.) PETRI NETS 2018. LNCS, vol. 10877, pp. 294–314. Springer, Cham (2018). https://doi.org/10.1007/978-3-319-91268-4_15
8. Valk, R.: Formal properties of circular traffic queues and cycloids (2019). http://uhh.de/inf-valk-traffic
9. Valk, R.: Formal properties of Petri's cycloid systems. Fundam. Informaticae **169**, 85–121 (2019)

PSPACE-Completeness of the Soundness Problem of Safe Asymmetric-Choice Workflow Nets

Guanjun Liu[✉]

Department of Computer Science, Tongji University, Shanghai 201804, China
liuguanjun@tongji.edu.cn

Abstract. Asymmetric-choice workflow nets (ACWF-nets) are an important subclass of workflow nets (WF-nets) that can model and analyse business processes of many systems, especially the interactions among multiple processes. Soundness of WF-nets is a basic property guaranteeing that these business processes are deadlock-/livelock-free and each designed action has a potential chance to be executed. Aalst *et al.* proved that the soundness problem is decidable for general WF-nets and proposed a polynomial algorithm to check the soundness for free-choice workflow nets (FCWF-nets) that are a special subclass of ACWF-nets. Tiplea *et al.* proved that for safe acyclic WF-nets the soundness problem is co-NP-complete, and we proved that for safe WF-nets the soundness problem is PSPACE-complete. We also proved that for safe ACWF-nets the soundness problem is co-NP-hard, but this paper sharpens this result by proving that for safe ACWF-nets this problem is PSPACE-complete actually. This paper provides a polynomial-time reduction from the acceptance problem of linear bounded automata (LBA) to the soundness problem of safe ACWF-nets. The kernel of the reduction is to guarantee that an LBA with an input string **does not accept** the input string if and only if the constructed safe ACWF-net is sound. Based on our reduction, we easily prove that the liveness problem of safe AC-nets is also PSPACE-complete, but the best result on this problem was NP-hardness provided by Ohta and Tsuji. Therefore, we also strengthen their result.

Keywords: Petri nets · Workflow nets · Asymmetric-choice nets · Soundness · Liveness · PSPACE-completeness

1 Introduction

Workflow nets (WF-nets) have been widely applied to the modelling and analysis of business process management systems [1–3, 21–24, 38]. To satisfy more design or analysis requirements, some extensions are proposed such as WFD-nets (workflow nets with data) [33, 36], WFT-nets (workflow nets with tables) [30] and PWNs (probabilistic workflow nets) [11, 27]. All of them use WF-nets to model execution logics of business processes including sequence, choice, loop, concurrence and/or interaction.

Soundness [1–3] is an important property required by these systems. Different types of models generally give different notions of soundness [30, 33], and even for classical

© Springer Nature Switzerland AG 2020
R. Janicki et al. (Eds.): PETRI NETS 2020, LNCS 12152, pp. 196–216, 2020.
https://doi.org/10.1007/978-3-030-51831-8_10

Table 1. Complexity results for some subclasses.

Petri net classes	Reachability	Liveness	WF-net classes	Soundness
Arbitrary	Decidable	Decidable	Arbitrary	Decidable
	EXPSPACE-hard	EXPSPACE-hard		EXPSPACE-hard
Safe	PSPACE-complete	PSPACE-complete	Safe	PSPACE-complete
Safe A.C.	**PSPACE-complete**	**PSPACE-complete**	Safe A.C.	**PSPACE-complete**
Safe F.C.	PSPACE-complete	Polynomial-time	Safe F.C.	Polynomial-time
Safe acyclic	NP-complete	Constant-time	Safe acyclic	co-NP-complete

A.C.: asymmetric-choice; F.C.: free-choice

WF-nets, there are different notions of soundness [3]. But most of them require that business processes have neither deadlock nor livelock and every designed action has a potential executed chance. This paper considers this classical notion of soundness.

It has been proven that the soundness problem is decidable and EXPSPACE-hard for general WF-nets [2,21,34]. These results are based on the fact that the problems of reachability[1], liveness, and deadlock of Petri nets are all EXPSPACE-hard [8,14]. Aalst [2] provided a polynomial-time algorithm to solve the soundness problem for free-choice WF-nets. For safe WF-nets we proved that this problem is PSPACE-complete [20]; and for safe acyclic WF-nets Tiplea *et al.* proved that this problem is co-NP-complete [32].

Asymmetric-choice workflow nets (ACWF-nets) [37] are an important subclass of WF-nets. Not only that, asymmetric-choice nets (AC-nets) [4,10] are also an important subclass of Petri nets. AC-nets not only own a stronger modelling power rather than their subclass FC-nets (free-choice nets), but also their special structures play an important role in analysing their properties such as liveness [5,10,15], response property [25] and system synthesis [35]. Free-choice structures can hardly model the interactions of multiple processes, while asymmetric-choice structures can usually handle them [1,19,26]. But so far, there are not too much theoretical results of complexity of deciding properties for AC-nets or ACWF-nets, and the existing ones are not too accurate, e.g. Ohta and Tsuji proved that the liveness problem of safe AC-nets is NP-hard [28], and we proved that the soundness problem of safe ACWF-nets is co-NP-hard [18].

Contributions. This paper sharpens these results by proving that the soundness problem of safe ACWF-nets and the liveness problem of safe AC-nets are both PSPACE-complete. We can construct a safe ACWF-net for any linear bounded automaton (LBA) with an input string by the following idea: if an LBA with an input string accepts the string, the constructed safe ACWF-net is not sound; if it does not accept the string, the constructed safe ACWF-net is sound. In addition, based on the PSPACE-completeness of the soundness problem of safe ACWF-nets and the equivalence between the soundness problem and the liveness and boundedness problem in WF-nets [2], we can infer

---

[1] The newest research shows that the reachability problem of general Petri nets is not elementary: a TOWER lower bound was found and proven by Czerwiński *et al.* [9], and an ACKERMANN upper bound was found and proven by Leroux and Schmitz [17].

that the liveness problem of safe AC-nets is PSPACE-complete. We also show that the reachability problem of safe AC-nets is PSPACE-complete.

The complexity results of reachability, liveness and soundness for some subclasses of Petri nets are summarised in Table 1 in which the results in bold type are obtained in this paper and others can be found in [2,10,20,32].

The rest of the paper is organised as follows. Section 2 reviews Petri nets, WF-nets and and the LBA acceptance problem. Section 3 introduces our results. Section 4 concludes this paper briefly.

2   Preliminary

In this section, we review some notions of Petri nets, WF-nets and the LBA acceptance problem. For more details, one can refer to [3,13,29].

2.1   Petri Nets

We denote $\mathbb{N} = \{0, 1, 2, \cdots\}$ as the set of nonnegative integers.

Definition 1 (Net). *A net is a 3-tuple* $N = (P, T, F)$ *where* P *is a set of places,* T *is a set of transitions,* $F \subseteq (P \times T) \cup (T \times P)$ *is a set of arcs,* $P \cup T \neq \varnothing$, *and* $P \cap T = \varnothing$.

A transition t is called an *input transition* of a place p and p is called an *output place* of t if $(t, p) \in F$. *Input place* and *output transition* can be defined similarly. Given a net $N = (P, T, F)$ and a node $x \in P \cup T$, the *pre-set* and *post-set* of x are defined as $^\bullet x = \{y \in P \cup T | (y, x) \in F\}$ and $x^\bullet = \{y \in P \cup T | (x, y) \in F\}$, respectively.

A *marking* of $N = (P, T, F)$ is a mapping $M: P \to \mathbb{N}$. A marking may be viewed as a $|P|$-dimensional nonnegative integer vector in which every element represents the number of tokens in the corresponding place at this marking, e.g. marking $M = (1, 0, 6, 0)$ over $P = \{p_1, p_2, p_3, p_4\}$ represents that at M, places p_1, p_2, p_3 and p_4 have 1, 0, 6, and 0 tokens, respectively. Note that we assume a total order on P so that the k-th entry in the vector corresponds to the k-th place in the ordered set. For convenience, a marking M is denoted as a multi-set $M = \{M(p) \cdot p | \forall p \in P\}$ in this paper. For the above example, it is written as $M = \{p_1, 6p_3\}$. A net N with an *initial marking* M_0 is called a *Petri net* and denoted as (N, M_0).

A place $p \in P$ is *marked* at M if $M(p) > 0$. A transition t is *enabled* at M if $\forall p \in {}^\bullet t: M(p) > 0$, which is denoted as $M[t\rangle$. *Firing* an enabled transition t leads to a new marking M', which is denoted as $M[t\rangle M'$ and satisfies that $M'(p) = M(p) - 1$ if $p \in {}^\bullet t \setminus t^\bullet$; $M'(p) = M(p) + 1$ if $p \in t^\bullet \setminus {}^\bullet t$; and $M'(p) = M(p)$ otherwise.

A marking M_k is *reachable* from a marking M if there is a firing sequence $\sigma = t_1 t_2 \cdots t_k$ such that $M[t_1\rangle M_1[t_2\rangle \cdots \rangle M_{k-1}[t_k\rangle M_k$. The above representation can be abbreviated to $M[\sigma\rangle M_k$, meaning that M reaches M_k after firing sequence σ. The set of all markings reachable from a marking M in a net N is denoted as $R(N, M)$. A marking M' is *always reachable* from a marking M in a net N if $\forall M'' \in R(N, M)$: $M' \in R(N, M'')$.

Given a Petri net $(N, M_0) = (P, T, F, M_0)$, a transition $t \in T$ is *live* if $\forall M \in R(N, M_0), \exists M' \in R(N, M): M'[t\rangle$. A Petri net is *live* if every transition in it is live.

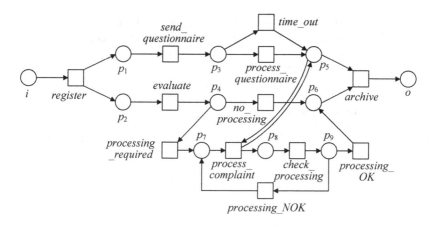

Fig. 1. A safe ACWF-net [2].

A Petri net $(N, M_0) = (P\ T, F, M_0)$ is *bounded* if $\forall p \in P, \exists k \in \mathbb{N}, \forall M \in R(N, M_0): M(p) \leq k$. A Petri net is *safe* if each place has at most 1 token in each reachable marking.

If a net $N = (P, T, F)$ satisfies $\forall p_1, p_2 \in P: p_1^\bullet \cap p_2^\bullet \neq \varnothing \Rightarrow p_1^\bullet \supseteq p_2^\bullet \vee p_1^\bullet \subseteq p_2^\bullet$, then it is called an *asymmetric-choice net* (AC-net). A *free-choice net* (FC-net) is a special AC-net such that $\forall p_1, p_2 \in P: p_1^\bullet \cap p_2^\bullet \neq \varnothing \wedge p_1 \neq p_2 \Rightarrow |p_1^\bullet| = |p_2^\bullet| = 1$.

2.2   WF-nets

Definition 2 (WF-net). *A net $N = (P, T, F)$ is a WF-net if*

1. *the net N has two special places i and o where $i \in P$ is called source place such that $^\bullet i = \varnothing$ and $o \in P$ is called sink place such that $o^\bullet = \varnothing$; and*
2. *the short-circuit $\overline{N} = (P, T \cup \{t_0\}, F \cup \{(t_0, i), (o, t_0)\})$ of N is strongly connected where $t_0 \notin T$.*

A WF-net is a special Petri net that has a source place representing the beginning of a task and a sink place representing the ending of the task, and for every node in the net there exists a directed path from the source place to the sink place through that node. A safe WF-net means that the WF-net is safe for the initial marking $\{i\}$, and an asymmetric-choice WF-net (ACWF-net) is a WF-net that is also an AC-net. For example, Fig. 1 shows a safe ACWF-net which is from literature [2].

Definition 3 (Soundness). *A WF-net $N = (P, T, F)$ is sound if the following conditions hold:*

1. $\forall M \in R(N, \{i\}): \{o\} \in R(N, M)$;
2. $\forall M \in R(N, \{i\}): M \geq \{o\} \Rightarrow M = \{o\}$; *and*
3. $\forall t \in T, \exists M \in R(N, \{i\}): M[t\rangle$.

The second condition in Definition. 3 can be removed because it is implied by the first one [3]. Weak soundness is a looser property than soundness since the former does not concern the firable chance of every transition. A WF-net $N = (P, T, F)$ is *weakly sound* if $\forall M \in R(N, \{i\})$: $\{o\} \in R(N, M)$.

2.3 LBA Acceptance Problem

An LBA is a Turing machine that has a finite tape containing initially an input string with a pair of bound symbols on either side.

Definition 4 (LBA). *A 6-tuple $\Omega = (Q, \Gamma, \Sigma, \Delta, \#, \$)$ is an LBA if*

1. $Q = \{q_1, \cdots, q_m\}$ *is a set of control states where $m \geq 1$, q_1 is the initial state and q_m is the acceptance state by default;*
2. $\Gamma = \{a_1, a_2, \cdots, a_n\}$ *is a tape alphabet where $n \geq 1$;*
3. $\Sigma \subseteq \Gamma$ *is an input alphabet;*
4. $\Delta \subseteq Q \times \Gamma \times \{R, L\} \times Q \times \Gamma$ *is a set of transitions where R and L represent respectively that the read/write head moves right or left by one cell; and*
5. $\# \notin \Gamma$ *and $\$ \notin \Gamma$ are two bound symbols that are next to the left and right sides of an input string, respectively.*

If an LBA is at state q_1, the read/write head is scanning a cell in which symbol a_1 is stored, and there is a transition $\delta = (q_1, a_1, R, q_2, a_2) \in \Delta$, then δ can be fired and firing it leads to the following consequences: 1) the read/write head erases a_1 from the cell, writes a_2 in the cell, and moves right by one cell; and 2) the state of the LBA becomes q_2. The case of left move can be understood similarly.

Given an LBA Ω with an input string S, an *instantaneous description* $[S_l q a S_r]$ means that the control state is q, the string in the tape is $S_l a S_r$ (note: the string $S_l a S_r$ contains the two bound symbols) and the read/write head is reading the symbol a. The *initial configuration* of Ω is an instantaneous description $[q_1 \# S\$]$, i.e., the LBA is at the initial state q_1, and the read/write head stays on the cell storing $\#$. A *final configuration* is an instantaneous description where the LBA is at the acceptance state. A *computation step* means that the LBA can go from an instantaneous description to another instantaneous description by firing a transition of Δ. A *computation* is a succession of computation steps from the initial configuration. Here, we do not consider whether a computation makes the LBA enter the acceptance state. The LBA *accepts* this input string if there is a computation making the LBA enter its acceptance state.

Those two cells storing bound symbols $\#$ and $\$$ are not allowed to store other symbols in any computation. Therefore, a transition of scanning bound symbol $\#$ (resp. $\$$) can only write $\#$ (resp. $\$$) into the cell and then the read/write head can only move right (resp. left) by one cell. Additionally, there is no transition leaving from the acceptance state q_m because the LBA halts correctly once q_m is reached.

LBA Acceptance Problem: Given an LBA with an input string, does it accept the string?

This problem is PSPACE-complete even for deterministic LBA [13].

3   PSPACE-completeness of the Soundness Problem of Safe ACWF-nets

First, we construct in polynomial time a safe ACWF-net for an LBA with an input string. Then, we prove that an LBA with an input string does not accept the string if and only if the constructed safe ACWF-net is sound. These results consequently indicate that the soundness problem of safe ACWF-nets is PSPACE-hard. Finally, an existing PSPACE algorithm that can decide the soundness problem for any safe WF-net means that this problem is PSPACE-complete. In order to construct a safe ACWF-net, we first construct a safe FC-net that can *weakly simulate* the computations of a given LAB with an input string.

3.1   Construct a Safe FC-net to Weakly Simulate the Computations of an LBA with an Input String

Given an LBA $\Omega = (Q, \Gamma, \Sigma, \Delta, \#, \$)$ with an input string S, we assume that the length of S is l ($l \geq 0$) and the j-th element of S is denoted as S_j. We denote $Q = \{q_1, \cdots, q_m\}$ and $\Gamma = \{a_1, \cdots, a_n\}$ where $m \geq 1$, $n \geq 1$, q_1 is the initial state and q_m is the acceptance state. Cells storing $\#S\$$ are labeled by 0, 1, $\cdots$, l, and $l + 1$, respectively.

We use places $A_{0,\#}$ and $A_{l+1,\$}$ to simulate tape cells 0 and $l + 1$, respectively. A token in $A_{0,\#}$ (resp. $A_{l+1,\$}$) means that tape cell 0 (resp. $l + 1$) stores $\#$ (resp. $\$$). Because the two bound symbols cannot be replaced by other symbols in the computing process, places $A_{0,\#}$ and $A_{l+1,\$}$ have been marked in the whole simulating process.

For each tape cell $k \in \{1, \cdots, l\}$, we construct a set of places $A_{k,a_1}, \cdots$, and A_{k,a_n} to simulate every possible stored symbol in that cell. They correspond to a_1, $\cdots$, and a_n, respectively. Therefore, a token in A_{k,a_j} means that the symbol in cell k is a_j. Obviously, one and only one of places $A_{k,a_1}, \cdots$, and A_{k,a_n} has a token in the simulating process since every tape cell only stores one symbol at any time in the computing process. These places are called *symbol places* for convenience.

For each tape cell $k \in \{0, 1, \cdots, l, l + 1\}$, we also construct a set of places $B_{k,q_1}, \cdots$, and $B_{k,q_{m-1}}$ to simulate every possible state the machine is keeping when the read/write head is scanning that cell. They correspond to q_1, $\cdots$, and q_{m-1}, respectively. A token in B_{k,q_j} means that the read/write head is on the cell k and the machine is at state q_j. Because an LBA halts correctly once it enters state q_m, we construct a place B_{q_m} to model this state. Because the LBA is only at one state and the read/write head is only scanning one tape cell at any time in the computing process, only one place in $\{B_{q_m}\} \cup \{B_{k,q_j} | k \in \{0, 1, \cdots, l, l + 1\}, j \in \{1, \cdots, m - 1\}\}$ is marked by one token in the simulating process. These places are called *state places* for convenience.

For each cell $k \in \{1, \cdots, l\}$ and each LBA transition $\delta \in \Delta$, if we use the method in [10, 20] to construct a net transition simulating δ, the constructed net cannot satisfy the *asymmetric-choice* requirement. Hence, we need some special treatments as follows.

For a given tape cell $k \in \{1, \cdots, l\}$ and a given transition $\delta = (q_{j_1}, a_{j_2}, D, q_{j_3}, a_{j_4})$ where $q_{j_1} \in Q \setminus \{q_m\}$ (i.e., $j_1 \in \{1, \cdots, m - 1\}$), $q_{j_3} \in Q$ (i.e., $j_3 \in \{1, \cdots, m\}$), $a_{j_2} \in \Gamma$ (i.e., $j_2 \in \{1, \cdots, n\}$), $a_{j_4} \in \Gamma$ (i.e., $j_4 \in \{1, \cdots, n\}$) and $D \in \{L, R\}$, we construct two places denoted as $\langle k, \delta, q_{j_1} \rangle$ and $\langle k, \delta, a_{j_2} \rangle$. In fact, place $\langle k, \delta, q_{j_1} \rangle$ is an *image* of state place $B_{k,q_{j_1}}$ w.r.t. δ, and place $\langle k, \delta, a_{j_2} \rangle$ is an *image* of symbol place $A_{k,a_{j_2}}$ w.r.t. δ. For k and $\delta = (q_{j_1}, a_{j_2}, D, q_{j_3}, a_{j_4})$, we now can construct a corresponding net transition which is denoted as $\langle k, \delta \rangle$ and satisfies that

$$- \ ^\bullet\langle k, \delta \rangle = \{\langle k, \delta, q_{j_1} \rangle, \langle k, \delta, a_{j_2} \rangle\}$$

$$- \ \langle k, \delta \rangle^\bullet = \begin{cases} \{A_{k,a_{j_4}}, B_{k+1,q_{j_3}}\} & q_{j_3} \neq q_m \wedge D = R \\ \{A_{k,a_{j_4}}, B_{k-1,q_{j_3}}\} & q_{j_3} \neq q_m \wedge D = L \\ \{B_{q_m}\} & q_{j_3} = q_m \end{cases}$$

When the read/write head is scanning tape cell k and LBA transition $\delta = (q_{j_1}, a_{j_2}, D, q_{j_3}, a_{j_4})$ can be fired, this means that the LBA is at state q_{j_1} and cell k stores symbol a_{j_2}. Therefore, the input places of net transition $\langle k, \delta \rangle$ should be $\langle k, \delta, q_{j_1} \rangle$ and $\langle k, \delta, a_{j_2} \rangle$. We should consider three cases for the output places of $\langle k, \delta \rangle$: 1) firing δ leads to a non-acceptance state and the read/write head moves right by one cell; 2) firing δ leads to a non-acceptance state and the read/write head moves left by one cell; and 3) firing δ leads to the acceptance state. The construction of the output places of $\langle k, \delta \rangle$ exactly match the three cases. The third case is noteworthy that: if firing δ leads to the acceptance state, then the output of the constructed transition $\langle k, \delta \rangle$ is exactly B_{q_m} (currently); in other words, the constructed transition does not consider the rewriting to the related cell or the moving of the read/write head since the machine halts correctly; but later, we will consider other outputs of this kind of constructed transition in order to reflect that the symbol in the related cell has been deleted.

Since places $\langle k, \delta, q_{j_1} \rangle$ and $\langle k, \delta, a_{j_2} \rangle$ are respectively an image of places $B_{k,q_{j_1}}$ and $A_{k,a_{j_2}}$ w.r.t. $\delta = (q_{j_1}, a_{j_2}, D, q_{j_3}, a_{j_4})$, we should construct some transitions, named *allocation transitions*, to connect them. For places $\langle k, \delta, q_{j_1} \rangle$ and $B_{k,q_{j_1}}$, we construct an allocation transition $\varepsilon_{k,\delta}^{q_{j_1}}$ such that

$$- \ ^\bullet\varepsilon_{k,\delta}^{q_{j_1}} = \{B_{k,q_{j_1}}\}$$

$$- \ \varepsilon_{k,\delta}^{q_{j_1}\bullet} = \{\langle k, \delta, q_{j_1} \rangle\}$$

and for places $\langle k, \delta, a_{j_2} \rangle$ and $A_{k,a_{j_2}}$, we construct an allocation transition $\varepsilon_{k,\delta}^{a_{j_2}}$ such that

$$- \ ^\bullet\varepsilon_{k,\delta}^{a_{j_2}} = \{A_{k,a_{j_2}}\}$$

$$- \ \varepsilon_{k,\delta}^{a_{j_2}\bullet} = \{\langle k, \delta, a_{j_2} \rangle\}$$

If a symbol or a state is associated with multiple LBA transitions (as their input), then a related symbol place or state place is associated with multiple allocation transitions. In other words, an allocation transition is to, in a free-choice form, allocate a state (if the related state place is marked) or a symbol (if the related symbol place is marked) to one of its images. For example, Fig. 2 shows the constructions for the following three LBA transitions: $\delta_1 = (q_1, a_1, R, q_2, a_2)$, $\delta_2 = (q_1, a_2, L, q_m, a_2)$ and $\delta_3 = (q_2, a_2, R, q_m, a_1)$ when considering the case of tape cell k. Note that q_m is the acceptance state. In the figure, allocation transitions are denoted by ε for simplification. Since symbol a_2 is associated with transitions δ_2 and δ_3, the symbol place A_{k,a_2} has two images: $\langle k, \delta_2, a_2 \rangle$ and $\langle k, \delta_3, a_2 \rangle$, that are an input place of transitions $\langle k, \delta_2 \rangle$ and $\langle k, \delta_3 \rangle$, respectively.

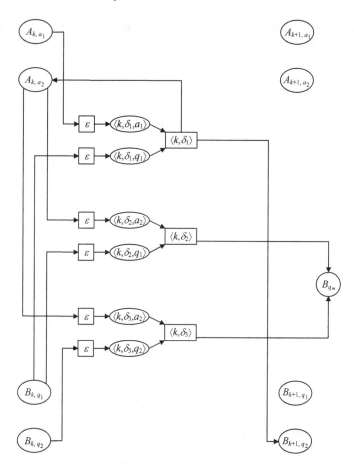

Fig. 2. Illustration of constructing a safe FC-net weakly simulating an LBA with an input string.

For tape cell 0 and an LBA transition $\delta = (q_{j_1}, \#, R, q_{j_3}, \#)$, we construct two places $\langle 0, \delta, \# \rangle$ and $\langle 0, \delta, q_{j_1} \rangle$ and a net transition $\langle 0, \delta \rangle$ such that

- ${}^\bullet \langle 0, \delta \rangle = \{ \langle 0, \delta, q_{j_1} \rangle, \langle 0, \delta, \# \rangle \}$
- $\langle 0, \delta \rangle^\bullet = \begin{cases} \{ A_{0,\#}, B_{1,q_{j_3}} \} & q_{j_3} \neq q_m \\ \{ B_{q_m} \} & q_{j_3} = q_m \end{cases}$

And we construct two allocation transitions $\varepsilon_{0,\delta}^{q_{j_1}}$ and $\varepsilon_{0,\delta}^{\#}$ to respectively connect $B_{0,q_{j_1}}$ and $\langle 0, \delta, q_{j_1} \rangle$ as well as $A_{0,\#}$ and $\langle 0, \delta, \# \rangle$, i.e.,

- ${}^\bullet \varepsilon_{0,\delta}^{q_{j_1}} = \{ B_{0,q_{j_1}} \}$
- $\varepsilon_{0,\delta}^{q_{j_1}}{}^\bullet = \{ \langle 0, \delta, q_{j_1} \rangle \}$
- ${}^\bullet \varepsilon_{0,\delta}^{\#} = \{ A_{0,\#} \}$
- $\varepsilon_{0,\delta}^{\#}{}^\bullet = \{ \langle 0, \delta, \# \rangle \}$

Similarly, we can construct places and transitions for tape cell $l + 1$ and LBA transition $\delta = (q_{j_1}, \$, L, q_{j_3}, \$)$. Here we do not introduce them any more. But, we should note that the above constructed net is free-choice.

Now, we finish the construction of an FC-net *weakly simulating* an LBA with an input string if we configure such an initial marking M_0: $M_0(A_{0,\#}) = M_0(A_{l+1,\$}) = M_0(B_{0,q_1}) = 1$, $M_0(A_{k,a_j}) = 1$ if cell k stores a_j ($k \in \{1, \cdots, l\}$ and $j \in \{1, \cdots, n\}$), and other places have no token. The token in B_{0,q_1}, i.e. $M_0(B_{0,q_1}) = 1$, means that the LBA is at the initial state q_1 and the read/write head stays on the tape cell 0. Other marked places represent the input string. Obviously, the constructed net is safe at M_0.

The reasons we call it *weakly simulating* are that:

– the following statement holds:
  an LBA with an input string accepts the string if and only if a marking in which place B_{q_m} has a token is **reachable** in the constructed Petri net,
– but the following one does not hold:
  an LBA with an input string accepts the string if and only if a marking in which place B_{q_m} has a token is **always reachable** in the constructed Petri net.

For an LBA accepting its input string, the above construction cannot guarantee that B_{q_m} will always be marked due to the free-choice-ness of those allocation transitions; in other words, those allocation transitions possibly result in a dead marking while B_{q_m} is not marked by the dead marking. Certainly, what we can make sure has that: 1) for an LBA not accepting its input string, state place B_{q_m} will never be marked in the constructed Petri net; and 2) for an LBA accepting its input string, there is a reachable marking at which B_{q_m} has a token. For example, we consider the case that A_{k,a_2} and B_{k,q_1} both have a token in Fig. 2. If the token in A_{k,a_2} is allocated to $\langle k, \delta_3, a_2 \rangle$ but the token in B_{k,q_1} is allocated to $\langle k, \delta_1, q_1 \rangle$ or $\langle k, \delta_2, q_1 \rangle$, then a deadlock occurs; only if the token in A_{k,a_2} is allocated to $\langle k, \delta_2, a_2 \rangle$ and the token in B_{k,q_1} is allocated to $\langle k, \delta_2, q_1 \rangle$, then transition $\langle k, \delta_2 \rangle$ can be fired and then puts a token into B_{q_m}.

Anyway, we can still utilise such a model to construct a safe ACWF-net satisfying the following requirements:

– if an LBA with an input string accepts the string, i.e., there is a reachable marking in the constructed safe ACWF-net that marks B_{q_m}, then a deadlock can be immediately caused after B_{q_m} is marked, i.e., the safe ACWF-net is not sound for this case;
– if an LBA with an input string does not accept the string, then the marking $\{o\}$ is always reachable and each transition has a potential chance to be executed, i.e., the constructed safe ACWF-net is sound for this case.

Simply speaking, an LBA with an input string does not accept the string if and only the constructed safe ACWF-net is sound. Our idea is to design a controller based on the above construction that makes the final net a safe ACWF-net and it satisfies the above requirements. As mentioned above, there is only one token in all state places and their images in the whole weakly-simulating process; for convenience, we call the token in state places or their images the *state token*.

3.2 Construct a Safe ACWF-net Based on the Constructed FC-net

In order to easily understand and clearly illustrate our control, we split it into three steps to introduce. The first step has two main purposes: 1) when the LBA accepts its input string, a deadlock is reachable; and 2) when the LBA does not accept the input string, the state token can freely move in state places and their images. The main purpose of the second control step is that all symbols can be deleted from the tape when the LBA does not accept its input string. After all symbols are deleted from the tape, our control either makes the net system enter the final marking $\{o\}$, or re-configures an arbitrary input string so that every transition has an enabled chance (and according to the second step, these symbols can be deleted again). These are the purposes of the third step.

3.2.1 The Control of Moving the State Token Freely or Entering a Deadlock

As shown in Fig. 3, we first construct the source place i, other 8 control places $\{v_1, \cdots, v_8\}$ and 5 transitions $\{t_1, \cdots, t_5\}$. Transition t_1 is constructed to initialise the LBA and put one token into control places v_1 and v_2 respectively, i.e.,

$$- \ {}^\bullet t_1 = \{i\}$$
$$- \ t_1^\bullet = \{v_1, v_2, B_{0,q_1}, A_{0,\#}, A_{l+1,\$}\} \cup \{A_{k,a_j} | S_k = a_j, k \in \{1, \cdots, l\}, j \in \{1, \cdots, n\}\}$$

One purpose of our control is that once a token enters B_{q_m} in the weakly-simulating process (i.e., the LBA accepts its input), the constructed ACWF-net can enter a deadlock (and thus it is not sound); therefore, we construct transition t_3 to realise this purpose:

$$- \ {}^\bullet t_3 = \{B_{q_m}, v_1\}$$
$$- \ t_3^\bullet = \{v_4\}$$

Obviously, if t_3 is fired before t_2, then only control places v_2 and v_4 are marked[2] and thus a deadlock occurs, where the construction of t_2 is:

$$- \ {}^\bullet t_2 = \{v_1, v_2\}$$
$$- \ t_2^\bullet = \{v_3, v_7\}$$

If no token enters B_{q_m} in the whole weakly-simulating process (i.e., the LBA does not accept its input), then t_3 can never be fired before t_2. For this case, the purpose of firing t_2 is to delete all symbols from the tape cells (and thus ensure the constructed ACWF-net is sound). In order to achieve this purpose, we should make the state token freely move in the state places and their images.

First, we need to take back the state token from images of state places so that the state token can be reallocated. Hence, for each image of each state place we construct a *recycling transition* whose input is that image and output is control place v_8 (i.e., the state token is always put into v_8 when it is taken back). For example in Fig. 3, $\langle k, \delta_1, q_1 \rangle$ is an image of state place B_{k,q_1}, and thus a recycling transition is constructed to connect $\langle k, \delta_1, q_1 \rangle$ and v_8. In Fig. 3, these recycling transitions are also denoted as ε for simplification. Note that although these recycling transitions can move the state

---

[2] Control place v_6 is also marked for this case and later we will introduce it.

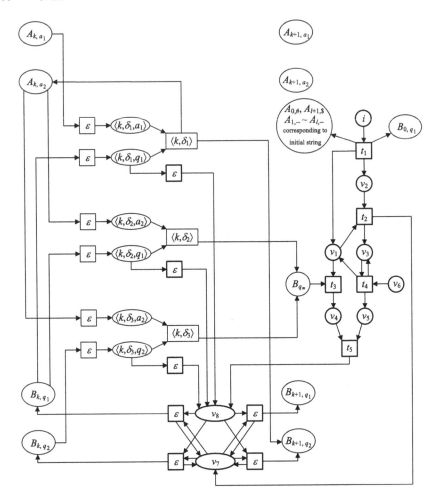

Fig. 3. Illustration of controlling the state token.

token into v_8, this does not destroy the fact that there exists a reachable marking which marks B_{q_m} if the LBA accepts its input. In order to re-allocate this state token in v_8 to a state place (in a free-choice form), say B_{k,q_j}, we construct a *re-allocation transition* $\varepsilon_{B_{k,q_j}}$ to connect v_8 and B_{k,q_j}. In order to ensure that a re-allocation cannot take place when the LBA accepts its input and t_3 is fired, each re-allocation transition is also connected with control place v_7 by a self-loop (since v_7 has a token only after firing t_2), i.e.,

- ${}^{\bullet}\varepsilon_{B_{k,q_j}} = \{v_7, v_8\}$
- $\varepsilon_{B_{k,q_j}}^{\bullet} = \{B_{k,q_j}, v_7\}$

Figure 3 illustrates these re-allocation transitions clearly and they are also denoted as ε for simplification. Note that when we add a re-allocation transition to a state place

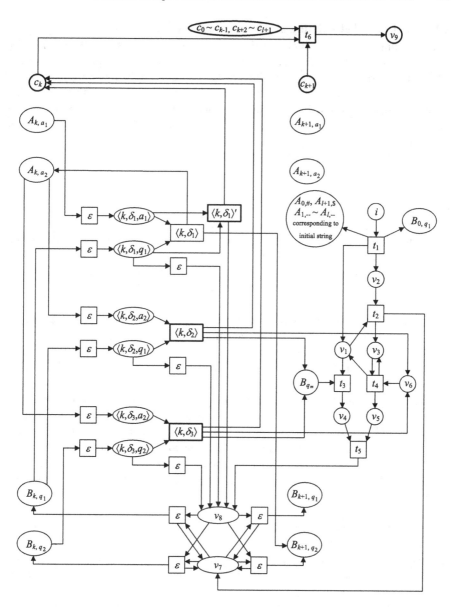

Fig. 4. Illustration of deleting symbols from the tape.

but the state place does not have an output, we should construct a recycling transition for this state place (i.e., the input of this recycling transition is the state place and its output is v_8) so that the state token re-allocated to it can be taken back again.

3.2.2 The Control of Deleting Symbols from the Tape

These recycling and re-allocation transitions guarantee that the state token can freely move in state places and their images when the LBA does not accept its input string, but their final purpose is to delete symbols from tape cells via the free movement of the state token matching that token in each tape cell. Therefore, for each net transition that corresponds to an LBA transition and does not output a token into B_{q_m}, say $\langle k, \delta \rangle$ with $\delta = (q_{j_1}, a_{j_2}, D, q_{j_3}, a_{j_4})$ and $q_{j_3} \neq q_m$, we construct a *deletion transition* $\langle k, \delta \rangle'$ that has the same inputs with $\langle k, \delta \rangle$ but has different outputs, i.e.,

- ${}^\bullet \langle k, \delta \rangle' = {}^\bullet \langle k, \delta \rangle = \{ \langle k, \delta, q_{j_1} \rangle, \langle k, \delta, a_{j_2} \rangle \}$
- $\langle k, \delta \rangle'^\bullet = \{ v_8, c_k \}$

where place c_k is used to represent whether the symbol in tape cell k is deleted or not, $\forall k \in \{0, 1, \cdots, l, l+1\}$. Because transitions $\langle k, \delta \rangle$ and $\langle k, \delta \rangle'$ are in a free-choice relation, the former is still enabled when the latter is enabled; in other words, we do not control the firing right of $\langle k, \delta \rangle$ since firing it does not disturb our purpose (this is because the state token in $B_{l+1,q_{j_3}}$ or $B_{l-1,q_{j_3}}$ can still be recycled after firing $\langle k, \delta \rangle$). For example, we construct such a deletion transition $\langle k, \delta_1 \rangle'$ in Fig. 4.

However, we should carefully deal with the case of such a net transition (say $\langle k, \delta \rangle$ with $\delta = (q_{j_1}, a_{j_2}, D, q_m, a_{j_4})$) that makes the LBA enter the acceptance state. In fact, we do not need to construct a deletion transition for such a transition; if we construct it, the original one is still enabled when it is enabled; consequently, firing the original one will still put a token into B_{q_m} so that we have to move the token from B_{q_m} to v_8. Therefore, we only need to modify the outputs of such a transition instead of constructing a deletion transition for it. In the previous construction of FC-net, the output of such a transition was B_{q_m}. Here, we add two outputs for it: c_k and v_6, e.g. the outputs of transitions $\langle k, \delta_2 \rangle$ and $\langle k, \delta_3 \rangle$ in Fig. 4. Place c_k as one output is used to represent that the symbol in the related tape cell has been deleted when such a transition is fired. Control place v_6 as one output is used to tell the controller to move the state token from B_{q_m} to v_8 in order to continuously delete other symbols in the tape. Firing the transition sequence $t_4 t_3 t_5$ will move the state token from B_{q_m} to v_8.

From Fig. 4 we can see that this part of the controller cannot only maintain the asymmetric-choice structure but also ensures that: 1) if t_3 can be fired previous to t_2 (i.e., the LBA accepts its input string), the net can reach a deadlock; 2) if t_2 is fired previous to t_3 (whether the LBA accepts its input string or not), all symbols in tape cells can finally be deleted; and 3) if the LBA does not accept its input string, then t_3 can never be fired previous to t_2.

When places $c_0, c_1, \cdots, c_l$ and c_{l+1} are all marked, symbols in tape cells are all deleted. Then, transition t_6 whose preset is $\{c_0, c_1, \cdots, c_l, c_{l+1}\}$ can be fired.

3.2.3 The Control of Reconfiguring an Arbitrary Input String or Entering the Final Marking

As shown in Fig. 5, after firing transition t_6, transition t_7 can be fired so that tokens in the control places v_7 and v_8 can finally be removed by firing transition t_8 and thus the token in control place v_3 is removed by firing transition t_9. Finally, only the sink place o is put one token into.

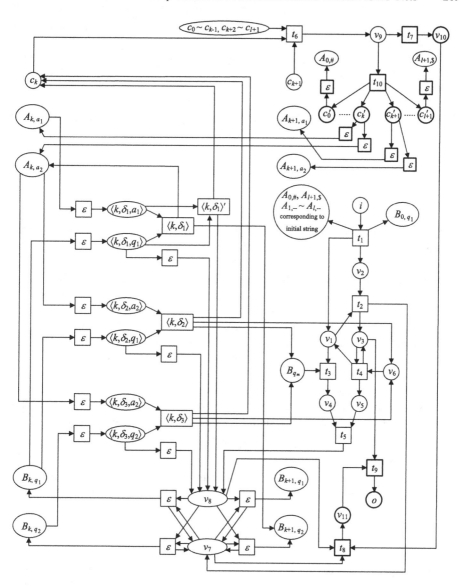

Fig. 5. Illustration of reconfiguring an arbitrary input string and entering the final marking.

Note that in the net constructed above, some symbol places possibly have no input transition and thus some constructed transitions as the outputs of these symbol places have no chance to be fired. Therefore, we construct a transition t_{10} and a group of places $\{c'_0, c'_1, \cdots, c'_l, c'_{l+1}\}$, as shown in Fig. 5, to reconfigure an arbitrary input string, i.e.,

- $\bullet t_{10} = \{v_9\}$
- $t_{10}^\bullet = \{c'_0, c'_1, \cdots, c'_l, c'_{l+1}\}$

And for each $k \in \{1, \cdots, l\}$ and $j \in \{1, \cdots, n\}$, we construct a *reconfiguration transition* $\varepsilon_{A_{k,a_j}}$ such that

- $^\bullet\varepsilon_{A_{k,a_j}} = \{c'_k\}$
- $\varepsilon^\bullet_{A_{k,a_j}} = \{A_{k,a_j}\}$

Specially, for c'_0 and c'_{l+1} we construct reconfiguration transitions $\varepsilon_{A_{0,\#}}$ and $\varepsilon_{A_{l+1,\$}}$ such that

- $^\bullet\varepsilon_{A_{0,\#}} = \{c'_0\}$
- $\varepsilon^\bullet_{A_{0,\#}} = \{A_{0,\#}\}$
- $^\bullet\varepsilon_{A_{l+1,\$}} = \{c'_{l+1}\}$
- $\varepsilon^\bullet_{A_{l+1,\$}} = \{A_{l+1,\$}\}$

Because those reconfiguration transitions associated with the same tape cell are in a free-choice relation, we can obtain an arbitrary input string. This can ensure that every transition has an enabled chance. Additionally, these symbols can still be deleted again from the tape according to the second control step.

Figure 5 shows the re-configuration clearly where re-configuration transitions are denoted as ε for simplification. After firing t_6, to fire t_{10} instead of t_7 can produce any possible input string and they all can be deleted again. Therefore, transition t_{10} has two functions: 1) it makes the net is a WF-net since for each node in the net there is a directed path from the source place i to the sink place o containing that node; and 2) it guarantees that each transition[3] has a potential chance to be fired (whether the LBA accepts the original input string or not).

Now, we finish our construction. From the above construction, we easily see that the constructed net is an ACWF-net and it is safe for the initial marking $\{i\}$.

3.2.4   An Example

We utilise a simple example to illustrate our constructions and the example comes from literature [20]. The LBA $\Omega_0 = (Q, \Gamma, \Sigma, \Delta, \#, \$)$ can produce the language

$$\{a^{i_1}b^{i_1}a^{i_2}b^{i_2} \cdots a^{i_m}b^{i_k} | i_1, i_2, \cdots, i_k, k \in \mathbb{N}\}$$

---

[3] Here, transition t_3 also has a potential chance to be fired when the LBA does not accept its input. The reason is that there is a reconfiguration enabling such a transition whose outputs are B_{q_m} and v_6, and thus t_3 has a fired chance. Maybe there exists such a case: the input string is empty and B_{q_m} is not an output of any transition in the constructed FC-net. For this case, we only needs a minor modification to our construction of FC-net: for tape cells 0 and $l + 1$, we also construct symbol places $A_{0,a_1}, \cdots, A_{0,a_n}, A_{l+1,a_1}, \cdots,$ and A_{l+1,a_n} besides $A_{0,\#}$ and $A_{l+1,\$}$; then for the two cells we continue to construct net transitions corresponding to Δ in which there is one whose output is B_{q_m}. Obviously, these symbol places and transitions do not disturb the original weak-simulation since these symbol places have no tokens in the weakly-simulating process. Certainly, our control of reconfiguration can connect them and provide them an enabled chance if we also construct reconfiguration transitions for them. Note that if the input string is nonempty, there is a transition whose output is B_{q_m} in the constructed FC-net.

where

- $Q = \{q_1, q_2, q_3, q_4, q_5\}$
- $\Gamma = \{a, b, X\}$
- $\Sigma = \{a, b\}$
- $\Delta = \{\delta_0 = (q_1, \#, R, q_2, \#),$
  $\delta_1 = (q_2, \$, L, q_5, \$), \delta_2 = (q_2, a, R, q_3, X), \delta_3 = (q_2, X, R, q_2, X),$
  $\delta_4 = (q_3, X, R, q_3, X), \delta_5 = (q_3, a, R, q_3, a), \delta_6 = (q_3, b, L, q_4, X),$
  $\delta_7 = (q_4, a, L, q_4, a), \delta_8 = (q_4, X, L, q_4, X), \delta_9 = (q_4, \#, R, q_2, \#)\}$

In Ω_0, q_1 is the initial state and q_5 is the acceptance state. Obviously, this LBA accepts the empty string. Figure 6 shows the constructed net corresponding to Ω_0 with the empty input string.

From Fig. 6 we can see that after firing transition t_1 at the initial marking $\{i\}$, transition $\langle 0, \delta_0 \rangle$ can be fired if the tokens in $A_{0\#}$ and B_{0,q_1} are allocated to the input places of transition $\langle 0, \delta_0 \rangle$. After we fire $\langle 0, \delta_0 \rangle$, if we allocate the tokens in $A_{1,\$}$ and B_{1,q_2} to the input places of transition $\langle 1, \delta_1 \rangle$, then $\langle 1, \delta_1 \rangle$ is enabled and firing it leads to a marked B_{q_5}. At this time, t_2 and t_3 are both enabled; if the latter is chosen to fire, then a **deadlock** occurs; if the former is chosen to fire, then from any succeeding marking, marking $\{o\}$ is always reachable.

After firing transition t_1 at the initial marking $\{i\}$, none of transitions $\langle 0, \delta_0 \rangle$, $\langle 0, \delta_9 \rangle$ and $\langle 1, \delta_1 \rangle$ can be fired if the token in symbol place $A_{0\#}$ is allocated to its image as an input place of transition $\langle 0, \delta_9 \rangle$ and the token in state place B_{0,q_1} is allocated to its image as an input place of transition $\langle 0, \delta_0 \rangle$. At this time, either the state token is recycled to v_8 or transition t_2 is fired. Only after we fire t_2, the token in v_8 can be re-allocated to state places. Similarly, from any succeeding marking, marking $\{o\}$ is always reachable.

In a word, this safe ACWF-net is not sound.

3.3 PSPACE-completeness

Our construction contains at most

$$(|S| + 2)(|Q| + |\Gamma| + 2|\Delta| + 2) + 13$$

places and at most

$$(|S| + 2)(|Q| + |\Gamma| + 5|\Delta|) + 10$$

transitions when we consider each tape cell corresponds to $|\Delta|$ LBA transitions. For an LBA with an input string, therefore, we can construct a safe ACWF-net in polynomial time. From the above introduction, we can know that an LBA with an input string does not accept the string if and only if the constructed safe ACWF-net is sound. Therefore, we have the following conclusion:

Lemma 1. *The soundness problem of safe ACWF-net is PSPACE-hard.*

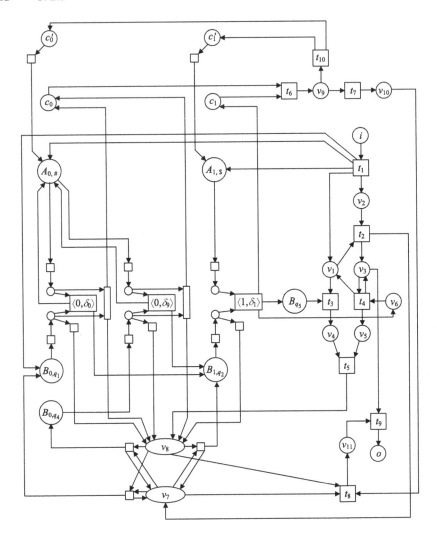

Fig. 6. Safe ACWF-net corresponding to the LBA Ω_0 with the empty input string.

In order to prove the PSPACE-completeness, we should provide a PSPACE algorithm to decide soundness for safe ACWF-nets. In fact, we have provided such an algorithm when we proved that the soundness problem is PSPACE-complete for bounded WF-nets in [20]. It mainly utilised the PSPACE algorithm of deciding the reachability problem for bounded Petri nets provided in [8], and it is omitted here.

Theorem 1. *The soundness problem of safe ACWF-net is PSPACE-complete.*

Obviously, this conclusion can be extended to the soundness problem of bounded ACWF-nets since the PSPACE algorithm in [20] is applicable to any bounded WF-net.

Corollary 1. *The soundness problem of bounded ACWF-net is PSPACE-complete.*

Since the definition of soundness is a special case of the definition of weak soundness, we have the following conclusion:

Corollary 2. *The weak soundness problem of bounded/safe ACWF-net is PSPACE-complete.*

Aalst [2] proved that the soundness of a WF-net is equivalent to the liveness and boundedness of its short-circuit.

Proposition 1 ([2]). *Let $N = (P, T, F)$ be a WF-net and $\overline{N} = (P, T \cup \{t_0\}, F \cup \{(t_0, i), (o, t_0)\})$ be the short-circuit of N where $t_0 \notin T$. Then, N is sound if and only if $(\overline{N}, \{i\})$ is live and bounded.*

If we consider the short-circuit of our constructed ACWF-net for an LBA with an input string, we easily know that 1) this short-circuit is still asymmetric-choice and it is still safe at the initial marking $\{i\}$, and 2) it is live at the initial marking $\{i\}$ if and only if the LBA does not accept the input string. Therefore, based on Proposition 1 or our construction, we can draw the following conclusion:

Lemma 2. *The liveness problem of safe AC-net is PSPACE-hard.*

Cheng, Esparza and Palsberg [8] introduced a PSPACE algorithm[4] to decide liveness of safe Petri nets, which is also applicable to safe AC-nets. Therefore, we have the following conclusion:

Theorem 2. *The liveness problem of safe AC-net is PSPACE-complete.*

Ohta and Tsuji [28] proved that the liveness problem of bounded AC-nets is NP-hard. Obviously, we sharpen their result.

We can know that the reachability problem for safe AC-nets is PSPACE-complete, which is based on the following facts: 1) the reachability problems for safe Petri nets and safe FC-nets are both PSPACE-complete [10], 2) safe AC-nets are a subclass of safe Petri nets, and 3) safe FC-nets are a subclass of safe AC-nets.

Theorem 3. *The reachability problem of safe AC-net is PSPACE-complete.*

4   Conclusion

In this paper, we first construct a safe FC-net to weakly simulate the computations of an LBA with an input string, and then add some controls into the FC-net so that the final net is a safe ACWF-net and the safe ACWF-net is sound if and only if the LBA does not accept the input string. We therefore prove that the soundness problem of safe ACWF-nets is PSPACE-hard, and this lower bound strengthens an existing result, i.e.,

---

[4] It is based on the proof technique of the liveness problem of 1-conservative nets provided by Jones *et al.* in [16].

NP-hardness provided in [18]. We have known that this problem is in PSPACE [20], and thus we obtain an accurate bound for this problem: PSPACE-complete. At the same time, we infer that the liveness problem of safe AC-nets is also PSPACE-complete. These results also provide a support to a rule of thumb proposed by Esparza in [12]: "All interesting questions about the behaviour of safe Petri nets are PSPACE-hard".

From Table 1 we can see that there is a large gap between safe FC-nets and safe AC-nets (resp. safe FCWF-nets and safe ACWF-nets) about their liveness (resp. soundness) problem's complexities: from polynomial-time to PSPACE-complete. These results also evidence that the computation power of AC-nets is much stronger than that of FC-nets. Although literatures [6,7,10,31] provide many theorems that hold for FC-nets but fail for AC-nets, we still plan to explore some subclasses of AC-nets and ACWF-nets in future, motivated by the large gap and focusing on the following questions:

1. whether there is a subclass such that it is larger than safe FC-net (resp. safe FCWF-net) and smaller than safe AC-net (resp. safe ACWF-net), but its liveness (resp. soundness) problem is below PSPACE (e.g. NP-complete[5])?
2. if there exists such a subclass, whether there is a good structure-based condition, just like that for FC-nets [6,7,10,31], to decide their liveness and soundness?

Acknowledgement. The author would like to thank the anonymous reviewers for their very detailed and constructive comments that help the author improve the quality of this paper. This paper was supported in part by the National Key Research and Development Program of China (grant no. 2018YFB2100801), in part by the Fundamental Research Funds for the Central Universities of China (grant no. 22120190198) and in part by the Shanghai Shuguang Program (grant no. 15SG18).

References

1. van der Aalst, W.M.P.: Interorganizational Workflows: an approach based on message sequence charts and petri nets. Syst. Anal. Model. **34**, 335–367 (1999)
2. van der Aalst, W.M.P.: Structural characterizations of sound workflow nets. Computing Science Report 96/23, Eindhoven University of Technology (1996)
3. van der Aalst, W.M.P., et al.: Soundness of workflow nets: classification, decidability, and analysis. Form. Aspects Comput. **23**, 333–363 (2011). https://doi.org/10.1007/s00165-010-0161-4
4. van der Aalst, W., Kindler, E., Desel, J.: Beyond asymmetric choice: a note on some extensions. Petri Net Newsl. **55**, 3–13 (1998)
5. Barkaoui, K., Pradat-Peyre, J.-F.: On liveness and controlled siphons in Petri nets. In: Billington, J., Reisig, W. (eds.) ICATPN 1996. LNCS, vol. 1091, pp. 57–72. Springer, Heidelberg (1996). https://doi.org/10.1007/3-540-61363-3_4
6. Best, E.: Structure theory of petri nets: the free choice hiatus. In: Brauer, W., Reisig, W., Rozenberg, G. (eds.) ACPN 1986. LNCS, vol. 254, pp. 168–205. Springer, Heidelberg (1987). https://doi.org/10.1007/978-3-540-47919-2_8
7. Best, E., Voss, K.: Free choice systems have home states. Acta Informatica **21**, 89–100 (1984). https://doi.org/10.1007/BF00289141

---

[5] We suppose NP $\neq$ PSPACE although nobody has proven NP $\subset$ PSPACE so far.

8. Cheng, A., Esparza, J., Palsberg, J.: Complexity results for 1-safe nets. Theoret. Comput. Sci. **147**, 117–136 (1995)
9. Czerwiński, W., Lasota, S., Lazić, R., Laroux, J., Mazowiecki, F.: The reachability problem for Petri nets is not elementary. In: Proceedings of the 51st Annual ACM SIGACT Symposium on Theory of Computing, pp. 24–33 (2019)
10. Desel, J., Esparza, J.: Free Choice Petri Nets. Cambridge Tracts in Theoretical Computer Science, vol. 40. Cambridge University Press, Cambridge (1995)
11. Esparza, J., Hoffmann, P., Saha, R.: Polynomial analysis algorithms for free choice probabilistic workflow nets. Perform. Eval. **117**, 104–129 (2017)
12. Esparza, J.: Decidability and complexity of Petri net problems — an introduction. In: Reisig, W., Rozenberg, G. (eds.) ACPN 1996. LNCS, vol. 1491, pp. 374–428. Springer, Heidelberg (1998). https://doi.org/10.1007/3-540-65306-6_20
13. Garey, M.R., Johnson, D.S.: Computer and Intractability: A Guide to the Theory of NP-Completeness. W. H. Freeman and Company, New York (1976)
14. Hack, M.: Petri Net languages. Technical report 159, MIT (1976)
15. Jiao, L., Cheung, T.Y., Lu, W.: On liveness and boundedness of asymmetric choice nets. Theoret. Comput. Sci. **311**, 165–197 (2004)
16. Jones, N.D., Landweber, L.H., Lien, Y.E.: Complexity of some problems in Petri nets. Theoret. Comput. Sci. **4**, 277–299 (1977)
17. Leroux, J., Schmitz, S.: Reachability in vector addition systems is primitive-recursive in fixed dimension. In: Proceedings of the 34th Annual ACM/IEEE Symposium on Logic in Computer Science, pp. 1–13 (2019)
18. Liu, G., Jiang, C.: Co-NP-hardness of the soundness problem for asymmetric-choice workflow nets. IEEE Trans. Syst. Man Cybern.: Syst. **45**, 1201–1204 (2015)
19. Liu, G., Jiang, C.: Net-structure-based conditions to decide compatibility and weak compatibility for a class of inter-organizational workflow nets. Sci. China Inf. Sci. **58**, 1–16 (2015). https://doi.org/10.1007/s11432-014-5259-5
20. Liu, G., Sun, J., Liu, Y., Dong, J.: Complexity of the soundness problem of workflow nets. Fundamenta Informaticae **131**, 81–101 (2014)
21. van Hee, K., Sidorova, N., Voorhoeve, M.: Generalised soundness of workflow nets is decidable. In: Cortadella, J., Reisig, W. (eds.) ICATPN 2004. LNCS, vol. 3099, pp. 197–215. Springer, Heidelberg (2004). https://doi.org/10.1007/978-3-540-27793-4_12
22. Kang, M.H., Park, J.S., Froscher, J.N.: Access control mechanisms for inter-organizational workflow. In: Proceedings of the Sixth ACM Symposium on Access Control Models and Technologies, pp. 66–74. ACM Press, New York (2001)
23. Kindler, E.: The ePNK: an extensible Petri net tool for PNML. In: Kristensen, L.M., Petrucci, L. (eds.) PETRI NETS 2011. LNCS, vol. 6709, pp. 318–327. Springer, Heidelberg (2011). https://doi.org/10.1007/978-3-642-21834-7_18
24. Kindler, E., Martens, A., Reisig, W.: Inter-operability of workflow applications: local criteria for global soundness. In: van der Aalst, W., Desel, J., Oberweis, A. (eds.) Business Process Management. LNCS, vol. 1806, pp. 235–253. Springer, Heidelberg (2000). https://doi.org/10.1007/3-540-45594-9_15
25. Malek, M., Ahmadon, M., Yamaguchi, S.: Computational complexity and polynomial time procedure of response property problem in workflow nets. IEICE Trans. Inf. Syst. **E101–D**, 1503–1510 (2018)
26. Martens, A.: On compatibility of web services. Petri Net Newsl. **65**, 12–20 (2003)
27. Meyer, P.J., Esparza, J., Offtermatt, P.: Computing the expected execution time of probabilistic workflow nets. In: Vojnar, T., Zhang, L. (eds.) TACAS 2019. LNCS, vol. 11428, pp. 154–171. Springer, Cham (2019). https://doi.org/10.1007/978-3-030-17465-1_9
28. Ohta, A., Tsuji, K.: NP-hardness of liveness problem of bounded asymmetric choice net. IEICE Trans. Fundam. **E85–A**, 1071–1074 (2002)

29. Reisig, W.: Understanding Petri Nets: Modeling Techniques, Analysis Methods, Case Studies. Springer, Heidelberg (2013). https://doi.org/10.1007/978-3-642-33278-4

30. Tao, X.Y., Liu, G.J., Yang, B., Yan, C.G., Jiang, C.J.: Workflow nets with tables and their soundness. IEEE Trans. Ind. Inf. **16**, 1503–1515 (2020)

31. Thiagarajan, P.S., Voss, K.: A fresh look at free choice nets. Inf. Control **61**, 85–113 (1984)

32. Tiplea, F.L., Bocaenealae, C., Chirosecae, R.: On the complexity of deciding soundness of acyclic workflow nets. IEEE Trans. Syst. Man Cybern.: Syst. **45**, 1292–1298 (2015)

33. Trčka, N., van der Aalst, W.M.P., Sidorova, N.: Data-flow anti-patterns: discovering data-flow errors in workflows. In: van Eck, P., Gordijn, J., Wieringa, R. (eds.) CAiSE 2009. LNCS, vol. 5565, pp. 425–439. Springer, Heidelberg (2009). https://doi.org/10.1007/978-3-642-02144-2_34

34. Verbeek, H.M.W., van der Aalst, W.M.P., Ter Hofstede, A.H.M.: Verifying workflows with cancellation regions and OR-joins: an approach based on relaxed soundness and invariants. Comput. J. **50**, 294–314 (2007)

35. Wimmel, H.: Synthesis of reduced asymmetric choice Petri nets. arXiv:1911.09133, pp. 1–27 (2019)

36. Xiang, D.M., Liu, G.J., Yan, C.G., Jiang, C.J.: A guard-driven analysis approach of workflow net with data. IEEE Trans. Serv. Comput. (2019, published online). https://doi.org/10.1109/TSC.2019.2899086

37. Yamaguchi, S., Matsuo, H., Ge, Q.W., Tanaka, M.: WF-net based modeling and soundness verification of interworkflows. IEICE Trans. Fundam. Electron. Commun. Comput. Sci. **E90-A**, 829–835 (2007)

38. van Zelst, S.J., van Dongen, B.F., van der Aalst, W.M.P., Verbeek, H.M.W.: Discovering workflow nets using integer linear programming. Computing **100**(5), 529–556 (2017). https://doi.org/10.1007/s00607-017-0582-5

Process Mining and Applications

Petri Nets Validation of Markovian Models of Emergency Department Arrivals

Paolo Ballarini[1(✉)], Davide Duma[2], Andras Horváth[2], and Roberto Aringhieri[2]

[1] Université Paris-Saclay, CentraleSupélec, MICS, 91190 Gif-sur-Yvette, France
paolo.ballarini@centralesupelec.fr
[2] Dipartimento di Informatica, Universitá di Torino, Turin, Italy
{duma,horvath,aringhier}i@di.unito.it

Abstract. Modeling of hospital's Emergency Departments (ED) is vital for optimisation of health services offered to patients that shows up at an ED requiring treatments with different level of emergency. In this paper we present a modeling study whose contribution is twofold: first, based on a dataset relative to the ED of an Italian hospital, we derive different kinds of Markovian models capable to reproduce, at different extents, the statistical character of dataset arrivals; second, we validate the derived arrivals model by interfacing it with a Petri net model of the services an ED patient undergoes. The empirical assessment of a few key performance indicators allowed us to validate some of the derived arrival process model, thus confirming that they can be used for predicting the performance of an ED.

Keywords: Stochastic petri nets · Markovian models · ED arrival process

1 Introduction

Optimisation of hospital's Emergency Departments (ED) is concerned with maximising the efficiency of health services offered to patients that shows up at an ED requiring treatments with different level of emergency. In this respect the ability to build formal models that faithfully reproduce the behaviour of an ED and to analyse their performances is vital. A faithful model of an ED must consist of at least two components: a model of the patients arrival and a model of the ED services (the various activities that an ED patient may undergo throughout his permanence at the ED). The composition of these two models must be such that the model of patients arrival is used to feed in the services model and the analysis of the performances of the composed model, through meaningful key performance indicators (KPIs), may yield essential indications as to where intervene so to improve the overall efficiency of the ED.

Contribution. In this paper we focus on the problem of deriving a stochastic model of the patient arrivals that is capable of correctly reproducing the statistical character of a given arrivals dataset. To this aim we considered a real

R. Janicki et al. (Eds.): PETRI NETS 2020, LNCS 12152, pp. 219–238, 2020.
https://doi.org/10.1007/978-3-030-51831-8_11

dataset relative to the activity of the ED of an hospital of the city of Cantù in northern Italy and which contains different data including the complete set of timestamps of the arrival of each patients over one year (i.e. the year 2015). Based on the statistical analysis of the dataset we derived different instances of Markovian models belonging to different classes (i.e. Markov renewal processes, Markov arrival processes, hidden Markov models) and we showed that only one model instance, amongst those derived, is capable of fully reproducing the statistical character of the dataset, including the periodicity of patients arrivals exhibited by the dataset. To further validate the derived models we developed a formal encoding in terms of stochastic Petri nets. This allowed us to compare the performances of each arrival models, coupled with a model of ED services, through assessment of a KPI formally encoded and accurately assessed through a statistical model checking tool. The obtained results evidenced that one amongst the derived arrival models faithfully match the dataset behaviour hence it can reliably be used for analysing the performances of different ED services.

Paper Organisation. The paper is organised as follows: In Sect. 2 we present the statistical analysis of the dataset and discuss the main statistical characteristics of the patient's arrival process by considering different perspectives. In Sect. 3 we describe the derivation of different kinds of Markovian models of the *patient arrival's* process, specifically we discuss the class of renewal Markov processes given in terms of continuous phase type (CPH) distributions, in Sect. 3.1, the class of Markov arrival process (MAP), in Sect. 3.2 and finally the class of hidden Markov models (HMM), in Sect. 3.3. In Sect. 4 we give the formal encoding of each of the derived models in terms of stochastic Petri nets. Finally in Sect. 5 we present the results of some experiments aimed at assessing a KPI against the model given by the composition of patient arrival models and a simple model of the ED services. Conclusive remarks wrap up the paper in Sect. 6.

Related Work. Literature on modeling of hospital EDs is very wide, faces different kinds of problems and cannot be reviewed in brief. We mention only some works to position ourselves in the literature and point out differences. Many works, such as [2,3,12,14,21,24], deal with forecasting the number of patients arrivals but do not experiment with Markovian models. The models proposed in this paper can also be applied to forecasting but we cannot elaborate on this issue due to space limits. Several works, e.g., [20,22], build a queuing model of ED but do not deal with data driven modeling of the arrival process. Few works, e.g., [31], are concerned with the data-driven development of stochastic models capturing both the patient arrival as well as the services received at the ED. Our paper falls in this category but with respect to other we experiment with a wide range of Markovian processes to model patient arrivals and provide also their Petri net counterpart which can be derived in an automatic manner.

2   Statistical Analysis of the Cantù Hospital Dataset

One way of representing the patient arrival flow is to register the time elapsed between consecutive show ups. To this end let $X_i, i = 1, 2, 3, \dots$ denote the ith inter-arrival time. Another possibility is to count the number of patients that arrive in an interval of a certain length. We will use intervals whose length is one hour, one day or one week. The associated series are $Y_{\mathcal{H},i}, Y_{\mathcal{D},i}$ and $Y_{\mathcal{W},i}$ with $i = 1, 2, 3, ...$ which denote the number of patient arrivals in the ith hour, day and week, respectively. Clearly, the second approach keeps less information with respect to the first but it has the advantage that it relates easily to the periodic nature of the patient arrival process. For example, the series $Y_{\mathcal{H},3+24j}$ with $j = 0, 1, 2, ...$ provides the number of arrivals between 2am and 3am during the 1st, the 2nd, the 3rd day, and so on.

We start off by analyzing the patient arrival pattern inside a day by considering the average and the variance of the number of patient arrivals during the ith hour of the day. I.e., we calculate $E_{\mathcal{H},i} = E\left[\{Y_{\mathcal{H},i+k}|k = 0, 24, 48, ...\}\right]$ and $Var_{\mathcal{H},i} = Var\left[\{Y_{\mathcal{H},i+k}|k = 0, 24, 48, ...\}\right]$ with $i = 1, 2, ..., 24$. In Fig. 1, $E_{\mathcal{H},i}$ and $Var_{\mathcal{H},i}$ are depicted as a function of i. The highest number of patient arrivals is registered between 9am and 10am with about 5.9 patients on average. Between 5am and 6am the average number of patients is instead less than 1. As expected, the time of the day has a strong impact on the number of patient arrivals. The variance, $Var_{\mathcal{H},i}$, has very similar values to those of the mean $E_{\mathcal{H},i}$, which is compatible with a well-known characteristic of the Poisson process, hence suggesting that the patient arrival process may be adequately modelled through a Poisson process with time inhomogeneous intensity. To further investigate this issue, we calculate the distribution of the number of arrivals during a given hour of the day and compare it the Poisson distribution with the same mean. The resulting probability mass function (pmf) is depicted in Fig. 2 for the interval 1pm-2pm and 5am-6am. The interval 1pm-2pm is the hour of the day where the Poisson distribution differs most from the actual distribution of the number of arrivals (the difference is calculated as the sum of absolute differences in the pmf) while the interval 5am-6am is the hour where the Poisson distribution is most similar to the experimental distribution.

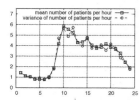
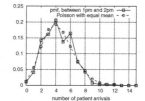
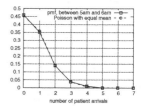
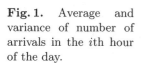

Fig. 1. Average and variance of number of arrivals in the ith hour of the day.

Fig. 2. Pmf of number of arrivals between 1pm and 2pm (left) and between 5am and 6am (right) and pmf of the Poisson distribution with equal mean.

The hourly autocorrelation function (acf) is defined as

$$R_{\mathcal{H},n} = \frac{E[(Y_{\mathcal{H},i} - E[Y_{\mathcal{H},i}])(Y_{\mathcal{H},i+n} - E[Y_{\mathcal{H},i}])]}{Var[Y_{\mathcal{H},i}]} \quad \text{with } n = 1, 2, 3, ...$$

and it is depicted in Fig. 3. The oscillation present in the autocorrelation is due to the changing patient arrival intensity during the day (Fig. 1) and it remains unaltered even for very large values of n.

We study now $Y_{\mathcal{D},i}$ in order to determine if there is a pattern in the patient arrivals during a week. In Fig. 4 we depict $E_{\mathcal{D},i} = E[\{Y_{\mathcal{D},i+k}|k = 0, 7, 14, ...\}]$ and $Var_{\mathcal{D},i} = Var[\{Y_{\mathcal{D},i+k}|k = 0, 7, 14, ...\}]$ with $i = 1, 2, ..., 7$ and observe that there are not large differences between days of the week. The variance differs to a large extent from the average on most days of the week. We calculated also the acf of $Y_{\mathcal{D},i}$ and saw that there is only a very mild correlation between number of arrivals of consecutive days. For this reason we do not aim to model daily correlation with our models.

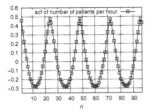

Fig. 3. $R_{\mathcal{H},n}$, i.e., acf of the sequence $Y_{\mathcal{H},i}$ as function of n.

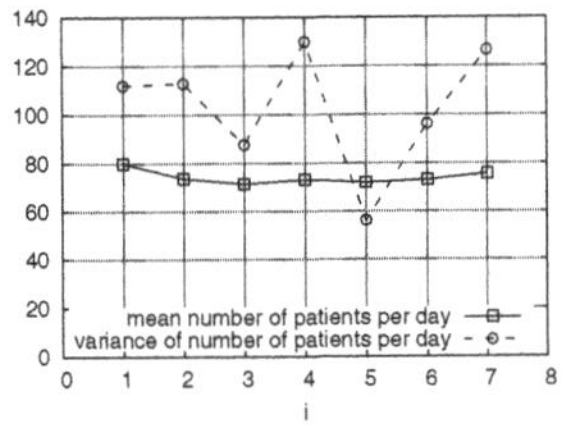

Fig. 4. Average and variance of number of arrivals in the ith day of the week.

Finally, we look at the series containing the inter-arrival times themselves, X_i with $i = 1, 2, 3...$. The average inter-arrival time is 19.43 and its variance is 483.18. The largest value in the series is 337 min. In Fig. 5 we provide the front and the tail of the probability density function (pdf) of the inter-arrival times, respectively. The autocorrelation present in X_i is depicted in Fig. 6 where there is oscillation due to the fact that a period with high arrival intensity (8am-22pm) are followed by a period with low arrival intensity (22pm-8am). This oscillation however is less easy to interpret than that in Fig. 3 and fades away as n grows because for large values of n there is no deterministic connection between the time of the day of the ith and the $(i + n)$th arrival.

3   Markovian Models of Patient Arrivals and Their Application to the Cantù Dataset

3.1   Markovian Renewal Process Approximation

The simplest arrival models are renewal processes, i.e., processes in which consecutive inter-arrival times are independent and share the same distribution. To

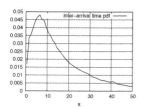
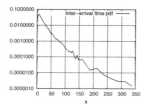
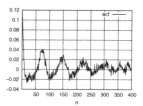

Fig. 5. Front (left) and tail (right) of the pdf of the inter-arrival times.

Fig. 6. Acf in the inter-arrival times series.

have a Markovian renewal process, we can approximate the observed inter-arrival time distribution by phase type (PH) distributions. A degree-n PH distributions is given by the distribution of time to absorption in a Markov chain (MC) with n transient states (the phases) and one absorbing state [25]. A PH distribution is continuous if a continuous time MC (CTMC) is used and it is discrete if a discrete time MC (DTMC) is applied. In this work we apply continuous PH (CPH) distributions.

In order to understand basic concepts about CPH models, consider the degree-3 CPH distribution depicted in Fig. 7 where numbers inside a state indicate the initial probability of the state and numbers on the arcs are transition intensities. The gray state is the absorbing one. A CPH distribution is conveniently described by the vector of initial probabilities of the transient states denoted by α (we assume in this work that the initial probability of the absorbing state is 0) and the matrix, denoted by Q, containing the transition intensities among the transient states and the opposite of the sum of the intensities of the outgoing transitions in the diagonal (this allows to determine the transition intensities toward the absorbing state). For the CPH in Fig. 7 we have

$$\alpha = (0.2,\ 0.8,\ 0),$$
$$Q = \begin{pmatrix} -3 & 2 & 0 \\ 2 & -3 & 1 \\ 0 & 2 & -8 \end{pmatrix} \tag{1}$$

The pdf of a CPH distribution can be obtained based on basic properties of CTMCs as $f(x) = \alpha e^{Qx}(-Q)\mathbb{1}$ where $\mathbb{1}$ is the column vector of 1s. From the previous the moments can be derived and have the form $m_i = i!\alpha(-Q)^{-i}\mathbb{1}$. We depicted the pdf of the above example in Fig. 7. Several well-known distributions are contained in the family of CPH distributions. The exponential distribution is a degree-1 CPH distribution. Also hyper-exponential and Erlang distributions are simple to represent as CPH distributions.

Our aim is to find α and Q such that the pdf of the associated CPH distribution is a good approximation of the pdf of the patient inter-arrival times. There are two families of approaches to face this problem. The first is based on the maximum likelihood principle, i.e., having set the degree n to a given value, α

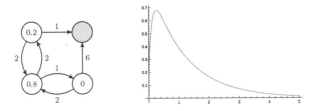

Fig. 7. Graphical representation of the degree-3 PH distribution used as example (left) and its pdf (right).

and Q are searched for such that it is most likely to reproduce the empirical value at hand. Techniques based on this approach suffers from two drawbacks: first, the number of parameters in α and Q grows quadratically with n; second, the vector-matrix representation provided by α and Q is redundant and this gives rise to difficult optimization problems. One of the first methods in this line is described in [9] and an associated tool is presented in [17]. The second family of approaches is based on matching a set of statistical parameters of the available data. The set usually includes only moments [5,18] but can refer also to characteristics of the pdf [4] or the cumulative distribution function at some points [15]. We apply here moment matching techniques which require much less computational effort, result in good fit of the inter-arrival distribution under study and have the advantage of considering a non-redundant representation (based on the moments every entry of α and Q is determined directly).

It is known that a degree-n CPH distribution is determined by $2n-1$ moments (see, e.g., [11]). Not any valid $2n-1$ moments can be realized however by a degree-n CPH distribution. The approaches proposed in [5,18] construct a degree-n CPH distribution matching $2n-1$ moments if the moments can be realized by a degree-n CPH distribution. The methods are implemented in the BuTools package (http://webspn.hit.bme.hu/~butools/) and we applied them with $n = 1, 2$ and 3. In the following we report the initial probabilities and the transition intensities of the resulting CPH distributions:

- $n = 1$ (one moment matched): $\alpha = (1)$ and $Q = (-0.051453)$
- $n = 2$ (three moments matched): $\alpha = (0.0409229, \ 0.959077)$ and

$$Q = \begin{pmatrix} -0.0213213 & 0.0213213 \\ 0 & -0.0570911 \end{pmatrix}$$

- $n = 3$ (five moments matched): $\alpha = (0.0284672, 0.881506, 0.0900266)$ and

$$Q = \begin{pmatrix} -0.699657 & 0 & 0 \\ 0.0667986 & -0.0667986 & 0 \\ 0 & 0.0260063 & -0.0260063 \end{pmatrix}$$

The pdf of the above CPH approximations is depicted in Fig. 8. With $n = 1$, the resulting exponential distribution, which captures only the first moment, is a bad approximation both of the front and of the tail. With $n = 2$ the tail is approximated well but the shape of the front is rather different. With $n = 3$ both the front and the tail are captured to a satisfactory extent.

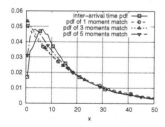 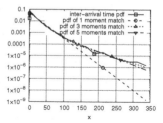

Fig. 8. Front (left) and tail (right) of pdf of CPH distribution approximations of the inter-arrival time distribution based on moment matching.

It is evident that, even if the inter-arrival time distribution is well approximated, a renewal process is significantly different from the data set under study because it cannot exhibit correlation and periodicity. Nevertheless, as we will show in Sect. 5, the renewal process with a proper PH distribution can give good approximation of some performance indices. In the following we propose models in which there is correlation in the inter-arrival time sequence.

3.2 Markovian Arrival Processes

Continuous time Markovian arrival processes (MAP), introduced in [26], are a wide class of point processes in which the events (arrivals) are governed by a background CTMC and they can be seen as the generalization of the Poisson process allowing for non-exponential and correlated inter-arrival times. We provide here a brief introduction to MAPs.

The infinitesimal generator matrix of the underlying CTMC of an n-state MAP will be denoted by the $n \times n$ matrix D. Arrivals can be generated in two ways. First, in each state i of the CTMC a Poisson process with intensity λ_i is active and can give rise to arrivals during a sojourn in the state. Second, when in the CTMC a transition from state i to j occurs an arrival is generated with probability $p_{i,j}$. A convenient and compact notation to describe a MAP is obtained by collecting all intensities in two matrices, $D^{(0)}$ and $D^{(1)}$, in such a way that $D^{(0)}$ contains the transition intensities that do not generate an event and $D^{(1)}$ those that give rise to one, including the intensities of the Poisson processes in the diagonal of $D^{(1)}$. Moreover, the diagonal entries of $D^{(0)}$ are set in such a way that we have $D^{(0)} + D^{(1)} = D$. Accordingly, the entries of $D^{(0)}$ and $D^{(1)}$ are

$$\forall i, j, i \neq j: \quad D^{(0)}_{i,j} = (1 - p_{i,j})D_{i,j}, \quad D^{(1)}_{i,j} = p_{i,j}D_{i,j}$$
$$\forall i: \quad D^{(0)}_{i,i} = -\sum_{\forall j, j \neq i} D_{i,j} - \lambda_i, \quad D^{(1)}_{i,i} = \lambda_i$$

In most of the literature, the vector of steady state probabilities of the background chain, denoted by $\gamma = (\gamma_1, ..., \gamma_n)$, is used as initial probability vector of the model[1]. In this paper we will do the same. As an example, consider the 2-state MAP described by

$$D^{(0)} = \begin{pmatrix} -3 & 1 \\ 0 & -7 \end{pmatrix} \quad D^{(1)} = \begin{pmatrix} 0 & 2 \\ 2 & 5 \end{pmatrix} \quad D = \begin{pmatrix} -3 & 3 \\ 2 & -2 \end{pmatrix} \tag{2}$$

which implies that during a sojourn in state 1 no arrivals are generated while during a sojourn in state 2 a Poisson process with intensity equal to 5 is active. Moreover, a transition from state 1 to state 2 generates an arrival with probability 2/3 while transitions from state 2 to 1 are always associated with an arrival. The steady state probabilities of the background chain are $\gamma = (0.4, 0.6)$.

The well-known Poisson process is a MAP with a single state ($n = 1$). The renewal process used in Sect. 3.1 can be expressed in terms of a MAP by setting $D^{(0)} = Q$, $D^{(1)} = (-Q)\mathbb{1}\alpha$. The Markov modulated Poisson process, in which a Poisson process is active in every state of a background CTMC, is a MAP whose $D^{(1)}$ matrix contains non-zero entries only in its diagonal.

As for CPH distributions, two families of parameter estimation methods have been developed in the literature: the first based on the maximum-likelihood principle ([13,27,29]) and the second based on matching a few statistical parameters of the arrival process. The representation given by $D^{(0)}$ and $D^{(1)}$ contains $2n^2 - n$ parameters (in every row of $D = D^{(0)} + D^{(1)}$ the sum of the entries must be zero) and it is redundant as an n-state MAP is determined by $n^2 + 2n - 1$ parameters ($2n - 1$ moments of the inter-arrival times and n^2 joint moments of consecutive inter-arrival times; see [30] for details). Maximum-likelihood based methods suffers from the same drawbacks described before in case of CPH distributions. In this paper we experiment with methods that belong to the second family of approaches.

A 2-state MAP is determined by 3 moments of the inter-arrival times and the lag-1 auto-correlation of the inter-arrival time sequence [10,30]. Our sequence has however such a lag-1 auto-correlation that cannot be realized with only 2 states. In [19] a method was proposed that creates a MAP with any 3 inter-arrival time moments and lag-1 auto-correlation. This method, implemented in the BuTools package, provides the following 6-state MAP:

---

[1] Notice that γ exists and is independent of the initial state as the background CTMC is, by definition, ergodic.

$$D^{(0)} = \begin{pmatrix} -0.0293 & 0.0293 & 0 & 0 & 0 & 0 \\ 0 & -0.1883 & 0 & 0 & 0 & 0 \\ 0 & 0 & -0.0103 & 0.0103 & 0 & 0 \\ 0 & 0 & 0 & -0.0989 & 0.0989 & 0 \\ 0 & 0 & 0 & 0 & -0.0989 & 0.0989 \\ 0 & 0 & 0 & 0 & 0 & -0.0989 \end{pmatrix} \quad (3)$$

$$D^{(1)} = \begin{pmatrix} 0. & 0. & 0. & 0. & 0. & 0. \\ 0.023 & 0.1288 & 0.0002 & 0.0362 & 0. & 0. \\ 0. & 0. & 0. & 0. & 0. & 0. \\ 0. & 0. & 0. & 0. & 0. & 0. \\ 0. & 0. & 0. & 0. & 0. & 0. \\ 0.0037 & 0.0207 & 0.0004 & 0.074 & 0. & 0. \end{pmatrix} \quad (4)$$

with which the steady state probability vector of the background Markov chain is $\gamma = (0.149503, 0.153343, 0.0126432, 0.22817, 0.22817, 0.22817)$.

As opposed to the approach used in Sect. 3.1, in the arrival process generated by a MAP there can be correlation between subsequent inter-arrival times. In Fig. 9 we depict the acf in the inter-arrival time sequence generated by the above 6-state MAP and that computed on the available dataset. As guaranteed by the applied method for $n = 1$ the auto-correlation is matched exactly but then it fails to follow the auto-correlation of the data. The acf of the sequence counting the number of arrival per hour generated by the same MAP is given instead in Fig. 10 which is very different from the that of the dataset. In the number of patients per day sequence the auto-correlation of the 6-state MAP with $n = 1$ is 0.00469433, i.e., it is negligible, as opposed to that in the data where it is 0.22.

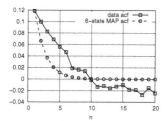

Fig. 9. Acf of the inter-arrival time sequence of the original data and of the MAP given in (3-4).

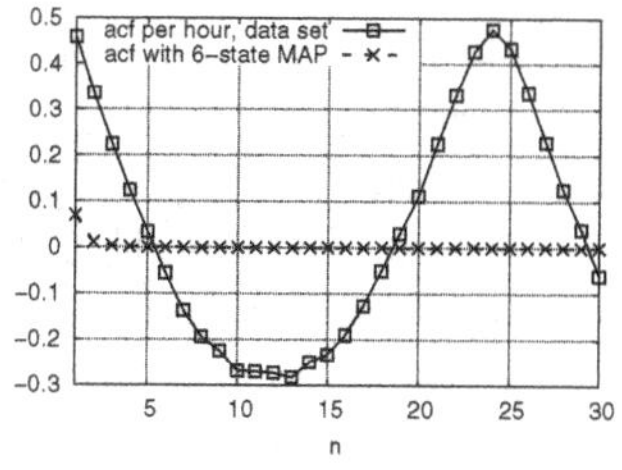

Fig. 10. Acf of the number of arrivals per hour sequence of the original data and of the MAP given in (3-4).

As seen above, 6 states are necessary to match three moments and just the lag-1 autocorrelation of the inter-arrival times. This indicates that a MAP with large number of states is necessary to capture the peculiar statistical features of the patient arrival process. General purpose MAP fitting techniques are not applicable however with such large number of states.

3.3 Hidden Markov Processes

A Hidden Markov model (HMM) [28] can be thought of as a generalisation of a DTMC for which an external observer cannot directly see the states but only observe some *output* whose probability to be emitted depends on the state. In practice an HMM is characterised by: *i)* a set of states $\mathcal{S} = \{s_1, \ldots, s_n\}$; *ii)* a set of possible observations $\mathcal{O} = \{o_1, \ldots, o_m\}$; *iii)* a $n \times n$ state-transition probability matrix $A = \{a_{ij}\}$ with a_{ij} being the probability of transitioning from state s_i to s_j; *iv)* a $n \times m$ state-observation probability matrix $B = \{b_{ij}\}$ where b_{ij} is the probability of observing o_j when the chain enters state s_i; *v)* an initial distribution over the set of states S denoted by α. Accordingly, an HMM is completely determined by the triple (α, A, B).

There exist three classical problems for HMM, all of which requiring a sequence of observations $O = (o_1, \ldots o_k)$. The first one is the *evaluation problem*: given an HMM by (α, A, B) and a sequence O, calculate the probability that the HMM produces O. The second, called *decoding problem*, consists of determining the most probable state sequence given a HMM and a sequence O. The third one, the *learning problem*, has only O as input and is about finding such triple (α, A, B) with which observing O is most probable.

In this paper we focus on the third type of problem and in particular we develop and study two HMMs aimed at reproducing the statistical characteristics of the patients' arrival dataset (described in Sect. 2). We start off with a rather coarse-grained (only 3-state) but general HMM, which reveals to be capable of reproducing only marginally the auto-correlation characteristics of the patients arrivals data. Then we propose a HMM with particular underlying DTMC that shows good agreement with the dataset from a statistical point of view.

A 3 States HMM. Here we consider a 3 state HMM in which every time slot corresponds to one hour and the possible observations are $\mathcal{O} = \{0, 1, \ldots, 14\}$ interpreted as the number of patient arrivals per hour[2]. This means that the sequence generated by the HMM has to be post-processed if we need to specify the exact arrival instance for each patient. This post-processing will consist of distributing the arrivals in uniform manner inside the hour.

There are 2 free parameters in π, $3 \times 2 = 6$ in A and $3 \times 14 = 42$ in B (because π must be normalized and also the rows both in A and B must be normalized). We applied the Baum-Welch algorithm [8] to determine the optimal parameters starting from several initial parameter sets chosen randomly. With this relatively small model, the final optimal parameters obtained by the Baum-Welch algorithm are independent of the initial values (apart from permutations of the states). The obtained HMM is with $\pi = (0, 1, 0)$ and

$$A = \begin{pmatrix} 0.865 & 0.135 & 0 \\ 0 & 0.835 & 0.165 \\ 0.134 & 0 & 0.866 \end{pmatrix} \quad B = \begin{pmatrix} 0.0007 & 0.0227 & 0.0807 & 0.1569 & 0.1756 & 0.1731 \\ 0.024 & 0.1103 & 0.2237 & 0.2249 & 0.19 & 0.1323 \\ 0.3533 & 0.3752 & 0.1989 & 0.0583 & 0.0116 & 0.0025 \end{pmatrix}$$

$$\begin{pmatrix} 0.1652 & 0.0991 & 0.0625 & 0.0292 & 0.0177 & 0.009 & 0.0058 & 0.0006 & 0.0006 \\ 0.0544 & 0.0303 & 0.0052 & 0.0025 & 0 & 0 & 0.0004 & 0 & 0 \\ 0 & 0 & 0 & 0 & 0 & 0 & 0 & 0 & 0 \end{pmatrix} \quad (5)$$

---

[2] More than 14 patients per hour are very rare in our dataset.

The mean number of arrivals per hour with the above HMM is 3.0862 and its variance is 5.67669 while in the data trace the mean and the variance are 3.08676 and 5.69061, respectively.

In Fig. 11 (left) we depict the autocorrelation of the number of arrivals per hour. As one can expect, the 3-state HMM is not able to reproduce the sustained oscillation present in the data shown in Fig. 3 (autocorrelation is negligible for $n \geq 10$) but does much better than the 6-state MAP given in (3-4) (see Fig. 10).

A 24 States HMM. In order to have a model that is able to exhibit oscillation in the autocorrelation function of the number of arrivals per hour, we define a 24-state HMM in which each state corresponds to an hour of the day and the transition probabilities are such that the process deterministically cycles through the 24 states. Accordingly, A is 24×24 and its entry in position (i, j) is 1 in $j = (i + 1) \pmod{24}$ and it is 0 otherwise. The initial probability vector is set to $\pi = (1, 0, ..., 0)$. As before, the possible observations are $\mathcal{O} = \{0, 1, ..., 14\}$.

The Baum-Welch algorithm is such that parameters set to 0 initially remain 0. Consequently, the algorithm does not change the matrix A and the vector π and has an effect only on the entries of B. The number of parameters is larger, it is $24 \times 14 = 336$, but thanks to the deterministic behavior of the underlying DTMC the Baum-Welch algorithm determines the parameters in a single iteration. With the resulting HMM the mean number of arrivals per hour is 3.08619 while the variance is 5.67465. The hourly autocorrelation is shown in Fig. 11 (center). This model provides very similar hourly autocorrelation to that of the data set.

We experimented also with a 24-state HMM letting the Baum-Welch algorithm to change any entry of the matrix A (i.e., the initial entries of A are random strictly greater than 0 and strictly smaller than 1). This way the number of free parameters is $24 \times 23 + 24 \times 14 = 888$. With this large number of parameters the Baum-Welch algorithm performs 4798 iterations and requires about 3 min of computation time on a standard portable computer. The resulting HMM has mean and variance equal to 3.08539 and 5.67468, respectively. The matrix A has a similar structure to the one before but the probabilities of going to the next state is not 1 but a value between 0.6 and 0.99. The resulting autocorrela-

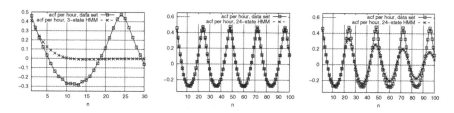

Fig. 11. Comparing the autocorrelation function (acf) of the dataset with that of the 3-state HMM (left), the 24-state HMM with deterministic (center) and non-deterministic (right) underlying Markov chain.

tion structure is depicted in 11 (right). With the non-deterministic underlying Markov chain the autocorrelation cannot exhibit sustained oscillation and the autocorrelation vanishes after about 150 hours (i.e., 6–7 days).

4   Petri Nets Model of the Patients Arrival Process

In order to assess the validity of the patients arrivals models presented in Sects. 3.1–3.3, we define a formal encoding of, respectively, the renewal Markov arrival model, the MAP and the HMM models, in terms of a superset of the Generalised Stochastic Petri Net (GSPN) [23] formalism, that we refer to as *extended* GSPN (eGSPN). Specifically, eGSPN is a class of stochastic Petri nets that, like GSPN, allows for combining immediate and stochastic timed transitions, but that, differently from the original GSPN, is not constrained to exponentially distributed timed transitions. We will use the eGSPN models described in this section to run a validation through the assessment of a number of key performance indicators (KPIs, see Sect. 5). For the sake of space we only give the syntactic elements necessary for characterizing an eGSPN, while we omit the formal semantics. We remark that if it is well known that the semantics of a GSPN can be reduced (by elimination of vanishing markings) to a CTMC process, equivalently it can be shown that the semantics of an eGSPN model corresponds to a generalised semi-Markov process (GSMP) [16], i.e. a larger class of stochastic processes that subsumes CTMCs.

Definition 1. *An eGSPN is defined by a 8-tuple* $(P, T, I, O, H, del, W, pol)$ *where:*

- P: set of places,
- T: set of transitions,
- $I, O, H : T \rightarrow Bag(P)$: are the input (I), output (O) and inhibition (H) arc functions (where $Bag(P)$ is the set of multi-sets built on P),
- $del : T \rightarrow dist(\mathbb{R}_{\geq 0})$: is a function that associates transitions with delay distributions (where $dist(\mathbb{R}_{\geq 0})$ is the set of probability distributions with non-negative real support),
- $W : T \rightarrow \mathbb{R}^+$: is the weight function that maps transitions onto positive real numbers,
- $pol : T \rightarrow \{single, infinite\}$ associates each transition with a semantics which, in this paper, is either *single server* or *infinite server*.

I, O and H give the multiplicities of arcs of connecting places of P with transitions of T. The function del allows to associate a generic probability distribution with the delay of firing of an enabled transition. Observe that since we put no restriction on the nature of the delay distributions, identical schedules for different transitions may have a positive probability (in case of non-continuous delay distributions): in this case the weight W is used to chose in a probabilistic manner the next transition to fire. In this paper, we will use four kinds of delay in

del. *Immediate* transitions are associated with constant zero firing time. *Expo-*
nential transitions fire after a delay described by the exponential distribution.
Deterministic transitions are associated with fixed positive delay. Finally, *uni-*
form transitions have a firing time according to the uniform distribution. For
what concerns *pol*, single server and infinite server semantics are as usual [23].

4.1 Petri Net Encoding of Markov Renewal Arrival Models

In the following we define a Petri net that represents the patient arrival process in
which inter-arrival times follow a degree-n PH distribution given by the vector-
matrix pair, (α, Q) (as described in Sect. 3.1).

The set of places is $P = \{P_0, P_1, ..., P_n, P_{n+1}, Arrs\}$ where P_0 will be used
to start the process according to α after each arrivals, $P_1, ..., P_n$ correspond to
the transient states of the PH distribution, P_{n+1} corresponds to the absorbing
state and $Arrs$ is the place where patient arrivals are accumulated. The set
of immediate transitions is $T_{\mathcal{I}} = \{t_{arr} \cup \{t_{0,i} : 1 \leq i \leq n, \alpha_i > 0\}\}$ where
the weight of transition t_{arr} is 1 and the weight associated with transitions
$t_i, 1 \leq i \leq n, \alpha_i > 0$ is α_i. Firing of t_{arr} will generate an arrival while firing
of a transition t_i will imply that phase i is chosen as initial state in the MC
associated with the PH distribution. Exponential transitions are used to model
the transitions among the phases. Their set is $T_{\mathcal{E}} = \{\{t_{i,j} : 1 \leq i, j \leq n, Q_{i,j} >$
$0\} \cup \{t_{i,n+1} : 1 \leq i \leq n, \sum_{j=1}^{n} Q_{i,j} < 0\}\}$ where the first set in the union
contains transitions among the transient states while the second those to the
absorbing state. Accordingly, the rate associated with these transitions is $Q_{i,j}$
for transition $t_{i,j}$ if $1 \leq i, j \leq n$ and it is $-\sum_{j=1}^{n} Q_{i,j}$ for transition $t_{i,n+1}$ with
$1 \leq i \leq n$. The set of transitions is $T = T_{\mathcal{I}} \cup T_{\mathcal{E}}$. Transition $t_{i,j}, 0 \leq i, j \leq n+1$,
if present, is with input bag $\{P_i\}$ and output bag $\{P_j\}$. Input bag of transition
t_{arr} is $\{P_{n+1}\}$ and its output bag is $\{P_0, Arrs\}$. In Fig. 12 we depict the PN
representation of the arrival process in which inter-arrival times are PH according
to the example proposed in Sect. 3.1. For immediate transitions (black bars) we
specify the weight while for exponential ones (white rectangles) the rate.

4.2 Petri Net Encoding of Markov Arrival Models

Now we turn our attention to MAPs defined by the pair of matrices $D^{(0)}$ and
$D^{(1)}$ as described in Sect. 3.2.

In case of an n-state MAP, the set of places is $P = \{P_0, P_1, ..., P_n, Arrs\}$
where P_0 is used to start to model, places $P_0, ..., P_n$ correspond to the states of
the MAP and place $Arrs$ is where the patients are gathered. A set of immediate
transitions is used to start the process according to the steady state probabilities,
γ. This set is $T_{\mathcal{I}} = \{t_{0,i} : 1 \leq i \leq n, \gamma_i > 0\}$ in which each transition $t_{0,i}$ is
associated with weight γ_i. In the set of exponential transitions, a transition is
associated with each positive entry of the matrices $D^{(0)}$ and $D^{(1)}$. Accordingly,
the set is $T_{\mathcal{E}} = \{\{t_{i,j} : 1 \leq i, j \leq n, D^{(0)}_{i,j} > 0\} \cup \{t^*_{i,j} : 1 \leq i, j \leq n, D^{(1)}_{i,j} > 0\}\}$
where the second set in the union is the set of transitions that generate arrival.
The rates of transitions $t_{i,j}$ and $t^*_{i,j}$ are $D^{(0)}_{i,j}$ and $D^{(1)}_{i,j}$, respectively. The set

of transitions is $T = T_{\mathcal{I}} \cup T_{\mathcal{E}}$. Transition $t_{i,j}, 0 \leq i,j \leq n$, if present, is with input bag $\{P_i\}$ and output bag $\{P_j\}$. Transition $t^*_{i,j}, 0 \leq i,j \leq n$, if present, is with input bag $\{P_i\}$ and output bag $\{P_j, Arrs\}$. Figure 13 shows the eGSPN encoding of the arrival process using the MAP specified in (2).

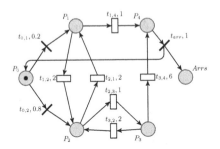 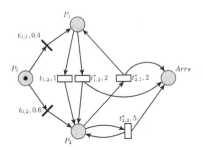

Fig. 12. eGSPN encoding of PH arrival model with PH specified in (1) in Sect. 3.1

Fig. 13. eGSPN encoding of MAP arrival model with MAP given in (2) in Sect. 3.2

4.3 Petri Net Encoding of HMM Arrival Models

In the following we provide the PN encoding of an n-state HMM given by the triple (π, A, B), assuming that the observations are interpreted as number of arrivals in an hour and distributing the arrivals within the hour in a uniform manner.

The set of places is $P = \{P_0, P_1, ..., P_n, O_1, ..., O_n, W, Arrs\}$ where P_0 is used to start the model according to the initial probabilities given in π, places $P_1, ..., P_n$ correspond to the states of the background chain, places $O_i, 1 \leq i \leq n$ are used to emit the observation when the background chain enters state i, place W collect arrivals that has to be distributed in a given hour, and place $Arrs$ gathers all arrivals. A set of immediate transitions, $T_{\mathcal{I},1} = \{t_{0,i} : 1 \leq i \leq n, \pi_i > 0\}$, is used to start the process according to π. A set of deterministic transitions models the background Markov chain: $T_{\mathcal{D}} = \{t_{i,j} : 1 \leq i, j \leq n, A_{i,j} > 0\}$ in which all transitions are associated with fixed delay of one time unit and $t_{i,j}$, if present, is with weight $A_{i,j}$. Another set of immediate transitions, $T_{\mathcal{I},2} = \{t^{(o)}_{i,j} : 1 \leq i \leq n, 1 \leq j \leq m, B_{i,j} > 0\}$ in which $t^{(o)}_{i,j}$ is with weight $B_{i,j}$, is used to emit observations. In order to distribute the arrivals inside an hour a uniform transition is used t_{dist} whose minimal firing time is 0 and maximal firing time is 1. The overall set of transitions is $T = T_{\mathcal{I},1} \cup T_{\mathcal{I},2} \cup T_{\mathcal{D}} \cup \{t_{dist}\}$. Input bag and output bag of transition $t_{0,i}$ is $\{P_0\}$ and $\{P_i, O_i\}$, respectively. Input bag and output bag of transition $t_{i,j}$ is $\{P_i\}$ and $\{P_j, O_j\}$, respectively. Input bag of transition $t^{(o)}_{i,j}$ is $\{O_i\}$ while its output bag contains $j-1$ times place W in order to generate the right number of arrivals. Input bag and output bag of transition t_{dist} is $\{W\}$ and $\{Arrs\}$, respectively. Transition t_{dist} is the only one associated with infinite server policy.

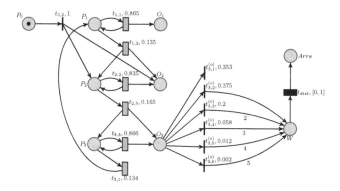

Fig. 14. Part of Petri net encoding of the 3-state HMM model given in (5).

In Fig. 14 we show a part of the PN that represents the 3-state HMM given in (5). Deterministic transitions are drawn as gray rectangles and are provided with their weight. The single uniform transition is drawn as a black rectangle and labeled with its firing interval. Number of occurrences of a place in a bag is written as a label to the corresponding arc if it is different from 1. In the figure we omitted transitions $t_{i,\bullet}^{(o)}$ with $i = 1, 2$ and connecting arcs.

5 Validation Through KPI Assessment

In order to validate the different models of patients arrival process we have coupled each of them with an elementary model of the ED services a patient may go through during his permanence at the ED and assessed meaningful KPI on the coupled models. For the sake of simplicity we assumed the ED patient flow consisting of a simple pipeline of 3 services: *triage*, *visit* and *discharge* for each of which we considered exponentially distributed service time with the following settings[3]: *triage*~Exp(12), the *visit*~Exp(6) and *discharge*~Exp(60) (i.e. visit is assumed to be the slowest while discharge the fastest service). Figure 15 shows the eGSPN encoding of the ED service model: notice that place *Arrs* is shared with the eGSPN models of the patient arrivals (representing the composition of the two models). To validate each arrival model we compared their performance with that resulting by the simulation of the ED service model (Fig. 15) with the actual arrivals of the dataset (for this we generated an eGSPN encoding of the dataset arrivals).

---

[3] The rates have been devised form statistical analysis of the Cantù dataset.

Fig. 15. A simple eGSPN model of ED services consisting of 3 pipelined services (to be composed by overlapping of place *Arrs* with an eGSPN model of the patients arrivals).

One relevant KPI that we considered for comparing the performances of the different patient arrival models coupled with the ED services model is:

$\phi_1 \equiv$ *pmf of the number of patients in the ED during a periodic time – window.*

We formally specified ϕ_1 by means of the Hybrid Automata Specification Language (HASL) [7] and assessed it through the statistical model checking platform Cosmos [1,6]. HASL model checking is a procedure that takes a stochastic model (in terms of an eGSPN) as input as well as a linear hybrid automaton (LHA) together with a list of quantities Z_i to be estimated. The procedure uses the LHA as a *monitor* (i.e., a filter) to select trajectories which are automatically sampled from the eGSPN. The statistics, which are stored in the LHA variables, are collected on the trajectories and used to obtain a confidence interval for each quantity Z_i associated with the LHA monitor (we refer the reader to [7] for more details).

Figure 16 depicts the HASL specification for ϕ_1. The LHA consists of 4 states $(l_0, l_{on}, l_{off}, l_{end})$, 1 clock variable (t), 2 stopwatches (t_a, t_p), $M+1$ real valued variables x_i $(0 \leq i \leq M$, whose final value is the probability that i patients have been observed in the ED during the periodic time windows), plus a number of auxiliary variables and parameters.

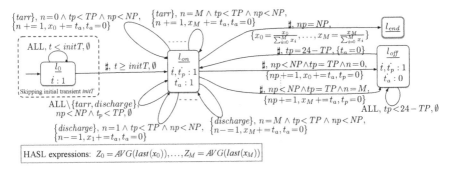

Fig. 16. Automaton for measuring the pmf of the number of patients in the ED.

The LHA is designed so that measuring, along a trajectory, is periodically switched ON/OFF (parameter *TP* being the ON period duration) and stops as soon as the NP-th period has occurred. In the initial state l_0 the occurrence of any transition is ignored ($l_0 \xrightarrow{ALL, t< initT, \emptyset} l_0$) for an initial transient of duration $initT$ ($initT$ being a parameter of the LHA) at the end of which the LHA moves to l_{on} ($l_0 \xrightarrow{\sharp, t \geq initT, \emptyset} l_{on}$). In l_{on} the LHA reacts to the occurrence of patients arrival as well as patients discharge events. When a patient arrives (resp. is discharged) while i patients are in the ED ($l_{on} \xrightarrow[\{n+=1, x_i +=t_a, t_a=0\}]{\{tarr\}, n=i \wedge tp< TP \wedge tp< TP \wedge np< NP} l_{on}$, resp. $l_{on} \xrightarrow[\{n-=1, x_i +=t_a, t_a=0\}]{\{discharge\}, n=i \wedge np< NP} l_{on}$) the patients counter n is incremented (resp. decremented), the duration of the last time interval on which the ED contained i patients is added up to x_i, and the arrival stopwatch t_a is reset. Any event different from a patient's arrival/discharge is ignored in l_{on} ($l_{on} \xrightarrow[tp< TP \wedge np< NP, t_p< TP, \emptyset]{ALL \setminus \{tarr, discharge\}} l_{on}$). As soon as the ON-period expires ($t_p = TP$) the LHA moves from l_{on} to l_{off} where it suspends registering the duration of different number of patients in the ED. In that respect observe that if the end of an ON-period corresponds with i patients being in the ED, the variable x_i (which accumulates the durations of i patients being in the ED) is added up with the time elapsed since the arrival of the last patient ($l_{on} \xrightarrow[np+=1, x_i +=t_a, t_p=0]{\sharp, np< NP \wedge tp=TP \wedge n=i} l_{off}$), hence the end of the ON-period is made corresponding with the end of the duration of having i patients in the ED. In l_{off} the automata ignores any event ($l_{off} \xrightarrow{ALL, tp< 24-TP, \emptyset} l_{off}$) while it switches back to l_{on} as soon as the reciprocal (w.r.t. 24 hours) of the ON-period duration elapsed ($l_{off} \xrightarrow{\sharp, tp=24-TP, \{t_a=0\}} l_{on}$) and in so doing the timer t_a (which stores the duration of the occupation at i patients since the last arrival/departure) is reset so that it can correctly be used in the freshly started ON-period. Finally, the LHA stops monitoring as soon as NP ON-periods have been observed along the monitored trajectory: at that moment each x_i is normalised w.r.t. the sum of all x_i, hence on ending the monitoring of trajectory x_i is assigned with the probability that the i patients have been observed in the ED during NP observed periods ($l_{on} \xrightarrow{\sharp, np=NP, \{x_0=\frac{x_0}{\sum_{i=0}^M x_i}, ..., x_M=\frac{x_M}{\sum_{i=0}^M x_i}\}} l_{end}$). Observe that such an LHA can also be used for "non-periodic" measures: it suffices to set $TP = 0$. The HASL specification for ϕ_1 is completed by the list of HASL expressions $AVG(last(x_i))$ which indicate that a confidence interval for average of the last value that x_i has at the end of an accepted trajectory is computed by COSMOS.

Figure 17 depicts plots computed with the COSMOS tool and resulting from assessing specification ϕ_1 (i.e., the pmf of the number of patients in the ED) against the eGSPN models of the patient arrivals coupled with the 3-services model of the ED (Fig. 15). Plotted results refer to a 1-year (365 days) observation window and have been computed as 99.99% confidence interval of 10^{-3} width. The plot on the left refers to the pmf measured without taking into account any specific time-window over the day (continuous measuring over 24 h for 365 days), while the plots in the center and on the right refer to periodic measuring over a 1-hour period, at low arrival intensity (from 2am to 3am, center picture)

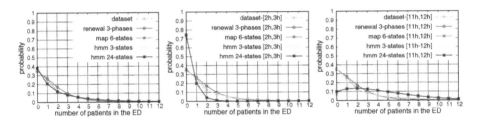

Fig. 17. Pmf of the num. of patients in the ED computed through 3-phases PH renewal model, 6-state MAP, 3-state HMM and 24-state HMM versus dataset model over the whole day (left), during a low-arrival hour (center) and high-arrivals hour (right).

and at high arrival intensity (from 11am to 12pm, right picture), respectively. The obtained results witness the clear advantage of the HMM24 model over the rest: if when no specific hour of the day is considered the pmf of the 3-phases renewal model and of the 6-states MAP provide an acceptable approximation of the pmf computed w.r.t. to dataset (red dashed plot) and the pmf of the HMM3 and HMM24 are essentially indistinguishable, this is no longer the case when the pmf is computed on a specific time window as shown in the center and right plots. Only the pmf of the HMM24 matches that of the dataset during a low-arrivals [2am,3am] window and a high-arrival [11am,12pm] window: the pmf of the renewal and map models instead are essentially identical to those measured continuously while that for the HMM3 exhibits a slight tendency towards matching the pmf of the dataset at low-arrivals but completely fails to do so when arrivals become intense.

6   Conclusion

In this paper we have considered the problem of deriving stochastic models that are capable to accurately reproduce the statistical nature of patients arrivals process observed on a real dataset of the ED of an Italian hospital. Starting from the statistical analysis of the dataset we figured out relevant characteristics of the patients arrival process, e.g., the hour of the day with highest/lowest number of arrivals, the moments, the variance and, most importantly, the periodic nature of number of arrivals-per-hour, the latter shown by auto-correlation function estimated on the dataset. In quest for a model capable of capturing all of these aspects we have considered different classes of Markovian models namely, renewal Markov processes, Markov arrival processes and hidden Markov models. Based on the considered dataset, we derived for each class a few model instances (1, 2 and 3-phase PH renewal process, a 6-state MAP, a 3-state and a 24-state HMM). We assessed the quality of each model by comparing their basic statistical characteristics (i.e., moments, auto-correlation) with those assessed initially on the dataset. To complete the validation process we then considered the coupling of patient's arrival model with a simple Petri net model of the ED services. To this aim we gave a formal encoding of the different kind of patients arrival

Markov models in form of stochastic Petri nets. We then evaluated each model by estimation of the *pmf of the number of patients in the ED during a given period of the day* which we formally encoded and accurately assessed through HASL statistical model checking. The obtained results showed that the 24-states HMM is the only model, amongst those considered, capable of reproducing the statistical characteristics of the dataset, including the hourly periodicity of the arrivals. Therefore such a model could be validly used for predicting the performances of a realistic ED design once coupled with a more realistic model of the internal ED services.

References

1. Cosmos home page. http://cosmos.lacl.fr
2. Abraham, G., Byrnes, G.B., Bain, C.A.: Short-term forecasting of emergency inpatient flow. IEEE Trans. Inf. Technol. Biomed. **13**(3), 380–388 (2009)
3. Afilal, M., Yalaoui, F., Dugardin, F., Amodeo, L., Laplanche, D., Blua, P.: Forecasting the emergency department patients flow. J. Med. Syst. **40**(7), 175 (2016)
4. Angius, A., Horváth, A., Halawani, S.M., Barukab, O., Ahmad, A.R., Balbo, G.: Constructing matrix exponential distributions by moments and behavior around zero. Math. Probl. Eng. **2014**, 1–13 (2014)
5. Avram, F., Chedom, D.F., Horváth, A.: On moments based Padé approximations of ruin probabilities. J. Comput. Appl. Math. **235**(10), 3215–3228 (2011)
6. Ballarini, P., Djafri, H., Duflot, M., Haddad, S., Pekergin, N.: COSMOS: a statistical model checker for the hybrid automata stochastic logic. In: Proceedings of QEST 2011, pp. 143–144. IEEE Computer Society Press, September 2011
7. Ballarini, P., Barbot, B., Duflot, M., Haddad, S., Pekergin, N.: HASL: a new approach for performance evaluation and model checking from concepts to experimentation. Perform. Eval. **90**, 53–77 (2015)
8. Baum, L.E., Petrie, T., Soules, G., Weiss, N.: A maximization technique occurring in the statistical analysis of probabilistic functions of Markov chains. Ann. Math. Stat. **41**(1), 164–171 (1970)
9. Bobbio, A., Telek, M.: A benchmark for PH estimation algorithms: results for Acyclic-PH. Stochast. Models **10**, 661–677 (1994)
10. Bodrog, L., Heindl, A., Horváth, G., Telek, M.: A Markovian canonical form of second-order matrix-exponential processes. Eur. J. Oper. Res. **190**, 459–477 (2008)
11. Bodrog, L., Horváth, A., Telek, M.: Moment characterization of matrix exponential and Markovian arrival processes. Ann. Oper. Res. **160**, 51–68 (2008). https://doi.org/10.1007/s10479-007-0296-8
12. Boyle, J., et al.: Regression forecasting of patient admission data. In: 2008 30th Annual International Conference of the IEEE Engineering in Medicine and Biology Society, pp. 3819–3822, August 2008
13. Buchholz, P.: An EM-algorithm for MAP fitting from real traffic data. In: Kemper, P., Sanders, W.H. (eds.) TOOLS 2003. LNCS, vol. 2794, pp. 218–236. Springer, Heidelberg (2003). https://doi.org/10.1007/978-3-540-45232-4_14
14. Carvalho-Silva, M., Monteiro, M.T., de Sa-Soares, F., Doria-Nobrega, S.: Assessment of forecasting models for patients arrival at emergency department. Oper. Res. Health Care **18**, 112–118 (2018). EURO 2016 - New Advances in Health Care Applications

15. Feldmann, A., Whitt, W.: Fitting mixtures of exponentials to long-tail distributions to analyze network performance models. Perform. Eval. **31**(3–4), 245–279 (1998)

16. Haas, P.J.: Stochastic Petri Nets - Modelling, Stability, Simulation. Springer Series in Operations Research and Financial Engineering. Springer, Heidelberg (2002). https://doi.org/10.1007/b97265

17. Horváth, A., Telek, M.: PhFit: a general phase-type fitting tool. In: Field, T., Harrison, P.G., Bradley, J., Harder, U. (eds.) TOOLS 2002. LNCS, vol. 2324, pp. 82–91. Springer, Heidelberg (2002). https://doi.org/10.1007/3-540-46029-2_5

18. Horváth, A., Telek, M.: Matching more than three moments with acyclic phase type distributions. Stochast. Models **23**(2), 167–194 (2007)

19. Horváth, G.: Matching marginal moments and lag autocorrelations with maps. In: Proceedings of the 7th International Conference on Performance Evaluation Methodologies and Tools, ValueTools 2013, ICST, Institute for Computer Sciences, Social-Informatics and Telecommunications Engineering, Brussels, Belgium, pp. 59–68 (2013)

20. Xia, H., Barnes, S., Golden, B.: Applying queueing theory to the study of emergency department operations: a survey and a discussion of comparable simulation studies. Int. Trans. Oper. Res. **25**(1), 7–49 (2018)

21. Jones, S.S.: A multivariate time series approach to modeling and forecasting demand in the emergency department. J. Biomed. Inform. **42**(1), 123–139 (2009)

22. Lin, D., Patrick, J., Labeau, F.: Estimating the waiting time of multi-priority emergency patients with downstream blocking. Health Care Manag. Sci. **17**(1), 88–99 (2013). https://doi.org/10.1007/s10729-013-9241-3

23. Marsan, M.A., Balbo, G., Conte, G., Donatelli, S., Franceschinis, G.: Modelling with Generalized Stochastic Petri Nets, 1st edn. Wiley, New York (1994)

24. Muthoni, G.J., Kimani, S., Wafula, J.: Review of predicting number of patients in the queue in the hospital using Monte Carlo simulation. IJCSI Int. J. Comput. Sci. Issues **11**(2), 219 (2014)

25. Neuts, M.F.: Probability distributions of phase type. In: Liber Amicorum Professor Emeritus H. Florin, pp. 173–206. University of Louvain (1975)

26. Neuts, M.F.: A versatile Markovian point process. J. Appl. Probab. **16**, 764–779 (1979)

27. Okamura, H., Dohi, T.: Faster maximum likelihood estimation algorithms for Markovian arrival processes. In: Sixth International Conference on the Quantitative Evaluation of Systems, QEST 2009, pp. 73–82. IEEE (2009)

28. Rabiner, L.R.: A tutorial on Hidden Markov models and selected applications in speech recognition. In: Readings in Speech Recognition, pp. 267–296. Morgan Kaufmann Publishers Inc., San Francisco (1990)

29. Ryden, T.: An EM algorithm for estimation in Markov-modulated poisson processes. Comput. Stat. Data Anal. **21**(4), 431–447 (1996)

30. Telek, M., Horváth, G.: A minimal representation of Markov arrival processes and a moments matching method. Perform. Eval. **64**(9–12), 1153–1168 (2007)

31. Whitt, W., Zhang, X.: A data-driven model of an emergency department. Oper. Res. Health Care **12**, 1–15 (2017)

Repairing Event Logs with Missing Events to Support Performance Analysis of Systems with Shared Resources

Vadim Denisov[1,3(✉)], Dirk Fahland[1], and Wil M. P. van der Aalst[1,2]

[1] Eindhoven University of Technology, Eindhoven, The Netherlands
{v.denisov,d.fahland}@tue.nl
[2] Process and Data Science (Informatik 9), RWTH Aachen University, Aachen, Germany
wvdaalst@pads.rwth-aachen.de
[3] Vanderlande Industries, Veghel, The Netherlands

Abstract. To identify the causes of performance problems or to predict process behavior, it is essential to have correct and complete event data. This is particularly important for distributed systems with shared resources, e.g., one case can block another case competing for the same machine, leading to inter-case dependencies in performance. However, due to a variety of reasons, real-life systems often record only a subset of all events taking place. For example, to reduce costs, the number of sensors is minimized or parts of the system are not connected. To understand and analyze the behavior of processes with shared resources, we aim to reconstruct bounds for timestamps of events that must have happened but were not recorded. We present a novel approach that decomposes system runs into token trajectories of cases and resources that may need to synchronize in the presence of many-to-many relationships. Such relationships occur, for example, in warehouses where packages for N incoming orders are not handled in a single delivery but in M different deliveries. We use linear programming over token trajectories to derive the timestamps of unobserved events in an efficient manner. This helps to complete the event logs and facilitates analysis. We focus on material handling systems like baggage handling systems in airports to illustrate our approach. However, the approach can be applied to other settings where recording is incomplete. The ideas have been implemented in ProM and were evaluated using both synthetic and real-life event logs.

Keywords: Log repair · Process mining · Performance analysis · Modeling · Material handling systems

1 Introduction

Precise knowledge about actual process behavior and performance is required for identifying causes of performance issues [16], as well as for predictive process monitoring of important process performance indicators [14]. For Material Handling Systems (MHS), such as Baggage Handling Systems (BHS) of airports, performance incidents are usually investigated offline, using recorded event data for finding root causes of

© Springer Nature Switzerland AG 2020
R. Janicki et al. (Eds.): PETRI NETS 2020, LNCS 12152, pp. 239–259, 2020.
https://doi.org/10.1007/978-3-030-51831-8_12

problems [10], while online event streams are used as input for predictive performance models [4]. Both analysis and monitoring heavily rely on the completeness and accuracy of input data. For example, events may not be recorded and, as a result, we do not know when they happened even though we can derive that they must have happened. Yet, when different cases are competing for shared resources, it is important to reconstruct the ordering of events and provide bounds for non-observed timestamps.

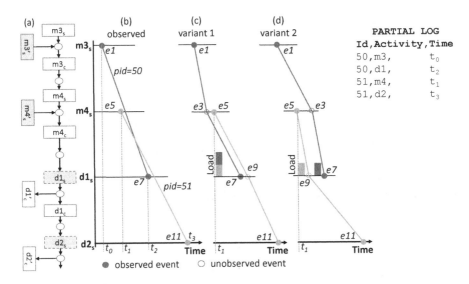

Fig. 1. An MHS model example (a), observed imprecise behavior for two cases 50 and 51 (b), possible actual behaviors (c,d).

However, in most real-life systems, items are not continuously tracked and not all events are stored for cost-efficiency, leading to incomplete performance information which impedes precise analysis. For example, an MHS tracks the location of an item, e.g., a bag or box, via hardware sensors placed throughout the system, generating tracking events for system control, monitoring, analysis, and prediction. Historically, to reduce costs, a tracking sensor is only installed when it is strictly necessary for the correct execution of a particular operation, e.g., only for the precise positioning immediately before shifting a bag from one conveyor onto another. Moreover, even when a sensor is installed, an event still can be discarded to save storage space. As a result, the recorded event data of an MHS are typically incomplete, hampering analysis based on such incomplete data. Therefore, it is essential to repair the event data before analysis. Figure 1 shows a simple MHS where events are not always recorded. The process model is given and for two cases the recorded incomplete sets of events are depicted using the so-called *Performance Spectrum* [10].

Figure 1(b) shows item pid = 50 entering the system via $m3$ at time t_0 (event e_1) and leaving the system via $d1$ at time t_2 (e_7), and item pid = 51 entering the system via $m4$ at time t_1 (e_5) and leaving the system via $d2$ at time t_3 (e_{11}). As only these four events are recorded, the event data do not provide information in which order both cases traversed the *segment* $m4 \rightarrow d1$. Naively interpolating the movement of both items, as shown in Fig. 1(b), suggests that item pid = 51 overtakes item pid = 50. This contradicts that all items are moved from $m4$ to $d1$ via a conveyor belt, i.e., a FIFO queue: item 51 cannot have overtaken item 50. In contrast, Fig. 1(c) and Fig. 1(d) show two possible behaviors that are consistent with our knowledge of the system. We know that a conveyor belt (FIFO queue) is a shared resource between $m4$ and $d1$. Both variants differ in the order in which items 50 and 51 enter and leave the shared resource, the speed with which the resource operated, and the load and free capacity the resource had during this time. In general, the longer the duration of naively interpolated segment occurrences, the larger the potential error. Errors in load, for example, make performance outlier analysis [10] or short-term performance prediction [9] rather difficult. Errors in order impede root-cause analysis of performance outliers, e.g., finding the cases that caused or were affected by outlier behavior.

Problem. In this paper, we address a novel type of problem as illustrated in Fig. 1 and explained above. The behavior and performance of the system cannot be determined by the properties of each case in isolation, but depends on the *behavior of other cases* and the *behavior of the shared resources* involved in the cases. Crucially, each case is handled by multiple resources and each resource handles multiple cases, resulting in *many-to-many* relations between them. The concrete problem we address is to reconstruct *unobserved* behavior and performance information of *each case* and each *shared resource* in the system that is *consistent* with both observed and reconstructed unobserved behavior and performance of all other cases and shared resources. More specifically, we consider the following information as given: (1) an event log L_1 containing the case identifier, activity and time for recorded events where intermediate steps are not recorded (i.e., the event log may be incomplete), (2) a model of the process (i.e., possible paths for handling each individual case), and (3) a description (model) of the resources involved in each step (e.g., queues, single server resources and their performance parameters such as processing and waiting time). Based on the above input, we want to provide a complete event log L_2 that describes (1) for each case the exact sequence of process steps, (2) and for each unobserved event a time-window of earliest and latest occurrence of the event so that (3) either all earliest or all latest timestamps altogether describe a consistent execution of the entire process over all shared resources.

Contribution. We propose a solution to this problem for a limited class of systems. We focus on processes where each step is served by one single-server resource and resources are connected by strict FIFO queues only. These assumptions are reasonable for a large class of MHSs. Our current solution formulation assumes the process to be *acyclic* which suffices for many real-life problem instances. Section 2 presents related work while elaborating on the problem. Specifically, prior work either only considers the case or the resource perspective explicitly, making implicit assumptions about their complex interplay. To overcome this limitation, we use *synchronous proclets* [11] in Sect. 3 to conceptually decompose a run of a system into individual *token trajectories* of cases, resources, and queues. Token trajectories synchronize when a resource or queue is involved in a case, allowing to explicitly describe their many-to-many relations in the run. Section 4 then formally captures token trajectories in terms of partial orders of events and defines the general problem. We solve the problem in Sect. 5 by formulating a Linear Programming (LP) problem [19] in terms of timestamps along the different token trajectories. To evaluate the approach, we compare the restored event logs with the ground truth for synthetic logs and estimate errors for real-life event logs for which the ground truth is unavailable (Sect. 6). We discuss our findings and future work in Sect. 7.

2 Related Work

In all operational processes (logistics, manufacturing, healthcare, education and so on) complete and precise event data, including information about workload and resource utilization, is highly valuable since it allows for process mining techniques uncovering compliance and performance problems. Event data can be used to replay processes on top of process models [2], to predict process behavior [5,9], or to visualize detailed process behavior using performance spectra [10]. All of these techniques rely on complete and correct event data. Since this is often not the case, we aim to transform *incomplete* event data into *complete* event data.

Various approaches exist for dealing with incomplete data of processes with non-isolated cases that compete for scarce resources. In call-center processes, thoroughly studied in [12], queueing theory models can be used for load predictions under assumptions about distributions of unobserved parameters, such as customer patience duration [6], while assuming high load snapshot principle predictors show better accuracy [21]. For time predictions in congested systems, the required features are extracted using congestion graphs [20] mined using queuing theory.

Techniques to repair, clean, and restore event data before analysis have been suggested in other works. An extensive taxonomy of quality issue patterns in event logs is presented in [22]. The taxonomy also lists approaches to repair inadvertent time intervals [22] in [8]. In [15] resource availability calendars are retrieved from event logs without the use of a process model, but assuming *start* and *complete* life-cycle transitions as well as a case arrival time present in a log. Using a process model, classical trace alignment algorithms [7] restore missing events but do not restore their timestamps. The authors conclude (see [7], p. 262) that incorporating other dimensions, e.g., resources, for multi-perspective trace alignment and conformance checking is an important challenge for the near future. Recently, also techniques for process discovery and conformance checking over uncertain event data were presented [17,18]. The output of our approach can provide the input needed for these techniques.

Our work contributes to the problem of reconstructing behavior of cases and limited shared resources for which the cases compete. We use the notion of *proclets* first introduced in [1] and adapted for process mining in [11] to approach the problem from control-flow and resource perspectives at once. We assume a system model given as a composition of a control-flow proclet (process) and resource/queue proclets. We restore missing events through classical trace alignments over control-flow proclets. The dynamic synchronization of proclets [11] allows us to infer how and when resource tokens must have traversed over the control-flow steps, which we express as a linear programming problem to compute timestamp intervals for the restored events. Event logs repaired in this way enable the use of analysis assuming complete event logs.

3 Modeling Inter-Case Behavior via Shared Resources

Prior work (cf. Sect. 2) approaches the problem of analyzing the performance of systems with shared resources primarily either from the control-flow perspective [5,9,15,17,18] or the resource/queuing perspective [6,12,20,21], leading to information loss about the other perspective. In the following, we show how to conceptualize the problem from both perspectives at once using *synchronous proclets* [11]. This way we are able to capture both control-flow and resource dynamics and their interaction as synchronizing token trajectories. We introduce the model in Sect. 3.1 and use it to illustrate how incomplete logging incurs information loss for performance analysis in Sect. 3.2.

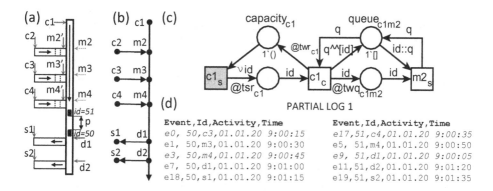

Fig. 2. A baggage handling system fragment (a) and its material flow diagram (b). Conveyor belts of check-in counters $c1 - c4$ merge at points $m2 - m4$, further downstream bags can divert at $d1$ and $d2$ to X-Ray security scanners $s1$ and $s2$. Red arrows show sensor (logging) locations. Conveyor $c1 : m2$ is modeled as a coloured Petri net model (c). An example of an incomplete event log of the system in (a) is shown in (d), where missing events are shown in the grey color. (Color figure online)

3.1 Processes-Aware Systems with Shared Resources

We explain the dynamics of process-aware systems over shared resources using a BHS handling luggage. The process control-flow takes a bag from a source (e.g., check-in or transfer from another flight), to a destination (e.g., the airplane, transfer) along intermediate process steps (e.g., baggage scanning, storage). BHS resources are primarily single-server machines (e.g., baggage scanners) connected via conveyor belts, i.e., FIFO queues. Figure 2(a) shows a typical system design pattern involving the control-flow and resource perspective: four parallel check-in desks (c1-c4) merge into one *linear conveyor* through *merge points* (m2-m4). *Divert points* (d1 and d2) can route bags from the linear conveyor to *scanners* (s1 and s2). Each merge point and scanner is preceded by a FIFO queue for buffering incoming cases (bags) in case the corresponding resource is busy. Figure 2(b) shows the plain control-flow of this BHS (also called Material Flow

Diagram (MFD)). A real-life BHS may contain hundreds of process steps and resources, and conveyors may also form loops.

Modeling with Coloured Petri Nets. Figure 2(c) shows a Coloured Petri Net (CPN) model for the segment $c1 \rightarrow m2$. In the model, transitions $c1_s$ and $c1_c$ describe *start* and *completion* of the check-in step $c1$. At the occurrence of $c1_s$ a new bag (v id) represented by a token with an id is inserted. Step $c1$ is served by a single resource (place $capacity_{c1}$) which has service-time tsr_{c1} to complete the step and waiting time trw_{c1} until the next bag can go through $c1$. All Resources in a BHS may require a waiting time to ensure sufficient "operating space" between two subsequent cases. After completion of $c1$, the bag enters a FIFO queue (modeling a conveyor belt) to the *start* of the merge step $m2_s$. Time annotation twq_{c1m2} models the minimum time it takes for a bag to travel from $c1$ to $m2$. Only then a bag may leave the queue at $m2_s$. The CPN model in Fig. 2(c) describes the impact of limited resource capacity and queues on the progress of a case, but does not model the resource itself as its own entity. The absence of the resource in the described behavior makes it impossible to reason about its behavior explicitly.

Modeling with Synchronous Proclets. The synchronous proclet system in Fig. 3 describes the entire BHS of Fig. 2(a) by using three types of proclets.

1. The *process proclet* (red border) is a Petri net describing the control-flow perspective of how bags may move through the system. It directly corresponds to the MFD of Fig. 2(b). It is transition-bordered and each occurrence of one of its initial transitions creates a new case identifier, see [11] for details.
2. Each *resource proclet* (green border) models a resource as its own entity with a cyclic behavior. For example, the *PassengerToSystemHandover* proclet (top left) identifies a concrete resource by token id $c1$; its life-cycle models that starting a task ($c1_s$) makes the resource *busy* and takes service time tsr_{c1}, after completing the task ($c1_c$) the resource has waiting time twr_{c1} before being *idle* again in the same way as Fig. 2(c). All other resource proclets follow the same pattern, though some resources such as *MergingUnit-m2* and *DivertingUnit-d1* may have two transitions to become *busy* or *idle*, respectively.
3. Each queue proclet (blue border) describes a FIFO queue as in Fig. 2(c). However, the queue state (the list) is accompanied by a queue identifier in place q. Items entering the queue are remembered by their number (generated from the *count* place).

The proclet system synchronizes process, resources, and queues via *synchronous channels* between transitions. Transitions linked via synchronous channels may only occur when all linked transitions are enabled; when they occur, they occur in a single synchronized event. For example, transition $c1_s$ is always enabled in *Process*, generating a new bag id, e.g., id $= 49$, but it may only occur together with $c1_s$ in *PassengerToSystemHandover*, i.e., when resource $c1$ is *idle*, thereby synchronizing the process case for bag $id = 49$ with the resource with identifier $c1$. By annotation *init c1,1:1 c1* is now correlated to $id = 49$. The subsequent correlation annotation $=c1, 1:1$ on the channel of the *complete* transition $c1_c$ ensures that resource $c1$ only synchronizes with the process case on which it started the step, i.e., $id = 49$; the next occurrence of $c1_s$ will create a new correlation to another process case, see [11] for details. In the example, each resource is statically linked to one process step, but the model also allows for one

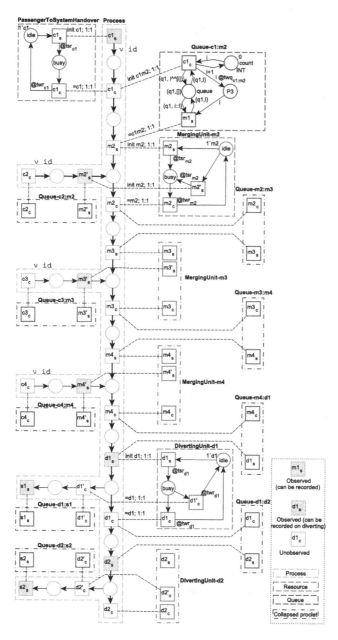

Fig. 3. The synchronous proclet model of the system shown in Fig. 2(a) consists of three types of proclets: *Process* for modeling a system layout and process control flow (red), *Resource* for modeling connector and sensor resources (services), and *Queue* for modeling conveyors transporting bags in the FIFO order. Only filled transitions can be observed in an event log. (Color figure online)

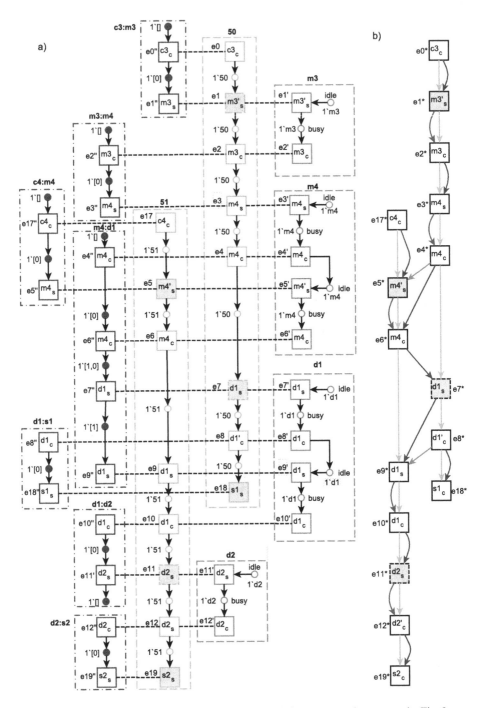

Fig. 4. Synchronization of multiple sub-runs of the synchronous proclet system in Fig. 3 over shared resources and queues (a), and a global partial order obtained by the union of partial orders of each sub-run (b) for synchronized events, shown by red, green and blue arrows for partial orders $<_{pid}$, $<_{rid}$ and $<_{qid}$ respectively. (Color figure online)

resource to participate in multiple different process steps, and multiple resources to be required for one process step. In the following, we call a proclet system that defines proclets for processes, queues, and resource that are linked via synchronous channels as described above, a *PQR system*.

Proclets Describe Synchronizing Token Trajectories. We now highlight how the partial-order semantics of synchronous proclets [11] preserves the identities of process, resources, and queues as "token trajectories". Figure 4(b) shows a partially-ordered run of the PQR system of Fig. 3 for two bags $id = 50$ and $id = 51$. The run in Fig. 4(b) can be understood as a synchronization of multiple runs of the process, resource, and queue proclets, one for each case, resource, or queue involved as shown in Fig. 4(a).

Bag 50 gets inserted via input transition $c3_c$ (event e_0^* in Fig. 4(b)). This event is a *synchronization* of events $e0$ ($c3_c$ occurs for bag 50 in the *Process* proclet) and $e0'$ ($c3_c$ occurs for the $c3{:}m3$ queue) in Fig. 4(a). The minimal waiting time twq_{c3m3} must pass before bag 50 reaches the end of the queue and process step $m3$ can start. The process step $m3$ merges bag 50 from the check-in conveyor $c3$ onto the main linear conveyor and may only start via transition $m3_s$ when *MergingUnit-m3* is *idle*. As this is the case, bag 50 leaves the queue ($e1''$ in $c3{:}m3$), $m3$ starts merging ($e1'$ in $m3$), the bag starts the merging step (event $e1$ in *Process*), resulting in the synchronized event $e1^*$ in Fig. 4(b).

By e_1', resource $m3$ switches from *idle* to *busy* and takes time tsr_{m3} before it can complete the merge step with $m3_c$ (event $e2'$) on bag 50 (event $e2$); this merge step also inserts bag 50 into queue $m3{:}m4$ ($e2''$) resulting in synchronized event $e2^*$. Subsequently, bag *50* leaves queue $m3{:}m4$ ($e3^*$) is pushed by merge unit $m4$ into queue $m4{:}d1$ ($e4^*$).

Concurrently, bag 51 is inserted via input transition $c4_c$ (event $e17^*$), moves via queue $c4{:}m4$ also to merge unit $m4$ to enter queue $m4{:}d1$, i.e., both bags 50 and 51 now compete for merge unit $m4$ and the order of entering $m4{:}d1$. In the run in Fig. 4, $m4$ executes $m4_s$ and $m4_c$ for bag 51 ($e5^*$ and $e6^*$) *after completing* this step for bag 50 ($e3^*$ and $e4^*$). Thus, 51 enters the queue ($e6^*$) after 50 entered the queue ($e5^*$) but before 50 leaves the queue $e7^*$. Consequently, divert unit $d1$ first serves 50 ($e7^*$ and $e8^*$) to reach scanner $s1$ ($e18^*$) before serving 51 ($e9^*$ and $e10^*$) to reach scanner $s2$ ($e19^*$).

Figure 4(b) shows how the process tokens of bag 50 and 51 synchronized with the resources and queue tokens along the run, forming sequences or *trajectories* of events where this token was involved. For example, bag 50 followed the trajectory $e0^*, e1^*, \ldots, e8^*, e18^*$ and queue $m4{:}d1$ followed trajectory $e4^*, e6^*, e7^*, e9^*$ thereby synchronizing with both bag 50 and bag 51.

3.2   Information Loss Because of Incomplete Logging

Although event data on objects that are tracked can be used for various kinds of data analysis [4,9], in practice sensors are placed only where it is absolutely necessary for correct operation of the system, e.g., for merge and divert operations, without considering data analysis needs. Applied to our example, only the transitions that are shaded in Fig. 3 would be logged, i.e., $c1_s, m2_s', m3_s', m4_s', d1_s, d2_s, s1_s, s2_2$ would be logged from the *control-flow* perspective only. The run of Fig. 4 would result in a "typical" but highly incomplete event log as shown in Fig. 2(d).

According to this incomplete log, bag 50 silently passes *m4* and is tracked again only at *d1* (*e7*) and finally at *s1* (*e18*) whereas 51 silently passes *d1* (as it moves further on the main conveyor) and is tracked again only at *d2* (*e11*). Based on this incomplete information the bags 50 and 51 may have traversed *m4:d1* in different orders and at different speeds resulting also in different loads as illustrated in Fig. 1. As a result, in case of congestion, we cannot determine the ordering of cases [10], cannot compute the exact load on each conveyor part for (predictive) process monitoring [5,9]. The longer an unobserved path (e.g., $c1 \rightarrow d2$), the higher the uncertainty about the actual behavior and the less accurate performance analysis outcome.

Although minimal (or even average) service and waiting times on conveyor belts and resource are known, we need to determine the *exact timestamps* of all missing events and their order to reconstruct for how long resources were occupied by particular cases and in which order cases were handled, e.g., did 50 precede 51 on *m4:d1* or vice versa?

The objective of this paper is to reconstruct from a subset of events logged from the control-flow perspective only the remaining events (including time information), so that the time order is consistent with a partially ordered run of the entire system, including resource and queue proclets. For example, from the recorded events of the event log in Fig. 2(d) we reconstruct the remaining events (Fig. 4(a)) with time information so that the resulting order (by time) is consistent with the partially ordered run in Fig. 4(b).

4   System Runs and Partial Event Logs

In Sect. 3, we showed how the behavior of resource and queue-aware processes can be modeled as a PQR system, a particular type of a synchronous proclet system. The partially-ordered run of a PQR system decomposes into token trajectories for process cases, resources, and queues. In this section, we first formalize this relation between a partially ordered run of a system and its token trajectories through projection on partially ordered sets. We then formalize *partial* and *complete* event logs of a system run within this model and state the formal problem we address.

We use the following notion. Let A be a set of *event classifiers*; A is usually the set of activity names or the set of locations in case of an MHS. Let T be the set of time durations and timestamps, e.g., the rational or real numbers. Let $\mathcal{E}$ be the *universe of unique events* with *attributes*, let AN be a set of attribute names. For any $e \in \mathcal{E}, n \in AN$, $\#_n(e)$ is the value of attribute n for event e ($\#_n(e) = \perp$ if attribute n is undefined for e). Each event has a mandatory attribute *act*, $\#_{act}(e) \in A$, a mandatory attribute *lt* for a life-cycle transition, $\#_{lt}(e) \in \{start, complete\}$ and an optional attribute *time*, $\#_{time}(e) \in T$. Finally, we allow events to be related to multiple case notions. Let $\mathcal{Z}$ be the *universe of case identifiers* and $ID \subset AN$ be a *set of case notions*. If $\#_{id}(e) = z$, then event e is related to case z under case notion $id \in ID$.

From Partial Orders to Token Trajectories. A run of a proclet system [11] can be observed in terms of a **Strict Partially Ordered Set** (SPOSET) $\pi = (E, <)$ of events $E \subseteq \mathcal{E}$. As usual, we write $e_1 < e_2$ if event e_1 *precedes* event e_2 and we write $e_1 \lessdot e_2$ iff e_1 *directly* precedes e_2, i.e., $e_1 < e_2$ and there is no other event e_3 with $e_1 < e_3 < e_2$. In a PQR system we can distinguish three case notions $ID = \{pid, rid, qid\}$ to distinguish cases of the process, resources, and queues. Each event $e \in \mathcal{E}$ in a run of a PQR system has one or more case notions from ID. For example, in Fig. 4, $\#_{pid}(e5^*) = 51, \#_{rid}(e5^*) = m4, \#_{qid}(e5^*) = c4{:}m4$.

Restricting $<$ of a system run to events of the same case notion $id \in ID$ results in a *case notion-specific* partial order $e_1 <_{id} e_2$ iff $e_1 < e_2$ and $\#_{id}(e_1) = \#_{id}(e_2) \neq \perp$. Only events which share the same case notion and case identifier are ordered by $<_{id}$ – events of different cases are unordered. For example, in Fig. 4(b) $e4^* <_{rid} e5^*$ but $e4^* \nless_{pid} e5^*$. Consequently, $<_{pid}$ orders all events wrt. the process perspective whereas $<_{rid}$ and $<_{qid}$ order all events wrt. the resource and queue perspective, respectively. For a given case notion $id \in ID$ and case identifier $z \in \mathcal{Z}$, the events $E_{id}^z = \{e \in E | \#_{id}(e) = z\}$ of case z and Strict Partial Order (SPO) $<_{id}^z |_{E_{id}^z \times E_{id}^z}$, restricted to the events of the same case, define a *sub-run* $\pi_{id}^z = (E_{id}^z, <_{id}^z)$. Each such sub-run formalizes one *token trajectory* in the system run π. For example, Fig. 4(a) shows the sub-runs, viz. token trajectories, of all cases of the run of Fig. 4(b), i.e., π_{pid}^{50} and π_{pid}^{51} from the perspective of the process, $\pi_{rid}^{m3}, \pi_{rid}^{m4}, \pi_{rid}^{d1}, \pi_{rid}^{d2}$ from the perspective of the resources, and $\pi_{qid}^{c3:m3}, \pi_{qid}^{m3:m4}, \pi_{qid}^{c4:m4}, \pi_{qid}^{m4:d1}, \pi_{qid}^{d1:d2}, \pi_{qid}^{d1:s1}, \pi_{qid}^{d2:s2}$ from the perspective of the queues. In this way, our model shows that events of different process cases ($pid = 50$ and $pid = 51$) are independent under the classical control-flow perspective $<_{pid}$, e.g., $e4^* \nless_{pid} e5^* \nless_{pid} e7^*$, but mutually depend on each other under $<_{rid}$ and $<_{qid}$, e.g., $e4^* <_{rid} e5^* <_{rid} e6^*$ and $e6^* <_{qid} e7^*$. Each sub-run π_{id}^z is a "proper" run of case z in the corresponding proclet (Lemma 2 in [11]).

Event Logs and Token Trajectories. Starting point for our analysis is the notion of a classical control-flow event log, which we express in our model of SPOSETs using *pid* as case notion. An *event log* $L = (E, <)$ is a finite set of events E where each event e has an activity $\#_{act}(e)$, a process case id $\#_{pid}(e)$. Note that e may have additional case identifiers $\#_{rid}(e)$ and $\#_{qid}(e)$ as attributes.

Adopting [13] to our setting, the optional timestamps $\#_{time}(e)$ induce the log's partial order, i.e., two events e_1 and e_2 are ordered if e_1 time-wise precedes e_2 and both are related in some case (for any $id \in ID$), i.e., $e_1 < e_2$ iff $\perp \neq \#_{time}(e_1) < \#_{time}(e_2) \neq \perp$ and there exists $id \in ID$ with $\#_{id}(e_1) = \#_{id}(e_2)$. If all events are only related to id cases, then $<$ and $<_{id}$ are identical. Further, each sub-run L_{id}^z (projection onto events with $\#_{id}(e) = z$) is a *trace* for case z under case notion id.

Event Logs and Token Trajectories. Given a model M of a PQR system, i.e., a system defining proclets for pid, rid, qid, we call log L *complete* wrt. events and ordering iff there is a run π of M such that L and M are isomorphic wrt. attributes act, pid, qid, rid. Note that the ordering in L is induced by event timestamps only, thus a complete log defines the "right" timestamps. Further note that in a complete log L, each trace L_{id}^z is also complete and describes a token trajectory, i.e., a sub-run π_{id}^z of π that fits the corresponding proclet. Further, all traces of process cases (pid) are ordered relative to each other via the shared resources and queues as described in M.

In reality often only a subset of activities $B \subseteq A$ and the control-flow case notion *pid* have been recorded in a log, making it *partial*. In this paper, we call a log L' *partial* if there exists a complete log $L = (E, <)$ of M (viz. system run π) such that $L' = (E', <')$ is the projection of L onto activities in B, $L|_B = (E_B, < |_{E_B \times E_B}), E_B = \{e \in E \mid \#_{act}(e) \in B\}$ such that additionally

1. each $e \in E'$ has only case notion pid, i.e., $\#_{pid}(e) \neq \bot, \#_{rid}(e) = \#_{qid}(e) = \bot$,
2. $\#_{time}(e)$ is defined, and
3. for each process case z occurring in L, L' contains at least the first and last event of the complete trace.

Thus, L' contains for each case z at least one *partial trace* L'^z_{pid} recording the entry and exit of the case and preserving the order of observed events, i.e., it can be completed to fit the model. An MHS typically records a partial log as defined above. Figure 2(d) shows a partial event log of the run on Fig. 4. In a partial event log, events of different process cases are less ordered, e.g., observed events $e1^*$ and $e5^*$ in Fig. 4 are unordered wrt. any resource or queue whereas they are ordered in the complete run. In the following, we investigate how to restore this lost ordering.

Problem Formulation. Reconstructing a complete log from a partial log as defined above requires to reconstruct all missing events, all missing case notion attributes, and their timestamp. Restoring the *exact* timestamp is generally infeasible and for most use cases also not required. We, therefore, formulate the problem as restoring *time-windows* providing minimal and maximal timestamps for each unobserved event.

Let M be a model of a PQR system defining life-cycles of process, resource, and queue proclets, which resources and queues synchronize on which process step, and for each resource the minimum service time tsr and waiting time twr and for each queue the minimum waiting time twq. Given M and a partial log $L_1 = (E_1, <_1)$ of M, we want to (1) reconstruct unobserved events E_u for all process cases in L_1 and their relations to queues and resources, and (2) for each unobserved event $e \in E_u$ a time-window of earliest and latest occurrence of the event $\#_{tmin}(e), \#_{tmax}(e) \in T$ so that (3) $L_2 = (E_1 \cup E_u, <_2)$ is a complete log of M when $<_2$ is inferred from $\#_{tmin}(e)$ or from $\#_{tmax}(e)$.

5   Inferring Timestamps Along Token Trajectories

In Sect. 4, we presented the problem of restoring missing events and time-windows for their timestamps from a partial event log L_1 such that the resulting log is consistent with resource and queueing behavior. In this section, we solve the problem for PQR systems with *acyclic* process proclets by casting it into a constraint satisfaction problem, that can be solved using Linear Programming (LP) [19]. In Sect. 5.1, we show how to infer unobserved events (from M) and how to infer resource and queue identifiers from M to construct an intermediate SPO $(E_2, <_2)$. All unobserved events $E_2 \setminus E_1$ have no timestamp, i.e., they are unordered in $<_2$. In Sect. 5.2 we then show how to determine minimal and maximal timestamps for each unobserved event (through a linear program) that preserves the already known ordering $<_2$. Inferring $<_3$ from the minimal (or maximal) timestamps refines $<_2$ and results in an SPO $L_3 = (E_2, <_3)$ which is a complete log of M and has L_1 as a partial log. We explain our approach using another (more compact running) example shown in Fig. 5(a) for two bags 53 and 54 processed in the system of Fig. 3. The events in grey italic (i.e., f3, f5, f6, f14) are unobserved.

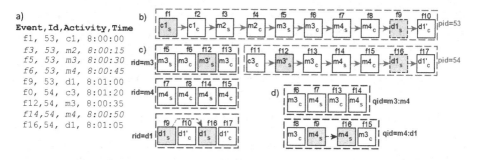

Fig. 5. Another partial event log of the system in Fig. 3 for bags 53 and 54 (a), partially complete traces of the Process (b), Resource (c) and Queue (d) proclets, restored by oracles O_1, O_2. Only observed events are ordered, e.g., $f9 <^{d1}_{rid} f16$, while the other events are isolated. (Color figure online)

5.1   Infer Potential Complete Runs from a Partial Run

We first derive for the partial log $L_1 = (E_1, <_1)$ an intermediate log $L_2 = (E_2, <_2)$ so that each trace $L^z_{2,pid}$ of a process case z is complete (i.e., fits the process proclet in M). In a second step, we relate each unobserved event $e \in E_u = E_2 \setminus E_1$ to a corresponding resource and/or queue case identifier which orders observed events wrt. $<_{rid}$ and $<_{qid}$, resulting in an SPO $\pi = (E_\pi, <_\pi)$ (with $E_\pi = E_2$). All unobserved events $e \in E_u$ lack a timestamp and hence are left unordered wrt. $<_{rid}$ and $<_{qid}$ in π; we later refine $<_\pi$ in Sect. 5.2.

We specify how to solve each of the steps in terms of two *oracles* O_1 and O_2 and describe concrete implementations for either. Oracle O_1 has to return $L_2 = (E_2, <_2) = O_1(E_1, <_1, M)$ by completing each partial trace $L^z_{1,pid}$ of some process case z into a complete trace $L^z_{2,pid}$ that fits the process proclet M. The restored *unobserved* events $E_u = E_1 \setminus E_1$ only have attributes *act, pid* and *lt* (life-cycle transition) and events are *totally ordered* along pid, i.e., $<_2=<_{pid}$ is a total order. O_1 can be implemented using well-known trace alignment [3]. For example, applying O_1 on the partial log of Fig. 5(a) results in the complete process traces of Fig. 5(b). Note that, slightly deviating from our model, O_1 constructs $<_2$ explicitly (not based on timestamps).

Oracle O_2 has to enrich events in L_2 with information about queues and resources so that for each $e \in E_2$ if resource r is involved in the step $\#_{act}(e)$, then $\#_{rid}(e) = r$ and if queue q was involved, then $\#_{qid}(e) = q$. Moreover, in order to formulate the linear program to derive timestamps in a uniform way, each event e has to be annotated with the performance information of the involved resource and/or queue. That is, if e is a start event and $\#_{rid}(e) = r \neq \perp$, then $\#_{tsr}(e)$ and $\#_{twr}(e)$ hold the minimum service and waiting time of r, and if $\#_{qid}(e) = q \neq \perp$, then $\#_{twq}(e)$ hold the minimum waiting time of q. For the concrete PQR systems considered in this paper, we set $\#_{rid}(e) = r$ based on the model M if r is the case id of the resource proclet that synchronizes with transition $t = \#_{act}(e)$ via a channel (there is at most one). Attributes $\#_{tsr}(e), \#_{twr}(e)$, can be set from the model as they are parameters of the resource proclet. To ease the LP formulation, if e is unrelated to a resource, we set $\#_{rid}(e) = r^*$ to fresh identifier and $\#_{tsr}(e) = \#_{twr}(e) = 0$; $\#_{qid}(e)$ and $\#_{twq}(e)$ are set correspondingly. By annotating the events in E_2, we obtain

SPO $\pi = (E_\pi, <_\pi)$, $E_\pi = E_2$ that also contains sub-runs for each queue and resource containing all events to be complete wrt. M but only observed events are ordered (due to their timestamps). For example, Fig. 5(d) shows the sub-run $\pi_{qid}^{m4:d1}$ containing events f_8, f_9, f_{16}, f_{15} with only $f_9 <_{qid} f_{16}$. Next, we define constraints based on the information in this intermediate log π to infer timestamps for all unobserved events.

5.2   Restoring Timestamps of Unobserved Events by Linear Programming

The SPO $\pi = (E, <)$ obtained in Sect. 5.1 from partial log $L_1 = (E_1, <_1)$ includes all unobserved events $E_u = E \setminus E_1$ of the complete log, but lacks timestamps for each $e \in E_u, \#_{time}(e) = \perp$. Each observed $e \in E_1$ has a timestamp $\#_{time}(e)$ and we also added minimum service time $\#_{tsr}$, waiting time $\#_{twr}(e)$ of the resource $\#_{rid}(e)$ involved in e and minimum waiting time $\#_{twq}(e)$ of the queue involved in e. We now define a constraint satisfaction problem that specifies the earliest $\#_{tmin}(e)$ and latest $\#_{tmax}(e)$ timestamps for each $e \in E_u$ so that all earliest (latest) timestamps yield a consistent ordering of all events in E wrt. $<_{pid}$ (events follow the process), $<_{rid}$ (events follow resource life-cycle), and $<_{qid}$ (events satisfy queueing behavior). The problem formulation propagates the known $\#_{time}(e)$ values along with the different case notions $<_{pid}$, $<_{rid}$, $<_{qid}$, using tsr, twr, twq. For that, we introduce variables x_e^{tmin}, $x_e^{tmax} \geq 0$ for representing event attributes $tmin, tmax$ of each $e \in E_u$. For all observed events $e \in E_1$, we set $x_e^{tmin} = x_e^{tmax} = \#_{time}(e)$ as here the correct timestamp is known. We now define two groups of constraints to constrain the x_e^{tmin} and x_e^{tmax} values for the unobserved events further. In the following, we assume for the sake of simpler constraints presented in this paper, that all observed events are start events (which is in line with logging in an MHS). The constraints can easily be reformulated to assume only complete events were observed (as in most business process event logs) or a mix (requiring further case distinctions).

The first group propagates constraints for $\#_{time}(e)$ along $<_{pid}$, i.e., for each token trajectory (viz. trace) π_{pid}^z of pid in π. By the steps in Sect. 5.1, events in π_{pid}^z are totally ordered and we write $\pi_{pid}^z = \langle e_1 ... e_m \rangle$ as a sequence of events. Each process step has a start and a complete event in π_{pid}^z, i.e., $m = 2 \cdot y, y \in \mathbb{N}$, odd events are start events and even events are complete events. For each process step $1 \leq i \leq y$, the time between start event e_{2i-1} and complete event e_{2i} is at least the service time of the resource involved (which we stored as $\#_{tsr}(e_{2i-1})$ in Sect. 5.1). Thus the following constraints must hold for the earliest and latest time of e_{2i-1} and e_{2i}.

$$x_{e_{2i}}^{tmin} = x_{e_{2i-1}}^{tmin} + \#_{tsr}(e_{2i-1}), \tag{1}$$

$$x_{e_{2i}}^{tmax} = x_{e_{2i-1}}^{tmax} + \#_{tsr}(e_{2i-1}). \tag{2}$$

For the remainder, it suffices to formulate constraints only for *start* events. We make sure that $tmin$ and $tmax$ define a proper interval for each start event:

$$x_{e_{2i-1}}^{tmin} \leq x_{e_{2i-1}}^{tmax}. \tag{3}$$

We write $e_i^s = e_{2i-1}$ for the start event of the i-th process step in π_{pid}^z and $\theta_{pid}^z = \langle e_1^s, ..., e_m^s \rangle$ for the sub-trace of start events of π_{pid}^z. Any event $e_i^s \in \theta_{pid}^z$ that was observed in L_1, i.e., $e_i^s \in E_1$, has $\#_{time}(e_i^s) \neq \perp$ defined. By the assumption in Sect. 4, π_{pid}^z as well as θ_{pid}^z always start and end with observed events, i.e., $e_1^s, e_y^s \in E_1$ and $\#_{time}(e_1^s), \#_{time}(e_y^s) \neq \perp$. An unobserved event e_i^s has no timestamp $\#_{time}(e_i^s) = \perp$ yet, but $\#_{time}(e_i^s)$ is bounded by $\#_{time}(e_1^s)$ (minimally) and $\#_{time}(e_y^s)$ (maximally). Furthermore, any two succeeding start events in $\theta_{pid}^z = \langle ..., e_{i-1}^s, e_i^s, ... \rangle$ are separated by the service time $\#_{tsr}(e_{i-1}^s)$ of step e_{i-1}^s and the waiting time $\#_{twq}(e_i)$ of the queue from e_{i-1} to e_i. Similar to Eq. 1 and 2, we formulate this constraint for both x_e^{tmin} and x_e^{tmax} variables:

$$x_{e_k^s}^{tmin} \geq x_{e_{k-1}^s}^{tmin} + (\#_{tsr}(e_{k-1}^s) + \#_{twq}(e_k^s)), \tag{4}$$

$$x_{e_k^s}^{tmax} \leq x_{e_{k+1}^s}^{tmax} - (\#_{tsr}(e_k^s) + \#_{twq}(e_{k+1}^s)). \tag{5}$$

Figure 6 uses the *Performance Spectrum* [10] to illustrate the effect of applying our approach step by step to the partially complete traces of Fig. 5 obtained in the steps of Sect. 5.1. The straight lines in Fig. 6(a) from f_1 to f_9 (for pid = 53) and from f_{12} to f_{16} (for pid = 54) illustrate that L_2 (after applying O_1) contains all intermediate steps that both process cases passed through but not their timestamps. Further (after applying O_2), we know for each process step the resources (i.e., c1, m2, m3, m4, d1) and the queues (c1:m2, m2:m3 etc.), and their minimum service and waiting times tsr, twr, twq. The sum $tsr + twq$ is visualized as bars on the time axis in Fig. 6(a), the duration of twr is shown in Fig. 6(b). We now explain the effect of applying Eq. 4 on pid = 53 for f_3, f_5 and f_7. We have $\theta_{pid}^{53} = \langle f_1, f_3, f_5, f_7, f_9 \rangle$ with f_1 and f_9 observed, thus $x_{f_i}^{tmin} = x_{f_i}^{tmax} = \#_{time}(f_i)$ for $i \in \{1, 9\}$. By Eq. 4, we obtain the lower-bound for the time for f_3 by $x_{f_3}^{tmin} \geq x_{f_1}^{tmin} + \#_{tsr}(f_1) + \#_{twq}(f_3)$ with $\#_{tsr}(f_1)$ and $\#_{twq}(f_3)$ the service time of resource c1 and waiting time of queue c1:m2. Similarly, Eq. 4 gives the lower bound for f_5 from the lower bound from f_3 etc. Conversely, the upper bounds $x_{f_i}^{tmax}$ are derived from f_9 "downwards" by Eq. 5. This way, we obtain for each $f_i \in \theta_{pid}^{53}$ an initial interval for the time of f_i between the bounds $x_{f_i}^{tmin} \leq x_{f_i}^{tmax}$ as shown by the intervals in Fig. 6(a). As $x_{f_1}^{tmin} = x_{f_1}^{tmax} = \#_{time}(f_1)$ and $x_{f_1}^{tmin} = x_{f_1}^{tmax} = \#_{time}(f_9)$, the lower and upper bounds for the unobserved events in θ_{pid}^{53} form a polygon as shown in Fig. 6(b). Case 53 must have passed over the process steps and resources as a path inside this polygon, i.e., the polygon contains all admissible solutions for the timestamps of the unobserved events of θ_{pid}^{53}; we call this polygon the *region* of case 53. The region for case 54 overlays with the region for case 53.

We now introduce a second group of constraints by which we infer more tight bounds for $x_{e_i}^{tmin}$ and $x_{e_i}^{tmax}$ based on the overlap with other regions. While the first group of constraints traversed token trajectories along *pid* (i.e., process traces), the second group of constraints traverses token trajectories for resources along *rid*. Each resource trace $\pi_{rid}^r = (E_{rid}^r, <_{rid}^r)$ in π, contains all events E_{rid}^r resource r was involved in - *across* multiple different process traces. The SPO $<_{rid}^r$ orders *observed* events of this resource trace due to their known timestamps; e.g. in Fig. 6(b) $f_9 <_{rid}^{m1} f_{16}$ with f_9 from pid = 53 and f_{16} from pid = 54. The order of the two events $e_{p1}^s <_{rid}^r e_{p2}^s$ for the *same* step $\#_{act}(e_{p1}^s) = \#_{act}(e_{p2}^s) = t_1$ in different cases $\#_{pid}(e_{p1}^s) = p1 \neq \#_{pid}(e_{p2}^s) = p2$ propagates "upwards" and "downwards" the process traces π_{pid}^{p1} and π_{pid}^{p2} as follows. Let events $f_{p1}^s \in E_{pid}^{p1}$ and $f_{p2} \in E_{pid}^{p2}$ be events in process traces π_{pid}^{p1} and π_{pid}^{p2} of the same

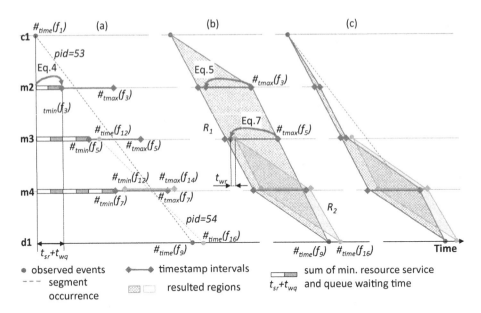

Fig. 6. Equations 1–5 define time intervals for unobserved events (a), defining regions for the possible traces (b). Equations 6–7 propagate orders of cases observed on one resource to other resources (b), resulting in tighter regions (c). (Color figure online)

step $\#_{act}(f_{p1}^s) = \#_{act}(f_{p2}^s) = t_n$. We say t_1 and t_n are *in FIFO relation* iff there is a unique path $\langle t_1 ... t_n \rangle$ between t_1 and t_n in the process proclet (i.e., no loops, splits, parallelism) so that between any two consecutive transitions t_k, t_{k+1} only synchronize with single-server resources or FIFO queues. If t_1 and t_n are in FIFO relation, then also $f_{p1}^s <_{rid}^{r2} f_{p2}^s$ on the resource r2 involved in t_n (as the case cannot overtake the case along this path). Thus $x_{f_{p1}^s}^{tmin} \leq x_{f_{p2}^s}^{tmin}$ must hold. More specifically, $x_{f_{p1}^s}^{tmin} + \#_{tsr}(f_{p1}^s) + \#_{twr}(f_{p1}^s) \leq x_{f_{p2}^s}^{tmin}$ must hold as the service time and waiting time of the resource involved in f_{p1}^s must elapse.

For any pair $e_{p1}^s, e_{p2}^s \in E_{rid}^r$ with $e_{p1}^s <_{rid}^r e_{p2}^s$ and any other trace θ_{rid}^{r2} for resource r2 and any pair $f_{p1}^s, f_{p2}^s \in E_{rid}^{r2}$ such that $\#_{pid}(e_{p1}^s) = \#_{pid}(f_{p1}^s)$, $\#_{pid}(e_{p2}^s) = \#_{pid}(f_{p2}^s)$ and transition $\#_{act} e_{p1}^s$ is in FIFO relation with $\#_{act}(f_{p1}^s)$, we generate the following constraint for *tmin*:

$$x_{f_{p1}^s}^{tmin} \leq x_{f_{p2}^s}^{tmin} - (\#_{tsr}(f_{p1}^s) + \#_{twr}(f_{p1}^s)), \tag{6}$$

and the following constraint for *tmax*:

$$x_{f_{p1}^s}^{tmax} \leq x_{f_{p2}^s}^{tmax} - (\#_{tsr}(f_{p1}^s) + \#_{twr}(f_{p1}^s)), \tag{7}$$

In the example of Fig. 6(b), we observe $f_9 <_{rid}^{d1} f_{16}$ (both of transition $d1_s$) along resource $d1$ at the bottom of Fig. 6(b). By Fig. 3, $d1_s$ and $m3_s$ are in FIFO-relation. Applying Eq. 7 yields $x_{f_5}^{tmax} \leq \#_{time}(f_{12}) - (\#_{tsr}(f_5) + \#_{twr}(f_5))$, i.e., f_5 occurs at latest before f_{12} minus the service and waiting time of $m3$. This operation significantly reduces the initial region R_1. By Eq. 5, the tighter upper bound for f_5 also propagates along the trace pid = 53 to f_3, i.e., $x_{f_3}^{tmax} \leq x_{f_5}^{tmax} - (\#_{tsr}(f_3) + \#_{twq}(f_5))$, resulting in a tighter region as shown in Fig. 6(c). If another trace $\langle m3_s, d1_s \rangle$ were present *before* trace 53, then this would cause reducing the *tmin* attributes of the events of trace 53 by Eq. 4, 6 in a similar way. In general, the more cases interact through shared resources, the more accurate timestamp intervals can be restored by Eq. 1–7 as we will show in Sect. 6.

To construct the linear program, we generate Eqs. 1 to 5 by iteration of each process trace in L_2. Further, iterate over each resource trace and for each pair of events $e_{p1} <_{rid}^{r} e_{p2}$ we generate Eqs. 6, 7 for each other pair of events $f_{p1} <_{rid}^{r2} f_{p2}$ that is in FIFO relation. The objective function to maximize is the sum of all intervals $\sum_{e \in E_2}(x_e^{tmax} - x_e^{tmin})$ to maximize the coverage of possible timestamp values by those intervals.

6   Evaluation

To evaluate our approach, we formulated the following questions. (Q1) Can timestamps be estimated in real-life settings and used to estimate performance reliably? (Q2) How accurately can the load (items per minute) be estimated for different system parts, using restored timestamps? (Q3) What is the impact of sudden deviations from the minimum service/waiting times, e.g., the unavailability of resource or stop/restart of an MHS conveyor, on the accuracy of restored timestamps and the computed load? For that, we extended the interactive ProM plug-in "Performance Spectrum Miner" with an implementation of our approach that solves the constraints using heuristics[1]. As input we considered the process of a part of real-life BHS shown in Fig. 7 and used Synthetic Logs (SL) (simulated from a model to obtain ground-truth timestamps) and Real-life Logs (RL) from a major European airport. Regarding Q3, we generated SL with regular performance and with *blockages* of belts (i.e., a temporary stand-still); the RL contained both performance characteristics. All logs were partial as described in Sect. 4. We selected the acyclic fragment highlighted in Fig. 7 for restoring timestamps of steps c_{1-4}, d_{1-2}, f, s.

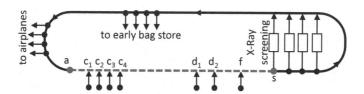

Fig. 7. In the BHS bags come from check-in counters c_{1-4} and another terminals d_{1-2}, f, go through mandatory screening and continue to other locations. (Color figure online)

We evaluated our technique against the ground truth known for SL as follows. For each event we measured the error of the estimated timestamp intervals $[t_{min}, t_{max}]$ against the actual time t as $\max\{|t_{max} - t|, |t_{min} - t|\}$ normalized over the sum of minimal service

---

[1] The simulation model, simulation logs, ProM plugin, and high-resolution figures are available on https://github.com/processmining-in-logistics/psm/tree/rel.

and waiting times of all involved steps (to make errors comparable). We report the Mean Absolute Error (MAE) and Root Mean Square Error (RMSE) of these errors. Applying our technique to SL with regular behavior, we observed very narrow time intervals for the estimated timestamps, shown in Fig. 8(a), and a MAE of <5%. The MAE of the estimated load (computed on estimated timestamps), shown in Fig. 8(e), was <2%. For SL with blockage behavior, the intervals grew proportionally with the duration of blockages (Fig. 8(b)), leading to a proportional growth of the MAE for the timestamps. However, the MAE of the estimated load (Fig. 8(f)) was at most 4%. The load MAE for different processing steps for both scenarios are shown in Table 1. Notably, both observed and reconstructed load showed load peaks each time the conveyor belt starts moving again.

When evaluating on the real-life event log, we measured errors of timestamps estimation as the length of the estimated intervals (normalized over the sum of minimal service and waiting times of all involved steps). Performance spectra built using the restored RL logs are shown in Fig. 8(c,d), and the load computed using these logs is shown in Fig. 8(g,h). The observed MAE was <5% in regular behavior and increased proportionally as observed on SL. The load error could not be measured, but similarly to synthetic data, it showed peaks after assumed conveyor stops.

The obtained results on SL show that the timestamps can be always estimated, and the actual timestamps are always within the timestamp intervals (Q1). When the system resources and queues operate close to the known performance parameters tsr, twr, twq, our approach restores accurate timestamps resulting in reliable load estimates in SL (Q2). During deviations in resource performance, the errors increase proportionally with performance deviation while the estimated load remains reliable (error <4% in SL) and shows known characteristics from real-life systems on SL and RL (Q3).

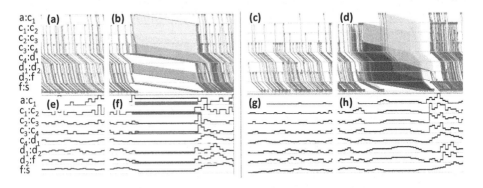

Fig. 8. Restored Performance Spectrum for synthetic (a,b) and real-life (c,d) logs. The estimated load (computed on estimated timestamps) for synthetic (e,f) and real-life (g,h) logs. For the synthetic logs, the load error is measured and shown in red (e,f). (Color figure online)

Table 1. The estimated load (computed on estimated timestamps) Root Mean Squared Error (RMSE) and Mean Absolute Error (MAE) are shown in % of max. load.

Scenario	MAE, $c_4 : d_1$	RMSE, $c_4 : d_1$	MAE, $d_1 : d_2$	RMSE, $d_1 : d_2$	MAE, $f : s$	RMSE, $f : s$
No blockages	0.16	1.01	0.22	1.66	0.17	0.89
Blockages	1.67	4.8	3.19	7.17	0.15	0.75

7 Conclusion

In this paper, we studied the problem of repairing a partial event log with missing events for the performance analysis of systems where case interact and compete for shared limited resources. We addressed the problem of repairing partial event logs that contain only a subset of events which impede the performance analysis of systems with shared limited resources and queues. To study and solve the problem, we used synchronous proclets [11] to model processes served by resources and queues (a PQR system). The model allows to decompose the interactions of resources and queues over multiple process cases into token trajectories for process cases, resources and queues that synchronize on shared events. We exploit the decomposition when restoring missing events along the process token trajectories using trace alignment [7]. We exploit the synchronization when formulating linear programming constraints over timestamps of restored events along, both, the process and the resource token trajectories. As a result, we obtain timestamps which are consistent for all events along the process, resource, and queue dimensions. The evaluation of our implementation in synthetic and real-life data shows errors of the estimated timestamps and of derived performance characteristics (i.e., load) of <5% under regular performance, while correctly restoring real-life dynamics (i.e. load peaks) after irregular performance behavior.

Limitations. The work made several limiting assumptions. (1) Although the proclet formalism allows for arbitrary, dynamic synchronizations between process steps, resources, and queues, we limited ourselves in this work to a static known resource/queue id per process step. The limitation is not severe for some use cases such as analyzing MHS, but generalizing oracle O_2 to a dynamic setting is an open problem. (2) The LP constraints to restore timestamps assume an acyclic process proclet without concurrency. Further, the LP constraints assume 1:1 interactions (at most one resource and/or queue per process step). Both assumptions do not hold in business processes in general; formulating the constraints for a more general setting is an open problem. (3) Our approach ensures consistency of either all earliest or all latest timestamps with the given model, it does not suggest how to select timestamps between the latest and earliest such that the consistency holds. (4) When the system performance significantly changes, e.g., due to sudden unavailability of resources, the error of restored timestamps is growing proportionally the duration of deviations. Points (3) and (4) require attention to further improve event log quality for performance analysis.

Acknowledgements. The research leading to these results has received funding from Vanderlande Industries in the project "Process Mining in Logistics". We also thank Mitchel Brunings for his comments that greatly improved our approach.

References

1. van der Aalst, W.M.P., Barthelmess, P., Ellis, C.A., Wainer, J.: Proclets: a framework for lightweight interacting workflow processes. Int. J. Coop. Inf. Syst. **10**(04), 443–481 (2001). https://doi.org/10.1142/S0218843001000412
2. van der Aalst, W.M.P.: Process Mining - Data Science in Action, 2nd edn. Springer, Heidelberg (2016). https://doi.org/10.1007/978-3-662-49851-4
3. Aalst, W.M.P., Adriansyah, A., Dongen, B.: Replaying history on process models for conformance checking and performance analysis. WIREs Data Min. Knowl. Discov. **2**, 182–192 (2012). https://doi.org/10.1002/widm.1045
4. Ahmed, T., Pedersen, T.B., Calders, T., Lu, H.: Online risk prediction for indoor moving objects. In: 2016 17th IEEE International Conference on Mobile Data Management (MDM), vol. 1, pp. 102–111, June 2016. https://doi.org/10.1109/MDM.2016.27
5. Senderovich, A., Francescomarino, C.D., Maggi, F.M.: From knowledge-driven to data-driven inter-case feature encoding in predictive process monitoring. Inf. Syst. **84**, 255–264 (2019). https://doi.org/10.1016/j.is.2019.01.007
6. Brown, L., et al.: Statistical analysis of a telephone call center. J. Am. Stat. Assoc. **100**(469), 36–50 (2005). https://doi.org/10.1198/016214504000001808
7. Carmona, J., van Dongen, B., Solti, A., Weidlich, M.: Conformance Checking - Relating Processes and Models. Springer, Heidelberg (2018). https://doi.org/10.1007/978-3-319-99414-7
8. Conforti, R., La Rosa, M., ter Hofstede, A.: Timestamp repair for business process event logs. Technical report (2018/04/05 2018). http://hdl.handle.net/11343/209011
9. Denisov, V., Fahland, D., van der Aalst, W.M.P.: Predictive performance monitoring of material handling systems using the performance spectrum. In: 2019 International Conference on Process Mining (ICPM), pp. 137–144, June 2019. https://doi.org/10.1109/ICPM.2019.00029
10. Denisov, V., Fahland, D., van der Aalst, W.M.P.: Unbiased, fine-grained description of processes performance from event data. In: Weske, M., Montali, M., Weber, I., vom Brocke, J. (eds.) BPM 2018. LNCS, vol. 11080, pp. 139–157. Springer, Cham (2018). https://doi.org/10.1007/978-3-319-98648-7_9
11. Fahland, D.: Describing behavior of processes with many-to-many interactions. In: Donatelli, S., Haar, S. (eds.) PETRI NETS 2019. LNCS, vol. 11522, pp. 3–24. Springer, Cham (2019). https://doi.org/10.1007/978-3-030-21571-2_1
12. Gans, N., Koole, G., Mandelbaum, A.: Telephone call centers: tutorial, review, and research prospects. Manuf. Serv. Oper. Manag. **5**, 79–141 (2003)
13. Lu, X., Fahland, D., van der Aalst, W.M.P.: Conformance checking based on partially ordered event data. In: Fournier, F., Mendling, J. (eds.) BPM 2014. LNBIP, vol. 202, pp. 75–88. Springer, Cham (2015). https://doi.org/10.1007/978-3-319-15895-2_7
14. Márquez-Chamorro, A.E., Resinas, M., Ruiz-Cortés, A.: Predictive monitoring of business processes: a survey. IEEE Trans. Serv. Comput. **11**(6), 962–977 (2018). https://doi.org/10.1109/TSC.2017.2772256
15. Martin, N., Depaire, B., Caris, A., Schepers, D.: Retrieving the resource availability calendars of a process from an event log. Inf. Syst. **88**, 101463 (2020). https://doi.org/10.1016/j.is.2019.101463. http://www.sciencedirect.com/science/article/pii/S0306437919305150
16. Maruster, L., van Beest, N.R.T.P.: Redesigning business processes: a methodology based on simulation and process mining techniques. Knowl. Inf. Syst. **21**(3), 267–297 (2009). https://doi.org/10.1007/s10115-009-0224-0
17. Pegoraro, M., Aalst, W.: Mining uncertain event data in process mining, pp. 89–96 (2019). https://doi.org/10.1109/ICPM.2019.00023

18. Pegoraro, M., Uysal, M.S., van der Aalst, W.M.P.: Discovering process models from uncertain event data. In: Di Francescomarino, C., Dijkman, R., Zdun, U. (eds.) BPM 2019. LNBIP, vol. 362, pp. 238–249. Springer, Cham (2019). https://doi.org/10.1007/978-3-030-37453-2_20

19. Schrijver, A.: Theory of Linear and Integer Programming. Wiley, Chichester (1986)

20. Senderovich, A., Beck, J., Gal, A., Weidlich, M.: Congestion graphs for automated time predictions. In: Proceedings of the AAAI Conference on Artificial Intelligence vol. 33, pp. 4854–4861 (2019). https://doi.org/10.1609/aaai.v33i01.33014854

21. Senderovich, A., Weidlich, M., Gal, A., Mandelbaum, A.: Queue mining – predicting delays in service processes. In: Jarke, M., Mylopoulos, J., Quix, C., Rolland, C., Manolopoulos, Y., Mouratidis, H., Horkoff, J. (eds.) CAiSE 2014. LNCS, vol. 8484, pp. 42–57. Springer, Cham (2014). https://doi.org/10.1007/978-3-319-07881-6_4

22. Suriadi, S., Andrews, R., ter Hofstede, A., Wynn, M.: Event log imperfection patterns for process mining: towards a systematic approach to cleaning event logs. Inf. Syst. **64**, 132–150 (2017). https://doi.org/10.1016/j.is.2016.07.011. http://www.sciencedirect.com/science/article/pii/S0306437915301344

Piecewise Affine Dynamical Models of Timed Petri Nets – Application to Emergency Call Centers

Xavier Allamigeon, Marin Boyet$^{(\boxtimes)}$, and Stéphane Gaubert

INRIA and CMAP, École Polytechnique, IP Paris, CNRS, Palaiseau, France
{xavier.allamigeon,marin.boyet,stephane.gaubert}@inria.fr

Abstract. We study timed Petri nets, with preselection and priority routing. We represent the behavior of these systems by piecewise affine dynamical systems. We use tools from the theory of nonexpansive mappings to analyze these systems. We establish an equivalence theorem between priority-free fluid timed Petri nets and semi-Markov decision processes, from which we derive the convergence to a periodic regime and the polynomial-time computability of the throughput. More generally, we develop an approach inspired by tropical geometry, characterizing the congestion phases as the cells of a polyhedral complex. We illustrate these results by a current application to the performance evaluation of emergency call centers in the Paris area.

Keywords: Timed Petri net · Performance evaluation · Markov decision process · Tropical geometry · Emergency call center

1 Introduction

Motivation. Emergency call centers exhibit complex synchronization and concurrency phenomena. Various types of calls induce diverse chains of actions, including reception of the call, instruction by experts, dispatch of emergency means and monitoring of operations in progress. The processing of calls is subject to priority rules, making sure that the requests evaluated as the most urgent are treated first. The present work originates from a specific case study, concerning the performance evaluation of the medical emergency call centers in Paris and its inner suburbs, operated by four *Services d'aide médicale urgente* (SAMU) of *Assistance Publique – Hôpitaux de Paris* (APHP). One needs to evaluate performance indicators, like the throughput (number of calls of different types that can be processed without delay). One also needs to optimize the resources (e.g., personnel of different kinds) to guarantee a prescribed quality of service for a given inflow of calls.

The second author is supported by a joint PhD grant of DGA and INRIA. The authors have been partially supported by the "Investissement d'avenir", référence ANR-11-LABX-0056-LMH, LabEx LMH and by a project of "Centre des Hautes Études du Ministère de l'Intérieur" (CHEMI).

© Springer Nature Switzerland AG 2020
R. Janicki et al. (Eds.): PETRI NETS 2020, LNCS 12152, pp. 260–279, 2020.
https://doi.org/10.1007/978-3-030-51831-8_13

Table 1. Correspondence between Petri nets and semi-Markov decision processes

Timed Petri nets	Semi-Markov decision processes
Transitions	States
Places	Actions
Physical time	Time remaining to live
Counter function	Finite horizon value function
Synchronization	Multiple actions
Preselection routing	Probabilistic moves
Priority routing	Negative probabilities
Throughput	Average cost
Bottleneck places	Optimal policies
Congestion phases	Cells of the average cost complex

Contribution. We develop a general method for the analysis of the timed behavior of Petri nets, based on a representation by piecewise linear dynamical systems. These systems govern *counter functions*, which yield the number of firings of transitions as a function of time. We allow routings based either on preselection or priority rules. Preselection applies to situations in which certain attributes of a token determine the path it follows, e.g. different types of calls require more or less complex treatments. Moreover, priority rules are used to allocate resources. We study a *fluid relaxation* of the model, in which the numbers of firings can take real values. Supposing the absence of priority routing, we establish a correspondence between timed Petri nets and semi-Markov decision processes. Table 1 provides the details of this correspondence that we shall discuss in the paper. Then, we apply methods from the theory of semi-Markov decision processes to analyze timed Petri nets. We show that the counter variables converge to a periodic orbit (modulo additive constants). Moreover, the throughput can be computed in polynomial time, by looking for affine stationary regimes and exploiting linear programming formulations. We also show that the throughput is given as a function of the resources (initial marking), by an explicit concave piecewise affine map. The cells on which this map is affine yield a polyhedral complex, representing the different "congestion phases". We finally discuss the extension of these analytic results to the case with priorities. The dynamics still has the form of a semi-Markov type Bellman equation, but with *negative* probabilities. Hence, the theoretical tools used to show the convergence to a periodic orbit do not apply anymore. However, we can still look for the affine stationary regimes, which turn out to be the points of a tropical variety. From this, we still obtain a phase diagram, representing all the possible throughputs of stationary regimes. Throughout the paper, these results are illustrated by the case study of emergency call centers. The final section focuses on the analysis of a policy proposed by the SAMU, involving a monitored reservoir.

Related Work. Our approach originates from the max-plus modeling of timed discrete event systems, introduced by Cohen, Quadrat and Viot and further investigated by Baccelli and Olsder and a number of authors. We refer the reader to the monographs [3,10] and to the survey of Komenda, Lahaye, Boimond and van den Boom [11]. The max-plus approach was originally developed for timed event graphs. Cohen, Gaubert and Quadrat extended it to fluid Petri nets with preselection routing [5,6]. Gaujal and Giua established in [9] further results on the model of [5,6]. Their results include a characterization of the throughput as the optimal solution of linear program. Recalde and Silva [14] obtained linear programming formulations for a different fluid model.

By comparison with [5,6,9], we use more powerful results on semi-Markov decision processes and nonexpansive mappings. This allows us, in particular, to deduce more precise asymptotic results, concerning the deviation $z(t) - \rho t$ between the counter function z at time t and its average growth ρt, instead of the mere existence of the limit $\lim_{t \to +\infty} z(t)/t = \rho$. We also establish the existence of the latter limit even in the case of irrational holding times. The present work is a follow-up of [2], in which Bœuf and two of the authors established an equivalence between timed Petri nets with priorities and a class of piecewise-linear models.

The present methods are complementary to probabilistic approaches [12]. Priority rules put our systems outside the classes of exactly solvable probabilistic models; only scaling limit type results on suitably purified models are known [4]. In contrast, fluid models allow one to compute analytically phase portraits. They lead to lower bounds of dimensioning which are accurate when the arrivals do not fluctuate, and which can subsequently be confronted with results of simulation.

2　Piecewise Affine Models of Timed Petri Nets

2.1　Preliminaries on Timed Petri Nets

A *timed Petri net* is given by a bipartite graph whose vertices are either *places* or *transitions*. We denote by $\mathcal{P}$ (resp. $\mathcal{Q}$) the finite set of places (resp. transitions). For two vertices x and y forming a place-transition pair, x is said to be an upstream (resp. downstream) vertex of y if there is an arc of the graph going from x to y (resp. from y to x). The set of upstream (resp. downstream) vertices of x is denoted by x^{in} (resp. x^{out}).

Every place p is equipped with an *initial marking* $m_p \in \mathbb{N}$, representing the number of tokens initially present in the place before starting the execution of the Petri net. The place p is also equipped with a holding time $\tau_p \in \mathbb{R}_{\geqslant 0}$, so that a token entering p must sojourn in this place at least for a time τ_p before becoming available for firing a downstream transition. In contrast, firing a transition is instantaneous. Every arc from a place p to a transition q (resp. from a transition q to a place p) is equipped with a positive and integer weight denoted by α_{qp} (resp. α_{pq}). Transition q can be fired only if each upstream place p contains α_{qp} tokens. In this case, one firing of the transition q consumes α_{qp} tokens in each upstream place p, and creates $\alpha_{p'q}$ in each downstream place p'. Unless specified, the weights are set to 1. The same transition can be fired as many times as necessary, as long as tokens in the upstream places are available. We shall assume that *transitions are fired as soon as possible*. By convention, the tokens of the initial marking are all available when the execution starts.

When a place has several downstream transitions, we must provide a *routing rule* specifying which transition is to be fired once a token is available. We distinguish two sets of rules: priority and preselection.

A *priority routing* on a place p is specified by a total order $\preceq_p$ over the downstream transitions of p. The principle of this routing rule is that a transition $q \in p^{\mathrm{out}}$ is fired only if there is no other fireable transition $q' \in p^{\mathrm{out}}$ with a higher priority, i.e. $q' \preceq_p q$. We represent the ordering of downstream transitions by a variable number of tips, like in Fig. 1, with the convention that the highest priority transition is the one pointed by the highest number of tips.

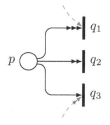

Fig. 1. Priority routing

Priority routing will be used in our model of monitored reservoir studied in Sect. 5. We denote by $\mathcal{Q}_{\mathrm{prio}}$ the subset of $\mathcal{Q}$ consisting of the downstream transitions of places subject to priority routing. We allow transitions in $\mathcal{Q}_{\mathrm{prio}}$ to admit multiple upstream places ruled by priority routings as long as the following compatibility condition is met.

Definition 1. *Let $\mathcal{P}_{\mathrm{prio}}$ denote the set of places subject to priority routing. We say that the rules $(\preceq_p)_{p \in \mathcal{P}_{\mathrm{prio}}}$ are compatible if their union (as binary relations) is acyclic.*

Acyclicity amounts to the extension of a linear extension of the partial orders $(\preceq_p)_{p \in \mathcal{P}_{\mathrm{prio}}}$. In consequence, our setting is equivalent to fixing a priority order over the set of all transitions.

The *preselection routing* on a place p is described by a partition $(m_{qp})_{q \in p^{\mathrm{out}}}$ of the initial marking and a collection of maps $(\Pi_q^p)_{q \in p^{\mathrm{out}}}$ satisfying the following properties:

$$\sum_{q \in p^{\mathrm{out}}} m_{qp} = m_p, \quad \text{and} \quad \forall n \in \mathbb{N}, \ \sum_{q \in p^{\mathrm{out}}} \Pi_q^p(n) = n \,.$$

For $q \in p^{\mathrm{out}}$, m_{qp} and $\Pi_q^p(n)$ represent the number of tokens, amongst the initial marking m_p and the n first tokens to enter place p, which are reserved to fire transition q. In other words, they cannot be used to fire any other transition of p^{out}. A natural example of preselection routing is the *proportional periodic routing*: if $p^{\mathrm{out}} = \{q_1, q_2, \ldots, q_k\}$, consider a positive integer L, a partition $(J_1, J_2, \ldots, J_k)$ of $\{1, 2, \ldots, L\}$ and define $\Pi_{q_k}^p(n) = \mathrm{card}(\{1, 2, \ldots, n\} \cap (J_k + L\mathbb{N}))$. For large values of n, we have $\Pi_{q_k}^p(n) \sim n \cdot \mathrm{card}(J_k)/L$.

In order to simplify the presentation of our following dynamical model, we assume that preselection routing is only allowed for places whose downstream transitions do not admit other upstream places. The general case can be reduced to this one by introducing extra places with holding time 0, as illustrated on Fig. 2.

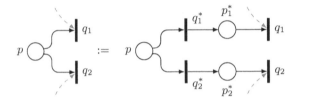

Fig. 2. Compact notation for preselection routing in case of multiple upstream places

Fig. 3. A synchronization pattern

We denote by $\mathcal{Q}_{\mathsf{psel}}$ the subset of $\mathcal{Q}$ consisting of the downstream transitions of places ruled by preselection routing. By construction, we have $\mathcal{Q}_{\mathsf{psel}} \cap \mathcal{Q}_{\mathsf{prio}} \neq \varnothing$. We define $\mathcal{Q}_{\mathsf{sync}} := \mathcal{Q} \setminus (\mathcal{Q}_{\mathsf{psel}} \cup \mathcal{Q}_{\mathsf{prio}})$, i.e. the set of transitions with no upstream place ruled by split or priority routing. As a result, we have a partition of $\mathcal{Q}$ into $\mathcal{Q}_{\mathsf{prio}}$, $\mathcal{Q}_{\mathsf{psel}}$, and $\mathcal{Q}_{\mathsf{sync}}$. Transitions of $\mathcal{Q}_{\mathsf{sync}}$ correspond to a synchronization pattern between several upstream places, as illustrated in Fig. 3. We point out that transitions with one upstream place can be of any of the three kind $\mathcal{Q}_{\mathsf{prio}}$, $\mathcal{Q}_{\mathsf{psel}}$, and $\mathcal{Q}_{\mathsf{sync}}$. The choice of their classification does not affect the analysis developed below.

Remark 1. Contrary to the preselection routing, the priority routing is essentially non-monotone. Indeed, a "young" token might activate some prioritized transition before some other "older" token activates a non-prioritized transition.

2.2 Dynamic Equations Governing Counter Functions

We associate with every transition $q \in \mathcal{Q}$ a *counter function* z_q from $\mathbb{R}$ to $\mathbb{R}_{\geqslant 0}$ such that $z_q(t)$ represents the number of firings of transition q that occurred up to time t included. Similarly, given a place $p \in \mathcal{P}$, we denote by $x_p(t)$ the number of tokens that have entered place p up to time t included, taking into account the tokens initially present in p. By construction, x_p and z_q are non-decreasing càdlàg functions. Given a càdlàg function f, we denote by $f(t^-)$ the left limit at the point t, possibly smaller than $f(t)$.

For each place $p \in \mathcal{P}$, $x_p(t)$ is given by the sum of the initial marking m_p and the number of firings of transitions $q \in p^{\mathrm{in}}$ weighted by α_{pq} (recall that one firing of transition q outputs α_{pq} tokens in p):

$$\forall p \in \mathcal{P}, \quad x_p(t) = m_p + \sum_{q \in p^{\mathrm{in}}} \alpha_{pq}\, z_q(t). \tag{P1}$$

For each transition $q \in \mathcal{Q}$, the equation satisfied by z_q depends on the routing policy of its upstream places. Suppose $q \in \mathcal{Q}_{\mathsf{sync}}$, so that its upstream places only admit q for downstream transition. Since transitions are fired as early as possible and must wait for all upstream tokens to be available, we have:

$$z_q(t) = \min_{p \in q^{\mathrm{in}}} \left\lfloor \alpha_{qp}^{-1} x_p(t - \tau_p) \right\rfloor. \tag{P2}$$

where $\lfloor \cdot \rfloor$ denotes the floor function (recall that z_q must be integer). Suppose now that $q \in \mathcal{Q}_{\mathsf{psel}}$. Because q admits only one upstream place p, we also have:

$$z_q(t) = \left\lfloor \alpha_{qp}^{-1} \left(m_{qp} + \Pi_q^p(x_p(t - \tau_p)) \right) \right\rfloor. \tag{P3}$$

Finally, suppose that $q \in \mathcal{Q}_{\mathsf{prio}}$. We have

$$z_q(t) = \min_{p \in q^{\mathrm{in}}} \left\lfloor \alpha_{qp}^{-1} \left(x_p(t - \tau_p) - \sum_{q' \prec_p q} \alpha_{q'p} z_{q'}(t) - \sum_{q' \succ_p q} \alpha_{q'p} z_{q'}(t^-) \right) \right\rfloor. \tag{P4}$$

This equation can be interpreted by examining $z_q(t) - z_q(t^-)$, which represents the number of tokens fired by q at time t. The amount of tokens available in place $p \in q^{\mathrm{in}}$ at time t^- is $x_p(t - \tau_p) - \sum_{q' \in p^{\mathrm{out}}} \alpha_{q'p} z_{q'}(t^-)$. However, transitions with higher priority than q relatively to p fire $\sum_{q' \prec_p q} \alpha_{q'p}(z_{q'}(t) - z_{q'}(t^-))$ of these tokens, leaving $x_p(t - \tau_p) - \sum_{q' \prec_p q} \alpha_{q'p} z_{q'}(t) - \sum_{q' \succ_p q} \alpha_{q'p} z_{q'}(t^-)$ available to fire q. Equation (P4) is obtained by packing these tokens in an integer number of groups of α_{qp} and taking the minimum of such terms over q^{in}.

The correspondence between the semantics of timed Petri net (expressed in terms of a transition system acting over states corresponding to timed markings) and the equations above has been proved in [2] in a more restricted model. It carries over to the current setting, allowing multiple levels of priority, preselection routings, and arcs with valuations.

It will be convenient to consider the continuous relaxation of the previous dynamics. This boils down to considering infinitely divisible tokens and real-valued counters functions. The weights α_{pq} or α_{qp} can now be allowed to take positive real values. The priority and preselection routing rules are not affected by the fluid approximation, though in what follows we choose to focus only on proportional preselection routing: if a place p is ruled by preselection, we fix a stochastic vector $(\pi_{qp})_{q \in p^{\mathrm{out}}}$ such that $m_{qp} = \pi_{qp} m_p$ and $\Pi_q^p(x) = \pi_{qp} x$. Equivalently, this corresponds to the continuous relaxation of a stochastic routing at place p, in which π_{qp} is the probability for a token to be routed to transition q. Finally, the continuous relaxation drops the floor functions. This leads to the dynamical system presented in Table 2, governing the counter functions z_q of the transitions.

Table 2. Dynamic equations followed by transitions counter functions

Type	Counter equation in the continuous model
$q \in \mathcal{Q}_{\mathsf{sync}}$	$z_q(t) = \min\limits_{p \in q^{\mathrm{in}}} \alpha_{qp}^{-1} \left(m_p + \sum\limits_{q' \in p^{\mathrm{in}}} \alpha_{pq'} z_{q'}(t - \tau_p) \right)$
$q \in \mathcal{Q}_{\mathsf{psel}}$	$z_q(t) = \pi_{qp} \cdot \alpha_{qp}^{-1} \left(m_p + \sum\limits_{q' \in p^{\mathrm{in}}} \alpha_{pq'} z_{q'}(t - \tau_p) \right)$
$q \in \mathcal{Q}_{\mathsf{prio}}$	$z_q(t) = \min\limits_{p \in q^{\mathrm{in}}} \alpha_{qp}^{-1} \left(m_p + \sum\limits_{q' \in p^{\mathrm{in}}} \alpha_{pq'} z_{q'}(t - \tau_p) - \sum\limits_{q' \prec_p q} \alpha_{q'p} z_{q'}(t) - \sum\limits_{q' \succ_p q} \alpha_{q'p} z_{q'}(t^-) \right)$

In order to prevent an infinite number of firings from occurring in a finite amount of time, we shall work with Petri nets whose underlying directed graph does not contain any circuit in which places have zero holding times. Such Petri nets are said to be *non-Zeno*.

Lemma 1. *Suppose that a Petri net is non-Zeno, and let T denote the maximum of the holding times of its different places. Then, the transition counter function $z : [-T, \infty) \to \mathbb{R}^{\mathcal{Q}}$, which follows the dynamics shown in Table 2, is uniquely determined by its restriction to the interval $[-T, 0]$.*

Proof (sketch). We can reduce to the case where the right-hand sides of the dynamic equations only involve counter-variables $z_{q'}(t - \tau_p)$ for which τ_p is positive. We obtain this reduction obtained by successive substitutions of the counter-variables $z_{q'}(t - \tau_p)$ such that $\tau_p = 0$. The non-Zeno assumption guarantees that finitely many substitutions are enough. Defining $\tau^* := \inf\{\tau_p : \tau_p > 0, \ p \in \mathcal{P}\}$, Lemma 1 can then be proved for all $t \in [-T, n\tau^*)$ by induction on $n \geqslant 0$. $\square$

3   Models of Medical Emergency Call Centers

We next present two models based on an ongoing collaboration with the Emergency Medical Services (EMS) of Paris and its inner suburbs (SAMU 75, 92, 93, and 94 of APHP). In France, the nation-wide phone number 15 is dedicated to medical distress calls, dispatched to regional call-centers. The calls are first answered by an operator referred to as a *medical regulation assistant* (MRA), who categorizes the request, takes note of essential personal information and transfers the call to one of the following two types of physicians, depending on the estimated severity of the case:

(i) an emergency doctor, able to dispatch Mobile Intensive Care Units or first-responding ambulances and to swiftly send the patient to the most appropriate hospital unit;

(ii) a general practitioner, who can dispatch ambulances and provide medical advice.

An MRA may also handle the call without transferring it if a conversation with a physician is not needed (report from a medical partner, dial error, etc.).

In case (i), the MRA must wait for an emergency physician to be available before transferring the call, in order to report the details of the request. In this way, the patient is constantly kept on line with an interlocutor. In case (ii), patients are left on hold, and dealt with by general practitioners who answer the calls in the order of arrival. As a first step, our main focus is the coupling between the answering operator and the emergency physician (which is a critical link of the system). Thus, for the sake of simplicity, we do not take into account what happens to calls in case (ii) after the MRA is released. In other words, we consider a simplified model in which only two types of inbound calls can occur: the ones which require the MRA to wait for an emergency physician and the ones which do not. We shall also consider that the patients do not leave the system before their call is picked up (infinite patience assumption).

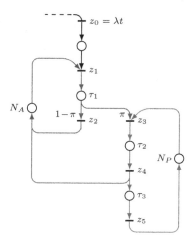

Fig. 4. A basic model of emergency call center (EMS-A)

We represent this emergency call-center by the (EMS-A) Petri net. Inbound calls arrive *via* the uppermost transition z_0. We may assume in what follows that $z_0(t) = \lambda t$ (arrivals at constant rate λ). The pool of MRAs is represented by the place with initial marking N_A. Transition z_1 is fired as soon as an MRA is available and a call is waiting for pick-up. Preliminary examination and information filling occur in place with holding time τ_1, ruled by preselection routing: a known fraction π of the patients are deemed to need the help of the emergency physician; for the complementary fraction $1 - \pi$ of the patients, the MRA is released at the firing of z_2. Transition z_3 is fired once a doctor is available from the pool of emergency physicians with initial marking N_P and an MRA waits for transfer. Summarizing the case takes a time τ_2 for both agents, then the firing of z_4 releases the MRA and the physician proceeds to the medical consultation with the patient for a time τ_3 before getting released by the firing of z_5. We use the color blue (resp. red) to highlight the circuits involving the MRA (resp. the emergency doctor). For the sake of readability, patient exits at transitions z_2 and z_5 are not depicted.

Applying the equations of the continuous relaxation of a timed Petri net recorded in Table 2, we obtain the following system of equations for the counter functions associated with transitions, where $x \wedge y$ stands for $\min(x, y)$.

$$\begin{cases} z_1(t) = & z_0(t) \quad \wedge \left(N_A + z_2(t) + z_4(t)\right) \\ z_2(t) = (1 - \pi)z_1(t - \tau_1) \\ z_3(t) = & \pi z_1(t - \tau_1) \wedge \left(N_P + z_5(t)\right) \qquad \text{(EMS-A)} \\ z_4(t) = & z_3(t - \tau_2) \\ z_5(t) = & z_4(t - \tau_3) \end{cases}$$

As we shall see in Sect. 4, a slowdown arising either in the MRAs circuit or in the physicians circuit causes a slowdown of the whole system, owing to the

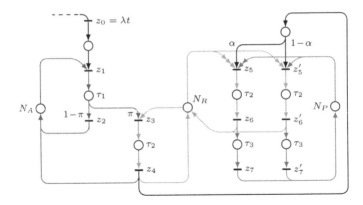

Fig. 5. Medical emergency call center with a monitored reservoir (EMS-B) (Color figure online)

synchronization step. To address this issue and still maintain the presence of an interlocutor with the patient and the brief oral summary told to physician, emergency doctors from the SAMU proposed to consider another model. One may create a new type of MRA, the *reservoir* assistant, who after a brief discussion with the MRA having answered the call, places the patient in a monitored reservoir. The answering MRA is released to pick-up other inbound calls. When an emergency physician becomes available, the reservoir assistant passes on the short briefing to the doctor and transfers the patient. While the queue of patients in the reservoir is non-empty, the reservoir assistant checks on the patients in the reservoir, and can call patients back in case they hung up. This replaces the synchronization between the physician and the answering MRA, enabling the first level MRAs to pick-up new calls more quickly. Another advantage of the reservoir mechanism is that a single reservoir assistant, having a consolidated vision of all the patients waiting for emergency physicians, can revise in real time their priority level if more severe cases arrive, whereas in the previous system the emergency physician may have had to ask each of the waiting MRAs.

The model (EMS-B), whose dynamics shall be introduced and studied in Sect. 5, implements these modifications; see Fig. 5. The reservoir assistant pool is a new place with initial marking N_R (not necessarily equal to 1). Reservoir assistants receive patients from the answering MRAs at transition z_3 and pass them to physicians at transitions z_5 and z_5', depending on the severity of the case. We denote by α the proportion of very urgent calls among patients who need to talk to an emergency physician. In case of conflict, reservoir assistants must first pass the calls already in the reservoir before placing other calls in, and should first handle very urgent calls. Release of the reservoir assistants happen at transitions z_4, z_6 and z_6'. Consultations with a physician take a time τ_3 after which transitions z_7 and z_7' can be fired. The circuits involving the reservoir assistant are depicted with color orange. It can be verified that the places standing for the pool of reservoir assistants and physicians have compatible priority rules.

4   Correspondence Between Fluid Petri Nets and Semi-Markov Decision Processes

4.1   Finite Horizon Problem

We next establish a correspondence between the Petri net dynamics and the Bellman equation of a semi-Markov decision process. Recall that *Markov Decision Processes* (MDPs) form a class of one-player games, in which one evolves through *states* by choosing *actions* at *discrete time instants*, which determine some *costs*. *Semi-Markov Decision Processes* (SMDPs, or Markov renewal programs) allow the time to take real values, while the state space remains discrete: between two successive moves, a *holding time* attached to states and actions must elapse. We refer for instance to [13,19] for background.

The finite set of states is denoted by S, and for all $i \in S$ the finite set of playable *actions* from state i is denoted by A_i. We denote $A := \bigcup_{i \in S} A_i$. As a result of playing action a from state i, the player incurs a cost r_i^a, is held in the state i for a non-negative time t_i^a, and finally goes to state $j \in S$ with probability P_{ij}^a (it is assumed that $\sum_{j \in S} P_{ij}^a = 1$ for all $i \in S$ and $a \in A_i$). Moreover, future costs are multiplied by a discount factor $\gamma_i^a \geqslant 0$. Given an initial state i and an horizon t, the *value* $v(i, t)$ is defined as the minimal expected cost incurred in horizon t. By convention, $v(\cdot, t) = 0$ when $t < 0$. Its formal definition in terms of *nonanticipative strategies* can be found in [13]. The value function satisfies the following Bellman-type dynamic programming equation (see for instance [19, §2, p. 800]):

$$v(i, t) = \inf_{a \in A_i} \left\{ r_i^a + \gamma_i^a \sum_{j \in S} P_{ij}^a v(j, t - t_i^a) \right\}. \tag{1}$$

Correspondence Theorem 1. *Consider a timed Petri net with no priority rules. Then, its dynamics is equivalent to the dynamic programming equation of a semi-Markov decision process with controlled discount factors.*

Proof. We extend the definition of proportions π_{qp} by letting $\pi_{qp} = 1$ if $q \in \mathcal{Q}_{\mathrm{sync}}$ and $p \in q^{\mathrm{in}}$. Similarly, we set the weights α_{qp} (resp. α_{pq}) to 0 if $p \notin q^{\mathrm{in}}$ (resp. $q \notin p^{\mathrm{in}}$). For all $q, q' \in \mathcal{Q}$ and $p \in \mathcal{P}$, we set

$$c_q^p := \begin{cases} \pi_{qp}\alpha_{qp}^{-1}m_p & \text{if } p \in q^{\mathrm{in}} \\ 0 & \text{otherwise,} \end{cases} \quad \text{and} \quad \tilde{\beta}_{qq'}^p := \begin{cases} \pi_{qp}\alpha_{qp}^{-1}\alpha_{pq'} & \text{if } p \in q^{\mathrm{in}} \\ 0 & \text{otherwise.} \end{cases}$$

By definition, for $q \in \mathcal{Q}$ and $p \in q^{\mathrm{in}}$, the nonnegative numbers $(\tilde{\beta}_{qq'}^p)_{q' \in \mathcal{Q}}$ are not all zero. We let $\kappa_q^p := \sum_{q' \in \mathcal{Q}} \tilde{\beta}_{qq'}^p$ and $\beta_{qq'}^p = \tilde{\beta}_{qq'}^p / \kappa_q^p$ so that $(\beta_{qq'}^p)_{q' \in \mathcal{Q}}$ is a probability vector. The dynamics summarized in Table 2 can then be written as

$$\forall q \in \mathcal{Q} \quad z_q(t) = \min_{p \in q^{\mathrm{in}}} \left\{ c_q^p + \kappa_q^p \sum_{q' \in \mathcal{Q}} \beta_{qq'}^p z_{q'}(t - \tau_p) \right\} \tag{2}$$

where we recognize the finite-horizon Bellman's equation of a discounted semi-Markovian decision process expressed in Eq. (1).   □

As we announced in Table 1, the states of the SMDP built in the proof corresponds to the transitions of the Petri net, and in each state $q \in \mathcal{Q}$ of the SMDP, the admissible actions are the upstream places $p \in q^{\text{in}}$. After playing action p from state q, the player incurs a cost $c_q^p = \pi_{qp}\alpha_{qp}^{-1}m_p$ and a discount factor $\kappa_q^p = \sum_{q' \in p^{\text{in}}} \pi_{qp}\alpha_{qp}^{-1}\alpha_{pq'}$. Then, the player is held for time τ_p, before moving to one of the states $q' \in p^{\text{in}}$ with probability $\beta_{qq'}^p = \pi_{qp}\alpha_{qp}^{-1}\alpha_{pq'}/\kappa_q^p$. In other words, the physical time of timed Petri nets is the backward time (time remaining to live) in semi-Markov decision processes.

In what follows, we shall interpret the other correspondences between these two families of models provided in Table 1.

4.2 Stationary Regimes and Average Cost Problem

We are interested in the long-run time behavior of Petri nets. For this purpose, we introduce a notion of affine stationary regime.

Definition 2. *We say that a trajectory z (counter functions of the transitions) of the Petri net is an* affine stationary regime *if there exists two vectors $\rho \in (\mathbb{R}_{\geqslant 0})^{\mathcal{Q}}$ and $u \in \mathbb{R}^{\mathcal{Q}}$ such that for all $t \geqslant -T$, $z(t) = \rho t + u$.*

The next proposition shows that, up to a shift in time, affine stationary regimes are characterized by a lexicographic system.

Proposition 1. *Given $\rho \in (\mathbb{R}_{\geqslant 0})^{\mathcal{Q}}$ and $u \in \mathbb{R}^{\mathcal{Q}}$, there exists a nonnegative number t_0 such that $z(t) := \rho(t + t_0) + u$ is a stationary regime if and only if*

$$\rho_q = \min_{p \in q^{\text{in}}} \left\{ \kappa_q^p \sum_{q' \in \mathcal{Q}} \beta_{qq'}^p \rho_{q'} \right\} \tag{L1}$$

$$u_q = \min_{p \in q_*^{\text{in}}} \left\{ c_q^p + \kappa_q^p \sum_{q' \in \mathcal{Q}} \beta_{qq'}^p (u_{q'} - \rho_{q'}\tau_p) \right\} \tag{L2}$$

where q_*^{in} is the subset of q^{in} where the minimum is achieved in (L1).

These equations are obtained by substituting $z(t) = \rho(t + t_0) + u$ in (2) and letting t tend to infinity.

There is a convenient and more abstract way to write this lexicographic system using germs of affine functions. A *germ* at infinity of a function f is an equivalence class for the relation which identifies two functions that coincide for sufficiently large values of their argument. The tuple $(\rho, u) \in \mathbb{R}^2$ will represent the germ of the affine function $t \mapsto \rho t + u$. The pointwise order on functions induces a total order on germs of affine functions, which coincides with the lexicographic order on the coordinates (ρ, u), the ρ coordinate being considered first. We complete $\mathbb{R}^2$ by introducing a greatest element $\top$ with respect to the lexicographic minimum. Then, $\mathbb{G} := \mathbb{R}^2 \cup \{\top\}$ equipped with the operations $\min^{\text{LEX}}$ and $+$ is a semifield (by convention, for all $g \in \mathbb{G}$, $g + \top = \top + g = \top$).

The multiplicative group $\mathbb{R}_{>0}$ acts on $\mathbb{G}$ by setting $a(\rho, u) := (a\rho, au)$, for $a > 0$ and $(\rho, u) \in \mathbb{R}^2$, and $a\top = \top$. If (ρ, u) is the germ of f, it is immediate to see that $(\rho, u - \rho\tau)$ is the germ of $t \mapsto f(t - \tau)$. The system (L1)–(L2) then becomes

$$\forall q \in \mathcal{Q} \quad (\rho_q, u_q) = \min_{p \in q^{\text{in}}}{}^{\text{LEX}}\left((0, c_q^p) + \kappa_q^p \sum_{q' \in \mathcal{Q}} \beta_{qq'}^p (\rho_{q'}, u_{q'} - \rho_{q'}\tau_p)\right) \quad (L)$$

In order to prove the existence of a stationary regime, we shall need the existence of a stoichiometric invariant:

Definition 3. *We say a vector* $(e_q)_{q \in \mathcal{Q}}$ *is a* stoichiometric invariant *of the Petri net whose dynamics is given by (2) if*

$$\forall q \in \mathcal{Q}, \quad \forall p \in q^{\text{in}}, \quad e_q = \kappa_q^p \sum_{q' \in \mathcal{Q}} \beta_{qq'}^p e_{q'}. \quad (3)$$

As an illustration, it can be checked that $(1, 1, 1-\pi, \pi, \pi, \pi)$ is a stoichiometric invariant of our model (EMS-A), with transitions indices in $\{0, \dots, 5\}$.

In the rest of the section, the following assumption is made:

Assumption A. *The Petri net is non-Zeno, has no priority rules and admits a positive stoichiometric invariant e.*

Correspondence Theorem 2. *Under Assumption A, the dynamics of the timed Petri net is equivalent to the dynamic programming equation of an undiscounted semi-Markov decision process.*

To see this, it suffices to observe that the transformed counters $\tilde{z}_q = z_q/e_q$ follow an equation of type (1), with $P_{qq'}^p := e_q^{-1}\kappa_q^p\beta_{qq'}^p e_{q'}$ and $\gamma_q^p = 1$, thanks to (3). We illustrate on Fig. 6 the construction of the undiscounted SMDP corresponding to the Petri net (EMS-A). Actions (depicted by squares) are labeled by pairs consisting of the associated cost and holding time, and probabilities are given along the arcs from actions to states (when non equal to 1). As we discussed after the proof of Correspondence Theorem 1, places and transitions of the Petri net are respectively mapped to the states and actions of the SMDP. The orientation of the arcs are flipped, and the time goes backward. For every state $q \in \mathcal{Q}$, the holding time of the action $p \in q^{\text{in}}$ is τ_p. Moreover, since all the arc weights in (EMS-A) are 0 or 1, the reward r_p^q reduces to m_p/e_q, which corresponds to a renormalization of the initial marking of the place p by the stoichiometric coefficient of the transition q.

Exploiting Correspondence Theorem 2, we arrive at our first main result.

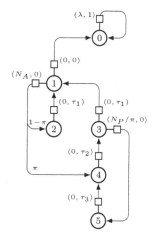

Fig. 6. The SMDP corresponding to the Petri net (EMS-A).

Theorem 3. *Under Assumption A,*

(i) *there exists an affine stationary regime, i.e. $(\rho, u) \in (\mathbb{R}_{\geqslant 0})^{\mathcal{Q}} \times \mathbb{R}^{\mathcal{Q}}$ such that, initializing the dynamics with $z(t) = \rho t + u$ for $t \in [-T, 0]$, we end up with $z(t) = \rho t + u$ for all $t \geqslant 0$.*
(ii) *the vector ρ in (i) is universal, i.e., for any initial condition, the solution $z(t)$ of the dynamics satisfies*

$$z(t) \underset{t \to \infty}{=} \rho t + O(1).$$

Theorem 3 relies on the equivalence with the undiscounted problem (Correspondence Theorem 2). Indeed, the term ρ_q/e_q can be interpreted as the optimal average cost per time unit of the associated undiscounted semi-Markov decision process. We refer to [8, 15, 16] for background on the average cost problem. Then, the lexicographic system (L1)–(L2) is equivalent to the optimality equation of the average cost problem. It is also known in the framework of undiscounted SMDPs that the optimal average cost is achieved by (stationary) *policies*. The existence of a solution to the optimality equations follows (constructively) from the termination of Howard's multichain policy iteration algorithm, proved by Denardo and Fox [8]. Finally, the universality result is proved by using the nonexpansiveness of the evolution semigroup of the time-delay system (2) with respect to an appropriate weighted sup-norm.

This prompts us to introduce policies on Petri nets: a map $\sigma: \mathcal{Q} \to \mathcal{P}$ is a *policy* if for all $q \in \mathcal{Q}$, $\sigma(q) \in q^{\text{in}}$. Given a policy σ and a stoichiometric invariant e, the $|\mathcal{Q}| \times |\mathcal{Q}|$ matrix P^σ with entries $(e_q^{-1} \kappa_q^{\sigma(q)} \beta_{qq'}^{\sigma(q)} e_{q'})_{q,q' \in \mathcal{Q}}$ is a probability matrix whose final classes are denoted by $\mathcal{F}(\sigma)$. We denote by μ_F^σ the unique invariant measure supported by the class $F \in \mathcal{F}$, and by $\phi_{F,q}^\sigma$ the probability of reaching F by applying policy σ starting from state q. The vectors m^σ (resp. τ^σ) stand for $(m_{\sigma(q)})_{q \in \mathcal{Q}}$ and $(\tau_{\sigma(q)})_{q \in \mathcal{Q}}$ and we finally define the diagonal matrix $D^\sigma := \text{diag}((e_q^{-1} \alpha_{q\sigma(q)}^{-1} \pi_{q\sigma(q)})_{q \in \mathcal{Q}})$. The fact that the average cost is achieved by stationary policies leads to the following explicit formula for the throughput.

Corollary 1 (Average cost complex). *Under Assumption A, the throughput vector ρ is given by*

$$\forall q \in \mathcal{Q}, \quad \rho_q = e_q \min_\sigma \sum_{F \in \mathcal{F}(\sigma)} \phi_{F,q}^\sigma \frac{\langle \mu_F^\sigma, D^\sigma m^\sigma \rangle}{\langle \mu_F^\sigma, \tau^\sigma \rangle}, \tag{4}$$

where the minimum is taken over all the policies.

This formula shows that the throughput ρ_q of the transition q is a concave piecewise affine function of the initial marking vector $m \in \mathbb{R}^{\mathcal{P}}$. As is customary in tropical geometry, we associate to this map a polyhedral complex (recall that a collection $\mathcal{L}$ of polyhedra is a *polyhedral complex* if for all $L \in \mathcal{L}$, any face F of L is also in $\mathcal{L}$ and for $L_1, L_2 \in \mathcal{L}$, the polyhedron $L_1 \cap L_2$ is a face of both L_1 and L_2, see [7]). If Σ is a set of policies, we define the polyhedral *cell* $\mathcal{C}_\Sigma$ to be the set of initial markings m such that the argument of the minimum in (4) is

Σ (note that the cell $\mathcal{C}_\Sigma$ may be empty for some choices of Σ). The space $\mathbb{R}^\mathcal{P}$ is covered by the cells $\mathcal{C}_\Sigma$ of maximal dimension, the latter can be interpreted as *congestion phases*. The policies σ which achieve the minimum for a given m determine bottleneck places.

We now consider the computational complexity of the problem of computing the throughput vector ρ.

Corollary 2. *Under Assumption A, the throughput vector ρ can be computed in polynomial time by solving the following linear program:*

$$\max \sum_{q \in \mathcal{Q}} \rho_q \quad s.t. \quad \begin{cases} \rho_q \leqslant \kappa_q^p \sum_{q' \in \mathcal{Q}} \beta_{qq'}^p \rho_{q'}, & \forall q \in \mathcal{Q}, \forall p \in q^{in} \\ u_q \leqslant c_q^p + \kappa_q^p \sum_{q' \in \mathcal{Q}} \beta_{qq'}^p (u_{q'} - \rho_{q'} \tau_{q'}), & \forall q \in \mathcal{Q}, \forall p \in q^{in} \end{cases}$$

in which $\rho, u \in \mathbb{R}^\mathcal{Q}$ are the variables. More precisely, if (ρ, u) is any optimal solution of this program, then ρ coincides with the throughput vector.

We refer the reader to the work of Gaujal and Giua [9] for an alternative linear programming approach. The asymptotic behavior of the value function in large horizon has been extensively studied [1,18]. As a corollary of these results, we arrive at:

Corollary 3 (Asymptotic Periodicity). *Suppose that Assumption A holds, and that the holding times are integer (so that $T \in \mathbb{N}$). Then, there exists an integer c, which is the order of an element of the symmetric group on nT letters, such that, for all $0 \leqslant r \leq c - 1$, $z(tc + r) - \rho(tc + r)$ converges as $t \to \infty$, for integer values of t.*

Whereas the earlier results of this section hold for irrational holding times, the integrality restriction in Corollary 3 is essential.

4.3   Application to Model (EMS-A)

We illustrate the above results on our running example (EMS-A). Since z_1 and z_3 have two upstream places each, there are a total of four policies. Though it is possible to use (4) to determine ρ, solving the lexicographic Eq. (L) turns out to be easier in practice. We also remark (for instance on Fig. 6) that z_2 (resp. z_4 and z_5) are always in the same recurrence class as z_1 (resp. z_3), in the sense of SMDP's chains. As a result, we shall just focus on the lexicographic optimality equation on ρ_1 and ρ_3:

$$\begin{cases} (\rho_1, u_1) = (\lambda, 0) \wedge ((1 - \pi)\rho_1 + \rho_3, N_A + (1 - \pi)(u_1 - \rho_1 \tau_1) + u_3 - \rho_3 \tau_2) \\ (\rho_3, u_3) = \pi(\rho_1, u_1 - \rho_1 \tau_1) \wedge (\rho_3, N_P + u_3 - \rho_3(\tau_2 + \tau_3)) \end{cases}$$

where $\wedge$ now stands for the $\min^{\text{LEX}}$ operation. Each policy (i.e. each choice of minimizing term in both equations) leads to a value of ρ_1 and ρ_3 and provides linear inequalities characterizing the associated validity domain. Eventually, we obtain $\rho_1 = \rho^*$ and $\rho_3 = \pi \rho^*$ with

$$\rho^* = \min\left(\lambda, \frac{N_A}{\tau_1 + \pi\tau_2}, \frac{N_P}{\pi(\tau_2 + \tau_3)}\right)$$

in which we retrieve the piecewise-affine form of ρ showed in Corollary 1.

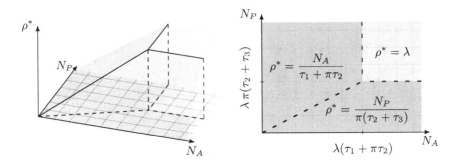

Fig. 7. The phase diagram of the (EMS-A) system

We interpret this result as follows: the "handling speed" ρ_1 of the MRAs and ρ_3 of the emergency physicians are always entangled and depend on three key dimensioning parameters: the arrival rate of inbound calls λ, the maximum MRA throughput $N_A/(\tau_1 + \pi\tau_2)$ and the maximum physician throughput $N_P/\pi(\tau_2 + \tau_3)$. We recognize in these last two terms a number of agents divided by a characteristic cycle time. Hence, if $N_A \geqslant N_A^* := \lambda(\tau_1 + \pi\tau_2)$ and $N_P \geqslant N_P^* := \lambda\pi(\tau_2 + \tau_3)$, we have $\rho^* = \lambda$ which means that all inbound calls are handled. If $N_A \leqslant N_A^*$ and $N_A/(\tau_1 + \pi\tau_2) \leqslant N_P/\pi(\tau_2 + \tau_3)$, there are too few MRAs, therefore they impose their maximum handling speed to the whole system (indeed emergency physicians wait for MRAs to pass them calls). Conversely, if $N_P \leqslant N_P^*$ and $N_P/\pi(\tau_2 + \tau_3) \leqslant N_A/(\tau_1 + \pi\tau_2)$, there are too few emergency physicians and they impose their handling speed to the whole system again (MRAs are waiting for doctors to take their calls and be released). This is illustrated by the phase diagram depicted on Fig. 7. We verify that its cells are the regions over which ρ^* is affine (as a function of N_A and N_P).

To sum it up, there are three different possible regimes, among which only one is fluid and guarantees that all calls are answered. This analysis can lead to minimal dimensioning recommendations: for such an emergency call-center and considering that calls arrive with rate λ, at least $\lceil\lambda(\tau_1 + \pi\tau_2)\rceil$ MRAs and $\lceil\lambda\pi(\tau_2 + \tau_3)\rceil$ emergency physicians are needed.

5   Case Study: Analysis of the Monitored Reservoir

In this section, we follow up on the analysis of the model (EMS-B) involving priority rules. Using Table 2, we can write the dynamics of the counter variable of the net. We present below a reduced system of equations where z_2, z_4, z_6, z_6', z_7 and z_7' have been substituted by expressions depending on z_1, z_3, z_5 and z_5' only. For the sake of readability, we denote $z|_{t_1}^{t_2} := z(t_2) - z(t_1)$, and $z|^t := z(t)$.

$$
\begin{aligned}
z_1(t) &= z_0|^t \quad \wedge \left(N_A + (1 - \pi) z_1|^{t-\tau_1} + z_3|^{t-\tau_2} \right) \\
z_3(t) &= \pi z_1|^{t-\tau_1} \quad\quad\quad\quad\quad \wedge \left(N_R + z_3|_t^{t-\tau_2} + z_5|_t^{t-\tau_2} + z_5'|_t^{t-\tau_2} \right) \\
z_5(t) &= \alpha z_3|^{t-\tau_2} \wedge \left(N_P + z_5|_t^{t-\tau_2-\tau_3} + z_5'|_{t-}^{t-\tau_2-\tau_3} \right) \wedge \left(N_R + z_3|_{t-}^{t-\tau_2} + z_5|_t^{t-\tau_2} + z_5'|_t^{t-\tau_2} \right) \\
z_5'(t) &= (1 - \alpha) z_3|^{t-\tau_2} \wedge \left(N_P + z_5|_t^{t-\tau_2-\tau_3} + z_5'|_t^{t-\tau_2-\tau_3} \right) \wedge \left(N_R + z_3|_{t-}^{t-\tau_2} + z_5|_t^{t-\tau_2} + z_5'|_t^{t-\tau_2} \right)
\end{aligned}
$$
$$\text{(EMS-B)}$$

If there are some priority routings, Correspondence Theorems 1 and 2 do not hold anymore: the dynamics has still the form of a Bellman equation, but the factors $\beta_{qq'}^p$ in (2) take negative values, implying that some "probabilities" are negative. However, it is still relevant to look for affine stationary regimes, and we next show that these regimes are the solutions of a lexicographic system over germs similar to (L). To do so, we need to derive other germ equations for transitions ruled by priority routing, whose dynamics is recalled in Table 2. One needs to address how the expressions of the form $z(t^-)$ behave when passing to germs. The problem may seem ill-posed since this value coincides with $z(t)$ for ultimately affine functions. Nonetheless, in [2], it has been shown that the problem of looking for ultimately affine stationary regimes on the δ-discretization of the fluid dynamics is well-posed. In this discretized model, the term $z(t^-)$ is replaced by $z(t - \delta)$. The detour via this discretized dynamics enables one to prove that, regardless of the choice of δ, small enough, some terms cannot achieve the minimum in the priority dynamic equations, and thus can be removed. This leads to the last equation of Table 3 and the following result.

Table 3. Dynamic equations followed by germs of transitions counter functions. When $q \in \mathcal{Q}_{\text{prio}}$ and $p \in q^{\text{in}}$, we denote by $p_{\text{max}}^{\text{out}}$ the transition of p^{out} with least priority and if $q \neq p_{\text{max}}^{\text{out}}$, $\sigma_p(q)$ denotes the successor of q relatively to the order of p^{out} induced by the priority rule, i.e. the transition with highest priority after q.

Type	Germs equation in stationary regime
$q \in \mathcal{Q}_{\text{sync}}$	$(\rho_q, u_q) =$ $$\min_{p \in q^{\text{in}}}{}^{\text{LEX}} \quad \alpha_{qp}^{-1}\left((0, m_p) + \sum_{q' \in p^{\text{in}}} \alpha_{pq'} (\rho_{q'}, u_{q'} - \rho_{q'}\tau_p) \right)$$
$q \in \mathcal{Q}_{\text{psel}}$	$(\rho_q, u_q) =$ $$\pi_{qp} \cdot \alpha_{qp}^{-1}\left((0, m_p) + \sum_{q' \in p^{\text{in}}} \alpha_{pq'} (\rho_{q'}, u_{q'} - \rho_{q'}\tau_p) \right)$$
$q \in \mathcal{Q}_{\text{prio}}$	$(\rho_q, u_q) =$ $$\min_{\substack{p \in q^{\text{in}} \\ q = p_{\text{max}}^{\text{out}} \text{ or } \rho_{\sigma_p(q)} = 0}}{}^{\text{LEX}} \quad \alpha_{qp}^{-1}\left((0, m_p) + \sum_{q' \in p^{\text{in}}} \alpha_{pq'} (\rho_{q'}, u_{q'} - \rho_{q'}\tau_p) - \sum_{q' \in p^{\text{out}} \setminus \{q\}} \alpha_{q'p}(\rho_{q'}, u_{q'}) \right)$$

Theorem 4. *The ultimately affine stationary regimes* $z(t) = \rho t + u$ *of the dynamics of Table 2 are solutions of the germ equations of Table 3.*

Applying Theorem 4 and the equations of Table 3 to the model (EMS-B) provide the following (reduced) system on the affine germs of counter variables:

$$(\rho_1, u_1) = (\lambda, 0) \wedge (\rho_1, N_A + (1 - \pi)(u_1 - \rho_1\tau_1) + u_3 - \pi\rho_1\tau_2)$$

$$(\rho_3, u_3) = (\pi\rho_1, \pi(u_1 - \rho_1\tau_1)) \wedge (\pi\rho_1, u_3 + N_R - \pi\rho_1\tau_2 - (\rho_5 + \rho_5')\tau_2)$$

$$(\rho_5, u_5) = \begin{cases} (\alpha\pi\rho_1, \alpha(u_3 - \pi\rho_1\tau_2)) \wedge (\rho_5, u_5 + N_P - \rho_5(\tau_2 + \tau_3)) & \text{if} \quad \rho_5' = 0 \\ (\alpha\pi\rho_1, \alpha(u_3 - \pi\rho_1\tau_2)) & \text{if} \quad \rho_5' > 0 \end{cases}$$

$$(\rho_5', u_5') = ((1 - \alpha)\pi\rho_1, (1 - \alpha)(u_3 - \pi\rho_1\tau_2)) \wedge (\rho_5', u_5' + N_P - (\rho_5 + \rho_5')(\tau_2 + \tau_3))$$

Note that the priority ruling the routing of tokens from the pool of reservoir assistants does not appear on the affine germs of z_5 and z_5'. This is an expected outcome since transitions z_5 and z_5' (high level of priority for the reservoir pool) can only receive tokens that have passed through transition z_3 (low level of priority for the reservoir pool) before, as a result z_5 and z_5' cannot ultimately inhibit themselves. Such a layout of priorities does remain appropriate to perform arbitration of tokens orientation in case of conflicts but is not captured in the scope of long-run time analysis of the system.

As in Sect. 4.3, a choice of policy (i.e. a choice of minimizing terms in the lexicographic system) provides affine equalities determining the throughput and regions on which it is an affine function of the parameters of the model. This leads to nine congestion phases (maximal cells of the throughput complex) covering $(\mathbb{R}_{\geqslant 0})^3$, depicted on Fig. 9. As expected, the introduction of a new type of resource agent (the reservoir assistant) introduces more slowdown phases if its initial marking N_R is insufficient. Therefore, to ensure the good behavior of the (EMS-B) model whose design relies substantially on the reservoir, one needs to take $N_R \geqslant N_R^* := 2\pi\lambda\tau_2$. Note that the minimum number of MRAs (resp. emergency physicians) to answer all the calls is not affected by the presence of the reservoir by comparison with (EMS-A) model, it is still $N_A^* := \lambda(\tau_1 + \pi\tau_2)$ (resp. $N_P^* := \pi\lambda(\tau_2 + \tau_3)$).

From there, two qualitative advantages of this new model must be noted. First, we focus on the effect of the priority level for very urgent calls (in proportion α among those passed to an emergency physician). Given an MRA throughput ρ_1, define the two functions $\underline{N}_P$ and $\overline{N}_P$ by

$$\underline{N}_P(\rho_1) := \pi\alpha(\tau_2 + \tau_3)\rho_1 \quad \text{and} \quad \overline{N}_P(\rho_1) := \pi(\tau_2 + \tau_3)\rho_1.$$

A minimum number of $\overline{N}_P$ physicians is needed to handle all the calls passed by the MRAs *via* the reservoir assistant. However, in case of a lack of physicians, the priority mechanism ensures that the very urgent calls remain handled as long as $N_P \geqslant \underline{N}_P$. Below the latter threshold, there are not enough physicians to handle even these very urgent calls only. Remark that in presence of priorities, the throughput function of transitions may not be concave anymore, see for instance ρ_5' as a function of N_A and N_P in Fig. 8 (supposing $N_R \geqslant \overline{N}_R$). In our

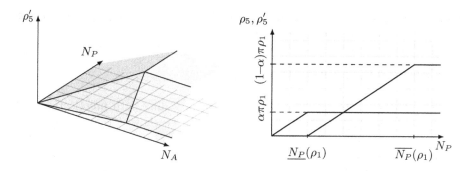

Fig. 8. The throughput ρ_5' is not concave, although ρ_5 and $\rho_5 + \rho_5'$ still are

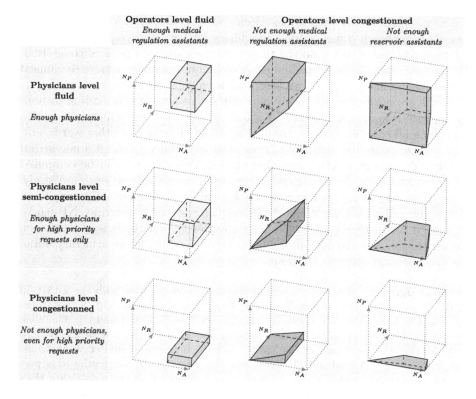

Fig. 9. Congestion phases of the model (EMS-B)

example though, note that in addition to ρ_1 and ρ_3, both ρ_5 and $\rho_5 + \rho_5'$ are still concave.

Contrary to the model (EMS-A), we also observe that a slowdown in the emergency physician circuit does not affect the throughput of the MRAs, as an effect of their desynchronization by the reservoir buffer. It may still happen that we encounter both a lack of MRAs and physicians, but the latter do not prevent the former to pick up inbound calls at their maximal possible throughput.

6   Concluding Remarks

We developed a model of fluid timed Petri net including both preselection and priority routings. In the absence of preselection, we showed that the dynamics of the net is equivalent to the Bellman equation of a semi-Markov decision problem, from which a number of properties follows: existence and universality of the throughput vector (independence from the initial condition), existence of stationary regimes by reduction to a lexicographic system, polynomial-time computability of the throughput by reduction to a linear program, and explicit representation of the throughput, as a function of resources, by a polyhedral complex. This approach provides tools to address further issues: e.g., an important practical problem is to bound the time needed to absorb a peak of congestion. We believe it can still be addressed using techniques of nonexpansive dynamical systems, along lines of [1,17], we leave this for a subsequent work.

In the presence of priority, only part of these results remain: finding stationary regimes is equivalent to solving a lexicographic system, which is a system of polynomial equations over a tropical semifield of germs. In other words, stationary regimes are the points of a tropical variety, and we still get a polyhedral complex, describing all the congestion phases. This complex can be computed in exponential time, by enumerating strategies, as we did on our case study. Whereas we do not expect worst-case polynomial-time computability results in such a generality (solving tropical polynomial systems is generally NP-hard), we leave it for further work to get finer complexity bounds. It is also an open problem to compare the asymptotic behavior of counters, for an arbitrary initial condition, with stationary solutions.

Acknowledgment. This work was done through a collaboration with the SAMU of APHP. We wish to thank especially, PR. P. Carli, DR. É. Chanzy, DR. É. Lecarpentier, DR. Ch. Leroy, DR. Th. Loeb, DR. J.-S. Marx, DR. N. Poirot and DR. C. Telion for making this work possible, for their support and for insightful comments. We also thank all the other personals of the SAMU, in particular DR. J. Boutet, J.-M. Gourgues, I. Lhomme, F. Linval and Th. Pérennou. The present work was developed as part of an ongoing study within a project between APHP and Préfecture de Police (PP) aiming at an interoperability among the different call centers. It strongly benefited from the experience acquired, since 2014, on the analysis of the new platform "PFAU" (answering to the emergency numbers 17, 18 and 112), developed by PP. We thank LCL S. Raclot and R. Reboul, in charge of the PFAU project at PP, for their constant support and insightful comments all along these years. This work also strongly benefited of scientific discussions with Ph. Robert, whom we thank. Finally, we thank the reviewers for their careful and detailed reviews.

References

1. Akian, M., Gaubert, S.: Spectral theorem for convex monotone homogeneous maps, and ergodic control. Nonlinear Anal. Theory Methods Appl. **52**(2), 637–679 (2003)
2. Allamigeon, X., Bœuf, V., Gaubert, S.: Performance evaluation of an emergency call center: tropical polynomial systems applied to timed Petri nets. In: Sankaranarayanan, S., Vicario, E. (eds.) FORMATS 2015. LNCS, vol. 9268, pp. 10–26. Springer, Cham (2015). https://doi.org/10.1007/978-3-319-22975-1_2
3. Baccelli, F., Cohen, G., Olsder, G.J., Quadrat, J.P.: Synchronization and Linearity. Wiley, Hoboken (1992)
4. Bœuf, V., Robert, P.: A stochastic analysis of a network with two levels of service. Queueing Syst. **92**, 203–232 (2019). https://doi.org/10.1007/s11134-019-09617-y
5. Cohen, G., Gaubert, S., Quadrat, J.: Asymptotic throughput of continuous timed Petri nets. In: Proceedings of the 34th Conference on Decision and Control, New Orleans, December 1995
6. Cohen, G., Gaubert, S., Quadrat, J.: Algebraic system analysis of timed Petri nets. In: Gunawardena, J. (ed.) Idempotency, pp. 145–170. Publications of the Isaac Newton Institute, Cambridge University Press (1998)
7. De Loera, J.A., Rambau, J., Santos, F.: Triangulations Structures for Algorithms and Applications. Springer, Heidelberg (2010). https://doi.org/10.1007/978-3-642-12971-1
8. Denardo, E., Fox, B.: Multichain Markov renewal programs. SIAM J. Appl. Math. **16**, 468–487 (1968)
9. Gaujal, B., Giua, A.: Optimal stationary behavior for a class of timed continuous Petri nets. Automatica **40**(9), 1505–1516 (2004)
10. Heidergott, G., Olsder, G.J., van der Woude, J.: Max Plus at Work. Princeton University Press, Princeton (2006)
11. Komenda, J., Lahaye, S., Boimond, J.L., van den Boom, T.: Max-plus algebra in the history of discrete event systems. Ann. Rev. Control **45**, 240–249 (2018)
12. L'Ecuyer, P., Gustavsson, K., Olsson, L.: Modeling bursts in the arrival process to an emergency call center. In: Rabe, M., Juan, A.A., Mustafee, N., Skoogh, A., Jain, S., Johansson, B. (eds.) Proceedings of the 2018 Winter Simulation Conference (2018)
13. Puterman, M.L.: Markov Decision Processes: Discrete Stochastic Dynamic Programming. Wiley, Hoboken (2014)
14. Recalde, L., Silva, M.: Petri net fluidification revisited: semantics and steady state. Eur. J. Autom. APII-JESA **35**(4), 435–449 (2001)
15. Schäl, M.: On the second optimality equation for semi-Markov decision models. Math. Oper. Res. **17**(2), 470–486 (1992)
16. Schweitzer, P.J., Federgruen, A.: The functional equations of undiscounted Markov renewal programming. Math. Oper. Res. **3**(4), 308–321 (1978)
17. Schweitzer, P., Federgruen, A.: The asymptotic behavior of undiscounted value iteration in Markov decision problems. Math. Oper. Res. **2**, 360–381 (1978)
18. Schweitzer, P., Federgruen, A.: Geometric convergence of value-iteration in multichain Markov decision problems. Adv. Appl. Prob. **11**, 188–217 (1979)
19. Yushkevich, A.: On semi-Markov controlled models with an average reward criterion. Theory Probab. Appl. **26**(4), 796–803 (1982)

Automated Repair of Process Models
Using Non-local Constraints

Anna Kalenkova[1] , Josep Carmona[2(✉)] , Artem Polyvyanyy[1] ,
and Marcello La Rosa[1]

[1] School of Computing and Information Systems, The University of Melbourne,
Parkville, VIC 3010, Australia
{anna.kalenkova,artem.polyvyanyy,marcello.larosa}@unimelb.edu.au
[2] Department of Computer Science, Polytechnic University of Catalonia,
C. Jordi Girona, 1-3, 08034 Barcelona, Spain
jcarmona@cs.upc.edu

Abstract. State-of-the-art process discovery methods construct free-choice process models from event logs. Hence, the constructed models do not take into account indirect dependencies between events. Whenever the input behavior is not free-choice, these methods fail to provide a precise model. In this paper, we propose a novel approach for the enhancement of free-choice process models, by adding non-free-choice constructs discovered a-posteriori via region-based techniques. This allows us to benefit from both the performance of existing process discovery methods, and the accuracy of the employed fundamental synthesis techniques. We prove that the proposed approach preserves fitness with respect to the event log, while improving the precision when indirect dependencies exist. The approach has been implemented and tested on both synthetic and real-life datasets. The results show its effectiveness in repairing process models discovered from event logs.

1 Introduction

Process mining is a family of methods used for the analysis of event data [1]. These methods include *process discovery* aimed at constructing process models from event logs; *conformance checking* applied for finding deviations between real (event logs) and expected (process models) behavior [13]; and *process enhancement* used for the enrichment of process models with additional data extracted from event logs. The latter also includes *process repair* applied to realign process models in accordance with the event logs. Event logs are usually represented as sequences of events (or traces). The main challenge of process discovery is to efficiently construct *fitting* (capturing traces of the event log), *precise* (not capturing traces not present in the event log) and simple process models.

Scalable process discovery methods, which are most commonly used for the analysis of real-life event data, either produce *directly follows graphs*, or use them as an intermediate process representation to obtain a Petri net or a BPMN model [24] (see e.g. *Inductive miner* [21] and *Split miner* [5]). Directly follows

© Springer Nature Switzerland AG 2020
R. Janicki et al. (Eds.): PETRI NETS 2020, LNCS 12152, pp. 280–300, 2020.
https://doi.org/10.1007/978-3-030-51831-8_14

graphs are directed graphs with nodes representing process activities and arcs representing the directly follows (successor) relation between them. Being simple and intuitive, these graphs considerably generalise process behaviour, e.g., they add combinations of process paths that are not observed in the event log. This is because they do not represent higher-level constructs such as parallism and long distance (i.e., non-local) dependencies. The above-mentioned discovery methods construct directly follows graphs from event logs and then recursively find relations between sets of nodes in these graphs, in order to discover a *free-choice* Petri net [15], which can then be seemlessly converted into a BPMN model – the industry language for representing business process models. In free-choice nets, the choice between conflicting activities (such that only one of them can be executed) is always "free" from additional preconditions. Although parallel activities can be modeled by free-choice nets, non-local choice dependencies are modeled by non-free-choice nets [30]. Several methods for the discovery of non-free-choice Petri nets exist. However, these methods are either computationally expensive [3,8,11,29,31], or heuristic in nature (i.e., the derived models may fail to replay the traces in the event log) [30]. Even these methods are not heuristic and demonstrate reasonable performance, they usually produce process models with complex structure [22,32]. In contrast, the approach proposed in this paper, starts with a simple free-choice "skeleton" enhancing it with additional modeling constructs.

In this paper, we propose a repair approach for the enhancement of free-choice nets by adding extra constructs to capture non-local dependencies. To find non-local dependencies, a transition system constructed from the initial event log is analyzed. This analysis checks whether all the free-choice constructs of the initial process model correspond to free-choice relations in the transition system. For process activities with non-free-choice relations in the transition system but with free-choice relation in the Petri net, region theory [7] is applied to identify, whenever possible, additional places and arcs to be added to the Petri net to ensure the non-local relations between the corresponding transitions. Remarkably, although we have implemented our approach over *state-based* region theory [3,11,29], the proposed approach can be also extended to *language-based* region theory [9,31], or to geometric or graph-based approaches that have been recently proposed [10,28].

Importantly, we apply a goal-oriented state-based region algorithm, to those parts of the transition system where the free-choice property is not fulfilled. This allows us to reduce the computation time, relegating region-theory to when it is really needed. We prove that important quality metrics of the initial free-choice (workflow) net are either preserved, or improved for those cases where non-local dependencies exist, i.e., *fitness* is never reduced and *precision* can increase. Hence, when using our approach on top of an automated discovery method that returns a free-choice Petri net, one can still keep the complexity of process discovery manageable, obtaining more precise process models that represent more faithfully the process behavior recorded in the event log.

In contrast to the existing process repair techniques, which change the structure of the process models by inserting, removing [4,17,25] or replacing tasks and sub-processes [23], the approach proposed in this paper only imposes additional restrictions on the process model behavior, preserving fitness and improving precision where possible.

We implemented the proposed approach as a plugin of Apromore [20][1] and tested it both on synthetic and real-world event data. The tests show the effectiveness of our approach within reasonable time bounds.

The paper is organized as follows. Section 2 illustrates the approach by a motivating example. Section 3 contains the main definitions used throughout the paper. The state-based region technique is introduced in Sect. 4. The proposed model repair approach is then described in Sect. 5. Additionally, Sect. 5 contains formal proofs of the properties of the repaired process model. High-level process modeling constructs, e.g., BPMN modeling elements representing non-free-choice routing are also discussed in Sect. 5. The results of the experiments are presented in Sect. 6. Finally, Sect. 7 concludes the paper.

2   Motivating Example

This section presents a simple motivating example inspired by real-life BPIC'2017 event log[2] and examples discussed in [30]. Consider a process of loan application. The process can be carried out by a client or by a bank employee on behalf of the client. Thus, this process can be described by two possible sequences of events (traces) which together can be considered as an event log: $L = \{\langle send\ application,\ check\ application, notify\ client,\ accept\ application\rangle,$ $\langle create\ application,\ check\ application,\ complete\ application,\ accept\ application\rangle\}$. According to one trace, the client sends a loan application to the bank, then this application is checked, after that the client is notified and the application is accepted. The other trace corresponds to a scenario when the application is initially created by a bank employee, then it is checked, after that, the bank employee contacts the client to complete the application, and finally, the application is accepted. Figure 1 presents a workflow net discovered by Inductive miner [21] and Split miner [5] from L. This model accepts two additional traces: $\langle send\ application,\ check\ application,\ complete\ application, accept\ application\rangle,$ $\langle create\ application,\ check\ application,\ notify\ client,\ accept application\rangle$ not presented in L. These traces violate the business logic of the process. If the application was sent by a client, it is completed, and there is no need to take *complete application* step. Also, if the application was initially created by a bank employee, the step *complete application* is mandatory.

This example demonstrates that the choice between *notify client* and *complete application* activities depends on the history of the trace. The transition system in Fig. 2 shows a behavior recorded in event log L (Fig. 2). State s_1 corresponds to a choice between activities *send application* and *create application*.

---

[1] https://apromore.org.
[2] https://data.4tu.nl/repository/uuid:5f3067df-f10b-45da-b98b-86ae4c7a310b.

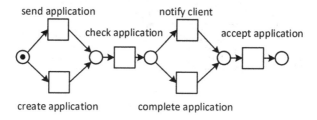

Fig. 1. A workflow net discovered from L by Inductive miner and Split miner.

This choice does not depend on any additional conditions. In contrast, for the system being in states s_4 and s_5 there is no free choice between *notify client* and *complete application* activities; in state s_4 only *notify client* step can be taken, in s_5 only *complete application* can be performed. This means that there are states in the transition system where activities *notify client* and *complete application* are not in a free-choice relation (the choice depends on additional conditions and is predefined), while they are in a free-choice relation within the discovered model (Fig. 1).

To impose additional restrictions on the process model the state-based region theory can be applied [12,14,29]. Figure 2 presents three regions $r_1 = \{s_4, s_5\}$, $r_2 = \{s_2, s_4\}$, and $r_3 = \{s_3, s_5\}$ with outgoing transitions labeled by *notify client* and *complete application* events discovered by the state-based region algorithm [29].

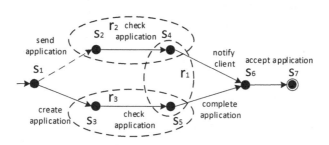

Fig. 2. Transition system that encodes event log L.

Figure 3 presents a target workflow net obtained from the initial workflow net (Fig. 1) by inserting places which correspond to the discovered regions. As one may note, in addition to r_1, two places r_2 and r_3 were added. These places impose additional constraints, such that the enhanced process model accepts event log L and does not support additional traces and, hence, is more precise.

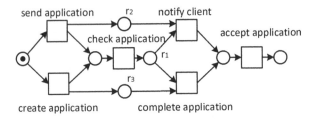

Fig. 3. A workflow net enhanced with additional regions (places) r_2 and r_3.

In the next sections, we formalise this technique and apply it to event data.

3   Preliminaries

In this section, we formally define event logs and process models, such as transition systems, Petri nets, and workflow nets.

3.1   Sets, Multisets, Event Logs

Let S be a finite set. A *multiset* m over S is a mapping $m : S \to \mathbb{N}_0$, where $\mathbb{N}_0$ is the set of all natural numbers (including zero), i.e., multiset m contains $m(s)$ copies of element $s \in S$.

For two multisets m, m' we write $m \subseteq m'$ iff $\forall s \in S : m(s) \leq m'(s)$ (the inclusion relation). The sum of two multisets m and m' is defined as: $\forall s \in S : (m+m')(s) = m(s)+m'(s)$. The difference of two multisets is a partial function: $\forall s \in S$, such that $m(s) \geq m(s')$, $(m - m')(s) = m(s) - m'(s)$.

Let E be a finite set of events. A *trace* σ (over E) is a finite sequence of events, i.e., $\sigma \in E^*$, where E^* is the set of all finite sequences over E, including the empty sequence of zero length. An *event log* L is a set of traces, i.e., $L \subseteq E^*$.

3.2   Transition Systems, Petri Nets, Workflow Nets

Let S and E be two disjoint non-empty sets of *states* and *events*, and $B \subseteq S \times E \times S$ be a *transition relation*. A *transition system* is a tuple $TS = (S, E, B, s_i, S_{fin})$, where $s_i \in S$ is an initial state and $S_{fin} \subseteq S$ – a set of final states. Elements of B are called *transitions*. We write $s \xrightarrow{e} s'$, when $(s, e, s') \in B$ and $s \xrightarrow{e}$, when $\exists s' \in S$, such that $(s, e, s') \in B$; $s \not\xrightarrow{e}$, otherwise.

A trace $\sigma = \langle e_1, \ldots, e_n \rangle$ is called *feasible* in TS iff $\exists s_1, \ldots, s_n \in S : s_i \xrightarrow{e_1} s_1 \xrightarrow{e_2} \ldots \xrightarrow{e_n} s_n$, and $s_n \in S_{fin}$, i.e., a *feasible* trace leads from the initial state to some final state. A *language accepted by* TS is defined as the set of all traces feasible in TS, and is denoted by $\mathcal{L}(TS)$.

We say that a transition system TS *encodes* an event log L iff each trace from L is a feasible trace in TS, and inversely each feasible trace in TS belongs to L. An example of a transition system is shown in Fig. 2. States and transitions are presented by vertices and directed arcs respectively. The initial state s_1 is marked by an additional incoming arrow, the only final state s_7 is indicated by a circle with double border.

Let P and T be two finite disjoint sets of *places* and *transitions*, and $F \subseteq (P \times T) \cup (T \times P)$ be a flow relation. Let also E be a finite set of events, and $l : T \to E$ be a labeling function, such that $\forall t_1, t_2 \in T, t_1 \neq t_2$, it holds that $l(t_1) \neq l(t_2)$, i.e., all the transitions are uniquely labeled. Then $N = (P, T, F, l)$ is a *Petri net*.

A *marking* in a Petri net is a multiset over the set of its places. A marked Petri net (N, m_0) is a Petri net N together with its *initial marking* m_0.

Graphically, places are represented by circles, transitions by boxes, and the flow relation F by directed arcs. Places may carry tokens represented by filled circles. A current marking m is designated by putting $m(p)$ tokens into each place $p \in P$. Marked Petri nets are presented in Figs. 1 and 3.

For a transition $t \in T$, an arc (p, t) is called an *input arc*, and an arc (t, p) an *output arc*, $p \in P$. The *preset* $\bullet t$ and the *postset* $t \bullet$ of transition t are defined as the multisets over P, such that $\bullet t(p) = 1$, if $(p, t) \in F$, otherwise $\bullet t(p) = 0$, and $t^\bullet(p) = 1$ if $(t, p) \in F$, otherwise $t^\bullet(p) = 0$. A transition $t \in T$ is *enabled* in a marking m iff $\bullet t \subseteq m$. An enabled transition t may *fire* yielding a new marking $m' =_{\text{def}} m - \bullet t + t^\bullet$ (denoted $m \xrightarrow{t} m'$, $m \xrightarrow{l(t)} m'$, or just $m \to m'$). We say that m_n is *reachable* from m_1 iff there is a (possibly empty) sequence of firings $m_1 \to \cdots \to m_n$ and denote this relation by $m_1 \xrightarrow{*} m_n$.

$\mathcal{R}(N, m)$ denotes the set of all markings reachable in Petri net N from marking m. A marked Petri net $(N, m_0), N = (P, T, F, l)$ is *safe* iff $\forall p \in P, \forall m \in \mathcal{R}(N, m_0) : m(p) \leq 1$, i.e., at most one token can appear in a place.

A *reachability graph* of a marked Petri net (N, m_0), $N = (P, T, F, l)$, with a labeling function $l : T \to E$, is a transition system $TS = (S, E, B, s_i, S_{fin})$ with the set of states $S = \mathcal{R}(N, m_0)$ and transition relation B defined by $(m, e, m') \in B$ iff $m \xrightarrow{t} m'$, where $e = l(t)$. The initial state in TS is the initial marking m_0. If some reachable markings in (N, m_0) are distinguished as final markings, they are defined as final states in TS. The *language* of a Petri net (N, m_0), denoted by $\mathcal{L}(N, m_0)$ is the language of its reachability graph, i.e., $\mathcal{L}(N, m_0) = \mathcal{L}(TS)$. We say that a Petri net (N, m_0) *accepts* a trace iff this trace is feasible in the reachability graph of (N, m_0); a Petri net *accepts* a language iff this language is accepted by its reachability graph.

Given a Petri net $N = (P, T, F, l)$, two transitions $t_1, t_2 \in T$ are in a *free-choice relation* iff $\bullet t_1 \cap \bullet t_2 = \emptyset$ or $\bullet t_1 = \bullet t_2$. Since we consider Petri nets with uniquely labeled transitions, we also say that events (or activities) $l(t_1)$ and $l(t_2)$ are in a *free-choice relation*. Petri net N is called *free-choice* iff for all $t_1, t_2 \in T$, it holds that t_1 and t_2 are in a free-choice relation. This is one of the several equivalent definitions for free-choice Petri nets presented in [15]. A Petri net is called *non-free-choice* iff it is not free-choice. Figure 4 presents an example of a non-free-choice Petri net, where for two transitions t_1 and t_2 holds that $\bullet t_1 \cap \bullet t_2 = \{p_1, p_2\} \neq \emptyset$ and $\bullet t_1 = \{p_1, p_2\} \neq \bullet t_2 = \{p_1, p_2, p_3\}$.

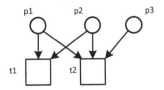

Fig. 4. A non-free-choice Petri net.

The choice of which transition will fire depends on an additional constraint imposed by place p_3. If $m(p_1) > 0$, $m(p_2) > 0$, and $m(p_3) = 0$, then only t_1 is enabled, thus there is no free-choice between t_1 and t_2. Another example of a non-free-choice Petri net was presented earlier in Fig. 3, where transitions labeled by *notify client* and *complete application* are not in a free-choice relation, thus the Petri net is not free-choice. An example of a free-choice Petri net is presented in Fig. 1.

Workflow nets is a special subclass of Petri nets designed for modeling workflow processes [2]. A workflow net has one initial and one final place, and every place or transition is on a directed path from the initial to the final place.

Formally, a marked Petri net $N = (P, T, F, l)$ is called a *workflow net* iff

1. There is one source place $i \in P$ and one sink place $o \in P$, such that i has no input arcs and o has no output arcs.
2. Every node from $P \cup T$ is on a directed path from i to o.
3. The initial marking contains the only token in its source place.

We denote by $[i]$ the initial marking in a workflow net N. Similarly, we use $[o]$ to denote the final marking in a workflow net N, defined as a marking containing the only token in the sink place o. The *language* of workflow net N is denoted by $\mathcal{L}(N)$.

A workflow net N with the initial marking $[i]$ and the final marking $[o]$ is *sound* iff

1. For every state m reachable in N, there exists a firing sequence leading from m to the final state $[o]$. Formally, $\forall m : [([i] \xrightarrow{*} m) \text{ implies } (m \xrightarrow{*} [o])]$;
2. The state $[o]$ is the only state reachable from $[i]$ in N with at least one token in place o. Formally, $\forall m : [([i] \xrightarrow{*} m) \wedge ([o] \subseteq m) \text{ implies } (m = [o])]$;
3. There are no *dead* transitions in N. Formally, $\forall t \in T \; \exists m, m' : ([i] \xrightarrow{*} m \xrightarrow{l(t)} m')$.

Note that both models presented in Figs. 1 and 3 are sound workflow nets.

4   Region State-Based Synthesis

In this section, we give a brief description of the well-known state-based region algorithm [14] applied for the synthesis of Petri nets from transition systems.

Let $TS = (S, E, T, s_i, S_{fin})$ be a transition system and $r \subseteq S$ be a subset of states. Subset r is a *region* iff for each event $e \in E$ one of the following conditions holds:

– all the transitions $s_1 \xrightarrow{e} s_2$ *enter* r, i.e., $s_1 \notin r$ and $s_2 \in r$,
– all the transitions $s_1 \xrightarrow{e} s_2$ *exit* r, i.e., $s_1 \in r$ and $s_2 \notin r$,
– all the transitions $s_1 \xrightarrow{e} s_2$ *do not cross* r, i.e., $s_1, s_2 \in r$ or $s_1, s_2 \notin r$.

In other words, all the transitions labeled by the same event are of the same type (*enter*, *exit*, or *do not cross*) for a particular region.

A region r' is said to be a *subregion* of a region r iff $r' \subseteq r$. A region r is called a *minimal region* iff it does not have any other subregions.

The state-based region algorithm covers the transition system by its minimal regions [16]. Figure 5 presents the transition system from Fig. 2 covered by minimal regions: $r_1 = \{s_4, s_5\}$, $r_2 = \{s_2, s_4\}$, $r_3 = \{s_3, s_5\}$, $r_4 = \{s_2, s_3\}$, $r_5 = \{s_6\}$, $r_6 = \{s_1\}$, and $r_7 = \{s_7\}$. According to the algorithm in [14], every minimal region is transformed to a place within the target Petri net and connected with transitions corresponding to the *exiting* and *entering* events by outgoing and incoming arcs respectively (refer to Fig. 6).

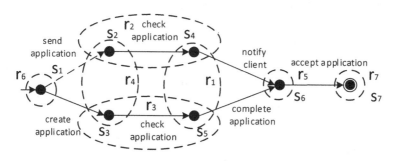

Fig. 5. Applying the state-based region algorithm to the transition system presented in Fig. 2.

Region r *separates* two different states $s, s' \in S$, $s \neq s'$, iff $s \in r$ and $s' \notin r$. Finding such a region is the *state separation problem* between s and s' and is denoted by $SSP(s, s')$. When an event e is not enabled in a state s, i.e., $s \not\xrightarrow{e}$, a region r, containing s may be found, such that e does not *exit* r. Finding such a region is known as the *event/state separation problem* between s and e and is denoted by $ESSP(s, e)$.

A well-known result in region theory establishes that if all *SSP* and *ESSP* problems are solved, then synthesis is exact [7]:

Theorem 1. *A TS can be synthesized into a safe Petri net N such that the reachability graph of N is isomorphic to TS if all SSP and ESSP problems are solvable.*

These problems are also known to be NP-complete [7]. In this paper, we reduce the size of the problem by constructing regions corresponding to particular events only.

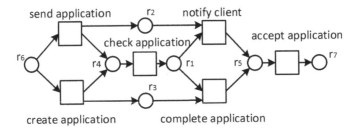

Fig. 6. A Petri net model synthesized from the transition system presented in Fig. 5.

5   Repairing Free-Choice Process Models

In this section, we describe our approach for repairing free-choice workflow nets using non-local constraints captured in the event logs. Additionally, we investigate formal properties of the repaired process models.

5.1   Problem Definition

Let N be a free-choice workflow net discovered from event log L and let TS be a transition system encoding L. Due to limitations of the automated discovery methods [5,21] that construct free-choice workflow nets, not all the places that correspond to minimal regions may have been derived, and therefore important $SSP/ESSP$ problems may not be solved in N, when considering $\mathcal{R}(N, [i])$ as the behavior to represent with N.

This brings us to the following characterization of the problem. Let $t_1, \ldots, t_n$ be transitions in N with ${}^{\bullet}t_1 = {}^{\bullet}t_2 = \cdots = {}^{\bullet}t_n$, i.e., $t_1, \ldots, t_n$ are in the free-choice relation in N, and let $TS = (S, E, T, s_i, S_{fin})$ be a minimal transition system encoding the event log L. If there exists a state $s \in S$, and $1 \le i < j \le n$ such that:

1. e_i, e_j correspond to transitions t_i, t_j, respectively,
2. $s \xrightarrow{e_i}$,
3. $s \not\xrightarrow{e_j}$

Then, the relation of $t_1, \ldots, t_n$ in N corresponds to a *false free-choice relation*, not observed in TS.

There is no place in N corresponding to a region that solves the $ESSP(s, e_j)$ problem, because $t_1, \ldots, t_n$ are in a free-choice relation in N. For instance, the Petri net in Fig. 1 contains places corresponding to regions r_1, r_4, r_5, r_6, and r_7 shown in Fig. 5, and none of those regions solves the $ESSP(s_4, complete\ application)$ and $ESSP(s_5, notify\ client)$ problems in the transition system.

Note that we define the notion of a false free-choice relation for a minimal transition system (transition system with a minimal number of states [19]) encoding the event log. This is done in order to avoid a case when there exists a state s' which is equivalent to s, such that $s' \xrightarrow{e_j}$. During the minimization these equivalent states will be merged into one state with outgoing transitions labeled by e_i and e_j showing that there is no false free-choice relation between corresponding transitions. Another reason to minimize the transition system is to reduce the number of states being analyzed.

Note that there is no guarantee that an *ESSP* problem can be solved. Nevertheless, in the running example, regions r_2 and r_3 solve $ESSP(s_4, complete\ application)$ and $ESSP(s_5, notify\ client)$ problems.

5.2   Algorithm Description

In this subsection, we present an algorithm for enhancement of a free-choice workflow net N with additional constraints from event log L (Algorithm 1). Firstly, by applying *ConstructMinTS*, a minimal transitional system encoding the event log L is constructed.[3] Then, false free-choice relations and corresponding *ESSP* problems are identified. According to the definition of a false free-choice relation presented earlier, procedure *FindFalseFreeChoiceRelations* is polynomial in time. Indeed, to find all the false free-choice relations one needs to check whether all the states of a transition system have none or all outgoing transitions labeled by events assumed to be in free-choice relations within the original workflow net N. When the false free-choice relations are discovered, for each corresponding *ESSP* problem function *ComputeRegionsESSP*, which finds regions solving the *ESSP* problem, is applied. Since the problem of finding minimal regions which solve *ESSP* problem is known to be NP-complete, this is the most time complex part of Algorithm 1. However, in contrast to the original synthesis approach, we do not solve *ESSP* problems for all the events in the net reducing the size of the problem. For instance, let $\overline{E} \subseteq E$ be the set of events that need to be checked. Let $e \in \overline{E}$, and S' be the states that have incoming or outgoing transitions labeled by e. Then we need to consider $O(2^{|S|-|S'|})$ regions, such that e enters them (for $|S'|$ states their inclusion to the region is predefined) and $O(2^{|S|-\lceil \frac{|S'|}{2} \rceil})$ regions, such that e does not cross them (for $\frac{|S'|}{2}$ states their inclusion to the region depends on the other $\frac{|S'|}{2}$ states). Hence we need to consider $O(|\overline{E}| \cdot 2^{|S|-\lceil \frac{|S'|}{2} \rceil})$ ($|S'|$ is the minimal for all the events from $\overline{E}$) possible regions in contrast to the original region-based approach which exhaustively considers all $O(2^{|S|})$ possible regions. Finally, if new regions solving *ESSP* problems are found, function *AddNewConstraints* is applied and corresponding constraints (places) are added to the target workflow net N'.

5.3   Formal Properties

In this subsection, we prove formal properties of Algorithm 1. Firstly, we study the relation between the languages of the initial and target workflow nets. Theorem 2 proves that if a trace *fits* the initial model (initial model accepts the

---

[3] Transition system can be constructed from the event log as a prefix-tree [3] with subsequent minimization [19].

Algorithm 1: RepairFreeChoiceWorkflowNet

Input: Free-choice workflow net N; Event log L.
Output: Repaired net N' obtained from N by inserting additional non-local
  constraints.

1 /* Construct minimal transition system */
2 $TS \leftarrow$ ConstructMinTS(L);

3 /* Compute ESSP problems */
4 $ESSPProblems \leftarrow$ FindFalseFreeChoiceRelations(N,TS);

5 $N' \leftarrow N$;

6 **foreach** (s, e) *from ESSPProblems* **do**
7   /* Solve $ESSP(s, e)$ */
8      $Y \leftarrow$ ComputeRegionsESSP(TS,s,e);

9   **if** $(Y \neq \emptyset)$ **then**
10   /* $ESSP(s, e)$ has been solved */
11          $N' \leftarrow$ AddNewConstraints(N', Y);

12 **end**
13 **return** N'

---

trace), it also *fits* the target model. Although the proof seems trivial, we need to consider different cases in order to verify that the final marking with only one token in the final place is reached.

Theorem 2 (Fitness). *Let $\sigma \in L$ be a trace of an event log $L \in E^*$, and $N = (P, T, F, l)$, $l : T \to E$ be a free-choice workflow net, such that its language contains σ, i.e., $\sigma \in \mathcal{L}(N)$. Workflow net $N' = (P \cup P', T, F', l)$, $l : T \to E$, is obtained from N and L using Algorithm 1. Then the language of N' contains σ, i.e., $\sigma \in \mathcal{L}(N')$.*

Proof. Let us prove that an insertion of a single place by Algorithm 1 preserves the ability of the workflow net to accept trace σ. Consider a place r (Fig. 7 b.) constructed from the corresponding region r (Fig. 7 a.) with entering events $b_1, ..., b_m$ and exiting events $a_1, ..., a_p$. Events $a_1, ..., a_p$ can belong to a larger set of events $a_1, ..., a_p, ..., a_k$ which are in a free-choice relation within N. Let us consider the workflow net N' with a new place r (the fragment of N' is presented in Fig. 7 b.). Next, we consider the following four cases:

1. Suppose $\sigma = \langle e_1, ..., e_l \rangle \in L$ does not contain events from $\{b_1, ..., b_m\}$ and $\{a_1, ..., a_p\}$ sets. Since $\sigma \in \mathcal{L}(N)$, there is a sequence of firings in N: $[i] \xrightarrow{e_1} m_1 \xrightarrow{e_2} ... \xrightarrow{e_l} [o]$, where $[i]$ and $[o]$ are the initial and final markings of the workflow net respectively. The same sequence of firings can be repeated within the target workflow net N', because σ does not contain events from the sets $\{b_1, ..., b_m\}$ and $\{a_1, ..., a_p\}$, and the place r is not involved in this sequence of firings.

2. Now let us consider trace $\sigma = \langle e_1, ..., b_i, ..., a_j, ..., e_l \rangle$ in which each occurrence of event b_i from the set $\{b_1, ..., b_m\}$ is followed by an occurrence

of event a_j from $\{a_1, ..., a_p\}$. Similarly, for the firing sequence within N: $[i] \xrightarrow{e_1} m_1 \xrightarrow{e_2} ... \xrightarrow{b_i} m_i \rightarrow ... \rightarrow m_j \xrightarrow{a_j} ... \xrightarrow{e_l} [o]$, there is a corresponding sequence $[i] \xrightarrow{e_1} m_1 \xrightarrow{e_2} ... \xrightarrow{b_i} m_i' \rightarrow ... \rightarrow m_j' \xrightarrow{a_j} ... \xrightarrow{e_l} [o]$ for N', such that $\forall p \in P : m_i'(p) = m_i(p)$, $m_i'(r) > 0$, $\forall p \in P : m_j'(p) = m_j(p)$, and $m_j'(r) > 0$.

3. Consider trace σ where an event from $\{b_1, ..., b_m\}$ is not followed by an event from $\{a_1, ..., a_p\}$. More precisely, there are two possible cases: (1) trace σ contains an event from $\{b_1, ..., b_m\}$ and does not contain an event from $\{a_1, ..., a_p\}$; (2) an occurrence of an event from set $\{b_1, ..., b_m\}$ is followed by another occurrence of an event from the same set $\{b_1, ..., b_m\}$ and only after that an event from the set $\{a_1, ..., a_p\}$ may follow. For the case (1), it is possible that the final state s_o belongs to the region r (Fig. 8a.).

Let us show that state s_o forms a region itself. Since the transition system constructed from the event log L was minimized, state s_o consolidates all the final states of the initial transition system. Let $f_1, .., f_s$ be events labeling incoming transitions (Fig. 9 a.). These events correspond to workflow net transitions connected with place o by outgoing arcs (Fig. 9 b.).

If transitions labeled by these events appear in other parts of the transition system (they are not final), then the initial workflow net N does not accept traces with these events. This can be proven by the fact that N is uniquely labeled and hence for marking m reachable by firing an event from $\{f_1, .., f_s\}$ it holds that $m(o) > 0$ and $\exists p \in P : p \neq o, m(p) > 0$. Obviously, from m the final marking $[o]$ cannot be reached in a workflow net. Thus, we have shown that there is another region $r' = \{s_0\} \subseteq r$, and r is not a minimal. This contradicts Algorithm 1 which builds minimal regions, and hence $s_o \notin r$.

The other possible scenario for the cases (1) and (2), is that the trace σ does not terminate inside region r. In both cases, there is a transition labeled by an event $c \notin \{a_1, ..., a_p\}$ which exits region r (Fig. 7 b.). While it is obvious for the case (1), for the case (2) this can be proven by the fact that there are two occurrences of events from $\{b_1, ..., b_m\}$ with no occurrences of events from $\{a_1, ..., a_p\}$ in between, and hence the trace σ leaves the region r in order to enter it again with a transition labeled by an event from $\{b_1, ..., b_m\}$. Having a new exiting event $c \notin \{a_1, ..., a_p\}$ contradicts the definition of the region r which has $\{a_1, ..., a_p\}$ as a set of exiting events. Thus, we have proven that there is no such a trace in the initial event log containing an event from the set $\{b_1, ..., b_m\}$ which is not followed by an event from the set $\{a_1, ..., a_p\}$.

4. Consider the last possible case when an event from $\{a_1, ..., a_p\}$ is not preceded by an event from $\{b_1, ..., b_m\}$ in trace σ. Here again we can distinguish two situations: (1) σ contains an event from $\{a_1, ..., a_p\}$ and does not contain an event from $\{b_1, ..., b_m\}$; (2) the occurrence of an event from $\{a_1, ..., a_p\}$ is firstly preceded by another occurrence of an event from $\{a_1, ..., a_p\}$ which in its turn can be preceded by an event from $\{b_1, ..., b_m\}$. Just like in the previous case, two scenarios are possible: the trace starts inside the region r (Fig. 8a.) or there is a transition entering r and labeled by an event $d \notin \{b_1, ..., b_m\}$ (Fig. 8 b). Similarly to the previous case, we can prove that in these scenarios

r is not a minimal region with entering and exiting events $\{b_1, ..., b_m\}$ and $\{a_1, ..., a_p\}$, respectively.

Thus, we have proven that if a place corresponding to a region constructed by the Algorithm 1 is added to the initial workflow net N then all the traces form L accepted by N are also accepted by the resulting workflow net N'.   □

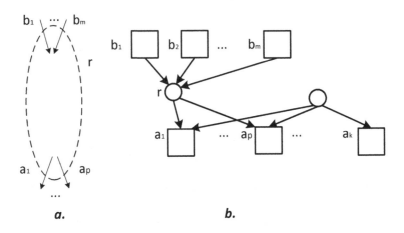

Fig. 7. a. A fragment of a transition system that encodes L. b. A fragment of N'.

The following theorem states that the resulting model cannot be less precise than the initial process model, i.e., it cannot accept new traces which were not accepted by the initial model.

Theorem 3 (Precision). *Let $N = (P, T, F, l)$, $l : T \rightarrow E$, be a free-choice workflow net and let L be an event log over set of events E. If workflow net N' is obtained from N and L by Algorithm 1, then the language of N contains the language of N', i.e., $\mathcal{L}(N') \subseteq \mathcal{L}(N)$.*

Proof. The proof follows from the well-known result that addition of new places (preconditions) can only restrict the behavior and, hence, the language of the Petri net [27].   □

Next, we formulate and prove a sufficient condition for the soundness of resulting workflow nets. This condition is formulated in terms of the state-based region theory.

Theorem 4 (Soundness).   *Let L be an event log over set E. Let $N = (P, T, F, l)$, $l : T \rightarrow E$ be a sound free-choice workflow net. Suppose that workflow net $N' = (P \cup P', T, F', l)$ is obtained from N and L by applying Algorithm 1 to one set of events in a free-choice relation within N. Suppose also*

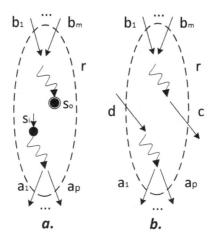

Fig. 8. Fragments of a transition system that encodes L.

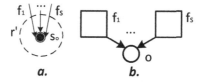

Fig. 9. Final state of N.

that $\{r^{(1)}, ..., r^{(n)}\}$ is a set of regions constructed at line of 8 Algorithm 1 in the transition system encoding L (Fig. 10 b.). Let $E_{ent}^{(1)} = \{b_1^{(1)}, ..., b_m^{(1)}\}, ..., E_{ent}^{(n)} = \{b_1^{(n)}, ..., b_t^{(n)}\}$ and $E_{exit}^{(1)} = \{a_1^{(1)}, ..., a_p^{(1)}\}, ..., E_{exit}^{(n)} = \{a_1^{(n)}, ..., a_k^{(n)}\}$ be sets of entering and exiting events for the regions $r^{(1)}, ..., r^{(n)}$ respectively. Consider unions of these sets: $E_{ent} = E_{ent}^{(1)} \cup ... \cup E_{ent}^{(n)}$ and $E_{exit} = E_{exit}^{(1)} \cup ... \cup E_{exit}^{(n)}$. If there exists a (not necessarily minimal) region r in the reachability graph of N (Fig. 10 a.) with entering and exiting sets of events E_{ent} and E_{exit}, respectively, which does not contain states corresponding to $[i]$ (initial) and $[o]$ (final) markings of N, then N' is sound.

Proof. Repeating the proof of Theorem 2 and taking into account that the initial and final states of the reachability graph of N do not belong to the region r, we can state that there is a following relation between E_{ent} and E_{exit} within $\mathcal{L}(N)$, i.e, for each trace, each occurrence of the event from E_{ent} is followed by an occurrence of the event from E_{exit} and there are no other occurrences of events from E_{ent} between them.

The firing sequences of N' which do not involve firings of transitions labeled by events from E_{ent} and E_{exit} repeat the corresponding firing sequences of N and do not violate the soundness of the model.

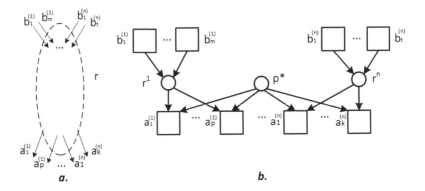

Fig. 10. a. A fragment of the reachability graph of N. b. A fragment of N'.

Let us consider a firing sequence of N' which involves firings of transitions labeled by events from E_{ent} and E_{exit}. Consider $b \in E_{ent}$, the firing sequence enabling and firing b in N': $[i] \xrightarrow{*} m'_1 \xrightarrow{b} m'_2$, corresponds to the firing sequence performed by N: $[i] \xrightarrow{*} m_1 \xrightarrow{b} m_2$, where $\forall p \in P : m_1(p) = m'_1(p)$, $m_2(p) = m'_2(p)$. Without loss of generality suppose that $b \in E_{ent}^{(i)}$, then $m'_2(r^i) = 1$, where r^i is a place constructed by Algorithm 1.

Since E_{ent} and E_{exit} events are in a following relation within $\mathcal{L}(N)$, they are in the following relation within $\mathcal{L}(N')$, because $\mathcal{L}(N') \subseteq \mathcal{L}(N)$. Consider sequences of steps leading to some of the events from E_{exit}. These firing sequences will be: $m'_2 \xrightarrow{*} m'_3$ and $m_2 \xrightarrow{*} m_3$, where $m_3(p) = m'_3(p)$ and $m'_3(r^i) = 1$, in N' and N respectively.

In model N' only transitions labeled by the events from E_{exit}^i will be enabled in m'_3, because according to Algorithm 1, the new preceding places are added only if they can be found for all the events from E_{exit}. Thus all other activities E_{exit} have their preceding places empty in the marking m'_3: $m'_3(r^j) = 0$, $i \neq j$.

In workflow net N' it holds that $m'_3(r^i) = 1$ and $m'_3(p^*) = 1$ (p^* is a choice place for the transitions in a free-choice relation within N, see Fig. 10 b.) and hence a step: $m'_3 \xrightarrow{a} m'_4$, where $a \in E_{exit}^{(i)}$ can be performed. After a is fired the place r^i is emptied. A corresponding firing step in N: $m_3 \xrightarrow{a} m_4$ can be taken, because $m_3(p^*) = 1$, and all the transitions labeled by events from E_{exit} are enabled in m_3. These steps lead models to the same markings: $\forall p \in P : m_4(p) = m'_4(p)$ from which firing the same transitions the final marking $[o]$ can be reached. If the rest sequence of firings contains events from E_{ent} and E_{exit}, we repeat the same reasoning.

Thus, we have shown that all the transitions within N' can be fired. Due to the soundness of N, since all the firing sequences of N' correspond to firing sequences of N, and the number of tokens in each place from P in corresponding markings of N' and N coincide, the final marking can be reached from any reachable marking of N' and there are no reachable markings in N' with tokens in the final place o and some other places.   □

5.4 Using High-Level Constructs to Model Discovered Non-local Constraints

In this subsection, we demonstrate how the discovered process models with non-local constraints can be presented using high-level modeling languages, such as BPMN (Business Process Model and Notation) [24]. Free-choice workflow nets can be modeled by a core set of process modeling elements that includes start and end events, tasks, parallel and choice gateways, and sequence flows. The equivalence of free-choice workflow nets and process models based on the core set of elements is studied in [2,18]. Most process modeling languages, such as BPMN, support these core elements. A BPMN model corresponding to the discovered free-choice workflow net (shown in Fig. 1) is presented in Fig. 11.

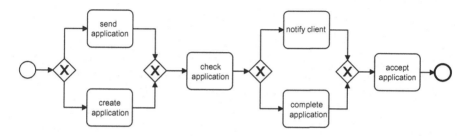

Fig. 11. A BPMN model that corresponds to the workflow net in Fig. 1.

If a workflow net is not free-choice, then it cannot be presented using core elements only [18]. However, BPMN language offers additional high-level modeling constructs which can be used to model non-free-choice constraints. Figure 12 demonstrates a BPMN model that corresponds to a non-free-choice net (in Fig. 3) constructed by Algorithm 1.

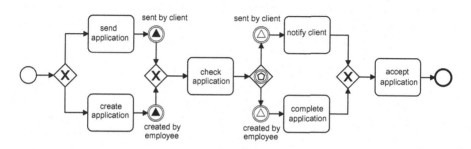

Fig. 12. A BPMN model that corresponds to the workflow net presented in Fig. 3.

In addition to core modeling elements, *signal events* and an *event-based gateway* are used. The signal events capture the discovered non-local dependencies. For instance, after *send application* task is performed, a signal *sent by client* is thrown. After that, an *event-based gateway* is used to select a branch depending on which of the catching signal event that immediately follows the gateway is fired. For example, if the type of the caught event is signal and its value is *sent by client*, then task *notify client* is performed.

6   Case Study

In this section, we demonstrate the results of applying our approach to synthetic and real-life event logs. The approach is implemented as an Apromore [20] plugin called *"Add long-distance relations"* and is available as part of Apromore Community Edition.[4] All the results were obtained in quasi real-time (in the order of milliseconds) using Intel(R) Core(TM) i7-8550U CPU @1.80 GHz with 16 GB RAM.

6.1   Synthetic Event Logs

To assess the ability of our approach to automatically repair process models we have built a set of workflow nets with non-local dependencies. An example of one of these workflow nets is presented in Fig. 13.

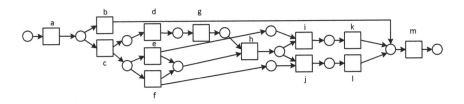

Fig. 13. A workflow net used for the synthesis of an event log.

We simulated each of the workflow nets and generated event logs containing accepted traces. After that, from each event log L we discovered a free-choice workflow net N using Split miner. Then our approach was applied to N and L producing an enhanced workflow net N' with additional constraints. To compare behaviors of N and N' workflow nets, conformance checking techniques [26] assessing fitness (the share of the log behavior accepted by a model) and precision (the share of the model behavior captured by the log) were applied. In all the cases, both models N and N' accept all the traces from L showing maximum fitness values of 1.0 (according to Theorem 2, if N accepts a trace, then N' also accepts this trace). Precision values as well as the structural characteristic of

---

4 https://apromore.org/platform/download/.

the workflow nets are presented in Table 1. These results demonstrate that our approach is able to automatically reveal hidden non-local constraints discovering precise workflow nets when applied to synthetic event logs.

Table 1. Structural (number of transitions and number of places) and behavioural characteristics (precision) of free-choice (N) and enhanced (N') workflow nets.

Event log	#Transitions/ #Places in N	#Transitions/ #Places in N'	Precision (N, L)	Precision (N', L)
1	18/14	18/18	0.972	1.0
2	13/12	13/14	0.945	1.0
3	10/9	10/11	0.899	1.0
4	12/13	12/15	0.911	1.0
5	6/4	6/6	0.841	1.0

At the same time, while other approaches for the discovery of non-free-choice workflow nets, such as α++ Miner [30] and the original Petri net synthesis technique [3] can also synthesize precise workflow nets form this set of simple event logs, they often either produce unsound workflow nets with dead transitions (in case of α++ Miner), or fail to construct a model in a reasonable time (in case of the original synthesis approach) when applied to real-life event logs. In the next subsection, we apply our approach to a real-life event log showing that our approach can discover a more precise and sound process model in a real-life setting. Note that α++ Miner produces a workflow net with dead transitions, which does not accept any of the event log traces, and the original synthesis approach fails to discover a Petri net from this real-life event log.

6.2 Real-Life Event Log

The proposed approach was applied to a real-life BPIC'2017 event log[5] of a loan application process. We analyzed car loan applications which had not been cancelled and had not passed the validation procedure at least once. The overall event log L after filtering contains 59 unique traces[6] and 12 events (each event can appear in the event log traces several times).

Figure 14 a. demonstrates a fragment of a workflow net N discovered from L by Split miner. N does not accept all the traces from L. The transition system was constructed only from those traces of L which are accepted by N. Algorithm 1 has revealed that there is a false free-choice relation between transitions labeled by *Accepted* and *Returned* events, and two additional places (regions) r_1 and r_2 were discovered and inserted in the repaired workflow net N' (Fig. 14 b.).

---

[5] https://data.4tu.nl/repository/uuid:5f3067df-f10b-45da-b98b-86ae4c7a310b.

[6] In contrast to the notion of event log used in this paper, a real-life event log can contain duplicate traces.

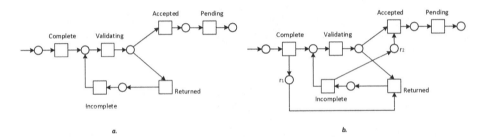

Fig. 14. a. A fragment of the workflow net N discovered from the real-life event log by Split miner. b. A fragment of the corresponding repaired workflow net N' with two places r_1 and r_2 added by Algorithm 1.

N and N' have the same fitness values (0.787), i.e., can accept the same share of traces from L, refer to Theorem 2. While their precision values are different (0.806 and 0.866, for N and N' respectively). Indeed, N' is more precise because it does not allow the sequence of events to be repeated more than once, and the repeating sequences are not presented in L traces accepted by N. The fulfillment of Theorem 4 conditions guarantees the soundness of N'.

7   Conclusion and Future Work

This paper presented an automated repair approach for obtaining precise process models under the presence of non-local dependencies. The approach identifies opportunities for improving the process model by analyzing the process behavior recorded in the input event log, and then uses goal-oriented region-based synthesis to discover new Petri net fragments that introduce non-local dependencies.

The theoretical contributions of this paper have been implemented as an open-source plugin of the Apromore process mining platform. This implementation has then been used to provide preliminary experiment results. Based on the experiments conducted so far, the proposed approach does not incur into significant performance penalties in practice. This is achieved by restricting the use of region theory to very specific situations.

We foresee different research directions arising from this work. First, implementing the proposed approach for alternative region techniques like language-based [9,31] or geometric [10,28], is an interesting avenue to explore. Second, evaluating the impact that well-known problems with event logs, like *noise* or *incompleteness*, may have on the approach, and proposing possible ways to alleviate/overcome these problems should be explored. Finally, in this paper, we only presented preliminary experimental results (e.g. we only tested the approach against a single, yet complex, real-life event log). Therefore, a concrete next step to extend this work is to perform more extensive experiments against automated discovery benchmarks such as [6].

Acknowledgments. This work was partly supported by the Australian Research Council Discovery Project DP180102839, and by MINECO and FEDER funds under grant TIN2017-86727-C2-1-R.

References

1. van der Aalst, W.: Process Mining: Data Science in Action. Springer, Heidelberg (2016). https://doi.org/10.1007/978-3-662-49851-4
2. van der Aalst, W.M.P., Hirnschall, A., Verbeek, H.M.W.: An alternative way to analyze workflow graphs. In: Pidduck, A.B., Ozsu, M.T., Mylopoulos, J., Woo, C.C. (eds.) CAiSE 2002. LNCS, vol. 2348, pp. 535–552. Springer, Heidelberg (2002). https://doi.org/10.1007/3-540-47961-9_37
3. Van der Aalst, W.M., Rubin, V., Verbeek, H., van Dongen, B.F., Kindler, E., Günther, C.W.: Process mining: a two-step approach to balance between underfitting and overfitting. Softw. Syst. Model. **9**(1), 87 (2010)
4. Armas Cervantes, A., van Beest, N.R.T.P., La Rosa, M., Dumas, M., García-Bañuelos, L.: Interactive and incremental business process model repair. In: Panetto, H., et al. (eds.) OTM 2017. LNCS, vol. 10573, pp. 53–74. Springer, Cham (2017). https://doi.org/10.1007/978-3-319-69462-7_5
5. Augusto, A., Conforti, R., Dumas, M., La Rosa, M., Polyvyanyy, A.: Split miner: automated discovery of accurate and simple business process models from event logs. Knowl. Inf. Syst. **59**(2), 251–284 (2018). https://doi.org/10.1007/s10115-018-1214-x
6. Augusto, A., et al.: Automated discovery of process models from event logs: review and benchmark. IEEE Trans. Knowl. Data Eng. **31**(4), 686–705 (2019)
7. Badouel, E., Bernardinello, L., Darondeau, P.: Petri Net Synthesis. TTCSAES. Springer, Heidelberg (2015). https://doi.org/10.1007/978-3-662-47967-4
8. Bergenthum, R.: Prime miner - process discovery using prime event structures. In: 2019 International Conference on Process Mining (ICPM), pp. 41–48, June 2019
9. Bergenthum, R., Desel, J., Lorenz, R., Mauser, S.: Synthesis of petri nets from finite partial languages. Fundam. Inform. **88**(4), 437–468 (2008)
10. Best, E., Devillers, R., Schlachter, U.: A graph-theoretical characterisation of state separation. In: Steffen, B., Baier, C., van den Brand, M., Eder, J., Hinchey, M., Margaria, T. (eds.) SOFSEM 2017. LNCS, vol. 10139, pp. 163–175. Springer, Cham (2017). https://doi.org/10.1007/978-3-319-51963-0_13
11. Carmona, J., Cortadella, J., Kishinevsky, M.: A region-based algorithm for discovering petri nets from event logs. In: Dumas, M., Reichert, M., Shan, M.-C. (eds.) BPM 2008. LNCS, vol. 5240, pp. 358–373. Springer, Heidelberg (2008). https://doi.org/10.1007/978-3-540-85758-7_26
12. Carmona, J., Cortadella, J., Kishinevsky, M.: New region-based algorithms for deriving bounded petri nets. IEEE Trans. Comput. **59**(3), 371–384 (2010)
13. Carmona, J., van Dongen, B.F., Solti, A., Weidlich, M.: Conformance Checking - Relating Processes and Models. Springer, Heidelberg (2018). https://doi.org/10.1007/978-3-319-99414-7
14. Cortadella, J., Kishinevsky, M., Lavagno, L., Yakovlev, A.: Deriving petri nets from finite transition systems. IEEE Trans. Comput. **47**(8), 859–882 (1998)
15. Desel, J., Esparza, J.: Free Choice Petri Nets. Cambridge University Press, Cambridge (1995)
16. Desel, J., Reisig, W.: The synthesis problem of petri nets. Acta Inf. **33**(4), 297–315 (1996). https://doi.org/10.1007/s002360050046

17. Fahland, D., van der Aalst, W.M.: Model repair – aligning process models to reality. Inf. Syst. **47**, 220–243 (2015)

18. Favre, C., Fahland, D., Völzer, H.: The relationship between workflow graphs and free-choice workflow nets. Inf. Syst. **47**, 197–219 (2015)

19. Hopcroft, J.E., Ullman, J.D.: An n log n algorithm for detecting reducible graphs. In: Proceedings of 6th Annual Princeton Conference on Information Sciences and Systems, pp. 119–122 (1972)

20. La Rosa, M., Reijers, H.A., Van Der Aalst, W.M., Dijkman, R.M., Mendling, J., Dumas, M., García-Bañuelos, L.: Apromore: an advanced process model repository. Expert Syst. Appl. **38**(6), 7029–7040 (2011)

21. Leemans, S.J.J., Fahland, D., van der Aalst, W.M.P.: Discovering block-structured process models from incomplete event logs. In: Ciardo, G., Kindler, E. (eds.) PETRI NETS 2014. LNCS, vol. 8489, pp. 91–110. Springer, Cham (2014). https://doi.org/10.1007/978-3-319-07734-5_6

22. Mannel, L.L., van der Aalst, W.M.P.: Finding uniwired petri nets using eST-miner. In: Di Francescomarino, C., Dijkman, R., Zdun, U. (eds.) BPM 2019. LNBIP, vol. 362, pp. 224–237. Springer, Cham (2019). https://doi.org/10.1007/978-3-030-37453-2_19

23. Mitsyuk, A., Lomazova, I., Shugurov, I., van der Aalst, W.: Process model repair by detecting unfitting fragments. In: AIST 2017, pp. 301–313. CEUR Workshop Proceedings (2017)

24. OMG: Business Process Model and Notation (BPMN), Version 2.0.2, December 2013. http://www.omg.org/spec/BPMN/2.0.2

25. Polyvyanyy, A., Aalst, W.M.P.V.D., Hofstede, A.H.M.T., Wynn, M.T.: Impact-driven process model repair. ACM Trans. Softw. Eng. Methodol. **25**(4) (2016). https://doi.org/10.1145/2980764

26. Polyvyanyy, A., Solti, A., Weidlich, M., Di Ciccio, C., Mendling, J.: Monotone precision and recall for comparing executions and specifications of dynamic systems. ACM Trans. Softw. Eng. Methodol. (TOSEM) (2020). https://doi.org/http://dx.doi.org/10.1145/3387909,inpress

27. Reisig, W.: Petri Nets: An Introduction. Springer, Heidelberg (1985). https://doi.org/10.1007/978-3-642-69968-9

28. Schlachter, U., Wimmel, H.: A geometric characterisation of event/state separation. In: Khomenko, V., Roux, O.H. (eds.) PETRI NETS 2018. LNCS, vol. 10877, pp. 99–116. Springer, Cham (2018). https://doi.org/10.1007/978-3-319-91268-4_6

29. Solé, M., Carmona, J.: Process mining from a basis of state regions. In: Lilius, J., Penczek, W. (eds.) PETRI NETS 2010. LNCS, vol. 6128, pp. 226–245. Springer, Heidelberg (2010). https://doi.org/10.1007/978-3-642-13675-7_14

30. Wen, L., Aalst, W.M., Wang, J., Sun, J.: Mining process models with non-free-choice constructs. Data Min. Knowl. Discov. **15**(2), 145–180 (2007)

31. van der Werf, J.M.E.M., van Dongen, B.F., Hurkens, C.A.J., Serebrenik, A.: Process discovery using integer linear programming. In: van Hee, K.M., Valk, R. (eds.) PETRI NETS 2008. LNCS, vol. 5062, pp. 368–387. Springer, Heidelberg (2008). https://doi.org/10.1007/978-3-540-68746-7_24

32. van Zelst, S.J., van Dongen, B.F., van der Aalst, W.M.P., Verbeek, H.M.W.: Discovering workflow nets using integer linear programming. Computing **100**(5), 529–556 (2017). https://doi.org/10.1007/s00607-017-0582-5

Extensions and Model Checking

Structural Reductions Revisited

Yann Thierry-Mieg[✉] [ID]

Sorbonne Université, CNRS, LIP6, 75005 Paris, France
yann.thierry-mieg@lip6.fr

Abstract. Structural reductions are a powerful class of techniques that reason on a specification with the goal to reduce it before attempting to explore its behaviors. In this paper we present new structural reduction rules for verification of deadlock freedom and safety properties of Petri nets. These new rules are presented together with a large body of rules found in diverse literature. For some rules we leverage an SMT solver to compute if application conditions are met. We use a CEGAR approach based on progressively refining the classical state equation with new constraints, and memory-less exploration to confirm counter-examples. Extensive experimentation demonstrates the usefulness of this structural verification approach.

1 Introduction

Structural reductions can be traced back at least to Lipton's transaction reduction [13] and in the context of Petri nets to Berthelot's seminal paper [1]. A structural reduction rule simplifies the structure of the net under study while preserving properties of interest. Structural reductions are complementary of any other verification or model-checking strategies, since they build a simpler net that can be further analyzed using other methods.

The main idea in reduction rules is either to discard parts of the net or to accelerate over parts of the behaviors by fusing adjacent transitions. Reduction rules exploit the locality property of transitions to define a reduction's effect in a small neighborhood. Most rules can be adapted to support preservation of stutter-invariant temporal logic.

Structural reductions have been widely studied with generalisations that apply to many other models than Petri nets e.g. [12]. The classical reduction rules [1] include pre and post agglomeration, for which [9, 11] give broad general definitions that can be applied also to colored nets. More recently, several competitors in the Model Checking Contest have worked on the subject, [5] defines 8 reduction rules used in the tool Tapaal and [2] defines very general transition-centric reduction rules used in the tool Tina.

In this paper, we develop a new framework that integrates an SMT solver, a memory-less pseudo random walk and structural reduction rules with the goal of jointly reducing a model and a set of properties expressed as invariants. The SMT constraints and the reduction rules we propose include classic ones as well as many contributions original to this paper. These components together form a powerful reduction engine, that can in many cases push reductions to obtain an empty net and only true or false properties.

© Springer Nature Switzerland AG 2020
R. Janicki et al. (Eds.): PETRI NETS 2020, LNCS 12152, pp. 303–323, 2020.
https://doi.org/10.1007/978-3-030-51831-8_15

2   Definitions

Petri Net Syntax and Semantics

Definition 1. *Structure. A Petri net* $N = \langle \mathcal{P}, \mathcal{T}, \mathcal{W}_-, \mathcal{W}_+, m_0 \rangle$ *is a tuple where* $\mathcal{P}$ *is the set of places,* $\mathcal{T}$ *is the set of transitions,* $\mathcal{W}_- : \mathcal{P} \times \mathcal{T} \mapsto \mathbb{N}$ *and* $\mathcal{W}_+ : \mathcal{P} \times \mathcal{T} \mapsto \mathbb{N}$ *represent the pre and post incidence matrices, and* $m_0 : \mathcal{P} \mapsto \mathbb{N}$ *is the initial marking.*

Notations: We use p (resp. t) to designate a place (resp. transition) or its index dependent on the context. We let markings m be manipulated as vectors of natural with $|\mathcal{P}|$ entries. We let $\mathcal{W}_-(t)$ and $\mathcal{W}_+(t)$ for any given transition t represent vectors with $|\mathcal{P}|$ entries. $\mathcal{W}_-^\top, \mathcal{W}_+^\top$ are the transposed flow matrices, where an entry $\mathcal{W}_-^\top(p)$ is a vector of $|\mathcal{T}|$ entries. We note $\mathcal{W}_e = \mathcal{W}_+ - \mathcal{W}_-$ the integer matrix representing transition *effects*.

In vector spaces, we use $v \geq v'$ to denote $\forall i, v(i) \geq v'(i)$, and offer sum $v + v'$ and scalar product $k \cdot v$ for scalar k with usual element-wise definitions.

We note $\bullet n$ (resp. $n \bullet$) the pre set (resp. post set) of a node n (place or transition). E.g. for a transition t its pre set is $\bullet t = \{p \in \mathcal{P} \mid \mathcal{W}_-(p,t) > 0\}$. A marking m is said to *enable* a transition t if and only if $m \geq \mathcal{W}_-(t)$. A transition t is said to *read* from place p if $\mathcal{W}_-(p,t) > 0 \wedge \mathcal{W}_e(p,t) = 0$.

Definition 2. *Semantics. The semantics of a Petri net are given by the firing rule* $\xrightarrow{t}$ *that relates pairs of markings: in any marking* $m \in \mathbb{N}^{|\mathcal{P}|}$*, if* $t \in \mathcal{T}$ *satisfies* $m \geq \mathcal{W}_-(t)$*, then* $m \xrightarrow{t} m'$ *with* $m' = m + \mathcal{W}_+(t) - \mathcal{W}_-(t)$*. The reachable set* $\mathcal{R}$ *is inductively defined as the smallest subset of* $\mathbb{N}^{|\mathcal{P}|}$ *satisfying* $m_0 \in \mathcal{R}$, *and* $\forall t \in \mathcal{T}, \forall m \in \mathcal{R}, m \xrightarrow{t} m' \Rightarrow m' \in \mathcal{R}$.

Properties of Interest. We focus on deadlock detection and verification of safety properties. A net contains a deadlock if its reachable set contains a marking m in which no transition is enabled. A safety property asserts an invariant I that all reachable states must satisfy. The invariant is given as a Boolean combination ($\vee, \wedge, \neg$) of atomic propositions that can compare ($\bowtie \in \{<, \leq, =, \geq, >\}$) arbitrary weighted sum of place markings to another sum or a constant, e.g. $\sum_{p \in \mathcal{P}} \alpha_p \cdot m(p) \bowtie k$, with $\alpha_p \in \mathbb{Z}$ and $k \in \mathbb{Z}$.

In the case of safety properties, the *support* of a property is the set of places whose marking is truly used in the predicate, i.e. such that at least one atomic proposition has a non zero α_p in a sum. The support $Supp \subseteq \mathcal{P}$ of the property defines the subset $S_t \subseteq \mathcal{T}$ of *invisible* or *stuttering* transitions t satisfying $\forall p \in Supp, \mathcal{W}_e(p,t) = 0$ [1]. For safety, we are only interested in the projection of reachable markings over the variables in the support, values of places in $\mathcal{P} \setminus Supp$ are not *observable* in markings. A small support means more potential reductions, as rules mostly cannot apply to observed places or their neighborhood.

3   Property Specific Reduction

We are given a Petri net N and either a set of safety invariants or a deadlock detection query. We consider the problem of building a structurally smaller net N' and/or simpler

---

[1] This sufficient condition for being stuttering might be relaxed by a more refined examination of the property and the effect of transitions on its truth value with respect to its value initially.

properties such that the resulting properties hold on the smaller net if and only if the original properties hold on the original net. In particular, properties that can be proven or disproven reduce to true or false, and when all properties are thus simplified, an empty system N' is enough to answer the problem.

In this paper we introduce a combination of three solution strategies: 1. we try to disprove an invariant by using a memory-less exploration that can randomly or with guidance encounter counter-example states thus *under-approximate* the behavior, 2. we try to prove an invariant holds using a system of SMT constraints to *over-approximate* reachable states and 3. we use structural reduction rules that preserve the properties of interest through a transformation. These approaches reinforce and complement each other and together provide a structural decision procedure often able to reduce the system to an empty net.

We consider an over-approximation of the state space, symbolically represented as set of constraints over a system of variables encoded in an SMT solver. We use this approximation to detect unfeasible behavior.

The SMT procedure while powerful is only a semi-decision procedure in the sense that UNSAT answers *prove that the invariant* holds ($\neg I$ is *not* reachable), but SAT answers are not trusted because we work with an over-approximation of the system. The now classic CEGAR scheme [6] proposes an elegant solution to this problem, consisting in replaying the abstract candidate counter-example on the original system, to try to exhibit a concrete counter-example thus proving the invariant *does not hold*. Similarly to [17] our constraint system is able to provide along with SAT answers a Parikh firing count (see Sect. 4) that can guide to a concrete counter-example.

We thus engineered a memory-less and fast transition engine able to explore up to millions of states per second to *under-approximate* behaviors by sampling. This engine can run in pseudo-random exploration mode or can be guided by a Parikh firing count coming from the SMT engine. If it can find a reachable marking that does not satisfy I the invariant is disproved.

We combine these solutions with a set of structural reduction rules (Sect. 5), that can simplify the net by examining its structure, and provide a smaller net where parts of the behavior are removed. The resulting simplified net and set of properties can then be exported in a format homogeneous to the initial input, typically for processing by a full-blown model-checker. If all properties have been reduced to true or false, the net will be empty.

4   Proving with SMT Constraints

In this section, we define an over-approximation of the state space, symbolically manipulated as a set of constraints over a system of variables encoded in a Satisfiability Modulo Theory (SMT) solver [15]. We use this approximation to detect unfeasible behavior. We present constraints that can be progressively added to a solver to over-approximate the state space with increasing accuracy. Structural reduction rules based on a behavioral characterisation of target conditions are also possible in this context (see Sect.7).

4.1   Approximating with SMT

SMT solvers are a powerful and relatively recent technology that enables flexible modeling of constraint satisfaction problems using first order logic and rich data types and theories, as well as their combinations. We use both linear arithmetic over reals and integers (LRA and LIA) to approximate the reachable set of states by constraints over variables representing the marking of places.

As first step in all approaches, we define for each place $p \in \mathcal{P}$ a variable m_p that represents its marking. These variables are initially only constrained to be positive: $\forall p \in \mathcal{P}, m_p \geq 0$. If we know that the net is one-safe (all place markings are at most one) e.g. because the net was generated, we add that information: $\forall p \in \mathcal{P}, m_p \leq 1$.

We then suppose that we are trying to find a reachable marking that *invalidates* a given invariant property I over a support. In other words we assert that the m_p variables satisfy $\neg I$. For deadlocks, we consider the invariant I asserting that at least one transition is enabled, expressed in negative form as $\neg I = \forall t \in \mathcal{T}, \exists p \in \bullet t, m_p < \mathcal{W}_-(p,t)$, and thus reduce the Deadlock problem to Safety.

An UNSAT answer is a definitive "NO" that ensures that I is indeed an invariant of the system. A SAT answer provides a candidate marking m_c with an assignment for the m_p variables, but is unreliable since we are considering an over-approximation of the state-space. To define the state equation constraint (see below) we also add a variable n_t for each transition t that counts the number of times it fired to reach the solution state. This Parikh count provides a guide for the random explorer.

Workflow. Because we are hoping for a definitive UNSAT answer, and that we have a large set of constraints, they are fed incrementally to the solver, hoping for an early UNSAT result. In practice we check satisfiability after every few assertions, and start with the simpler constraints that do not need additional variables.

Real arithmetic is much easier to solve than integer arithmetic, and reals are an over-approximation of integers since if no solution exists in reals (UNSAT), none exists in integers either. We therefore always first incrementally add constraints using the real domain, then at the end of the procedure, if the result is still SAT, we test if the model computed (values for place markings and Parikh count) are actually integers. If not, we escalate the computation into integer domain, restarting the solver and introducing again from the simplest constraints. At the end of the procedure we either get UNSAT or a "likely" Parikh vector that can be used to cheaply test if the feasibility detected by SAT answer is actually doable.

4.2   Incremental Constraints

We now present the constraints we use to approximate the state space of the net, in order of complexity and induced variables that corresponds to the order they are fed to our solver. We progressively add generalized flow constraints, trap constraints, the state equation, read arc constraints, and finally add new "causality" constraints.

Generalized Flows. A flow $\mathcal{F}$ is a weighted sum of places $\sum_{p \in \mathcal{P}} \alpha_p \cdot m_p$ that is an invariant of the state space. A semi-flow only has positive integers as $\alpha_i \in \mathbb{N}$ coefficients while generalized flows may have both positive and negative integer coefficients $\alpha_i \in \mathbb{Z}$.

It is possible to compute all generalized flows of a net in polynomial time and space, and the number of flows is itself polynomial in the size of the net. We use for this purpose a variant of the algorithm described in [7], initially adopted from the code base of [3] and then optimized for our scenario. This provides a polynomial number of simple constraints each only having as variables a subset of places (the α_i are fixed). The constant value of the flow in any marking can be deduced from the initial marking of the net. We do not attempt to compute semi-flows as there can be an exponential number of them, but we still first assert semi-flow constraints (if any were found) before asserting generalized flows as these have far fewer solutions due to markings being positive.

Trap Constraints. In [8], to reinforce a system of constraints on reachable states, "trap constraints" are proposed. Such a constraint asserts that an initially marked trap must remain marked.

Definition 3. *Trap. A trap S is a subset of places such that any transition consuming from the set must also feed the set. $S \subseteq \mathcal{P}$ is a trap iff.*

$$\forall p \in S, \forall t \in p\bullet, \exists p' \in t \bullet \wedge p' \in S.$$

The authors show that traps provide constraints that are a useful complement to state equation based approaches, as they can discard unfeasible behavior that is otherwise compatible with the state equation. The problem is that in general these constraints are worst case exponential in number. Leveraging the incremental nature of SMT solvers, we therefore propose to only introduce "useful" trap constraints, that actually contradict the current candidate solution.

We consider the candidate state m_c produced as SAT answer to previous queries, and try to contradict it using a trap constraint: we look for an initially marked trap that is empty in the solution. The search for such a potential trap can be done using a separate SMT solver instance.

For each place p in the candidate state m_c, we introduce a Boolean variable b_p that will be true when p is in the trap. We then add the trap constraints:

$\exists p \in \mathcal{P}, m_0(p) > 0 \wedge b_p$ trap was initially marked
$\forall p \in \mathcal{P}, m_c(p) > 0 \Rightarrow \neg b_p$ no finally marked places
$b_p \Rightarrow \forall t \in p\bullet, \exists p' \in t \bullet \wedge b_{p'}$ trap definition

If this problem is SAT, we have found a trap S from which we can derive a constraint expressed as $\bigvee_{p \in S} m_p > 0$ that can be added to the main solution procedure. Otherwise, no trap constraint existed that could contradict the given witness state. The procedure is iterated until no more useful trap constraints are found or UNSAT is obtained.

State Equation. The state equation [16] is one of the best known analytical approximations of the state space of a Petri net.

Definition 4. *State equation. We define for each transition $t \in \mathcal{T}$ a variable $n_t \in \mathbb{N}$. We assert that*

$$\forall p \in P, m_p = m_0(p) + \sum_{t \in p \bullet \cup \bullet p} n_t \cdot \mathcal{W}_e(p,t)$$

The state equation constraint is thus implemented by adding for each transition $t \in \mathcal{T}$ a variable $n_t \geq 0$ and then asserting for each place $p \in \mathcal{P}$ a linear constraint. Instead of considering all transitions and adding a variable for each of them, we can limit ourselves to one variable per possible transition *effect*, thus having a single variable for transitions t, t' such that $\mathcal{W}_e(t) = \mathcal{W}_e(t')$. Care must be taken when interpreting the resulting Parikh vectors however. In the worst case, the state equation adds $|\mathcal{T}|$ variables and $|\mathcal{P}|$ constraints, which can be expensive for large nets. We always start by introducing the constraints bearing on places in the support.

A side effect of introducing the state equation constraints is that we now have a candidate Parikh firing vector, in the form of the values taken by n_t variables. The variables in this Parikh vector can now be further constrained, as we now show.

Read $\Rightarrow$ Feed Constraints. A known limit of the state equation is the fact that it does not approximate read arc behavior very well, since it only reasons with actual effects $\mathcal{W}_e$ of transitions. However we can further constrain our current solution to the state equation by requiring that for any transition t used in the candidate Parikh vector, that reads from an initially insufficiently marked place p, there must be a transition t' with a positive Parikh count that feeds p.

Definition 5. *Read Arc Constraint. For each transition $t \in \mathcal{T}$, for every initially insufficiently marked place it reads from i.e. $\forall p \in \bullet t$, such that $\mathcal{W}_e(p,t) = 0 \wedge \mathcal{W}_-(p,t) > m_0(p)$, we assert that:*

$$n_t > 0 \Rightarrow \bigvee_{\{t' \in \bullet p \setminus \{t\} \,|\, \mathcal{W}_e(p,t') > 0\}} n_{t'} > 0$$

These read arc constraints are easy to compute and do not introduce any additional variables so they can usually safely be added after the problem with the state equation returned SAT, thus refining the solution.

While in practice it is rare that these additional constraints allow to conclude UNSAT, they frequently improve the feasibility of the Parikh count solution on nets that feature a lot of read arcs (possibly due to reductions), going from an unfeasible solution to one that is in fact realizable.

Causality Constraints. Solutions to the state equation may contain transition cycles, that "borrow" non-existing tokens and then return them. This leads to spurious solutions that are not in fact feasible. However, we can break such cycles of transitions if we consider the partial order that exists over the first occurrence of each transition in a potential concrete trace realizing a Parikh vector.

Indeed, any time a transition t consumes a token in place p, but p is not initially sufficiently marked, it must be the case that there is another transition t' that feeds p, and that t' *precedes* t in the trace.

We thus consider a precedes $\prec\subseteq\mathcal{T}\times\mathcal{T}$ relation between transitions that is: non-reflexive $\forall t\in\mathcal{T},\neg(t\prec t)$, transitive $\forall t_1,t_2,t_3\in\mathcal{T},t_1\prec t_2\wedge t_2\prec t_3\Rightarrow t_1\prec t_3$, anti-symmetric $\forall t_1,t_2\in\mathcal{T},t_1\prec t_2\Rightarrow\neg(t_2\prec t_1)$. This relation defines a strict partial order over transitions.

Definition 6. *Causality Constraint. We add the definition of the precedes relation to the solver. For each transition $t\in\mathcal{T}$, for each input of its input places that is insufficiently marked i.e. $\forall p\in\bullet t,\mathcal{W}_-(p,t)>m_0(p)$, we assert that*

$$n_t>0\Rightarrow\bigvee_{\{t'\in\bullet p\setminus\{t\}\,|\,\mathcal{W}_e(p,t')>0\}}(n_{t'}>0\wedge t'\prec t)$$

These constraints reflect the fact that insufficiently marked places must be fed *before* a continuation can take place. These constraints offer a good complement to the state equation as they forbid certain Parikh solutions that use a cycle of transitions and "borrow" tokens: such an Ouroboros-like cycle now needs a causal predecessor to be feasible. Our solutions still over-approximate the state-space as we are only reasoning on the first firing of each transition, and we construct conditions for each predecessor place separately, so we cannot guarantee that all input places of a transition have been *simultaneously* marked.

Notes: The addition of causal constraints forming a partial order to refine state equation based reasoning has not been proposed before in the literature to our knowledge.

To encode the *precedes* constraints in an SMT solver the approach we found most effective in practice consists in defining a new integer (or real) variable o_t for each transition t, and use strict inferior $o_{t_1}<o_{t_2}$ to model the precedes relation $t_1\prec t_2$. This remains a partial order as some o_t variables may take the same value, and avoids introducing any additional theories or quantifiers.

5 Structural Reduction Rules

This section defines a set of structural reduction rules. For each rule, we give a name and identifier; whether it is applicable to deadlock detection, safety or both; an informal description of the rule; a formal definition of the rule; a sketch of correctness where $\Rightarrow$ proves that states observably satisfying the property (or deadlocks) are not removed by the reduction, $\Leftarrow$ proves that new observable states (or deadlocks) are not added.

Deadlock detection can be stated as the invariant "at least one transition is enabled". But this typically implies that all places are in the support, severely limiting rule application. So instead we define deadlock specific reductions that consider that the support is empty, and that are mainly concerned with preserving divergent behavior (loops).

5.1 Elementary Transition Rules

Rule 1. Equal transitions modulo k
Applicability: Safety, Deadlock
Description: When two transitions are equal modulo k, the larger one can be discarded.

Definition: If $\exists t, t' \in \mathcal{T}, \exists k \in \mathbb{N}, W_-(t) = k \cdot W_-(t') \wedge W_+(t) = k \cdot W_+(t')$, discard t.
Correctness: $\Rightarrow$: In any state where t is enabled, t' must be enabled, and firing k times t' leads to the same state as firing t. $\Leftarrow$: Discarding transitions cannot add states. For deadlocks, any state enabling t in the original net still has a successor by t'.

Rule 2. Dominated transition
Applicability: Safety, Deadlock
Description: When a transition t has the same effect as t' but more preconditions, which can happen due to read arc behavior, t can be discarded.
Definition: If $\exists t, t' \in \mathcal{T}, W_e(t) = W_e(t') \wedge W_-(t) \geq W_-(t')$, discard t.
Correctness: $\Rightarrow$: any state that enables t also enables t', and the resulting state is the same. $\Leftarrow$: Discarding transitions cannot add states. For deadlocks, any state enabling t in the original net still has successors by t' in the resulting net.

Rule 3. Redundant Composition
Applicability: Safety, Deadlock
Description: When a transition t has the same effect and more input places than a composition $t_1.t_2$, where t_1 enabled implies $t_1.t_2$ is enabled, t can be discarded.
Definition: If $\exists t, t_1, t_2 \in \mathcal{T}, W_-(t) \geq W_-(t_1), W_+(t_1) \geq W_-(t_2), W_e(t) = W_e(t_1) + W_e(t_2)$, discard t.
Correctness: $\Rightarrow$: any state that enables t also enables t_1, t_2 is necessarily enabled after firing t_1, and the state reached by the sequence $t_1.t_2$ is the same as reached by t. $\Leftarrow$: Discarding transitions cannot add states. For deadlocks, any state enabling t in the original net still has successors by t_1 in the resulting net.
Notes: This pattern could be extended to compositions of more transitions, but there is a risk of explosion as we end up exploring intermediate states of the trace. [2] notably proposes a more general version that subsumes the four first rules presented here but is more costly to evaluate.

Rule 4. Neutral transition
Applicability: Safety
Description: When a transition has no effect it can be discarded.
Definition: If $\exists t \in \mathcal{T}, W_-(t) = W_+(t)$, discard t.
Correctness: $\Rightarrow$: any state reachable by a firing sequence using t can also be reached without firing t. $\Leftarrow$: Discarding transitions cannot add states.

Rule 5. Sink Transition
Applicability: Safety
Description: A transition t that has no outputs and is stuttering can be discarded.
Definition: If $\exists t \in \mathcal{S}_t, t\bullet = \emptyset$, discard t.
Correctness: $\Rightarrow$: any state reached by firing t enables less subsequent behaviors since tokens were consumed by t. Any firing sequence of the original net using t is still possible in the new net if occurrences of t are removed, and leads to a state satisfying the same properties because t stutters. $\Leftarrow$: Discarding transitions cannot add states.

Rule 6. Source Transition
Applicability: Deadlock
Description: If a transition has no input places then the net has no reachable deadlocks.

Definition: If $\exists t \in \mathcal{T}, \bullet t = \emptyset$, discard all places and discard all transitions except t.
Correctness: $\Rightarrow$: Since t is fireable in any reachable state, there cannot be reachable deadlock states. $\Leftarrow$: The resulting model has no deadlocks, like the original model.

5.2 Elementary Place Rules

Rule 7. Equal places modulo k
Applicability: Deadlock, Safety
Description: When two places are equal modulo k (flow matrices and initial marking), either one can be discarded (we discard the larger one).
Definition: If $\exists p, p' \in \mathcal{P} \setminus Supp, \exists k \in \mathbb{N}, m_0(p) = k \cdot m_0(p'), \mathcal{W}_-^\top(p) = k \cdot \mathcal{W}_-^\top(p') \wedge \mathcal{W}_+^\top(p) = k \cdot \mathcal{W}_+^\top(p')$, discard p.
Correctness: $\Rightarrow$: Removing a place cannot remove any behavior. $\Leftarrow$: Inductively we can show that $m(p) = k \cdot m(p')$ in any reachable marking m. Thus enabling conditions on output transitions t of $p\bullet = p'\bullet$ are always equivalent: either p and p' are both insufficiently marked or both are sufficiently marked to let t fire. Removing one of these two conditions thus does not add any behavior.

Rule 8. Sink Place
Applicability: Deadlock, Safety
Description: When a place p has no outputs, and is not in the support of the property, it can be removed.
Definition: If $\exists p \in \mathcal{P} \setminus Supp, p\bullet = \emptyset$, discard p
Correctness: $\Rightarrow$: Removing a place cannot remove any behavior. $\Leftarrow$: Since the place had no outputs it already could not enable any transition in the original net.

Rule 9. Constant place
Applicability: Deadlock, Safety
Description: When a place's marking is constant (typically because of read arc behavior), the place can be removed and the net can be simplified by "evaluating" conditions on output transitions.
Definition: If $\exists p \in \mathcal{P}, \mathcal{W}_-^\top(p) = \mathcal{W}_+^\top(p)$, discard p and any transition $t \in \mathcal{T}$ such that $\mathcal{W}_-(p,t) > m_0(p)$.
Correctness: $\Rightarrow$: Removing a place cannot remove any behavior. The transitions discarded could not be enabled in any reachable marking so no behavior was lost. $\Leftarrow$: The remaining transitions of $p\bullet$ have one less precondition, but it evaluated to true in all reachable markings, so no behavior was added. Discarding transitions cannot add states.
Notes: This reduction also applies to places in the support, leading to simplification of the related properties.

Rule 10. Maximal Unmarked Siphon
Applicability: Deadlock, Safety
Description: An unmarked *siphon* is a subset of places that are not initially marked and never will be in any reachable state. These places can be removed and adjacent transitions can be simplified away.

Definition: A maximal unmarked siphon $S \subseteq \mathcal{P}$ can be computed by initializing with the set of initially unmarked places $S = \{p \in \mathcal{P} \mid m_0(p) = 0\}$, and $T \subseteq \mathcal{T}$ with the full set $\mathcal{T}$ then iterating:

- Discard from T any transition that has no outputs in S, $t \in T, t \bullet \cap S = \emptyset$,
- Discard from T any transition t that has no inputs in S and discard all of t's output places from S. So $\forall t \in T$, if $\bullet t \cap S = \emptyset$, discard t from T and discard $t \bullet$ from S,
- iterate until a fixed point is reached.

If S is non-empty, discard any transition t such that $\bullet t \cap S \neq \emptyset$ and all places in S.

Correctness: $\Rightarrow$: The discarded transitions were never enabled so no behavior was lost. Removing places cannot remove behavior. $\Leftarrow$: Removing transitions cannot add behavior. The places removed were always empty so they could not enable any transition.

Notes: Siphons have been heavily studied in the literature [14]. This reduction also applies to places in the support, leading to simplification of the related properties.

Rule 11. Bounded Marking Place

Applicability: Deadlock, Safety

Description: When a place p has no true inputs, i.e. all transitions effects can only reduce the marking of p, $m_0(p)$ is an upper bound on its marking that can be used to reduce adjacent transitions.

Definition: If $\exists p \in \mathcal{P}, \forall t \in \mathcal{T}, \mathcal{W}_e(p,t) \leq 0$, discard any transition $t \in \mathcal{T}$ such that $\mathcal{W}_-(p,t) > m_0(p)$

Correctness: $\Rightarrow$: Since the transitions discarded were never enabled in any reachable marking, removing them cannot lose behaviors. $\Leftarrow$: Discarding transitions cannot add behaviors.

Rule 12. Implicit Fork/Join place

Applicability: Deadlock, Safety

Description: Consider a place p not in the support that only touches two transitions: t_{fork} with two outputs (of which p) and t_{join} with two inputs (of which p). If we can prove that the only tokens that can mark the other input p' of t_{join} must result from firings of t_{fork}, p is implicit and can be discarded.

Definition: If $\exists p \in \mathcal{P} \setminus Supp, \exists t_f, t_j \in \mathcal{T}, \bullet p = \{t_f\} \wedge p \bullet = \{t_j\}, \mathcal{W}_+(p, t_f) = \mathcal{W}_-(p, t_j) = 1, t_f \bullet = \{p, p''\} \wedge \mathcal{W}_+(p'', t_f) = 1, \bullet t_j = \{p, p'\} \wedge \mathcal{W}_-(p', t_j) = 1$, then if p' is *induced by* t_f, discard p.

We use a simple recursive version providing sufficient conditions for the test "is p induced by t":

- If p has t as single input and $\mathcal{W}_+(p,t) = 1$, return true.
- If p has a single input t' and $\mathcal{W}_+(p,t') = 1$, if there exists any input p' of t', such that $\mathcal{W}_+(p', t') = 1$, and (recursively) p' is induced by t return true. Else false.

Correctness: $\Rightarrow$: Removing a place cannot remove any behavior. $\Leftarrow$: In any marking that disabled t_j, either both p and p' were unmarked, or only p' was unmarked. Removing the condition on p thus does not add behavior.

Notes: There are many ways we could refine the "induced by" test, and widen the application scope, but the computation should remain fast. We opted here for a reasonable complexity vs. applicability trade-off. The implementation further bounds recursion depth (to 5 in the experiments), and protects against recursion on a place already in the stack. Implicit places are studied in depth in [10], the concept is used again in SMT backed Rule 21.

Rule 13. Future equivalent place
Applicability: Deadlock, Safety
Description: When two places p and p' enable isomorphic behaviors up to permutation of p and p', i.e. any transition consuming from p has an equivalent but that consumes from p', the tokens in p and p' enable the same future behaviors. We can fuse the two places into p, by redirecting arcs that feed p' to instead feed p.
Definition: We let $v \equiv_{p|p'} v'$ denote equality under permutation of elements at index p and p' of two vectors v and v'.

If $\exists p, p' \in \mathcal{P} \setminus Supp, p \bullet \cap p' \bullet = \emptyset, \forall t \in p\bullet, \mathcal{W}_-(p,t) = 1$
$\wedge \exists t' \in p'\bullet, \mathcal{W}_-(t) \equiv_{p|p'} \mathcal{W}_-(t') \wedge \mathcal{W}_+(t) \equiv_{p|p'} \mathcal{W}_+(t')$,
then $\forall t \in \bullet p'$, set $\mathcal{W}'_+(p,t) = \mathcal{W}_+(p,t) + \mathcal{W}_+(p',t)$, update initial marking to $m'_0(p) = m_0(p) + m_0(p')$, and discard p' and transitions in $p'\bullet$.
Correctness: $\Rightarrow$: Any firing sequence of the original net using transitions consuming from p' still have an image using the transitions feeding from p. These two traces are observation equivalent since neither p nor p' are in the support, so no behavior was lost. This transformation does not preserve the bounds on p's marking however. $\Leftarrow$: The constraints on having only arcs with value 1 feeding from p and not having common output transitions feeding from both p and p' ensure there is no confusion problem for the merged tokens in the resulting net; a token in p' of the original net allowed exactly the same future behaviors (up to the image permutation) as any token in p of the resulting net. Merging the tokens into p thus did not add more behaviors. Discarding transitions cannot add states.
Notes: This test can be costly, but sufficient conditions for non-symmetry allow to limit complexity and prune the search space, e.g. we group places by number of output transitions, use a sparse "equality under permutation test"... The effect can be implemented as simply moving the tokens in p' to p and redirecting arcs to p, other rules will then discard the now constant place p' and its outputs.

5.3 Agglomeration Rules

Agglomeration in p consists in replacing p and its surrounding transitions (feeders and consumers) to build instead a transition for every element in the Cartesian product $\bullet p \times p\bullet$ that represents the effect of the sequence of firing a transition in $\bullet p$ then immediately a transition in $p\bullet$. This "acceleration" of tokens in p reduces interleaving in the state space, but can preserve properties of interest if p is chosen correctly. This type of reduction has been heavily studied [11,12] as it forms a common ground between structural reductions, partial order reductions and techniques that stem from transaction reduction.

Definition 7. *Agglomeration of a place $p \in \mathcal{P}$:*
$\forall h \in \bullet p, \forall f \in p\bullet,$ *define a new transition t:*

$$\begin{cases} Let\ k = \frac{\mathcal{W}_+(p,h)}{\mathcal{W}_-(p,f)}, k \in \mathbb{N}, k \geq 1, \\ \mathcal{W}_-(t) = \mathcal{W}_-(h) + k \cdot \mathcal{W}_-(f) \wedge \mathcal{W}_+(t) = \mathcal{W}_+(h) + k \cdot \mathcal{W}_+(f) \end{cases} \tag{1}$$

Discard transitions in •p and p•. Discard place p.

Note the introduction of the k factor, reflecting how many times f can be fed by one firing of h. This factor should be a natural number for the agglomeration to be well defined. As a post processing, it is recommended to apply identity reduction Rule 1 to the set of newly created transitions, this set is much smaller than the full set of transitions but often contains duplicates.

Rule 14. Pre Agglomeration

Applicability: Deadlock, Safety

Description: Basically, we assert that once an $h \in •p$ transition becomes enabled, it will stay enabled until some tokens move into the place p by actually firing h. Transition h cannot feed any other places, so the only behaviors it enables are continuations f that feed from p. So we can always "delay" the firing of h until it becomes relevant to enable an f transition. We fuse the effect of tokens exiting p using f with its predecessor action h, build a set of $h.f$ agglomerate actions and discard p.

Definition:

$\exists p \in \mathcal{P} \setminus Supp$ p not in support
$m_0(p) = 0$ initially unmarked
$•p \cap p• = \emptyset$ distinct feeders and consumers
$•p \subseteq S_t$ feeders are stuttering

$\forall h \in •p,$

$$\begin{cases} \mathcal{W}_+(p,h) = 1, & \text{feed arc weights are one} \\ h• = \{p\} & p \text{ is the single output of } h \\ \exists p_1 \in \mathcal{P}, \mathcal{W}_+(p_1,h) < \mathcal{W}_-(p_1,h) & h \text{ is divergent free} \\ \forall p_2 \in •h, p_2• = \{h\} & h \text{ is strongly quasi-persistent} \end{cases}$$

$\forall f \in p•,$

$\{\mathcal{W}_-(p,f) = 1$ consume arc weights are one

Then perform a pre agglomeration in p.

Correctness: $\Rightarrow$: Any sequence using an h and an f still has an image using the agglomerated transition $h.f$ in the position f was found in the original sequence. Because h is invisible and only feeds p that is itself not in the support, delaying it does not lose any observable behaviors. $\Leftarrow$: Any state that is reachable in the new net by firing an agglomerate transition $h.f$ was already reachable by firing the sequence h then f in the original net, so no behavior is added.

Notes: Pre agglomeration is one of the best known rules, this version is generalized to more than one feeder or consumer and uses terminology taken from [11].

Rule 15. Post Agglomeration

Applicability: Deadlock, Safety

Description: Basically, we assert that once p is marked, it fully controls its outputs, so the tokens arriving in p necessarily have the choice of when and where they wish to go to. Provided p is not in the support and its output transitions are invisible, we can fuse the effects of feeding p with an immediate choice of what happens to those tokens after that. We fuse the effect of tokens entering p using h with a successor action f, build a set of $h.f$ agglomerate actions and discard p.

Definition: If
$\exists p \in \mathcal{P} \setminus Supp$ p not in support
$m_0(p) = 0$ initially unmarked
$\bullet p \cap p\bullet = \emptyset$ distinct feeders and consumers
$p\bullet \subseteq S_t$ consumers are stuttering

$\forall f \in p\bullet,$
$\begin{cases} \bullet f = \{p\} & \text{no other inputs to } f \\ \forall h \in \bullet p, \mathcal{W}_-(p,h)/\mathcal{W}_+(p,f) \in \mathbb{N} & \text{natural ratio constraint} \end{cases}$

Then perform a post agglomeration in p.

Correctness: $\Rightarrow$: Any sequence using an h still has an image using the agglomerated transition $h.f$ in the position h was found in the original sequence, that leads to a state satisfying the same propositions since f transitions stutter. Any sequence using an f transition must also have an h transition preceding it, and the same trace where the f immediately follows the h is feasible in both nets and leads to a state satisfying the same propositions. It is necessary that the f transitions stutter so that moving them in the trace to immediately follow h does not lead to observably different states. $\Leftarrow$: Any state that is reachable in the new net by firing an agglomerate transition $h.f$ was already reachable by firing the sequence h then f in the original net, so no behavior is added.

Notes: Post agglomeration has been studied a lot in the literature e.g. [11], this version is generalized to an arbitrary number of consumers and feeders and a natural ratio constraint on arc weights. This procedure can grow the number of transitions when both $|\bullet p|$ and $|p\bullet|$ are greater than one, which becomes more likely as agglomeration rules are applied, and can lead to an explosion in the number of transitions of the net. In practice we refuse to agglomerate when the Cartesian product size is larger than 32.

Rule 16. Free Agglomeration
Applicability: Safety
Description: Basically, we assert that all transitions h that feed p *only* feed p and are invisible. It is possible that the original net lets h fire but never enables a continuation f, these behaviors are lost since resulting $h.f$ is never enabled, making the rule only valid for safety. In the case of safety, firing h makes the net lose tokens, allowing *less* observable behaviors until a continuation $f \in p\bullet$ is fired, so the lost behavior leading to a dead end was not observable anyway. We agglomerate around p.
Definition:
$\exists p \in \mathcal{P} \setminus Supp$ p not in support
$m_0(p) = 0$ initially unmarked
$\bullet p \cap p\bullet = \emptyset$ distinct feeders and consumers
$\bullet p \subseteq S_t$ feeders are stuttering
$\forall h \in \bullet p,$

$\begin{cases} h\bullet = \{p\} & p \text{ is the single output of } h \\ \mathcal{W}_+(p,h) = 1 & \text{feed arc weights is one} \end{cases}$

$\forall f \in p\bullet,$

$\begin{cases} \mathcal{W}_-(p,f) = 1 & \text{consume arc weights is one} \end{cases}$

Then perform a free agglomeration in p.

Correctness: ⇒: If there exists a sequence using one of the f transitions in the original system, it must contain an h that precedes the f. The trace would also be possible if we delay the h to directly precede the f, because h only stores tokens in p, it cannot causally serve to mark any other place than p, and since h transitions are stuttering it leads to the same observable state in the new system. Traces that do not use an f transition are not impacted. Sequences that use an h but not an f are no longer feasible, but because h transitions stutter, the same sequence without the h that is still possible in the new system would lead to the same observable states. So no observable behavior is lost as sequences and behaviors that are lost were not observable. ⇐: Any state that is reachable in the new net by firing an agglomerate transition $h.f$ was already reachable by firing the sequence h then f in the original net, so no behavior is added.

Notes: Free agglomeration is a new rule, original to this paper that can be understood as relaxing conditions on pre-agglomeration in return for less property preservation. It is a reduction that may remove deadlocks, as it is no longer possible to fire h without f, which forbids having tokens in p that could potentially be stuck because no f can fire. After firing h the net is less powerful since we took tokens from it and placed them in p, these situations are no longer reachable.

Rule 17. Controlling Marked Place

Applicability: Deadlock, Safety

Description: A place p that is initially marked and which is the only input of its single stuttering output transition t, can be emptied using t. Since p controls its output, once it is emptied it will be post-agglomerable.

Definition: If $\exists p \in P, p\bullet = \{t\}, \bullet t = \{p\}, p \notin t\bullet, \exists k \in \mathbb{N}, m_0(p) = k \cdot W_-(p,t)$, update $m_0' = m_0 + k \cdot W_e(t)$.

Correctness: ⇒: Since t is stuttering and only consumes from p, firing it at the beginning of any firing sequence will not change the truth value of the property in the reached state. ⇐: the new initial state was already reachable in the original model.

Notes: This is the first time to our knowledge that a structural reduction rule involving token movement is proposed. This reduction also may also consume some prefix behavior as long as a single choice is available.

6   Graph-Based Reduction Rules

In this section we introduce a set of new rules that reason on a structural over approximation of the net behavior to quickly discard irrelevant behavior. The main idea is to study variants of the token flow graph underlying the net to compute when sufficient conditions for a reduction are met.

In these graphs, we use places as nodes and add edges that partly abstract away the transitions. Different types of graphs considered, all are abstractions of the structure of the net. A graph is a tuple $G = (N,E)$ where nodes N are places $N \subseteq P$ and edges E in $P \times P$ are oriented. We can notice these graphs are small, at most $|P|$ nodes, so these approaches are structural.

We consider that computing the prefix of a set of nodes, and computing strongly connected components (SCC) of the graph are both solved problems. The prefix of S is the least fixed point of the equation $\forall s \in S, \exists s', (s',s) \in E \Rightarrow s' \in S$. The SCC of a graph form a partition of the nodes, where for any pair of nodes (p, p') in a subset, there exists a path from p to p' and from p' to p. The construction of the prefix is trivial; decomposition into SCC can be computed in linear time with Tarjan's algorithm.

Rule 18. Free SCC

Applicability: Deadlock, Safety

Description: Consider a set of places P not in the support are linked by elementary transitions (one input, one output). Tokens in any of these places can thus travel freely to any other place in this SCC. We can compute such SCC, and for each one replace all places in the SCC by a single "sum" place that represents it.

Definition: We build a graph that contains a node for every place in $\mathcal{P} \setminus Supp$ and an edge from p to p' iff. $\exists t \in \mathcal{T}, \bullet t = \{p\} \wedge W_-(p, t) = 1, t \bullet = \{p'\} \wedge W_+(p', t) = 1$.

For each SCC S of size 2 or more of this graph, we define a new place p such that: $\forall t \in \mathcal{T}, W_-(p, t) = \sum_{p' \in S} W_-(p', t) \wedge W_+(p, t) = \sum_{p' \in S} W_+(p', t)$, and $m_0(p) = \sum_{p' \in S} m_0(p')$. Then we discard all places in the SCC S.

Correctness: $\Rightarrow$: Any scenario that required to mark one or more places of the SCC is still feasible (more easily) using the "sum" place; no behavior has been removed. $\Leftarrow$: The sum place in fact represents any distribution of the tokens it contains within the places of the SCC in the original net. Because these markings were all reachable from one another in the original net, the use of the abstract "sum" place does not add any behavior.

Notes: This powerful rule is computationally cheap, provides huge reductions, and is not covered by classical pre and post agglomerations. [5] has a similar rule limited to fusing two adjacent places linked by a pair of elementary transitions. This rule (and a generalization of it) is presented using a different formalization in [2].

Rule 19. Prefix of Interest: Deadlock

Applicability: Deadlock

Description: Consider the graph that represents all potential token flows, i.e. it has an edge from every input place of a transition to each of its output places. Only SCC (lakes) in this token flow graph can lead to absence of deadlocks in the system, if the net flows has no SCC (like a river), it must eventually must lose all its tokens and deadlock. Tokens and places that are initially above (feeding streams) or in an SCC are relevant, as well as tokens that can help empty an SCC (they control floodgates). The rest of the net can simply be discarded.

Definition: We build a graph G that contains a node for every place in $\mathcal{P}$ and an edge from p to p' iff. $\exists t \in \mathcal{T}, p \in \bullet t \wedge p' \in t \bullet$

We compute the set of non trivial SCC of this graph G: SCC of size two or more, or a consisting of a single place p but only if it has a true self-loop $\exists t \in \mathcal{T}, \bullet t = t \bullet = \{p\} \wedge W_-(p, t) = W_+(p, t)$. We let S contain the union of places in these non trivial SCC. We add predecessors of output transitions of this set to the set, $S \leftarrow S \cup \{\bullet t \mid \exists p \in S, \exists t \in p \bullet\}$. This step not iterated. We then compute in the graph the nodes in the prefix of S add them to this set S.

We finally discard any places that do not belong to Prefix of Interest S, as well as any transition fed by such a place. Discard all places $p, p \notin S$, and transitions in $p \bullet$.

Correctness: $\Rightarrow$: The parts of the net that are removed inevitably led to a deadlock (ending in a place with no successor transitions or being consumed) for the tokens that entered them. These tokens now disappear immediately upon entering the suffix region, correctly capturing the fact this trace would eventually lead to a deadlock in the original net. So no deadlocks have been removed. $\Leftarrow$: Any scenario leading to a deadlock must now either empty the tokens in the SCC or consist in interlocking the tokens in the SCC. Such a scenario using only transitions that were preserved was already feasible in the original net, reaching a state from which a deadlock was inevitable once tokens had sufficiently progressed in the suffix of the net that was discarded.

Notes: This very powerful rule is computationally cheap and provides huge reductions. The closest work we could find in the literature was related to program slicing rather than Petri nets. The main strength is that we ignore the structure of the discarded parts, letting us discard complex (not otherwise reducible) parts of the net. The case where S is empty because the net contains no SCC is actually relatively common in the MCC and allows to quickly conclude.

Rule 20. Prefix of Interest: Safety

Applicability: Safety

Description: Consider the graph that represents all actual token flows, i.e. it has an edge from every input place p of a transition t to each of its output places p' distinct from p, but only if t is not just reading from p'. This graph represents actual token movements and takes into account read arcs with an asymmetry. A transition consuming from p_1 to feed p_2 under the control of reading from p_3 would induce an edge from p_1 to p_2 and from p_3 to p_2, but not from p_1 to p_3. Indeed p_1 is not causally responsible for the marking in p_3 so it should not be in its prefix.

We start from places in the support of the property, which are interesting, as well as all predecessors of transitions consuming from them (these transitions are visible by definition). These places and their prefix in the graph are interesting, the rest of the net can simply be discarded.

Definition: We build a graph G that contains a node for every place in $\mathcal{P}$ and an edge from p to p' iff.

$$\exists t \in \mathcal{T}, p \in \bullet t \wedge p' \in t \bullet \wedge p \neq p' \wedge \mathcal{W}_-(p',t) \neq \mathcal{W}_+(p',t)$$

We let S contain the support of the property $S = Supp$.

We add predecessors of output transitions of this set to the set,

$S \leftarrow S \cup \{\bullet t \mid \exists p \in S, \exists t \in p\bullet\}$. This step not iterated.

We then add any place in the prefix of S to the interesting places S. We finally discard all places $p \in \mathcal{P} \setminus S$, and for each of them the transitions in $p\bullet$.

Correctness: $\Rightarrow$: The parts of the net that are removed are necessarily stuttering effects, leading to more stuttering effects. The behavior that is discarded cannot causally influence whether a given marking of the original net projected over the support is reachable or not. Any trace of the original system projected over the transitions that remain in new net is still feasible and leads to a state having the same properties as the original net. So no observable behavior has been removed. $\Leftarrow$: Any trace of the new system is also feasible in the original net, and leads to a state satisfying the same properties as in the original net. So no behavior has been added.

Notes: This very powerful rule is computationally cheap, provides huge reductions, and is not otherwise covered in the literature. Similarly to the rule for Deadlock, it can discard complex (not otherwise reducible) parts of the net. The refinement in the graph for read arcs allows to reduce parts of the net (including SCC) that are *controlled by* the places of interest, but do not themselves actually feed or consume tokens from them.

7   SMT-backed Behavioral Reduction Rules

Leveraging the over-approximation of the state space defined in Sect. 4, we now define reduction rules that test behavioral application conditions using this approximation.

Rule 21. Implicit place
Applicability: Deadlock, Safety
Description: An implicit place p never restricts any transition t in the net from firing: if t is disabled it is because some other place is insufficiently marked, never because of p. Such a place is therefore not useful, and can be discarded from the net.
Definition: Implicit place: a place p is *implicit* iff. for any transition t that consumes from p, if t is otherwise enabled, then p is sufficiently marked to let t fire. Formally,

$$\forall m \in \mathcal{R}, \forall t \in p\bullet, (\forall p' \in \bullet t \setminus \{p\}, m(p') \geq \mathcal{W}_-(p',t)) \Rightarrow m(p) \geq \mathcal{W}_-(p,t)$$

To use our SMT engine to determine if a place $p \in \mathcal{P} \setminus Supp$ is assuredly implicit, we assert:

$$\exists t \in p\bullet, m_p < \mathcal{W}_-(p,t) \wedge \forall p' \in \bullet t \setminus \{p\}, m_{p'} \geq \mathcal{W}_-(p',t)$$

If the result in UNSAT, we have successfully proved p is implicit and can discard it.

Correctness: $\Rightarrow$: Removing a place cannot remove any behavior. $\Leftarrow$: Removing the place p does not add behavior since it could not actually disable any transition.
Notes: The notion of implicit place and how to structurally or behaviorally characterise them is discussed at length in [10], but appears already in [1].

We recommend to heuristically start by testing the places that have the most output transitions, as removing them has a larger impact on the net. The order is important when two (or more) places are mutually implicit, so that each of them satisfies the criterion, but they share an output transition t that only consumes from them (so both cannot be discarded).

Rule 22. Structurally Dead Transition
Applicability: Deadlock, Safety
Description: If in any reachable marking t is disabled, it can never fire and we can discard t.
Definition: For each transition $t \in \mathcal{T}$, we use our Safety procedure to try to prove the invariant "t is disabled" negatively expressed by asserting:

$$\bigwedge_{p \in \bullet t} m_p \geq \mathcal{W}_-(p,t)$$

If the result is UNSAT, we have successfully proved t is never enabled and can discard it.

Correctness: $\Rightarrow$: t was never enabled even in the over-approximation we consider, therefore discarding it does not remove any behavior. $\Leftarrow$: Removing a transition cannot add any behavior.

Notes: Because of the refined approximation of the state space we have, this test is quite strong in practice at removing otherwise reduction resistant parts of the net.

8 Evaluation

8.1 Implementation

The implementation of the algorithms described in this paper was done in Java and relies on Z3 [15] as SMT solver. The code is freely available under the terms of Gnu GPL, and distributed from http://ddd.lip6.fr as part of the ITS-tools. Using sparse representations everywhere is critical; we work with transition based column sparse matrix (so preset and postset are sparse), and transpose them when working with places. The notations we used when defining the rules in this paper deliberately present immediate parallels with an efficient sparse implementation of markings and flow matrices. For instance, because we assume that $\forall p \in \bullet t$ is a sparse iteration we always prefer it to $\forall p \in P$ in rule definitions.

Our random explorer is also sparse, can restart (in particular if it reaches a deadlock, but not only), is more likely to fire a transition again if it is still enabled after one firing (encouraging to fully empty places), can be configured to prefer newly enabled transitions (pseudo DFS) or a contrario transitions that have been enabled a long time (pseudo BFS). It can be guided by a Parikh firing count vector, where only transitions with positive count in the vector are (pseudo randomly) chosen and counts decremented after each firing. For deadlock detection, it can also be configured to prefer successor states that have the least enabled events. These various heuristics are necessary as some states are exponentially unlikely to be reached by a pure random memory-less explorer. Variety in these heuristics where each one has a strong bias in one direction is thus desirable. After each restart we switch heuristic for the next run.

The setting of Sect. 2 is rich enough to capture the problems given in the Model Checking Contest in the *Deadlock, ReachabilityFireability* and *ReachabilityCardinality* examinations. We translate fireability of a transition t to the state based predicate $m \geq W_-(t)$. We also negate reachability properties where appropriate so that all properties are positive invariants that must hold on all states. To perform a reduction of a net and a set of safety properties, we iterate the following steps, simplifying properties and net as we progress:

1. We perform a random run to see if we can visit any counter-example markings within a time bound.
2. We perform structural reductions preserving the union of the support of remaining properties.
3. We try to prove that the remaining properties hold using the SMT based procedure.
4. If properties remain, we now have a candidate Parikh vector for each of them that we try to pseudo-randomly replay to contradict the properties that remain.

5. If the computation has not progressed yet in this iteration, we apply the more costly SMT based structural reduction rules (see Sect. 7)
6. If the problem has still not progressed in this iteration, we do a refined analysis to simplify atoms of the properties, reducing their support
7. As long as at least one step in this process has made progress, and there remain properties to be checked, we iterate the procedure.

Step 6 is trying to prove for every atomic proposition in every remaining property that the atom is invariant: its value in all states is the same as in the initial state. Any atom thus proved to be constant can then be replaced by its value in the initial state to simplify the properties and their support. This procedure is also applicable to arbitrary temporal logic formulas, as shown in [4]. We can thus in some cases even solve CTL and LTL logic formulas by reducing some of their atomic propositions to true or false.

8.2 Experimental Validation

We used the MCC2019 models and formulas, limiting ourselves to examinations where all formulas were solved in 2019. All the formulas solved by our tool agree with the control values from the contest[2]. An examination consists in a model instance and either 16 safety predicates (cardinality or fireability) or a single deadlock detection task. Model instances come from 90 distinct families of Petri nets, some of which features colors.

For deadlock detection, the approach was able to fully solve 902 (536 true, 366 false) out of 932 deadlock problem instances (96.8%), where true means a deadlock was found. For safety properties, we fully solved 1634 out 1748 examinations (93.5%), and in total reduced 27594 out of 27968 formulas to true or false (98.6%).

We limited our experiments to 12 minutes of runtime and $8GB$ of RAM. We feel this is a reasonable timeout for a filter in front of an exhaustive model-checker since the contest gives 1 hour per examination. 21 of the 2680 examinations timed out; the total runtime was $82k$ seconds thus averaging at 31 seconds per examination overall.

Of the 28496 formulas solved, the solutions were due to pseudo-random exploration or Parikh guided exploration in 17829(62%) of formulas, to structural reductions and immediate simplification for 4720(17%) formulas and the SMT procedure proved 5947(21%) formulas in total. These statistics reveal a bias in the benchmark in favor of invariants that can be disproved by a counter-example.

The reduction rules presented in this paper are all relevant on this benchmark for more than one model, and combine on top of each other to achieve superior reductions.

The problems that are not fully solved often have small state spaces on which invariants do hold, but which the SMT constraints fail to prove and our memory-less explorer cannot prove. The SMT solver is particularly useful; besides proving properties to be true, in many cases the reduction becomes stuck until SMT can prove some places to be implicit, which starts another round of reductions. The random memory-less walker also benefits hugely from reductions since they make it increasingly likely that we can observe the target situation within a reasonable number of steps.

---

[2] The raw logs and procedure to reproduce the experiment are available on https://lip6.github.io/ITSTools-web/structural.html as well as graphically rendered examples.

9   Conclusion

The approach presented in this paper combines over-approximation using an SMT solver, under-approximation by sampling with a random walk, and works with a system that is progressively simplified by property preserving structural reduction rules.

Structural approaches are a strong class of techniques to analyse Petri nets, that bypass state space explosion in many cases. Structural reductions are particularly appealing because any gain in structural complexity usually implies an exponential state space reduction. The approach presented in this paper is complementary of other verification strategies as the behavior for the given properties is preserved by the transformations: it can act as an elaborate filter in front of any verification tool.

The choice of using an SMT based solver rather than a more classical ILP engine gives us flexibility and versatility, so that extending the refinement with new constraints is relatively easy. Our current directions include some new partial agglomeration rules where only some transitions stutter, investigating other more complete ways of replaying a Parikh candidate such as the approach proposed in [17], and extending our set of reduction rules to cover more fully the new advanced rules presented in [2].

References

1. Berthelot, G.: Checking properties of nets using transformation. In: Rozenberg, G. (ed.) APN 1985. LNCS, vol. 222, pp. 19–40. Springer, Heidelberg (1985). https://doi.org/10.1007/BFb0016204
2. Berthomieu, B., Le Botlan, D., Dal Zilio, S.: Counting Petri net markings from reduction equations. Int. J. Softw. Tools Technol. Transf. **22**(2), 163–181 (2019). https://doi.org/10.1007/s10009-019-00519-1
3. Best, E., Schlachter, U.: Analysis of Petri nets and transition systems. In: ICE. EPTCS, vol. 189, pp. 53–67 (2015)
4. Bønneland, F., Dyhr, J., Jensen, P.G., Johannsen, M., Srba, J.: Simplification of CTL formulae for efficient model checking of petri nets. In: Khomenko, V., Roux, O.H. (eds.) PETRI NETS 2018. LNCS, vol. 10877, pp. 143–163. Springer, Cham (2018). https://doi.org/10.1007/978-3-319-91268-4_8
5. Bønneland, F.M., Dyhr, J., Jensen, P.G., Johannsen, M., Srba, J.: Stubborn versus structural reductions for Petri nets. J. Log. Algebr. Meth. Program. **102**, 46–63 (2019)
6. Clarke, E., Grumberg, O., Jha, S., Lu, Y., Veith, H.: Counterexample-guided abstraction refinement. In: Emerson, E.A., Sistla, A.P. (eds.) CAV 2000. LNCS, vol. 1855, pp. 154–169. Springer, Heidelberg (2000). https://doi.org/10.1007/10722167_15
7. D'Anna, M., Trigila, S.: Concurrent system analysis using Petri nets: an optimized algorithm for finding net invariants. Comput. Commun. **11**(4), 215–220 (1988)
8. Esparza, J., Melzer, S.: Verification of safety properties using integer programming: beyond the state equation. Formal Methods Syst. Des. **16**(2), 159–189 (2000)
9. Evangelista, S., Haddad, S., Pradat-Peyre, J.-F.: Syntactical colored petri nets reductions. In: Peled, D.A., Tsay, Y.-K. (eds.) ATVA 2005. LNCS, vol. 3707, pp. 202–216. Springer, Heidelberg (2005). https://doi.org/10.1007/11562948_17
10. García-Vallés, F., Colom, J.M.: Implicit places in net systems. In: PNPM, pp. 104–113. IEEE Computer Society (1999)
11. Haddad, S., Pradat-Peyre, J.: New efficient Petri nets reductions for parallel programs verification. Parallel Process. Lett. **16**(1), 101–116 (2006)

12. Laarman, A.: Stubborn transaction reduction. In: Dutle, A., Muñoz, C., Narkawicz, A. (eds.) NFM 2018. LNCS, vol. 10811, pp. 280–298. Springer, Cham (2018). https://doi.org/10.1007/978-3-319-77935-5_20

13. Lipton, R.J.: Reduction: a method of proving properties of parallel programs. Commun. ACM **18**(12), 717–721 (1975)

14. Liu, G., Barkaoui, K.: A survey of siphons in Petri nets. Inf. Sci. **363**, 198–220 (2016)

15. de Moura, L., Bjørner, N.: Z3: an efficient SMT solver. In: Ramakrishnan, C.R., Rehof, J. (eds.) TACAS 2008. LNCS, vol. 4963, pp. 337–340. Springer, Heidelberg (2008). https://doi.org/10.1007/978-3-540-78800-3_24

16. Murata, T.: State equation, controllability, and maximal matchings of Petri nets. IEEE Trans. Autom. Control **22**, 412–416 (1977)

17. Wimmel, H., Wolf, K.: Applying CEGAR to the Petri net state equation. In: Abdulla, P.A., Leino, K.R.M. (eds.) TACAS 2011. LNCS, vol. 6605, pp. 224–238. Springer, Heidelberg (2011). https://doi.org/10.1007/978-3-642-19835-9_19

Efficient Unfolding of Coloured Petri Nets Using Interval Decision Diagrams

Martin Schwarick[1], Christian Rohr[1], Fei Liu[2], George Assaf[1], Jacek Chodak[1], and Monika Heiner[1(✉)]

[1] Brandenburg Technical University (BTU), Cottbus, Germany
monika.heiner@b-tu.de
[2] South China University of Technology, Guangzhou, China
https://www-dssz.informatik.tu-cottbus.de

Abstract. We consider coloured Petri nets, qualitative and quantitative ones alike, as supported by our PetriNuts tool family, comprising, among others, Snoopy, Marcie and Spike. Currently, most analysis and simulation techniques require to unfold the given coloured Petri net into its corresponding plain, uncoloured Petri net representation. This unfolding step is rather straightforward for finite discrete colour sets, but tends to be time-consuming due to the potentially huge number of possible transition bindings. We present an unfolding approach building on a special type of symbolic data structures, called Interval Decision Diagram, and compare its runtime performance with an unfolding engine employing an off-the-shelf library to solve constraint satisfaction problems. For this comparison we use the 22 scalable coloured models from the MCC benchmark suite, complemented by a few from our own collection.

Keywords: Coloured Petri nets · Unfolding · Symbolic data structures · Interval decision diagrams

1 Motivation

We consider coloured Petri nets, comprising coloured qualitative (i.e., time-free) Petri nets ($\mathcal{PN}^\mathcal{C}$) as well as coloured quantitative ones, consisting of coloured stochastic Petri nets ($\mathcal{SPN}^\mathcal{C}$), coloured continuous Petri nets ($\mathcal{CPN}^\mathcal{C}$) and coloured hybrid Petri nets ($\mathcal{HPN}^\mathcal{C}$), as supported by our PetriNuts tool family, including, among others, the modelling and simulation tool Snoopy [9], the analysis and simulation tool Marcie [10], and the simulation tool Spike [2].

Coloured Petri nets are a powerful modelling formalism, which has been used in all sorts of applications in various fields, covering natural sciences, engineering sciences and life sciences alike. But modelling is only one side of the coin. The other side calls for exploration of the model behaviour. Petri nets come generally along with a wide range of analysis techniques, extending from structural to behavioural analysis, and may include model animation to gain some initial trust in the model behaviour, as well as a variety of sophisticated simulation techniques in the case of stochastic, continuous or hybrid Petri nets.

© Springer Nature Switzerland AG 2020
R. Janicki et al. (Eds.): PETRI NETS 2020, LNCS 12152, pp. 324–344, 2020.
https://doi.org/10.1007/978-3-030-51831-8_16

However, most analysis and simulation techniques require currently an unfolding of the given coloured Petri net into its corresponding plain, uncoloured Petri net representation. That's also the case for our tool family. Thus, $\mathcal{PN}^C$ are unfolded to $\mathcal{PN}$, $\mathcal{SPN}^C$ to $\mathcal{SPN}$, $\mathcal{CPN}^C$ to $\mathcal{CPN}$, and $\mathcal{HPN}^C$ to $\mathcal{HPN}$; see [9] for the supported modelling paradigms and the relations among them.

When we confine ourselves to coloured Petri nets with finite discrete colour sets, this unfolding step is rather straightforward from an algorithmic point of view, but tends to be fairly time-consuming due to the potentially huge number of possible transition bindings. This situation may even get worse for quantitative Petri nets, because the transitions' rate functions can be colour-dependent, too.

Initially, we followed the unfolding approach proposed in [20], which gave us our first unfolding engine, which we called *Generic unfolding*. This engine applies templates and basically uses a similar pattern matching mechanism as CPN tools. But with increasing size of the generally scaleable coloured models, we observed an annoying increase of the unfolding runtime. Thus, we suggested in [22] an unfolding approach which deploys an off-the-shelf Constraint Satisfaction Problem (CSP) solver by means of the constraint solver library Gecode [6]. Its implementation, let's call it *Gecode unfolding*, brought some relief by a notable acceleration of the unfolding step.

However, there were always some doubts whether the Gecode-based CSP approach would be the ideal solution for the problem on hand. A CSP solver must generally find *one* solution for a given CSP. In contrast, unfolding of coloured Petri nets requires to iterate efficiently over the complete solution space.

Motivated by our long-term overwhelmingly positive experience with Interval Decision Diagrams (IDD) [11,12,30,34] and inspired by the idea of IDD-based creation of net transitions defined by guarded reward structures [32], we designed an IDD-based representation of the unfolding solution space. Currently, the IDD unfolding engine is integrated into Snoopy, Marcie and Spike. All three tools deal with coloured qualitative and/or quantitative Petri Nets according to [25] and communicate via files in CANDL (Coloured Abstract Net Description Language), a human-readable proprietary text format used within the PetriNuts tool family.

In this paper we explain step by step the general approach of IDD-based unfolding; it is organised as follows. We start off with recalling the basic principles of coloured Petri nets, followed by a gentle introduction into IDDs, where we illustrate IDD-based unfolding by an easy-to-follow simple example. For those interested in more algorithmic details, we describe next the implementation strategy in pseudocode notation. We evaluate our IDD unfolding engine by comparing its runtime with the one required by the Gecode unfolding engine. For this purpose, we consider coloured models representing different application scenarios, comprising all scalable (qualitative) coloured models from the PNML benchmark suite collected over the years for the Model Checking Contest (MCC) [15], complemented by quantitative coloured models from our own collection. We conclude the paper with a brief summary and outlook on future work.

We have deliberately written this paper in an informal way, avoiding any formal definitions and thus (self-) plagiarism; instead we provide pointers to literature, where the interested reader will find the formalisation of the concepts applied in this paper. This not only allows us to cope with the given page limit, but hopefully also contributes to the paper's readability.

2   Coloured Petri Nets

We briefly recall in an informal way some of the core ideas paving the way for the success story of coloured Petri nets; see [22] for more details and formal definitions of the coloured Petri nets as applied in this paper.

Coloured Petri nets [7,18] build on plain (uncoloured) Petri nets; thus they are equally made of places, transitions and arcs, follow the same basic principles of bipartite graphs, and enjoy an execution semantics, just as their uncoloured predecessors do. Additionally, coloured Petri nets are enriched with a tailored, rather sophisticated concept of (finite) discrete data types, called colour sets and inspired by high-level programming languages. Typical colour sets provided include integer subranges and enumeration sets. These colour sets induce the necessity for a couple of related net annotations, which in turn require some basic programming skills. Annotations often make use of guards, which are technically Boolean expressions over colour variables and constants, possibly involving user-defined functions.

- *Places* - get assigned colour sets and may contain a multiset of distinguish-able tokens, each coloured with a colour of the place's colour set. The initial marking definition may involve guards to specify subsets of the place's colour set, where each colour of a given subset shall get the same amount of tokens.
- *Transitions* - get assigned guards, which must be evaluated to true for enabling the transition. The trivial guard *true* is usually not explicitly given.
- *Arcs* - get assigned colour expressions, over colour variables and constants, possibly involving user-defined functions – just like guards, but the result type of a colour expression is a multiset over the colour set of the connected place. This colour expression may also incorporate guards to specify different colour expressions for different conditions.
- *Rate functions* - Quantitative nets involve transition rate functions, which may be colour-dependent. This requires again guards to assign different rate functions for different colours.

These colour-related annotations permit, among others, to conveniently encode a regular grid where each grid position represents a specific point in space; see Fig. 1 for a related introductory example, encoding diffusion in 3D space as $\mathcal{CPN}^{\mathcal{C}}$. Here, colours serve as addresses for grid positions, or to put it differently: movement in space boils down to recolouring of tokens; see [8] for details.

Ideally, all colour-related definitions are specified in a scaleable way, so we can easily adjust a model, here the spatial resolution, by just changing a few constants. In the diffusion example, we have just one scaling constant, which

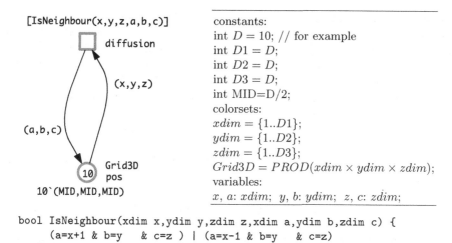

```
[IsNeighbour(x,y,z,a,b,c)]
        □  diffusion

            (x,y,z)

(a,b,c)

        ⑩  Grid3D
           pos
     10`(MID,MID,MID)
```

constants:
int $D = 10$; // for example
int $D1 = D$;
int $D2 = D$;
int $D3 = D$;
int MID=$D/2$;
colorsets:
$xdim = \{1..D1\}$;
$ydim = \{1..D2\}$;
$zdim = \{1..D3\}$;
$Grid3D = PROD(xdim \times ydim \times zdim)$;
variables:
x, a: $xdim$; y, b: $ydim$; z, c: $zdim$;

```
bool IsNeighbour(xdim x,ydim y,zdim z,xdim a,ydim b,zdim c) {
    (a=x+1 & b=y   & c=z )  | (a=x-1 & b=y   & c=z)
  | (a=x   & b=y+1 & c=z)  | (a=x   & b=y-1 & c=z)
  | (a=x   & b=y   & c=z+1)| (a=x   & b=y   & c=z-1) };
```

Fig. 1. Diffusion in 3D space as $\mathcal{CPN}^{\mathcal{C}}$. The transition guard (given in square brackets) determines by means of the colour function *IsNeighbour()*, if (a,b,c) is one of the six neighbours of (x,y,z).

is D, the size of the spatial cube. By unfolding we obtain a $\mathcal{CPN}$ made of D^3 places, $6 \cdot D^3 - 6 \cdot D^2$ transitions, and twice as many arcs.

Coloured Petri nets can be constructed from uncoloured Petri nets by folding, when the partitions of places and transitions are given. Then, these partitions define the colour sets of the coloured net. Vice versa, coloured Petri nets with finite colour sets can be automatically unfolded into uncoloured Petri nets, which then allows the application of all analysis and simulation techniques and related tools available for the corresponding unfolded Petri net class. See [26] for a formal definition of the unfolding problem. In the next section we show how unfolding can take advantage of symbolic data structures.

3   Interval Decision Diagrams

We first briefly recall in an informal way the core principles of Interval Decision Diagrams (IDD) as applied in this paper; see [31] for more details and formal definitions. Next, we illustrate the IDD-based unfolding by a simple example.

3.1   General Principles

IDDs have been first proposed in [21] and [33], probably independently. IDDs belong to the symbolic data structures and can be seen as a generalisation of the popular Binary decision diagrams (BDD). BDDs became a widely used data

structure to encode Boolean functions, while IDDs encode interval logic functions, induced by expressions of the interval logic, originally defined in [21] to
describe marking sets of P/T nets. Interval logic functions are, loosely speaking,
Boolean expressions involving atomic predicates defining integer intervals, such
as $x_1 \in [6, 8)$ or $x_2 > 0$.

Like BDDs, IDDs are Directed acyclic graphs (DAGs) with two types of nodes
– terminal and non-terminal ones. There are two terminal nodes (typically represented as boxes), labelled with 0 and 1, and the non-terminal nodes (typically
represented as circles or ellipses) are labelled with the variables occurring in the
interval logic function to be encoded. These variables have to be totally ordered,
i.e., they occur in the same order and at most once along each path from the
root to one of the two terminal nodes.

In contrast to BDDs, non-terminal nodes in IDDs may have an arbitrary
number of outgoing arcs labelled with intervals of natural numbers (including
zero) partitioning the set of natural numbers. We consider intervals which have
the form $[a, b)$; the lower bound a is included in the interval $[a, b)$, the upper
bound b not. Note that intervals of the form $[a, \infty)$ are allowed as well, see Fig. 2
for two examples.

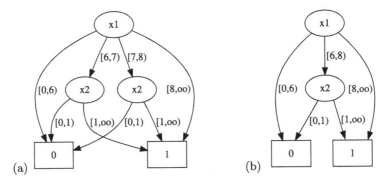

Fig. 2. Two IDDs representing $f = (x_1 \geq 8) \vee (x_1 \in [6, 8) \wedge x_2 > 0)$; **(a)** not reduced;
(b) reduced.

Reduced ordered interval decision diagrams (ROIDD) are a canonical representation for interval logic functions and often provide a compact representation
in many application areas. An IDD is called reduced if three conditions hold.

– The interval partitions labelling the outgoing arcs of each non-terminal node
  are reduced.
– Each non-terminal node has at least two different children.
– There exist no two nodes with isomorphic subgraphs.

The IDD in Fig. 2(a) is obviously not reduced. The third rule asks to merge
the two nodes labelled with $x2$, which yields, having applied the first rule as

well, the ROIDD in Fig. 2(b). These reduction rules are a straightforward generalisation of the rules for reduced ordered BDDs (ROBBD). As IDDs are a generalisation of BDDs, it does not come as a big surprise that the same issues hold with respect to the variable ordering.

- The variable ordering can have a great impact on the size of an ROIDD.
- In general, finding an optimal ordering is infeasible, even checking if a particular ordering is optimal is NP-complete.
- There exist interval logic functions that have ROIDD representations of exponential size for any variable ordering.
- Heuristics taking into consideration that variables which depend on each other should be close together in the ordering bring often good results.

Nevertheless, IDDs allow to define and implement efficient algorithms for the manipulation of interval logic functions [31,34]. All our algorithms use shared ROIDDs, an implementation principle to keep several ROIDDs within one data structure. Technically speaking, a shared ROIDD is a single multi-rooted DAG representing a collection of interval logic functions. All functions in the collection must be defined over the same set of variables, using the same variable ordering. Thanks to the canonicity of ROIDDs, two functions in the collection are identical if and only if the ROIDDs representing these functions have the same root in the shared ROIDD.

Our IDD implementations often perform fairly well, as we have seen for Marcie's model checkers for CTL [12,34] and CSL [11,30,32]. Thus, we will explore next how IDDs may possibly be of help for the unfolding problem.

3.2   IDD-Based Unfolding

All our unfolding engines proceed basically in three steps.

1. *Unfolding of coloured places* – generates for each coloured place as many unfolded places as we have colours in the place's colour set, which is also reflected in the applied naming convention for the generated unfolded places. If the initial marking of a coloured place p contains n tokens of the colour c, then the unfolded place p_c has initially n (black) tokens.
2. *Unfolding of coloured transitions* – generates an unfolded transition (transition instance) for every variable binding and connects this unfolded transition with those unfolded places which correspond to the binding. The naming convention for the generated unfolded transition reflects the variable binding.
3. *Deleting any isolated unfolded places* – Colours which are never used yield isolated places, which will never influence the net behaviour, even if initially holding tokens; thus they can be safely removed.

The first and last step are relatively easy. The core problem of efficient unfolding is to determine the transition instances, i.e. all bindings of values to the variables involved, potentially enabling coloured transitions. Fortunately, we can consider each coloured transition t separately, and the problem can be formulated as a constraint satisfaction problem (CSP), defined by

- *the set of variables* – all variables occurring on any arc adjacent to t;
- *the domain of each variable* – given by its (finite, discrete) colour set;
- *the constraints* – any guards involved, which are all Boolean expressions.

To solve the CSP, we build step-wise bottom-up a corresponding IDD, the so-called *constraint IDD*. First, the domain of each individual variable is represented as IDD; the only colour set type causing here problems is *union*. Next, the constraint IDD is constructed using standard IDD algorithms. The set of all paths going from the root to the terminal node 1 describes all solutions of the given constraint problem; typically, one path encodes more than one solution. Thus, we can easily pick all CSP solutions from the constraint IDD.

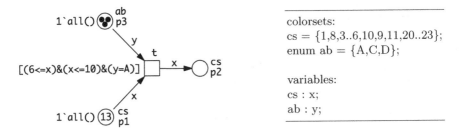

colorsets:
cs = {1,8,3..6,10,9,11,20..23};
enum ab = {A,C,D};

variables:
cs : x;
ab : y;

Fig. 3. A simple coloured Petri nets.

For illustration let's consider the simple coloured Petri net given in Fig. 3. The stepwise computation of the constraint IDD to find all instances for the single transition t is documented in Fig. 4 and comprises the following steps.

(a) Encoding the entire colour set cs, specified in the coloured Petri net by the (deliberately) unordered list $\{1, 8, 3..6, 10, 9, 11, 20..23\}$. The IDD's basic principles yield automatically an ordered list; compare all paths going to the terminal node 1.
(b) Constraining the colour set cs given in (a) to $6 \leq x$ yields the subrange comprising $\{6, 8..11, 20..23\}$.
(c) Constraining the colour set cs given in (a) to $x \leq 10$ yields the subrange comprising $\{1, 3..6, 8..10\}$.
(d) Combining (b) and (c) by the logical operator & yields the IDD for the interval logical expression $6 \leq x$ & $x \leq 10$; the corresponding subrange now comprises $\{6, 8..10\}$.
(e) Encoding the entire enumeration colour set $ab = \{A, C, D\}$ automatically involves a mapping of the enumerated constants to integer identifiers; thus, the integer identifier 0 represents A, 1 stands for C, and 2 for D.
(f) Constraining the colour set ab given in (e) to $y = A$ yields the subrange comprising the single value A, represented by its identifier 0.
(g) The final result is computed by combining (d) and (f) by the logical operator & defining all possible bindings for the transition t according to its guard

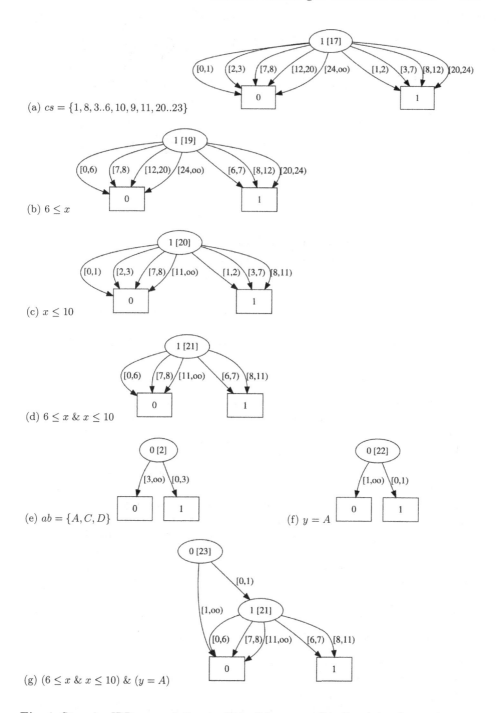

Fig. 4. Stepwise IDD computation to find all instances (bindings) for the single transition t of the coloured Petri net given in Fig. 3.

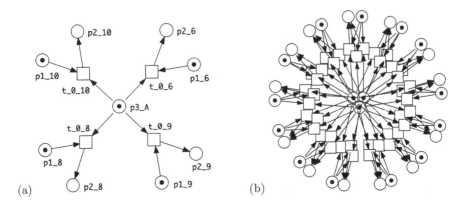

Fig. 5. **(a)** The unfolded Petri net corresponding to the coloured Petri net in Fig. 3. Transition bindings determined by the constraint IDD given in Fig. 4(g). **(b)** The unfolded Petri net for the coloured Petri net in Fig. 3 without transition guard. Layouts automatically generated by Snoopy's built-in layout library.

$(6 \leq x \ \& \ x \leq 10) \ \& \ (y = A)$. The (automatically) chosen variable order is $y \succ x$. There are two paths going to the terminal node 1. The first path encodes the value binding (A,6), and the second path encodes three value bindings: (A,8), (A,9), (A,10). Thus, in total we obtain four bindings, giving us four uncoloured transition instances for the coloured transition t.

All IDDs in Fig. 4 have been generated by a logging mechanism integrated in the unfolding engine for debugging purposes and visualised with Graphviz [4]. Non-terminal nodes are labelled with the variable index and the node number in the shared IDD (in square brackets).

Initial Marking. In Fig. 3, the two coloured places $p1$ and $p3$ are initialised with one token of each colour of the places' colour sets (by help of the pre-defined function all()). Thus, all places which we obtain by unfolding $p1$ and $p3$ are initialised with one token, and all unfolded places obtained by unfolding $p2$ remain empty. The finally generated unfolded Petri net is given in Fig. 5(a).

If the guard in the coloured Petri net in Fig. 3 were empty (meaning being simply *true*), then the cross product of the colour sets of x and y would be computed, and the unfolded Petri net would comprise $2 \cdot 13 + 3$ places and $13 \cdot 3$ transitions, each having three adjacent arcs; compare Fig. 5(b).

To deal with variables of union type, we need to consider – alternatively – all different data types subsumed by the union type, each yielding one constraint IDD. With other words: if we have two variables of a union type subsuming three types (colour sets), we obtain the solution by considering nine constraint IDDs. This might explain somehow why the union type is not really popular (anymore) in living programming languages.

As we have seen, guards define constraints. Guards, which may be arbitrarily complex, may not only serve as transition guards, as in the example just discussed, but also help to conveniently define colour sets as subsets of previously defined colour sets or to specify the initial marking in a concise and scalable way. Both needs to be considered when unfolding places. Likewise, guards also permit to specify colour-dependent transition rate functions or the arcs' conditional colour expressions. The algorithms given in the next section will take care of all of them.

4   Algorithms

This section sketches our implementation of the IDD unfolding engine by a pseudocode description; see Algorithms 1–4. The pseudocode is meant to be self-explanatory; we spend a few complementary remarks anyway.

Algorithm 1. The main procedure of the IDD unfolding engine follows the basic steps outlined previously in Sect. 3.2. But before unfolding the coloured places (line 17) and unfolding the coloured transitions (line 18), all colour-related net annotations have to be registered (lines 5–16). This comprises four categories of declarations: constants, colour sets, variables, and colour functions. Constants are crucial to design scalable and easily adjustable coloured Petri nets; thus they are often used in colour sets and colour functions.

Algorithm 1. Unfold CPN

```
 1   Net unfoldedNet
 2   placeRefTable ⊂ String × Int × Int = ∅ // (name, tokens, number of references)
 3   Environment env // some kind of registry
 4   proc UNFOLDNET (CPN net)
 5       forall c ∈ net.constants do
 6           env.registerConstant(c.name, c.expr)
 7       od
 8       forall cs ∈ net.colorsets do
 9           env.registerColorset(cs.name, cs.expr)
10       od
11       forall v ∈ net.variables do
12           env.registerVariable(v.name, v.colorset)
13       od
14       forall cf ∈ net.colorfunctions do
15           env.registerColorFunction(cf)
16       od
17       unfoldPlaces(net) // Algorithm 2
18       unfoldTransitions(net) // Algorithm 3
19       forall (place, tokens, ref) ∈ placeRefTable do
20           if ref > 0 then unfoldedNet.addPlace(place, tokens) fi
21       od
22   end
```

The actual unfolding happens in Algorithms 2 and 3, which involves setting up and solving a CSP for every place and every transition, respectively. This is here done by help of IDDs, but could be equally achieved by any other appropriate data structure. Algorithm 2 creates unfolded places, but does not add them to the unfolded net. Algorithm 3 creates unfolded transitions and their unfolded adjacent arcs and does indeed add them to the unfolded net. Afterwards, all unfolded places, which are involved in the unfolding of transitions, are actually added to the unfolded net in the final step (lines 19–21), which implicitly prunes the unfolded net by ignoring isolated places.

Algorithm 2. Unfold Places

```
1   proc UNFOLDPLACES (CPN net )
2      forall p ∈ net.places do
3         substituteColorFunctions(p.markingExpr, env) // replace fun call by its body
4         // guard used to describe subsets
5         Guard g_cs = env.implicitGuard(p.colorset)
6         Set G_p = {g_cs}
7         forall expr ∈ p.markingExpr do// separated by '++'
8            Set vars = collectVariables(markingExpr, env)
9            IDDSolutionSpaceRepr S(vars, expr.guard, env)
10           createPlaces(p, S, expr.value, expr.color)
11           G_p = G_p ∪ {expr.guard ∩ g_cs}
12        od
13        // remaining places are empty
14        IDDSolutionSpaceRepr S(vars, ⋂_{g∈G_p} ¬g, env)
15        createPlaces(p, S, 0, expr.color)
16     od
17   end
18
19   proc CREATEPLACES (Place p, IDDSolutionSpaceRepr S, ColExpr value, ColExpr color)
20      forall sol ∈ S do
21         place_s = createPlace(p, color, sol, env)
22         value_s = createValue(value, sol, env)
23         placeRefTable = placeRefTable ∪ {(place_s, value_s, 0)}
24      od
25   end
```

Algorithm 2. The unfolding of places can be done place by place and requires to determine all colours of a place's colour set. Thus, the computational load for this unfolding step depends on the kind of colour sets supported. Colour sets known by our tool family include the following.

- *Integer sets.* All colour sets are based on integer sets (to be precise: natural numbers). An integer colour set can be specified by a set of single elements or valid ranges, and may incorporate the usual set operations.
- *Enumeration types* are treated as integer sets, where all elements are given by constants.
- *Product sets.* Building on previously defined colour sets more complex, compound colour sets can be defined by means of the Cartesian product.
- *Subsets.* Given a previously defined colour set, it is possible to select specific elements characterised by a Boolean expression (*guard*). These guards are treated as implicit guards during the unfolding (line 5).

The computation of all colours for the colour set of a given place is achieved by constructing an IDD for the solution space (lines 9, 14). We obtain the solutions by following all path to the IDD's terminal node 1 (supported by a corresponding iterator concept); each solution generates an unfolded place (lines 20–24).

The creation of unfolded places includes the generation of their initial marking according to the given marking expression (lines 7–12). Places which remain empty are created afterwards (lines 14–15). Please note, places are created, but not added yet to the unfolded net.

Algorithm 3. The unfolding of transitions can be done transition by transition and requires to determine for every transition all variable bindings. To set up the corresponding CSP, the algorithm first iterates over all adjacent arcs (line 5), which are grouped into *conditions* (read arcs, inhibitory arcs, equal arcs, reset arcs) and *updates* (standard arcs connecting pre- and post-places). A transition guard may be additionally restricted by the implicit guards of any adjacent places with a subset colour set. Thus, those implicit guards have to be collected (lines 6–9). Next, all variables involved in any adjacent arc or transition guard

---

Algorithm 3. Unfold Transitions

```
 1  proc UNFOLDTRANSITIONS (CPN net )
 2    forall t ∈ net.transitions do
 3      Guard G_a = ∅
 4      // preparation step;
 5      forall (p, arcType, arcExpr) ∈ t.conditions ∪ t.updates do
 6        // guard used to describe subsets
 7        Guard g_p = env.implicitGuard(p.colorset)
 8        substituteColorFunctions(arcExpr, env) // replace fun call by its body
 9        G_a = G_a ∪ {arcExpr.guards ∩ g_p}
10      od
11      Set vars = collectVariables(t.conditions ∪ t.updates ∪ t.guard, env)
12      IDDSolutionSpaceRepr S(vars, t.guard ∩ (⋃_{g ∈ G_a} g), env)
13      // creation step
14      forall sol ∈ S do
15        Set arcs
16        forall (p, arcType, arcExpr) ∈ t.conditions ∪ t.updates do
17          arc = createArc(p, arcType, sol, arcExpr.guard, arcExpr.value, arcExpr.color)
18          if arc ≠ null then arcs = arcs ∪ {arc} fi
19        od
20        if arcs ≠ ∅ then unfoldedNet.addTransition(createTransition(t, sol, env), arcs) fi
21      od
22    od
23  end
24
25  func CREATEARC (Place p, ArcType arcType, Solution sol, Guard guard, ColExpr value, ColExpr color)
26    // guard used to describe subsets
27    Guard g_p = env.implicitGuard(p.colorset)
28    if sol ⊨ guard ∩ g_p then
29      place_s = createPlace(p, color, sol, env)
30      value_s = createValue(value, sol, env)
31      placeRefTable = placeRefTable − {(place_s, value_s, n)} ∪ {(place_s, value_s, n + 1)}
32      return Arc(place_s, arcType, value_s)
33    else
34      return null
35    fi
36  end
```

are collected (line 11), which then permits to create the IDD representation of the solution space of the given CSP (line 12).

Next, the CSP solutions are evaluated by iterating over the solution space, i.e., following all path to the IDD's terminal node 1 (lines 14–21). Every solution generally generates a set of arcs, whereby the unfolding of arcs always preserves the arc type; e.g., a coloured read arc will always be unfolded to read arcs. If there are no arcs for a given CSP solution, no unfolded transition is created (line 20).

Unfolded places will be ignored in Algorithm 1, if they are never connected to any transition. Thus, the entry in the *placeRefTable* is updated by removing the previous tuple and adding a new tuple with the number of references (i.e., usage of this place) increased by 1 (line 31).

Algorithm 4. This pseudocode is actually a data structure and provides the algorithm to construct an IDD representing the solution space for a given CSP,

---

Algorithm 4. IDD solution space representation

```
1   struct IDDSolutionSpaceRepr // Class
2       IDD solutions
3       proc constructor(Set vars, Guard guard, Environment env)
4           createVariableOrder(vars, guard)
5           solutions = 1 // the universe
6           // create the potential solution space
7           forall v ∈ vars do
8               Intset is = env.getIntColorset(v)
9               solutions = solutions ∩ makeIDD(is,env)
10          od
11          // create the actual solution space
12          solutions = solutions ∩ makeIDD(guard,env)
13      end
14
15      func makeIDD(Guard g, Environment env)
16          if g ≡ g₁ ∧ g₂ then return makeIDD(g₁,env) ∩ makeIDD(g₂,env) fi
17          if g ≡ g₁ ∨ g₂ then return makeIDD(g₁,env) ∪ makeIDD(g₂,env) fi
18          if g ≡ ¬g₁ then return 1− makeIDD(g₁,env) fi
19          if g ≡ f₁ ∘ f₂ : ∘ ∈ {=,≠,≤<,>,≥} then
20              Set vars = collectVariables(f1) ∪ collectVariables(f2)
21              IDD S = 0 // empty set
22              forall v ∈ vars do
23                  Intset is = env.getIntColorset(v)
24                  S = S ∪ makeIDD(is,env)
25              od
26              return ExtractAP(S, g)
27          fi
28      end
29  end
```

characterised by a set of variables and a guard in the context of the coloured net to be unfolded. The algorithm starts with choosing the variable order (line 4). A good variable order often depends on the specific guard involved; thus the guard occurs as parameter and is evaluated by the procedure *createVariableOrder* to determine which variables are close to each other. Next, the potential solution space is constructed by combining the colour sets of all variables involved (lines 7–10), which is afterwards restricted to the actual solution space by considering the guard (line 12). The actual IDD construction (lines 15–28) follows the standard IDD algorithms, see [31,32]. ExtractAP (Extract atomic proposition, line 26) extracts all states in S fulfilling the guard g, represented as IDD; for details see [32], Algorithm 6.

Our implementation is equipped with an iterator enabling the efficient iteration over all solutions, which is used in Algorithms 2 and 3. As a special feature, the iterator automatically updates the environment only with regard to changed variable values.

This algorithm is not IDD-specific. One could replace the type IDD just by some set type and the algorithm will work. Although IDDs often yield a very compact representation of sets and permit very efficient manipulation algorithms, it may be worth considering explicit or other symbolic data structures.

5 Experiments

The IDD-based unfolding is integrated in our `dssd_util` library used by Snoopy, Marcie and Spike. This section compares its runtime performance with our Gecode unfolding engine, i.e., with an unfolding engine employing the off-the-shelf library Gecode, which is a C++ toolkit for developing constraint-based systems [6]. Both unfolding engines are supported by our command-line tool Marcie. All experiments were done running scripts on a Linux machine in our computer lab (2.83 GHz, 32 GB RAM, running Ubuntu 18.04 LTS).

There is a growing library of (qualitative) benchmark models gathered over the years by the Petri net community for the yearly Model Checking Contest (MCC) [15]; among them are 22 scaleable coloured PNML models, for which different model versions are provided. We run our scripts over all of them. There is one model (BART) exceeding Marcie's unfolding skills; we did not get any results at all. All other models can be clearly classified into two categories.

- *No substantial runtime.* There are 13 $\mathcal{PN}^C$ with no substantial runtime for any model version; some even require a runtime below 1 sec, i.e., below a reliable measuring accuracy. A good example to illustrate this category is the *Bridges and vehicles* model.
- *Exponential runtime increase.* There are 8 $\mathcal{PN}^C$ yielding all qualitatively equal results, which is best seen in the *Family reunion* model.

To cover different application scenarios we extend our benchmark suite by two examples from our own case study collection; both are quantitative $\mathcal{PN}^C$ modelling biological processes and are scaleable with regard to some constant(s).

Thus, we consider here the following four models for our performance experiments.

Bridges and Vehicles (MCC). This $\mathcal{PN}^{\mathcal{C}}$ models an one-lane bridge of limited capacity, used by two types of vehicles. $\mathcal{PN}^{\mathcal{C}}$ size: 15 places, 11 transitions, and 57 arcs. There are three scaling parameters: the number of vehicles in each class, the bridge capacity, and the maximum number of vehicles of the same type being allowed to pass the bridge; see MCC website for details. There are 20 model versions, none requires substantial runtime, the entire experiment is completed in less than a minute; compare Fig. 6.

Family Reunion (MCC). This $\mathcal{PN}^{\mathcal{C}}$ models the reunification process, which legal permanent residents have to follow to reunite with their families. $\mathcal{PN}^{\mathcal{C}}$ size: 104 places, 66 transitions, and 198 arcs. The model is parameterised by the number of legal residents, which in turn govern four further parameters; see MCC website for details. The unfolding time is short for smaller model versions, but quickly explodes for larger model versions; compare Fig. 7.

Diffusion in 3D. Diffusion is a basic mechanism underlying many biological processes and is thus crucial for many case studies undertaken by help of our unifying framework, building on scaleable $\mathcal{SPN}^{\mathcal{C}}$, $\mathcal{CPN}^{\mathcal{C}}$ or $\mathcal{HPN}^{\mathcal{C}}$, as introduced in [8]. Here, we consider the $\mathcal{CPN}^{\mathcal{C}}$ given in Fig. 1, where each grid element in the 3D cube has six neighbours. The scaling factor is the grid size, i.e., the edge length of the 3D cube; compare Fig. 8.

Brusselator. The Brusselator [29] is a popular reaction diffusion system, typically modelled as partial differential systems (PDE). Using this case study, we have shown in [23] how to systematically encode PDEs as $\mathcal{CPN}^{\mathcal{C}}$, which then can be equally read as $\mathcal{SPN}^{\mathcal{C}}$ or $\mathcal{HPN}^{\mathcal{C}}$. The coloured Petri net modelling the Brusselator involves diffusion in 2D with four neighbours and yields exactly the same results as obtained in [29] using PDEs. $\mathcal{CPN}^{\mathcal{C}}$ size: 2 places, 6 transitions, and 11 arcs. The nodes can either be read as discrete or continuous nodes. However, this does not have any effect on the unfolding efficiency. The scaling factor is the edge length of the 2D grid; compare Fig. 9.

Discussion. The IDD engine seems to come with some initial overhead; thus it is not advisable to use it for models, not requiring some substantial runtime (more than a few seconds), see Fig. 6. This also applies for the smaller versions of all models with an exponential increase of the runtime. But IDD unfolding outperforms Gecode unfolding for larger model versions, and we observe clear margins with increasing runtime, see Fig. 7. The margins in favour of the IDD unfolding turn out to be specifically dramatically for those case studies involving diffusion, see Figs. 8 and 9. These results equally apply to all our corresponding biological case studies, see, e.g., [5,13,17,24,28], to mention just a few. Scaling is crucial for case studies modelling biological processes evolving in time and space as it brings a higher resolution of the final results. In contrast, the scaling of all other examples shown here is more an academic exercise to challenge the unfolding engines. Based on these experiments, we recommend to use Gecode unfolding for such models, using only fairly simple and smallish colour sets and

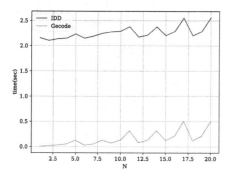

N	$\|P\|$	$\|T\|$	$\|A\|$
1	28	52	326
2	48	288	2090
4	78	968	7350
5	108	2228	17 190
9	128	1328	10 010
10	138	2348	18 090
11	168	5408	42 330
15	188	2108	15 950
16	198	3728	28 830
20	228	8588	67 470

Fig. 6. Bridges and vehicles (MCC); N – sequential model version number; Table shows selective results.

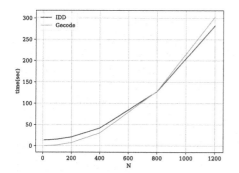

N	$\|P\|$	$\|T\|$	$\|A\|$
10	1475	1234	3799
20	3271	2753	8446
50	12 194	10 560	32 238
100	40 605	36 871	112 728
200	143 908	134 279	411 469
400	537 708	508 489	1 558 729
800	2 075 308	1 976 909	6 061 249
1200	4 612 908	4 405 329	13 507 769

Fig. 7. Family Reunion (MCC); N – legal residents.

N	$\|P\|$	$\|T\|$	$\|A\|$
5	125	600	1200
10	1000	5400	10 800
15	3375	18 900	37 800
20	8000	72 998	145 996
25	15 625	90 000	180 000
30	27 000	156 600	313 200
35	42 875	249 900	499 800
40	64 000	374 400	748 800
45	91 125	534 600	1 069 200
50	125 000	735 000	1 470 000

Fig. 8. 3D Diffusion with six neighbours; N – grid size of a 3D cube.

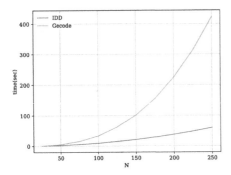

N	$|P|$	$|T|$	$|A|$
25	1250	11 908	23 191
50	5000	48 808	95 116
75	11 250	110 708	215 791
100	20 000	197 608	385 216
125	31 250	309 508	603 391
150	45 000	446 408	870 316
175	61 250	608 308	1 185 991
200	80 000	795 208	1 550 416
225	101 250	1 007 108	1 963 591
250	125 000	1 244 008	2 425 516

Fig. 9. Brusselator; N – grid size of a 2D square.

guards, and the IDD unfolding for models with more complex and larger colour sets and guards.

Reproducibility. All Petri nets used in our computational experiments are publicly available. The test cases taken from the MCC collection of coloured PNML files can be found at the MCC's website[1]. Our complementary set of test cases is provided as Snoopy and CANDL files[2]. In the same folder, one also finds a brief report documenting all computational experiments which we performed while writing this paper, and the Python scripts to repeat our computational experiments. All diagrams were generated with Python's Matplotlib [16].

6   Conclusions

In this paper we presented an approach building on symbolic data structures to efficiently unfold coloured Petri nets, qualitative and quantitative alike, into their plain, i.e. uncoloured counterparts. Our implementation builds on Interval Decision Diagrams and is available in the PetriNuts tool family for the modelling and simulation tool Snoopy, the analysis and simulation tool Marcie, and the simulation tool Spike, tailored to support reproducible simulation experiments.

We compared the runtime performance of the IDD unfolding engine with our Gecode unfolding engine; both are supported by Marcie. Our comparison incorporates all scalable coloured PNML models of the MCC benchmark suite. To the best of our knowledge, this has not been done before. We presented results of four case studies; two of the PNML benchmark suite and two of our own collection. There is no clear winner; which unfolding engine to use best seems to depend on the application scenario. The performance gain by IDD-based unfolding is specifically substantial for case studies involving non-trivial (implicit or explicit) guards. We appreciate that the exact runtime figures may

---

[1] https://mcc.lip6.fr/models.php.
[2] https://www-dssz.informatik.tu-cottbus.de/DSSZ/Software/Examples?
dir=IddUnfolding.

depend on some implementation details, however we do generally not expect a dramatic change in the total order in which the two compared unfolding methods performed in our experiments. Due to the given page constraints, we had to confine ourselves to present four case studies only; results for more case studies are available as supplementary material on our website, including all scalable coloured PNML models of the MCC. Actually, Marcie reads PNML files; thus external users can test the IDD unfolding engine themselves without having to learn first Snoopy's coloured Petri nets or our exchange format CANDL.

Related work. Another idea to exploit symbolic data structures for the unfolding problem is to represent the unfolded net symbolically, as done in [19] using Data Decision Diagrams (DDD) [3]. First, the coloured net is unfolded in a straightforward manner and represented symbolically. Next, a couple of reductions of the unfolded net are performed on its symbolic representation. These reductions include the a posteriori removal of any false-guarded transitions, which should not have been generated in the first place. This requires, according to [19], some non-trivial DDD variable reordering and turned out to be less efficient than the traditional transition enabling test as done, e.g., by the tool Maria [27], which was one of the inspirational sources for our Generic unfolding engine. If required, the symbolic representation of the reduced unfolded net can be finally transformed into an explicit representation.

It would be interesting to explore whether our IDD-based unfolding strategy could be appropriately combined with a symbolic representation of the generated unfolded net. This would be specifically helpful if the analysis and simulation algorithms could directly deal with a symbolic net representation. Such a combination might help to efficiently deal with much larger unfolded nets as we are currently able to cope with.

Future Work. Useful extensions of the current functionality of coloured Petri nets in the PetriNuts tool family include place-dependent arc expressions (self-modifying $\mathcal{PN}^\mathcal{C}$), as exploited in [14] to capture cell growth and division in the eukaryotic cell cycle.

In our experiments, we compared the competing unfolding engines w.r.t. runtime, neither memory nor power consumption have been considered so far. This might not be appropriate for some specific use cases.

Envisaged performance-related improvements include:

- *Multi-threading exploiting state-of-the-art multi-core computers.* So far, the unfolding considers coloured places/transitions sequentially. one after the other. Obviously, this could be done in parallel, as each coloured place/transition defines its own CSP, which can be all solved independently. On top, the constraint IDD for a given CSP could be computed in a data-flow driven partial order.
- *Reuse of already computed solutions.* Often, variables, guards and colour sets are the same for several places and/or transitions. In this case, a recomputation of the actual solution space is a waste of time. Instead, the IDD unfolding should reuse previously computed IDD instances.

– *Choosing a variable order strategy.* It is well known that the efficiency of any algorithm relying on symbolic data structures generally depends on the chosen variable order; compare also Sect. 3.1. The current implementation deploys a simple dependency analysis of the guards. More sophisticated heuristics may be necessary for models with higher numbers of variables; we consider the option to explore some of the ordering strategies discussed in [1].

Acknowledgement. Fei Liu has been supported by National Natural Science Foundation of China (61873094), Science and Technology Program of Guangzhou, China (201804010246), and Natural Science Foundation of Guangdong Province of China (2018A030313338).

References

1. Amparore, E., Donatelli, S., Beccuti, M., Garbi, G., Miner, A.: Decision diagrams for Petri nets: which variable ordering? In: Proceedings of PNSE 2017. CEUR Workshop Proceedings, vol. 1846, pp. 31–50. CEUR-WS.org (2017)
2. Chodak, J., Heiner, M.: Spike – reproducible simulation experiments with configuration file branching. In: Bortolussi, L., Sanguinetti, G. (eds.) CMSB 2019. LNCS, vol. 11773, pp. 315–321. Springer, Cham (2019). https://doi.org/10.1007/978-3-030-31304-3_19
3. Couvreur, J.-M., Encrenaz, E., Paviot-Adet, E., Poitrenaud, D., Wacrenier, P.-A.: Data decision diagrams for Petri net analysis. In: Esparza, J., Lakos, C. (eds.) ICATPN 2002. LNCS, vol. 2360, pp. 101–120. Springer, Heidelberg (2002). https://doi.org/10.1007/3-540-48068-4_8
4. Gansner, E., North, S.: An open graph visualization system and its applications to software engineering. Softw.: Practice Experience **30**(11), 1203–1233 (2000)
5. Gao, Q., Gilbert, D., Heiner, M., Liu, F., Maccagnola, D., Tree, D.: Multiscale modelling and analysis of planar cell polarity in the drosophila wing. IEEE/ACM Trans. Comput. Biol. Bioinform. **10**(2), 337–351 (2013)
6. Gecode Team: Gecode: Generic constraint development environment. http://www.gecode.org
7. Genrich, H.J., Lautenbach, K.: The analysis of distributed systems by means of predicate/transition-nets. In: Kahn, G. (ed.) Semantics of Concurrent Computation. LNCS, vol. 70, pp. 123–146. Springer, Heidelberg (1979). https://doi.org/10.1007/BFb0022467
8. Gilbert, D., Heiner, M., Liu, F., Saunders, N.: Colouring space - a coloured framework for spatial modelling in systems biology. In: Colom, J.-M., Desel, J. (eds.) PETRI NETS 2013. LNCS, vol. 7927, pp. 230–249. Springer, Heidelberg (2013). https://doi.org/10.1007/978-3-642-38697-8_13
9. Heiner, M., Herajy, M., Liu, F., Rohr, C., Schwarick, M.: Snoopy – a unifying Petri net tool. In: Haddad, S., Pomello, L. (eds.) PETRI NETS 2012. LNCS, vol. 7347, pp. 398–407. Springer, Heidelberg (2012). https://doi.org/10.1007/978-3-642-31131-4_22
10. Heiner, M., Rohr, C., Schwarick, M.: MARCIE – model checking and reachability analysis done efficiently. In: Colom, J.-M., Desel, J. (eds.) PETRI NETS 2013. LNCS, vol. 7927, pp. 389–399. Springer, Heidelberg (2013). https://doi.org/10.1007/978-3-642-38697-8_21

11. Heiner, M., Rohr, C., Schwarick, M., Streif, S.: A comparative study of stochastic analysis techniques. In: Proceedings of CMSB 2010, pp. 96–106. ACM Digital Library (2010)

12. Heiner, M., Rohr, C., Schwarick, M., Tovchigrechko, A.A.: MARCIE's secrets of efficient model checking. In: Koutny, M., Desel, J., Kleijn, J. (eds.) Transactions on Petri Nets and Other Models of Concurrency XI. LNCS, vol. 9930, pp. 286–296. Springer, Heidelberg (2016). https://doi.org/10.1007/978-3-662-53401-4_14

13. Herajy, M., Liu, F., Rohr, C., Heiner, M.: Coloured hybrid Petri nets: an adaptable modelling approach for multi-scale biological networks. Comput. Biol. Chem. **76**, 87–100 (2018)

14. Herajy, M., Schwarick, M., Heiner, M.: Hybrid Petri nets for modelling the eukaryotic cell cycle. In: Koutny, M., van der Aalst, W.M.P., Yakovlev, A. (eds.) Transactions on Petri Nets and Other Models of Concurrency VIII. LNCS, vol. 8100, pp. 123–141. Springer, Heidelberg (2013). https://doi.org/10.1007/978-3-642-40465-8_7

15. Hillah, L.M., Kordon, F.: Petri nets repository: a tool to benchmark and debug Petri net tools. In: van der Aalst, W., Best, E. (eds.) PETRI NETS 2017. LNCS, vol. 10258, pp. 125–135. Springer, Cham (2017). https://doi.org/10.1007/978-3-319-57861-3_9

16. Hunter, J.: Matplotlib: a 2D graphics environment. Comput. Sci. Eng. **9**(3), 90–95 (2007)

17. Ismail, A., Herajy, M., Heiner, M.: A graphical approach for hybrid modelling of intracellular calcium dynamics based on coloured hybrid Petri nets. In: Liò, P., Zuliani, P. (eds.) Automated Reasoning for Systems Biology and Medicine. CB, vol. 30, pp. 349–367. Springer, Cham (2019). https://doi.org/10.1007/978-3-030-17297-8_13

18. Jensen, K.: Coloured Petri nets and the invariant-method. Theoret. Comput. Sci. **14**(3), 317–336 (1981)

19. Kordon, F., Linard, A., Paviot-Adet, E.: Optimized colored nets unfolding. In: Najm, E., Pradat-Peyre, J.-F., Donzeau-Gouge, V.V. (eds.) FORTE 2006. LNCS, vol. 4229, pp. 339–355. Springer, Heidelberg (2006). https://doi.org/10.1007/11888116_25

20. Kristensen, L., Christensen, S.: Implementing coloured Petri nets using a functional programming language. High.-Order Symb. Comput. **17**(3), 207–243 (2004)

21. Lautenbach, K., Ridder, H.: A completion of the S-invariance technique by means of fixed point algorithms. Technical report 10–95, Universität Koblenz-Landau (1995)

22. Liu, F.: Colored Petri nets for systems biology. Ph.D. thesis, BTU Cottbus, Department of CS (2012)

23. Liu, F., Blätke, M., Heiner, M., Yang, M.: Modelling and simulating reaction-diffusion systems using coloured Petri nets. Comput. Biol. Med. **53**, 297–308 (2014)

24. Liu, F., Heiner, M.: Multiscale modelling of coupled Ca2+ channels using coloured stochastic Petri nets. IET Syst. Biol. **7**(4), 106–113 (2013)

25. Liu, F., Heiner, M., Rohr, C.: Manual for colored Petri nets in Snoopy. Technical report 02–12, BTU Cottbus, Department of Computer Science (2012)

26. Liu, F., Heiner, M., Yang, M.: An efficient method for unfolding colored Petri nets. In: Proceedings of WSC 2012. 978-1-4673-4781-5/12. IEEE (2012). http://informs-sim.org

27. Mäkelä, M.: Optimising enabling tests and unfoldings of algebraic system nets. In: Colom, J.-M., Koutny, M. (eds.) ICATPN 2001. LNCS, vol. 2075, pp. 283–302. Springer, Heidelberg (2001). https://doi.org/10.1007/3-540-45740-2_17

28. Pârvu, O., Gilbert, D., Heiner, M., Liu, F., Saunders, N., Shaw, S.: Spatial-temporal modelling and analysis of bacterial colonies with phase variable genes. ACM TOMACS **25**(2), 25p (2015)

29. Peña, B., Pérez-García, C.: Stability of Turing patterns in the Brusselator model. Phys. Rev. E **64**, 056213 (2001)

30. Schwarick, M., Heiner, M.: CSL model checking of biochemical networks with interval decision diagrams. In: Degano, P., Gorrieri, R. (eds.) CMSB 2009. LNCS, vol. 5688, pp. 296–312. Springer, Heidelberg (2009). https://doi.org/10.1007/978-3-642-03845-7_20

31. Schwarick, M., Tovchigrechko, A.: IDD-based model validation of biochemical networks. Theoret. Comput. Sci. **412**(26), 2884–2908 (2011)

32. Schwarick, M.: Symbolic on-the-fly analysis of stochastic Petri nets. Ph.D. thesis, BTU Cottbus, Department of CS (2014)

33. Strehl, K., Thiele, L.: Symbolic model checking using interval diagram techniques. Technical report, Computer Engineering and Networks Lab (TIK), ETH Zurich (1998)

34. Tovchigrechko, A.: Efficient symbolic analysis of bounded Petri nets using Interval Decision Diagrams. Ph.D. thesis, BTU Cottbus, Department of CS (2008)

Dynamic Recursive Petri Nets

Serge Haddad and Igor Khmelnitsky[✉]

LSV, Université Paris-Saclay, ENS Paris-Saclay, CNRS, Inria, Gif-sur-Yvette, France
{haddad,khmelnitsky}@lsv.fr

Abstract. In the early two-thousands, Recursive Petri nets (RPN) have been introduced in order to model distributed planning of multi-agent systems for which counters and recursivity were necessary. While having a great expressive power, RPN suffer two limitations: (1) they do not include more general features for transitions like reset arcs, transfer arcs, etc. (2) the initial marking associated with the recursive "call" only depends on the calling transition and not on the current marking of the caller. Here we introduce Dynamic Recursive Petri nets (DRPN) which address these issues. We show that the standard extensions of Petri nets for which decidability of the coverability problem is preserved are particular cases of DPRN. Then we establish that w.r.t. coverability languages, DRPN are strictly more expressive than RPN. Finally we prove that the coverability problem is still decidable for DRPN.

Keywords: Recursive Petri nets · Expressiveness · Coverability · Decidability

1 Introduction

Limitations of Petri Nets (PN). When modelling dynamic systems, formalisms that can handle infinite transition systems are required in several contexts among them: the concurrent execution of parallel sequential processes that produce and consume resources and the dynamical creation of processes. While Petri nets are well suited for specifying the first pattern, their static structure forbids the modelling of the second pattern. Furthermore even the management of an unbounded number of resources by PNs suffers some limitations: the reset operation that cleans a buffer or the transfer operation of a set of resources from one buffer to another one cannot be performed in an atomic way.

Recursive Petri Nets (RPN). This formalism has been introduced in order to address the issue of modelling dynamic structures with PN [3] (see also [13,14] for similar models). Roughly speaking, a state of an RPN consists of a tree of *threads* where the local state of each thread is a marking. Any thread fires an *elementary* or *abstract* transition. When the transition is elementary, the firing updates its marking as in Petri nets; when it is abstract, this only consumes the tokens specified by the input arcs of the transition and creates a child thread initialised with the *starting marking* of the transition. When a marking of a

© Springer Nature Switzerland AG 2020
R. Janicki et al. (Eds.): PETRI NETS 2020, LNCS 12152, pp. 345–366, 2020.
https://doi.org/10.1007/978-3-030-51831-8_17

thread covers one of the *final markings*, it may perform a *cut* transition pruning its subtree and producing in its parent the tokens specified by the output arcs of the abstract transition that created it. In RPN, reachability, is decidable [9,10] by reducing this property to several reachability problems of PNs. Furthermore, the coverability and termination problems of RPNS have the same complexity as the ones of PNs (EXPSPACE-complete, see [4]). In [11], several additional features are proposed while preserving the decidability of the verification problems.

Static Extensions of Petri Nets. In another direction, PNs have been extended by adding capabilities of transitions while the static structure given by the set of places is unchanged. The reset and transfer arcs allow to perform the corresponding operations by a single transition [2]. The arcs of self-modifying nets are labelled by expressions in such a way that the numbers of tokens consumed or produced by the transitions depend on the current marking [15]. Affine Petri nets unifies the previous extensions with a concise syntax [6]. While reachability becomes undecidable, depending on the model several properties remain decidable including coverability (implying sometimes weak restrictions).

Our Contribution. While having a great expressive power, RPN suffer two limitations: (1) they do not include more general features for transitions like reset arcs, transfer arcs, etc. (2) the initial marking associated the recursive "call" only depends on the calling transition and not on the current marking of the caller. So we introduce Dynamic Recursive Petri nets (DRPN) which address these issues. We show that the extensions of Petri nets (discussed above) for which decidability of the coverability problem is preserved are particular cases of DPRN. Then we establish that w.r.t. coverability languages, DRPN are strictly more expressive than RPN. Finally we prove that the coverability problem is still decidable for DRPN.

Outline. In Sect. 2 we introduce DRPN, illustrate their modelling capabilities and define a quasi-order between states of DRPN. In Sect. 3, we study from a theoretical point of view, expressiveness of DRPNs. In Sect. 4, we establish that the coverability problem is decidable. Finally in Sect. 5, we conclude and give some perspectives to this work. All missing proofs can be found in [8].

2   Dynamic Recursive Petri Nets

Well Quasi-Ordered Sets. A quasi-ordered set $(X, \leq)$ is *well quasi-ordered* if given any infinite sequence $(x_n)_{n \in \mathbb{N}} \in X^\omega$, there exist $i < j$ such that $x_i \leq x_j$. For instance, $\mathbb{N}^P$, where P is a finite set, equipped with the component order is well-ordered. From a computability point of view, we assume that the representation of items of X allows to decide whether $x \leq x'$ which is obviously the case for $\mathbb{N}^P$. Well quasi-ordered sets fulfill properties that we exploit here:

- Let $Y \subseteq X$. Then Y is *upward closed* if for all $x \leq x' \in X$, $x \in Y$ implies $x' \in Y$. Given an arbitrary set Y, the *upward closure* of Y, denoted $Y^\uparrow$ is defined by $Y^\uparrow = \{x' \mid \exists x \in Y \ x \leq x'\}$. The set of minimal elements of an upward closed set Y, denoted $\min(Y)$, is finite and fulfills $Y = \min(Y)^{\uparrow}$. So whenever we handle upward closed sets, they are implicitly defined by their set of minimal elements.

- Given any infinite sequence of upward closed sets $(Y_n)_{n \in \mathbb{N}} \in (2^X)^{\omega}$ where for all n, $Y_n \subseteq Y_{n+1}$, there exists n_0 such that $\bigcup_{n \in \mathbb{N}} Y_n = \bigcup_{n \leq n_0} Y_n$.
- Let f be a partial non decreasing function from X to X, f is *effective* if (1) there is an algorithm that takes as input $x \in X$, decides whether x belongs to the domain of f and, in the positive case, computes $f(x)$ and (2) there is an algorithm that takes as input $x \in X$ and computes $\min(f^{-1}(\{x\}^{\uparrow}))$.
- Let $\mathcal{F}$ be a finite set of effective functions and Y be an upward closed set. Define $Cov(\mathcal{F}, Y)$ as the smallest set C that contains Y and fulfills $\bigcup_{f \in \mathcal{F}} f^{-1}(C) \subseteq C$. Then $Cov(\mathcal{F}, Y)$ can be computed the following *backward exploration*:

$$C \leftarrow Y; \textbf{ repeat } oldC \leftarrow C; \; C \leftarrow C \cup \bigcup_{f \in \mathcal{F}} f^{-1}(C) \textbf{ until } C = oldC$$

For instance, coverability in Petri nets can be decided using this backward exploration (see for instance [7]) and we will apply it in several contexts.

Notation. Let $X \subseteq Y$ be two sets. The mapping **Id** denotes the identity mapping from X to Y where X and Y should be clear from the context.

Let us introduce dynamic recursive Petri nets (DRPN). Like a Petri net, a DRPN has a set of places P and a set of transitions T partitioned in elementary and abstract transitions (resp. T_{el} and T_{ab}). A state s of an SRPN is a tree whose vertices are labelled by markings (defined by the mapping M) and edges are labelled by transitions (defined by the mapping Λ). A transition t may fire in any vertex u provided that the marking of this vertex $M(u)$ belongs to an upward closed set Grd_t. If t is elementary then $M(u)$ is updated by applying an effective function Upd_t. If t is abstract then (1) a vertex v is created as a child of u with $\Lambda(u, v) = t$, (2) marking $M(v)$ is defined by $Beg_t(M(u))$ where Beg_t is an effective function, and (3) $M(u)$ is updated by applying an effective function $Upd_t^- \leq \textbf{Id}$. A DRPN is equipped with End, an upward closed set of $\mathbb{N}^P$. When for some vertex v, $M(v) \in End$ then τ, the *cut transition*, can be fired whose effect consists to (1) delete the subtree rooted v and (2) when $v \neq r$ where r denotes the root of the state to update $M(u)$ where u is the parent of v by applying the effective function $Upd_t^+ \geq \textbf{Id}$. The state consisting in the empty tree is denoted $\bot$.

Definition 1 (DRPN). *A Dynamic Recursive Petri Net is a 6-tuple* $\mathcal{N} = \langle P, T, Grd, Upd, Upd^-, Upd^+, Beg, End \rangle$ *where:*

- *P is a finite set of places;*
- *$T = T_{el} \uplus T_{ab}$ with $P \cap T = \emptyset$ is a finite set of transitions;*
- *$Grd = \{Grd_t\}_{t \in T}$ is a family of upward closed sets of $\mathbb{N}^P$;*
- *$Upd = \{Upd_t\}_{t \in T_{el}}$ is a family of effective functions with $Upd_t \in (\mathbb{N}^P)^{Grd_t}$;*
- *$Upd^- = \{Upd_t^-\}_{t \in T_{ab}}$ is a family of effective functions with $Upd_t^- \in (\mathbb{N}^P)^{Grd_t}$;*

- $Upd^+ = \{Upd_t^+\}_{t \in T_{ab}}$ is a family of effective functions with $Upd_t^+ \in (\mathbb{N}^P)^{\mathbb{N}^P}$;
- For all $t \in T_{ab}$, $Upd_t^- \leq \mathbf{Id}$ and $\mathbf{Id} \leq Upd_t^+$;
- $Beg = \{Beg_t\}_{t \in T_{ab}}$ is a family of effective functions with $Beg_t \in (\mathbb{N}^P)^{Grd_t}$;
- End is an upward closed set of $\mathbb{N}^P$.

As discussed above, a state of a DRPN is a labelled tree.

Definition 2 (State). *Let $\mathcal{N}$ be a DRPN. Then a state $s = \langle V, M, E, \Lambda \rangle$ of $\mathcal{N}$ is defined by:*

- *V its finite set of vertices;*
- *$M : V \to \mathbb{N}^P$, a function that labels vertices with markings;*
- *$E \subseteq V \times V$, a set of edges such that (V, E) is a directed tree;*
- *$\Lambda : E \to T_{ab}$, a function that labels edges with abstract transitions.*

One denotes by $Des_s(v)$(respectively $Anc_s(v)$) the set of descendants (respectively ancestors) of $v \in V$ in the underlining tree of s (including v itself). If $v \neq r$ then $prd(v)$ is the parent of v in the tree. Given a $U \subseteq V$ we will denote by $Anc_s(U) = \cup_{v \in U} Anc_s(v)$. The *depth* of s is the depth of its tree. Given $\mathbf{m} \in \mathbb{N}^P$, $s_{\mathbf{m}}$ denotes a tree consisting of a single vertex r with marking $M(r) = \mathbf{m}$. Let us formally define the firing of elementary, abstract and cut transitions.

Definition 3. *Let $\mathcal{N}$ be a DRPN, s a state of $\mathcal{N}$, $v \in V$ and $t \in T \cup \{\tau\}$. t is fireable by v from s if either $t \neq \tau$ and $M(v) \in Grd_t$ or $t = \tau$ and $M(v) \in End$. In this case, its firing leads to the state $s' = \langle V', M', E', \Lambda' \rangle$, defined below:*

- *If $t \in T_{el}$ then $s' = \langle V, M', E, \Lambda \rangle$ where $M'(u) = M(u)$ for all $u \in V \setminus \{v\}$ and $M'(v) = Upd_t(M(v))$;*
- *If $t \in T_{ab}$ then:*
  - *$V' = V \cup \{w\}$ where w is a fresh identifier;*
  - *$M'(u) = M(u)$ for all $u \in V \setminus \{v\}$, $M'(v) = Upd_t^-(M(v))$ and $M'(w) = Beg_t(M(v))$;*
  - *$E' = E \cup \{(v, w)\}$ and $\Lambda'(e) = \Lambda(e)$ for all $e \in E$ and $\Lambda'((v, w)) = t$.*
- *If $t = \tau$ and $v = r$ then $s' = \perp$;*
- *If $t = \tau$ and $v \neq r$ then let $w = prd(v)$:*
  - *$V' = V \setminus Des_s(v)$;*
  - *for all $u \neq w$, $M'(u) = M(u)$ and $M'(w) = Upd_{\Lambda(w,v)}^+(M(w))$;*
  - *$E' = E \cap (V' \times V')$ and Λ' is the restriction of Λ on E'.*

The transition firing is denoted $s \xrightarrow{(v,t)} s'$ and when there are several nets, $s \xrightarrow{(v,t)}_{\mathcal{N}} s'$. A *firing sequence* is a sequence of transition firings, written in detailed way: $s_0 \xrightarrow{(v_1,t_1)} s_1 \xrightarrow{(v_2,t_2)} \cdots \xrightarrow{(v_n,t_n)} s_n$, or when the context allows it, in a more concise way like $s_0 \xrightarrow{\sigma} s_n$ for $\sigma = (v_1, t_1)(v_2, t_2) \ldots (v_n, t_n)$. The *length* of σ, denoted $|\sigma|$, is n. The *abstract length* of σ, denoted $|\sigma|_{ab}$, is $|\{i \leq n \mid t_i \in T_{ab}\}|$. The *depth* of σ is the maximal depth of states $s_0, \ldots, s_n$. A *closing sequence* is a firing sequence that reaches $\perp$. Given a firing sequence that includes the firing of an abstract transition t in vertex v creating vertex w and followed later by the cut transition in w, we say that $(v,t), (w,\tau)$ are *matched* in σ. The *reachability* set $Reach(\mathcal{N}, s_0)$ is defined by: $Reach(\mathcal{N}, s_0) = \{s \mid \exists \sigma\ s_0 \xrightarrow{\sigma} s\}$.

Discussion. The main limitations of the modelling power of DRPN are the requirements that (1) sets like Grd_t must be upward closed and (2) functions like Upd_t must be monotonic. Despite these limitations, DRPNs include many models like:

- *Petri nets* (PN) that can be defined without abstract transitions and such that for all $t \in T$, $Grd_t = \{\mathbf{m} \mid \mathbf{m} \geq \mathbf{Pre}(t)\}$ and $Upd_t = \mathbf{Id} + \mathbf{C}(t)$ where $\mathbf{Pre}(t)$ (resp. $\mathbf{C}(t)$) is the column vector indexed by t of the backward incidence matrix $\mathbf{Pre}$ (resp. incidence matrix $\mathbf{C}$).
- *Affine Petri nets* ([6]) that can also be defined without abstract transitions and such that for all $t \in T$, there exist a matrix $\mathbf{A}_t \in \mathbb{N}^{P \times P}$ and a vector $\mathbf{B}_t \in \mathbb{Z}^P$ with $Grd_t = \{\mathbf{m} \mid \mathbf{A}_t\mathbf{m} + \mathbf{B}_t \geq 0\}$ and $Upd_t(\mathbf{m}) = \mathbf{A}_t\mathbf{m} + \mathbf{B}_t$.
- *Recursive Petri nets* (RPN) ([10]) such that for all $t \in T$, $Grd_t = \{\mathbf{m} \mid \mathbf{m} \geq \mathbf{Pre}(t)\}$ and when $t \in T_{el}$, $Upd_t = \mathbf{Id} + \mathbf{C}(t)$ and when $t \in T_{ab}$, $Upd_t^- = \mathbf{Id} - \mathbf{Pre}(t)$, $Upd_t^+ = \mathbf{Id} + \mathbf{Post}(t)$ and Beg_t is some constant. Here $\mathbf{Post}(t)$ is the column vector indexed by t of the forward incidence matrix $\mathbf{Post}$.

Graphical Representation. For modelling purposes, we equip DRPN with a graphical representation based on net representations. Places (resp. transitions) are depicted by circles (resp. rectangles). However a transition does not have input arcs but only output arcs represented by double-headed arrows and labelled by expressions where a place represents the current value of its marking. The guard of an elementary transition is also represented by a boolean expression inside the rectangle. For instance, the elementary transition t figured below is defined by: $Grd_t = \{\mathbf{m} \mid \mathbf{m}(p_1) > 2\}$ and $Upd_t(\mathbf{m}) = (\mathbf{m}(p_1) + \mathbf{m}(p_2))\overrightarrow{p_1} + \lfloor\sqrt{\mathbf{m}(p_2)}\rfloor\overrightarrow{p_2}$ where $\overrightarrow{p}$ denotes the vector defined by $\overrightarrow{p}[p] = 1$ and for all $p' \neq p$, $\overrightarrow{p}[p'] = 0$.

$$p_2\ \bigcirc \xleftarrow{\lfloor\sqrt{p_2}\rfloor} \boxed{p_1 > 2} \xrightarrow{p_1 + p_2} \bigcirc\ p_1$$

The rectangle of an abstract transition t is divided into several parts: on the top corner left $(-)$ starts the edges representing Upd_t^-, on the top center Grd_t is represented, on the bottom corner left Beg_t is represented, and on the bottom corner right $(+)$ start the edges representing Upd_t^+. There are no edges for unchanged place markings. For instance, the abstract transition figured below is defined by: $Grd(t) = \{\mathbf{m} \mid \mathbf{m}(p_1) > 2 \vee \mathbf{m}(p_3) > 1\}$, $Upd_t^-(\mathbf{m}) = (\mathbf{m}(p_1) - 1)\overrightarrow{p_1} + \lfloor 0.5\mathbf{m}(p_2)\rfloor\overrightarrow{p_2} + \mathbf{m}(p_3)\overrightarrow{p_3}$, $Upd_t^+(\mathbf{m}) = \mathbf{m}(p_1)\overrightarrow{p_1} + 2\mathbf{m}(p_2)\overrightarrow{p_2} + \mathbf{m}(p_3)^2\overrightarrow{p_3}$, and $Beg_t = \mathbf{m}(p_1)\overrightarrow{p_2}$. Observe that $Upd_t^- \leq \mathbf{Id}$ and $\mathbf{Id} \leq Upd_t^+$.

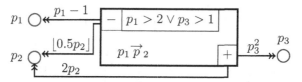

Example 1 (Hiring an assassin). In order to illustrate the modelling capabilities of DRPN, we present an example of distributed planning. The DRPN $\mathcal{N}_{Jaqen}$ of Fig. 1 represents the possible behaviour of an assassin hired for a job. The transitions filled in black are Petri net transitions and so are presented with input and output arcs as usual.

The assassin is given 3 days (3 tokens in p_{time}), an advance of 20 bit-coins (20 tokens in p_{adv}), and is promised to get a reward of 20 bitcoins after the job is done (20 tokens in p_{reward}). In order to try catching their target he needs to devote one bitcoin and one day of his time. After this day either the assassin is successful (t_{found}) or fails (t_{lost}) and needs to spend another day. When successful, the assassin can collect the reward ($t_{collect}$). However, the assassin has also another strategy which consists of hiring another assassin by giving him a quarter of his advance money and promise him an equal reward, telling him the number of days left (t_{hire} where $f_{pay}(\mathbf{m}) = \mathbf{m}(p_{time})\overrightarrow{p}_{time} + \lfloor 0.5 \lceil 0.5\mathbf{m}(p_{adv}) \rceil \rfloor \overrightarrow{p}_{adv} + \lceil 0.5 \lceil 0.5\mathbf{m}(p_{adv}) \rceil \rceil \overrightarrow{p}_{reward}$). If some hired assassin is successful then he can report his success to the hiring guy by firing the cut transition (due to the specification *End*). The state presented on the right of Fig. 1 consists of three assassins where the last hired one has killed the target. Observe that as long as a guy has money he can hire several assassins and that even after hiring he can still try to kill the target by himself.

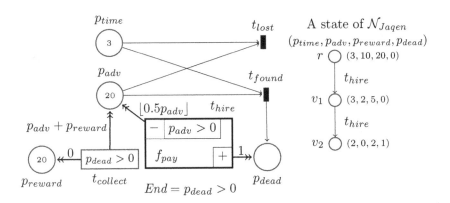

Fig. 1. A DRPN with a state

A firing sequence of $\mathcal{N}_{Jaqen}$ is presented in Fig. 2 where the vertex who fires the transition is filled in black. The initial assassin first tries to find the target but fails ((r, t_{lost})), losing one bitcoin and one day. Then he hires another assassin by firing the abstract transition (r, t_{hire}), losing half of his advance money and creating a new vertex v), where the hired assassin has two days, an advance of five bitcoins and a promised reward of five bitcoins for completing the job ($M(v) = 2\overrightarrow{p}_{time} + 5\overrightarrow{p}_{adv} + 5\overrightarrow{p}_{reward}$). This assassin kills the target and collects the reward ((v, t_{lost}) followed by ($v, t_{collect}$)). Then using the cut transition he

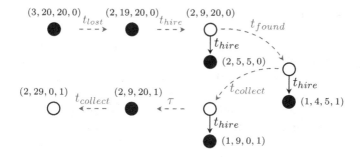

Fig. 2. Firing sequence

reports it to his employer $((v, \tau))$, which removes v and adds one token to p_{dead} in $M(r)$. Finally the original assassin can collect his money by firing $t_{collect}$.

Ordering States of a DRPN. We now define a quasi-order on the states of a DRPN. Given two states s, s' of $\mathcal{N}$ we say that s is smaller or equal than s' or equivalently that s' *covers* s if (without considering labels) there is a subtree in s' isomorphic to s (by some matching) such that (1) given any pair of matched vertices (u, u'), $M(u) \leq M'(u')$ and (2) given any pair of transitions (t, t') labelling matched edges, $Upd_t^+ \leq Upd_{t'}^+$.

Definition 4. *Let $\mathcal{N}$ be DRPN and $s = \langle V, M, E, \Lambda \rangle$, $s' = \langle V', M', E', \Lambda' \rangle$ of a DRPN $\mathcal{N}$, be two states of $\mathcal{N}$ and φ be an injective mapping from V to V'. We say that s' covers s by φ, denoted $s \preceq_\varphi s'$ if:*

1. *For all $v \in V$, $M(v) \leq M'(\varphi(v))$;*
2. *For all $(u, v) \in E$, $(\varphi(u), \varphi(v)) \in E'$ and $Upd_{\Lambda(u,v)}^+ \leq Upd_{\Lambda'(\varphi(u), \varphi(v))}^+$.*

We say that s' covers s denoted $s \preceq s'$ if there exists φ such that $s \preceq_\varphi s'$.

Given states s, s', deciding whether s' covers s is a necessary condition for designing algorithms related to the coverability relation. So we assume that given a DRPN, for all pairs $t, t' \in T_{ab}$ one can decide whether $Upd_t^+ \leq Upd_{t'}^+$. This hypothesis is satisfied by all "reasonable" effective functions.

The *coverability set* $Cov(\mathcal{N}, s_0)$ is defined as the upward closure of the reachability set: $Cov(\mathcal{N}, s_0) = Reach(\mathcal{N}, s_0)^\uparrow$. As for recursive Petri nets (see [4]) this quasi-order is strongly compatible (and even more): for all $s \preceq_\varphi s'$ and $s \xrightarrow{(v,t)} s_1$ there exists $s_1' {}_{\varphi_1} \!\succeq s_1$ such that $s' \xrightarrow{(\varphi(v),t)} s_1'$ and φ and φ_1 coincide on the intersection of their domain. However this quasi-order is not a well quasi-order (see also [4]) and thus in order to solve the coverability problem, one cannot apply the backward exploration.

Notation. Let S_t be a finite set of states. We call a sequence σ such that $s \xrightarrow{\sigma} s' \succeq s'' \in S_t$ a *covering sequence*.

3   Expressiveness

Expressiveness of a formalism may be defined by the family of languages that
it can generate. In [4], expressiveness of RPNs was studied using coverability
languages. In order to compare RPN and DRPN we need to define coverability
languages of DRPNs and so we equip any transition $t \in T \cup \tau$ with a *label*
$\lambda(t) \in \Sigma \cup \{\varepsilon\}$ where Σ is an alphabet and ε is the empty word of Σ^* fulfilling
$\lambda(\tau) = \varepsilon$. The labelling is extended to transition sequences in the usual way.
Thus given a labelled marked DRPN $(\mathcal{N}, s_{init})$ and a finite set of states S_t, the
coverability language $\mathcal{L}(\mathcal{N}, s_{init}, S_t)$ is defined by:

$$\mathcal{L}(\mathcal{N}, s_{init}, S_t) = \{\lambda(\sigma) \mid \exists \; s_0 \xrightarrow{\sigma} s \succeq s' \wedge s' \in S_t\}$$

i.e. the set of labellings for sequences covering some state $s \in S_t$ of $\mathcal{N}$. We say
that $\mathcal{L} \subseteq \Sigma^*$ is a *coverability language* if $\mathcal{L} = \mathcal{L}(\mathcal{N}, s_{init}, S_t)$ for some $\mathcal{N}$, s_{init}
and S_t. We also introduce $\mathcal{L}_B(\mathcal{N}, s_{init}, S_t)$ the B-bounded coverability language
with $B \in \mathbb{N}$.

$$\mathcal{L}_B(\mathcal{N}, s_{init}, S_t) = \{\lambda(\sigma) \mid \exists \; s_0 \xrightarrow{\sigma} s \succeq s' \wedge s' \in S_t \wedge |\sigma|_{ab} \leq B\}$$

We say that $\mathcal{L} \subseteq \Sigma^*$ is a *B-bounded coverability language* if $\mathcal{L} = \mathcal{L}_B(\mathcal{N}, s_{init}, S_t)$
for some $\mathcal{N}$, s_{init} and S_t.

The next proposition is an important ingredient for our expressiveness result
(see proof in [8]). The main idea of this proof consists in considering a PN with
enough copies of P and T such that each copy mimicks the behaviour of a vertex
in a state of the RPN with height at most B and outgoing degree at most B for
every vertex. Additional places and transitions allow to express the child relation
between vertices and the existence of the vertices in the current mimicked state
of the RPN.

Proposition 1. *Let $\mathcal{L}$ be a B-bounded coverability language. Then $\mathcal{L}$ is a PN
coverability language.*

We say that a function $f : \mathbb{N} \to \mathbb{N}$ is *sublinear* if $\lim_{n \to \infty} \frac{f(n)}{n} = 0$. Let f
be sublinear non decreasing with $\lim_{n \to \infty} f(n) = \infty$, we define on the alphabet
$\{a, b\}$ the language $\mathcal{L}_f = \{a^k b^m \mid m \leq f(k)\}$. Examples of such functions are
$\lfloor \log(1 + n) \rfloor$ or $\lfloor \sqrt{n} \rfloor$. As an immediate consequence of the properties of f, one
defines by induction the strictly increasing sequence $(\alpha(n))_{n \in \mathbb{N}}$: $\alpha(0) = 0$ and
$\alpha(n + 1) = \min(m \mid \alpha(n) < m \wedge f(\alpha(n)) < f(m))$. Note that α depends on f,
but since in the sequel we consider a single arbitrary f, for sake of readability
we write α instead of α_f.

The next proposition establishes that $\mathcal{L}_f$ is a DRPN coverability language.
Indeed the coverability language of the DRPN below (without abstract transitions) such that the initial state consists of a single vertex with one token in p_{w_a}
and the final state consists of a single vertex with one token in p_{w_b} is $\mathcal{L}_f$ (see
the full proof in [8]). So we will use $\mathcal{L}_f$ for witnessing that DRPN coverability
languages strictly include RPN coverability languages.

$p_a \quad t_a, a \quad p_{w_a} \qquad t_{w_b}, \varepsilon \qquad \qquad p_{w_b} \quad t_b, b$

$0 \qquad \boxed{p_{w_a} > 0} \quad f(p_a)+1 \qquad \qquad \boxed{End = \emptyset}$

Proposition 2. *For all f, $\mathcal{L}_f$ is a DRPN coverability language.*

The remainder of this section consists in showing that $\mathcal{L}_f$ is not an RPN coverability language. Let us pick an arbitrary labelled RPN $\mathcal{N}$ with an initial state s_{init}, a finite set of states S_t such that $\mathcal{L}_f \subseteq \mathcal{L}(\mathcal{N}, s_{init}, S_t)$. Let $\{\sigma_n\}_{n \in \mathbb{N}}$ be a family of sequences covering some state of S_t such that $\lambda(\sigma_n) = a^{\alpha(n)} b^{f(\alpha(n))}$ where among the possible σ_n's we pick one with the minimal depth and among those with minimal depth one with the minimal length (i.e. $\min |\sigma_n|$). The skeleton of the proof is as follows.

- $\mathcal{L}_f$ is a not a PN coverability language (Proposition 3 proved in [8]). Therefore there does not exist B such that $\mathcal{L}_f$ is a B-bounded language.
- If the depth of $\{\sigma_n\}_{n \in \mathbb{N}}$ is bounded then $\mathcal{L}(\mathcal{N}, s_{init}, S_t)$ is a B-bounded language (Proposition 4) which shows that $\mathcal{L}_f \subsetneq \mathcal{L}(\mathcal{N}, s_{init}, S_t)$.
- If the depth of $\{\sigma_n\}_{n \in \mathbb{N}}$ is unbounded $\mathcal{L}(\mathcal{N}, s_{init}, S_t)$ contains words that do not belong to $\mathcal{L}_f$, (Proposition 5) which shows that $\mathcal{L}_f \subsetneq \mathcal{L}(\mathcal{N}, s_{init}, S_t)$.

The following proposition is obtained by an adaptation of a result related to the (non) weak computability of sublinear functions in PN [12].

Proposition 3. *$\mathcal{L}_f$ is not a PN coverability language.*

Remark 1. In the appendix of [5] Lemma 6 shows that for any RPN language there exists a marked RPN with an initial state consisting of only one vertex. Therefore we will assume in the following that s_{init} consists of a single vertex.

In order to alleviate notations in RPN and to be consistent with the notations of DRPN, for all $t \in T$, we denote $\mathbf{Pre}(t)$, $\mathbf{Post}(t)$ and $\mathbf{C}(t)$ respectively by $\mathbf{Pre}_t$, $\mathbf{Post}_t$ and $\mathbf{C}_t$.

Given a labelled RPN $\mathcal{N}$ and an abstract transition t, we introduce the following languages:

$$\mathcal{L}_{\mathcal{N}}(t) = \{\lambda(\sigma) \mid Beg_t \xrightarrow{\sigma}\} \; ; \; \mathcal{L}_{\mathcal{N}}^{\perp}(t) = \{\lambda(\sigma) \mid Beg_t \xrightarrow{\sigma} \perp\}$$

Proposition 4. *Let $\mathcal{N}$ be a labelled RPN, s_{init} be its initial state and S_t be a finite set of states such that $\mathcal{L}_f \subseteq \mathcal{L}(\mathcal{N}, s_{init}, S_t)$. Assume that the depths of $\{\sigma_n\}_{n \in \mathbb{N}}$ are bounded. Then $\mathcal{L}_f \subsetneq \mathcal{L}(\mathcal{N}, s_{init}, S_t)$.*

Proof. Let D denote a bound of the depths of the family of $\{\sigma_n\}_{n=1}^{\infty}$, and let $\mathcal{L}_{\mathcal{N}}$ denote more concisely $\mathcal{L}(\mathcal{N}, s_{init}, S_t)$. We are going to build a net $\mathcal{N}''$ and some $\mathcal{L}''$, a B-bounded language of $\mathcal{N}''$ such that: $\mathcal{L}_f \subseteq \mathcal{L}''$. Due to Proposition 1 and 3, $\mathcal{L}_f \subsetneq \mathcal{L}''$. We stop the construction earlier if we can conclude that $\mathcal{L}_f \subsetneq \mathcal{L}_{\mathcal{N}}$. Otherwise the relation between $\mathcal{L}''$ and $\mathcal{L}_{\mathcal{N}}$ will allow us to conclude. We first build an RPN $\mathcal{N}'$ that fulfills $\mathcal{L}(\mathcal{N}', s_{init}, S_t) = \mathcal{L}_{\mathcal{N}}$ as follows. For all $t \in T_{ab}$, one adds places and transitions according to $\mathcal{L}_{\mathcal{N}}(t)$ and $\mathcal{L}_{\mathcal{N}}^{\perp}(t)$:

- If $a^+b^+ \cap \mathcal{L}_\mathcal{N}(t) = \emptyset$ and $z_t = \max\{m \mid b^m \in \mathcal{L}_\mathcal{N}(t)\} < \infty$ then one adds elementary transitions t^-, t^b and a place p_t (see left side of the figure below), where:

$$\begin{aligned}
\mathbf{Pre}_{t^-} &= \mathbf{Pre}_t, & \mathbf{C}_{t^-} &= -\mathbf{Pre}_t + z_t \cdot \overrightarrow{p_t}, & \lambda(t^-) &= \lambda(t); \\
\mathbf{Pre}_{t^b} &= \overrightarrow{p_t}, & \mathbf{C}_{t^b} &= -\overrightarrow{p_t}, & \lambda(t^b) &= b;
\end{aligned}$$

- $a^+b^+ \cap \mathcal{L}_\mathcal{N}^\perp(t) = \emptyset$ and $\max\{m \mid b^m \in \mathcal{L}_\mathcal{N}^\perp(t)\} < \infty$ then one adds, elementary transitions $t_\perp^-, t_\perp^x, t_\perp^+$ and places $p_{t^-}^\perp, p_{t^+}^\perp$ such that where the y_t and x are defined below:

$$\begin{aligned}
\mathbf{Pre}_{t_\perp^-} &= \mathbf{Pre}_t, & \mathbf{C}_{t_\perp^-} &= y_t \cdot \overrightarrow{p_{t^-}^\perp} - \mathbf{Pre}_{t_\perp^-}, & \lambda(t_\perp^-) &= \lambda(t); \\
\mathbf{Pre}_{t_\perp^x} &= \overrightarrow{p_{t^-}^\perp}, & \mathbf{C}_{t_\perp^x} &= \overrightarrow{p_{t^+}^\perp} - \overrightarrow{p_{t^-}^\perp}, & \lambda(t_\perp^x) &= x; \\
\mathbf{Pre}_{t_\perp^+} &= y_t \cdot \overrightarrow{p_{t^+}^\perp}, & \mathbf{C}_{t_\perp^+} &= \mathbf{Post}_t - y_t \cdot \overrightarrow{p_{t^+}^\perp}, & \lambda(t_\perp^+) &= \varepsilon
\end{aligned}$$

  ○ If $b^m \in \mathcal{L}_t^\perp$ for $m > 0$ then $y_t = \max\{m \mid b^m \in \mathcal{L}_t^\perp\}$ and $x = b$;
  ○ Else if $a^\ell \in \mathcal{L}_t^\perp$ for $\ell > 0$ then $y_t = \min\{\ell \mid a^\ell \in \mathcal{L}_t^\perp\}$ and $x = a$;
  ○ Otherwise, $y_t = 1$ and $x = \varepsilon$.

On the one hand $\mathcal{L}_\mathcal{N} \subseteq \mathcal{L}(\mathcal{N}', s_{init}, S_t)$ since any firing sequence in $\mathcal{N}$ can be performed in $\mathcal{N}'$. On the other hand, the new transitions are built according to $\mathcal{L}_\mathcal{N}(t)$ and $\mathcal{L}_\mathcal{N}^\perp(t)$ in such a way that every firing of a new transition can be replaced by a firing of a sequence of transitions with the same produced label. Hence $\mathcal{L}(\mathcal{N}', s_{init}, S_t) = \mathcal{L}_\mathcal{N}$.
We now show that for $\mathcal{N}'$ there exists some B, such for all $n \in \mathbb{N}$ there is a firing sequence σ_n' in $\mathcal{N}'$ with $|\sigma_n'|_{ab} \leq B$ and $\lambda'(\sigma_n') = \lambda(\sigma_n)$. Denote by $G = \max\{|V_{s_t}| \mid s_t \in S_t\}$. Pick an arbitrary $n \in \mathbb{N}$ and denote more explicitly the covering sequence $s_{init} \xrightarrow{\sigma_n} s \nsucceq s_t \in S_t$. Assume there is an occurrence $t \in T_{ab}$ by the vertex u in σ_n creating a vertex v. We transform σ_n according to whether the firing (u, t) has a matching cut transition (v, τ) in σ_n:

The firing (u, t) Does Not Have a Matching Cut.

- If $a^+b^+ \cap \mathcal{L}_t \neq \emptyset$ then let us suppose that there are more than $2D + G$ occurrences of t without a matching cut. Then there are two vertices v, v' created by t in σ_n which are not: (1) in $\varphi(V_{s_t})$, (2) in the branch leading to it, and (3) both in the same branch. Therefore, one can build a covering sequence σ with $\ell_i, m_i > 0$ such that $\lambda(\sigma) = \ldots a^{\ell_1} b^{m_1} \ldots a^{\ell_2} b^{m_2} \ldots \in \mathcal{L}_\mathcal{N}$. So $\mathcal{L}_f \subsetneq \mathcal{L}_\mathcal{N}$ and we are done.
- If $\max\{m \mid b^m \in \mathcal{L}_\mathcal{N}(t)\} = \infty$ then let us suppose that there are more than $D + G$ occurrences of t without a matching cut. There is a vertex v created by t which is neither in $\varphi(V_{s_t})$ nor in the branch leading to it. Then one can build a covering sequence σ with $k \leq \alpha(n)$ and $m > f(k)$ such that $\lambda(\sigma) = a^k b^m \in \mathcal{L}_\mathcal{N}$. So $\mathcal{L}_f \subsetneq \mathcal{L}_\mathcal{N}$ and we are done.

- Otherwise (i.e. $a^+b^+ \cap \mathcal{L}_{\mathcal{N}}(t) = \emptyset$ and $z_t = \max\{m \mid b^m \in \mathcal{L}_{\mathcal{N}}(t)\} < \infty$), we replace the firing of (u,t) and all firings from $Des(v)$ by:

$$(u,t^-)\overbrace{(u,t^b)\ldots(u,t^b)}^{z_t \text{ times}}$$

which will give us a covering sequence σ_n' such that $\lambda(\sigma_n') = a^\ell b^m$ with $\ell \leq \alpha(n)$ and $m \geq f(\alpha(n))$. If $\ell < \alpha(n)$, and $m > f(\alpha(n))$ then $\mathcal{L}_f \subsetneq \mathcal{L}_{\mathcal{N}}$ and we are done. Otherwise $\lambda'(\sigma_n') = \lambda'(\sigma_n')$ and $|\sigma_n'|_{ab} < |\sigma_n|_{ab}$. We can repeat this process until either one concludes that $\mathcal{L}_f \subsetneq \mathcal{L}_{\mathcal{N}}$ or there is no more firing of t without a matching cut transition in σ_n'.

The Firing (u,t) Has a Matching Cut.

- Assume that $a^+b^+ \cap \mathcal{L}_{\mathcal{N}}^\perp(t) \neq \emptyset$. If there are more than D occurrences of t with a matching cut in σ_n then there are two occurrences (w_1,t) and (w_2,t) where w_2 is neither a descendent nor an ascendent of w_1. So one could build a covering sequence σ with $\ell_i, m_i > 0$ such that $\lambda(\sigma) = \ldots a^{\ell_1}b^{m_1}a^{\ell_2}b^{m_2}\ldots \in \mathcal{L}_{\mathcal{N}}$. So $\mathcal{L}_f \subsetneq \mathcal{L}_{\mathcal{N}}$ and we are done.
- If $\max\{m \mid b^m \in \mathcal{L}_{\mathcal{N}}^\perp(t)\} = \infty$. Consider m the occurrences of b in σ_n produced at the subtree rooted in v then there exists $m' > m$ such that one can build a covering sequence σ with $m > f(\alpha(n)) + 1$ such that $\lambda(\sigma) = a^{\alpha(n)}b^{f(\alpha(n))+m'-m} \in \mathcal{L}_{\mathcal{N}}$ So $\mathcal{L}_f \subsetneq \mathcal{L}_{\mathcal{N}}$ and we are done.
- Otherwise (i.e. $a^+b^+ \cap \mathcal{L}_{\mathcal{N}}^\perp(t) = \emptyset$ and $\max\{m \mid b^m \in \mathcal{L}_{\mathcal{N}}^\perp(t)\} < \infty$), we replace the firing of (u,t) by the sequence below and remove all firings from $Des(v)$,

$$(u,t_\perp^-)\overbrace{(u,t_\perp^x)\ldots(u,t_\perp^x)}^{y_t \text{ times}}(u,t_\perp^+)$$

and obtain a covering sequence σ_n' such that $\lambda(\sigma_n') = a^\ell b^m$ with $\ell \leq \alpha(n)$ and $m \geq f(\alpha(n))$. If $\ell < \alpha(n)$, and $m > f(\alpha(n))$ then $\mathcal{L}_f \subsetneq \mathcal{L}_{\mathcal{N}}$ and we are done. Otherwise $\lambda'(\sigma_n') = \lambda'(\sigma_n')$ and $|\sigma_n'|_{ab} < |\sigma_n|_{ab}$. We can repeat this process until either one concludes that $\mathcal{L}_f \subsetneq \mathcal{L}_{\mathcal{N}}$ or there is no more firing of t with a matching cut transition in σ_n'.

If we are not yet done, we have built a sequence σ_n' with $\lambda'(\sigma_n') = \lambda'(\sigma_n')$ and such that $|\sigma_n'|_{ab} \leq |T_{ab}|(2D + G)$. So we choose $B = |T_{ab}|(2D + G)$.

In order that for all $a^\ell b^m \in \mathcal{L}_f$ there is a covering sequence σ with $|\sigma|_{ab} \leq B$, we build $\mathcal{N}''$ from $\mathcal{N}'$. We observe that the definition of α implies that: $\mathcal{L}_f = \{a^\ell b^m \mid \exists n, \delta^-, \delta^+ \ \ell = \alpha(n) + \delta^+ \wedge m = f(\alpha(n)) - \delta^-\}$. Due to the observation about RPN languages, the initial state of $\mathcal{N}'$ consists of only one vertex, whose initial marking is denoted $\mathbf{m}_{ini}$. So one builds the RPN $\mathcal{N}''$ with initial marking $\overrightarrow{p}_{ini}$ from $\mathcal{N}'$ as follows.

- Add elementary transitions t_a, t_{run}, t_{ab} and places p_{ini}, p_{run} such that:

$$\begin{aligned}
\mathbf{Pre}_{t_a}'' &= \overrightarrow{p}_{ini}, & \mathbf{C}_{t_a}'' &= 0, & \lambda''(t_a) &= a; \\
\mathbf{Pre}_{t_r}'' &= \overrightarrow{p}_{ini}, & \mathbf{C}_{t_r}'' &= \mathbf{m}_{ini} - \overrightarrow{p}_{ini} + \overrightarrow{p}_{run}, & \lambda''(t_r) &= \varepsilon; \\
\mathbf{Pre}_{t_{ab}}'' &= \overrightarrow{p}_{run}, & \mathbf{C}_{t_{ab}}'' &= \overrightarrow{p}_{run}, & \lambda''(t_{ab}) &= \varepsilon.
\end{aligned}$$

- For all $t \in T$, we set:

$$\mathbf{Pre}''_t = \mathbf{Pre}'_t + \vec{p}_{run}, \mathbf{C}''_t = \mathbf{C}'_t - \vec{p}_{run}, Beg''_t = Beg'_t + \vec{p}_{run} \text{ when } t \in T_{ab}$$

- For any transition $t \in T$ with $\lambda'(t) = b$, we add the transition t_ε which is a copy of t with $\lambda''(t_\varepsilon) = \varepsilon$.
  Let $a^\ell b^m \in \mathcal{L}_f$. Then there exist n, δ^-, δ^+ such that $\ell = \alpha(n) + \delta^+$ and $m = f(\alpha(n)) - \delta^-$. Let σ_n be a covering sequence in $\mathcal{N}'$ such that $\lambda'(\sigma_n) = a^{\alpha(n)} b^{f(\alpha(n))}$. We define σ a covering sequence of $\mathcal{N}''$ as follows.
  σ starts by $(r, t_a)^{\delta^+}(r, t_r)(r, t_{ab})^{|\sigma_n|}$. Then σ is completed by $\hat{\sigma}_n$ where $\hat{\sigma}_n$ is obtained from σ_n by:
  - changing δ^- occurrences of transitions with label b by their copy;
  - whenever σ_n creates a new vertex v, one inserts $(v, t_{ab})^{|\sigma_n|}$ firings.
  Observe that $\lambda''(\sigma) = a^\ell b^m$. Let w be a word. Define $w_{\downarrow b}$ as the set of words obtained from w by omitting some occurrences of b. Define $\mathcal{L}'' = \{w \mid \exists w' \in \mathcal{L}_\mathcal{N} \ w \in a^* w'_{\downarrow b}\}$. Therefore $\mathcal{L}_f \subseteq \mathcal{L}_B(\mathcal{N}'', \vec{p}_{ini}, S_t) \subseteq \mathcal{L}''$. Thus $\mathcal{L}_f \subsetneqq \mathcal{L}''$. If $\mathcal{L}_\mathcal{N} = \mathcal{L}_f$ then $\mathcal{L}'' = \mathcal{L}_f$ which concludes the proof. ∎

We are now in position to conclude.

Proposition 5. $\mathcal{L}_f$ *is not an RPN coverability language.*

Proof. Let $\mathcal{N}$ be an RPN with initial state s_{init} and final states S_t such that $\mathcal{L}_f \subseteq \mathcal{L}(\mathcal{N}, s_{init}, S_t)$. Let us denote more concisely $\mathcal{L}(\mathcal{N}, s_{init}, S_t)$ by $\mathcal{L}_\mathcal{N}$. By Proposition 4 if $\{\sigma_n\}_{n \in \mathbb{N}}$ are bounded then $\mathcal{L}_f \subsetneqq \mathcal{L}_\mathcal{N}$ and we are done. So assume that the depths of $\{\sigma_n\}_{n \in \mathbb{N}}$ are unbounded. Denote by $G = \max\{|s_t| \mid s_t \in S_t\}$. Let $h = (5|T_{ab}| + 1)G$. There is some $\sigma_n = \sigma'_n \sigma''_n$ such that $s_0 \xrightarrow{\sigma'_n} s'$ where the depth of s' is greater than h. There exists $\{v_i\}_{0 < i \leq 5} \subseteq V_{s'}$ on the same branch which were created by the firing of the same abstract transition with corresponding depths are $\{d_i\}$ and such that $d_{i+1} - d_i > G$. Denote by T_f the subtree of the final state of σ_n which matches the state to be covered and by Br the branch leading to it. Due to the choice of G there exist three consecutive vertices v_i, v_{i+1} and v_{i+2} such that:

  — The branch Br_i between v_i and v_{i+2} does not intersect with T_f;
  — Either Br_i does not intersect with Br or Br_i is included in Br.

Each of these vertex either may or may not have a matching cut in σ_n. We pick two of these vertices v, v' ($v' \in Des(v)$) such that either both of them have a matching cut or both of them do not. Denote by w and w' the labellings of the sequences performed in the subtree rooted in v and v' along σ_n, we split the proof in two cases:

- **Case $w \neq w'$.** We denote by $\hat{w} \neq \epsilon$ the trace of the sequence performed in the subtree rooted in v without the one performed in the subtree rooted in v':

$\circ$ $\widehat{w} = a^{\ell}$, for $\ell > 0$. Then one can build another covering sequence by mimicking the behavior of v' from v. But then the trace of the new covering sequence will be $a^{\alpha(n)-\ell}b^{f(\alpha(n))}$ and since $\ell > 0$ we get that $a^{\alpha(n)k-\ell}b^{f(\alpha(n))} \notin \mathcal{L}_f$, from which we conclude that $\mathcal{L}_f \subsetneq \mathcal{L}_{\mathcal{N}}$.

$\circ$ $\widehat{w} = b^m$, for $m > 0$. Then one can build another covering sequence by mimicking the behavior of v from v'. But then the trace of the new covering sequence will be $a^{\alpha(n)}b^{f(\alpha(n))+m} \notin \mathcal{L}_f$ from which we conclude that $\mathcal{L}_f \subsetneq \mathcal{L}_{\mathcal{N}}$.

$\circ$ $\widehat{w} = a^{\ell}b^m$, for $\ell, m > 0$. Then one can build a family of covering sequences $\{\widehat{\sigma}_x\}_{x \in \mathbb{N}}$ by mimicking the behavior of v_i from v_{i+1} recursively x times. We would get that $\lambda(\widehat{\sigma}_x) = a^{\alpha(n)+x\ell}b^{f(\alpha(n))+xm}$ for any $x \in \mathbb{N}$. But that would give us that:

$$\frac{f(\alpha(n)) + xm}{\alpha(n) + x\ell} \xrightarrow{x \to \infty} \frac{m}{\ell} > 0$$

Since f is sublinear there exists $x \in \mathbb{N}$ such that $\lambda(\widehat{\sigma}_x) \notin \mathcal{L}_f$, from which we conclude that $\mathcal{L}_f \subsetneq \mathcal{L}_{\mathcal{N}}$.

• **Case $w = w'$.** Then one can build another covering sequence with the trace $a^{\alpha(n)}b^{f(\alpha(n))}$ where we mimic the behavior of v' in v. By doing so we get a covering sequence σ'_n not deeper then σ_n but which is shorter then σ_n, i.e. $|\sigma_n| \geq |\sigma'_n|$ which is a contradicts to our assumption about σ_n.   ∎

4   Decidability of the Coverability Problem

The *coverability problem* takes as input a DRPN $\mathcal{N}$, and two states s_0, s and asks whether there exists a sequence $s_0 \xrightarrow{\sigma} s' \succeq s$. Before developing the proof that the coverability problem is decidable let us describe its scheme.

– The algorithm builds a DRPN $\widehat{\mathcal{N}}$ by adding elementary transitions to $\mathcal{N}$ (Definition 6). The DRPN $\widehat{\mathcal{N}}$ is equivalent to $\mathcal{N}$ w.r.t. coverability. Furthermore for any firing sequence σ of $\mathcal{N}$, there is an equivalent firing sequence $\widehat{\sigma}$ of $\widehat{\mathcal{N}}$ such that in $\widehat{\sigma}$ there is no occurrence of an abstract transition followed later by a matching cut step.

– The definition of $\widehat{\mathcal{N}}$, is based on several upward closed sets of $\mathbb{N}^P$: (1) *Endable($\mathcal{N}$)*, the set of markings $\mathbf{m}$ such that from $s_{\mathbf{m}}$ one can reach $\perp$ and (2) for all $t \in T_{ab}$, *$Closed_t$* the set of markings from which one can fire t and create a vertex whose marking belongs to *Endable($\mathcal{N}$)*. So we establish that one can compute these sets (Proposition 6).

– Afterwards we successively define and solve two intermediate particular coverability problems: (1) the *restricted rooted coverability problem* which takes as input a DRPN $\mathcal{N}$, a marking $\mathbf{m}_0$, and a state s and asks whether there exists a sequence $s_{\mathbf{m}_0} \xrightarrow{\sigma} s'_{\mathbf{Id}} \succeq s$ and (2) the *restricted coverability problem* which takes as input a DRPN $\mathcal{N}$, a marking $\mathbf{m}_0$, and a state s and asks whether there exists a sequence $s_{\mathbf{m}_0} \xrightarrow{\sigma} s' \succeq s$. The decidability of the latter problem (Theorem 2) is partially based on the decidability of the former one (Theorem 1).

– Finally we solve the coverability problem (Theorem 3) by a case based analysis of the covering sequence which, depending on the case, is based on either Theorem 1 or Theorem 2.

As announced above, building the following sets is a key ingredient for the decidability of the coverability problem. Due to the properties of $\preceq$, these sets are upward closed.

Definition 5. *Let $\mathcal{N}$ be a DRPN. Then $Endable(\mathcal{N}) \subseteq \mathbb{N}^P$ is defined by:*

$$Endable(\mathcal{N}) = \{\mathbf{m} \mid \exists \sigma \; s_{\mathbf{m}} \xrightarrow{\sigma} \perp\}$$

and for all $t \in T_{ab}$, $Closed_t$ is defined by:

$$Closed_t = Beg_t^{-1}(Endable(\mathcal{N}))$$

Example 2. Consider the DRPN of Fig. 1. With one token in p_{time} and in p_{adv} one fires t_{found} producing a token in p_{dead} which allows to fire a cut transition. Furthermore firing the abstract transition t_{hire} does not help to reach $\perp$ since in the new vertex, the marking of p_{time} is equal to the marking of p_{time} in its parent vertex and the marking p_{adv} is smaller than the markingof p_{adv} in in its parent vertex. Thus $Endable(\mathcal{N}) = \{p_{time} + p_{adv}\}^\uparrow$. The marking of a vertex created by t_{hire} is greater or equal than $p_{time} + p_{adv}$ if in its parent vertex there is at least one token in p_{time} and three tokens in p_{adv}. Thus $Closed_{t_{hire}} = \{p_{time} + 3p_{adv}\}^\uparrow$.

If we are able to compute *Endable* and thus $\{Closed_t\}_{t \in T_{ab}}$ then we can build a DRPN $\widehat{\mathcal{N}}$ which in some sense is equivalent to $\mathcal{N}$. The interest of $\widehat{\mathcal{N}}$ is that for any firing sequence σ of $\mathcal{N}$, there is an equivalent firing sequence $\widehat{\sigma}$ where the firing of an abstract transition t followed later by a matching cut transition can be replaced by the firing of an elementary t^- followed later by the firing of another elementary transition t^+. So the set of transitions is extended with $T_r = \{t^-, t^+ \mid t \in T_{ab}\}$. To ensure the sequentiality between these firings, the set of places is extended with $P_r = \{p_t \mid t \in T_{ab}\}$ with one token produced (resp. consumed) in p_t by t^- (resp. t^+). To ensure that the firing of t^- is performed when the corresponding firing of t can be matched later by the firing of the cut transition, the guard of t^- is the guard of t intersected with $Closed_t$. In order to formally define $\widehat{\mathcal{N}}$ and to exhibit relations between states and thus markings of $\mathcal{N}$ and $\widehat{\mathcal{N}}$, we introduce the projection Proj from $\mathbb{N}^{\widehat{P}}$ to $\mathbb{N}^P$ where $\widehat{P} = P \cup P_r$. In addition $\widehat{\mathcal{N}}_{el}$, obtained from $\widehat{\mathcal{N}}$ by deleting T_{ab}, allows to track the evolution of the marking of a vertex in $\widehat{\mathcal{N}}$ when no firing of abstract transitions occurs in this vertex.

Definition 6. *Let $\mathcal{N}$ be a DRPN.*
Then $\widehat{\mathcal{N}} = \langle \widehat{P}, \widehat{T}, \widehat{Grd}, \widehat{Upd}, \widehat{Upd}^-, \widehat{Upd}^+, \widehat{Beg}, \widehat{End} \rangle$ is a DPRN defined by:

– $\widehat{P} = P \cup P_r$ and $\widehat{T} = T \cup T_r$;
– for all $t \in T$, $\widehat{Grd}_t = \mathsf{Proj}^{-1}(Grd_t)$
   for all $t^- \in T_r$, $\widehat{Grd}_{t^-} = \mathsf{Proj}^{-1}(Grd_t \cap Closed_t)$
   for all $t^+ \in T_r$, $\widehat{Grd}_{t^+} = p_t > 0$;

- *for all $t \in T_{el}$, all $p \in P$ and all $p_{t'} \in P_r$,*
  $\widehat{Upd}_t(p) = Upd_t(p) \circ \mathsf{Proj}$ *and* $\widehat{Upd}_t(p_{t'}) = p_{t'}$;
- *for all $t^- \in T_r$, all $p \in P$ and all $p_{t'} \in P_r$,*
  $\widehat{Upd}_{t^-}(p) = Upd_t^-(p) \circ \mathsf{Proj}$ *and* $\widehat{Upd}_t(p_{t'}) = p_{t'} + 1_{t=t'}$;
- *for all $t^+ \in T_r$, all $p \in P$ and all $p_{t'} \in P_r$,*
  $\widehat{Upd}_{t^+}(p) = Upd_t^+(p) \circ \mathsf{Proj}$ *and* $\widehat{Upd}_t(p_{t'}) = p_{t'} - 1_{t=t'}$;
- *for all $t \in T_{ab}$, all $p \in P$ and all $p_{t'} \in P_r$,*
  1. $\widehat{Upd}_t^-(p) = Upd_t^-(p) \circ \mathsf{Proj}$, $\widehat{Upd}_t^+(p) = Upd_t^+(p) \circ \mathsf{Proj}$,
     $\widehat{Beg}_t(p) = Beg_t(p) \circ \mathsf{Proj}$;
  2. $\widehat{Upd}_t^-(p_{t'}) = p_{t'}$, $\widehat{Upd}_t^+(p_{t'}) = p_{t'}$, $\widehat{Beg}_t(p_{t'}) = 0$;
- $\widehat{End} = \mathsf{Proj}^{-1}(End)$;

$\widehat{\mathcal{N}}_{el}$ *is obtained from* $\widehat{\mathcal{N}}$ *by deleting* T_{ab}.

The following lemma states the correspondence between $\mathcal{N}$, $\widehat{\mathcal{N}}$ and $\widehat{\mathcal{N}}_{el}$. We extend Proj to states of $\widehat{\mathcal{N}}$ by applying it to the marking of vertices and furthermore to sets of states by the standard set extension. In the reverse direction, given a marking $\mathbf{m} \in \mathbb{N}^P$ we define $\widehat{\mathbf{m}} \in \mathbb{N}^{\widehat{P}}$ the extended marking with no token in P_r. Similarly, given s a state of $\mathcal{N}$, we define $\widehat{s}$ a state of $\widehat{\mathcal{N}}$ by extending the marking of vertices of s with no token in P_r.

Lemma 1. *Let $\mathcal{N}$ be a DRPN and s_0 be a state. Then:*

1. *For all $\widehat{s}_0 \xrightarrow{\sigma}_{\widehat{\mathcal{N}}} s$, there exists $s_0 \xrightarrow{\sigma'}_{\mathcal{N}} s'$ with $\mathsf{Proj}(s) \preceq_{\mathbf{Id}} s'$*
   and for all $s_0 \xrightarrow{\sigma'}_{\mathcal{N}} s$, $\widehat{s}_0 \xrightarrow{\sigma}_{\widehat{\mathcal{N}}} \widehat{s}$;
2. $Cov(\mathcal{N}, s_0) = \mathsf{Proj}(Cov(\widehat{\mathcal{N}}, \widehat{s}_0))$;
3. *For all $s \in Reach(\widehat{\mathcal{N}}, \widehat{s}_0)$, there exists a sequence $\widehat{s}_0 \xrightarrow{\sigma}_{\widehat{\mathcal{N}}} s$ such that no firing of abstract transition is matched by a cut transition.*
4. $\mathsf{Proj}(Endable(\widehat{\mathcal{N}}) \cap \bigcap_{t \in T_{ab}} p_t = 0) = \mathsf{Proj}(Endable(\widehat{\mathcal{N}}_{el}) \cap \bigcap_{t \in T_{ab}} p_t = 0)$
   $= Endable(\mathcal{N})$.

Proof.

- Let $\widehat{s}_0 \xrightarrow{\sigma}_{\widehat{\mathcal{N}}} s$. Consider successively all $t \in T_{ab}$. Observe that due to the presence of place p_t every occurrence of (v, t^+) in σ' can be matched with an occurrence of (v, t^-). The unmatched occurrences of (v, t^-) can be omitted since they only produce useless tokens in p_t and do not increase the marking of any other place. So we get a new firing sequence $\widehat{s}_0 \xrightarrow{\sigma^*}_{\widehat{\mathcal{N}}} s^*$ with $\mathsf{Proj}(s) \preceq_{\mathbf{Id}}$ $\mathsf{Proj}(s^*)$. We transform this sequence in a sequence $s_0 \xrightarrow{\sigma'}_{\mathcal{N}} \mathsf{Proj}(s^*)$ as follows. For every pair of matching firings $(v, t^-), (v, t^+)$, we substitute to (v, t^-) the firing (v, t) creating the vertex w with an initial $\mathbf{m} \in Endable(\mathcal{N})$ due to the guard of t^-. Then we substitute to (v, t^+) a sequence $s_{\mathbf{m}} \xrightarrow{\sigma_{\mathbf{m}}} \bot$ applying it to w. The "reverse" direction is immediate since for all $t \in T \cup \{\tau\}$ and all states s and s', $s \xrightarrow{(v,t)}_{\mathcal{N}} s'$ implies $\widehat{s} \xrightarrow{(v,t)}_{\widehat{\mathcal{N}}} \widehat{s}'$.
- Assertion 2 is an immediate consequence of Assertion 1 of the lemma.

- Let $\widehat{s}_0 \xrightarrow{\sigma'} s$ be a firing sequence of $\widehat{\mathcal{N}}$. Consider successively all $t \in T_{ab}$ and all matching pairs $(v, t), (w, \tau)$ occurring in σ' where w is the vertex created by the firing of (v, t). Let σ^* be the subsequence of σ of firings in the subtree rooted at w ended by (w, τ). Then one substitutes to (v, t), the firing (v, t^-) which is fireable as witnessed by σ^* and one deletes σ^* substituting (w, τ) by (v, t^+). Iterating this process, one gets the sequence we are looking for.

- Let $\mathbf{m} \in Endable(\widehat{\mathcal{N}}) \cap \bigcap_{t \in T_{ab}} p_t = 0$. Consider a sequence $s_{\mathbf{m}} \xrightarrow{\sigma}_{\widehat{\mathcal{N}}} \bot$. Using Assertion 3, one can assume that in σ, no firing of an abstract transition is matched with a cut transition. Consider (r, t) the firing of an abstract transition in the root occurring in σ. Since it is not matched by a cut transition one can delete it and all the firings in the subtree rooted at the created vertex and still reaches $\bot$. Such a sequence is thus a firing sequence of $\widehat{\mathcal{N}}_{el}$. The other direction is immediate since the set of transitions of $\widehat{\mathcal{N}}_{el}$ is included in the one of $\widehat{\mathcal{N}}$.

The inclusion $Endable(\mathcal{N}) \subseteq \mathsf{Proj}(Endable(\widehat{\mathcal{N}}) \cap \bigcap_{t \in T_{ab}} p_t = 0)$ is immediate since for all $t \in T \cup \{\tau\}$, $s \xrightarrow{(v, t)}_{\mathcal{N}} s'$ implies $\widehat{s} \xrightarrow{(v, t)}_{\widehat{\mathcal{N}}} \widehat{s}'$.

Let $\mathbf{m} \in \mathsf{Proj}(Endable(\widehat{\mathcal{N}}) \cap \bigcap_{t \in T_{ab}} p_t = 0)$. One applies Assertion 1 of the lemma with $s_0 = s_{\mathbf{m}}$. Let $\widehat{s}_{\mathbf{m}} \xrightarrow{\sigma}_{\widehat{\mathcal{N}}} s \xrightarrow{(r, \tau)}_{\widehat{\mathcal{N}}} \bot$. Then there exists $s_0 \xrightarrow{\sigma}_{\mathcal{N}} s'$ with $\mathsf{Proj}(s) \preceq_{\mathbf{Id}} s'$. Thus $s' \xrightarrow{(r, \tau)}_{\mathcal{N}} \bot$ which establishes that $\mathbf{m} \in Endable(\mathcal{N})$. ∎

Let us describe how Algorithm 1 computes $Endable$ and $\{Closed_t\}_{t \in T_{ab}}$. During lines 2–14, it builds a version of $\widehat{\mathcal{N}}_{el}$ where for every t, $Closed_t$ is replaced by $Closed[t]$. Since $Closed[t]$ will be updated during the loop of lines 15–23, the definition of Grd_{t^-} is performed at line 16. Still in this loop, using a standard backward exploration, during lines 17–20, it computes in variable X, $Endable$ for this version of $\widehat{\mathcal{N}}_{el}$. Afterwards still in this loop, it updates Y by restricting X to the markings with no token in P_r and then projecting it to P. Then using Y, updates for every t, $Closed[t]$. The algorithm terminates when Y is no more enlarged.

Proposition 6. *Algorithm 1 terminates and upon termination $Y = Endable$ and for all $t \in T_{ab}$, $Closed[t] = Closed_t$.*

Algorithm 1: Computing the closure of abstract transitions

Closing($\mathcal{N}$)
Input: $\mathcal{N}$ a DRPN
/* $\widehat{P} = P \cup \{p_t \mid t \in T_{ab}\}$, $\widehat{T}_{el} = T_{el} \cup \{t^-, t^+ \mid t \in T_{ab}\}$ */
/* Proj is the projection from $\mathbb{N}^{\widehat{P}}$ to $\mathbb{N}^P$ */
Data: $X, oldX$ subsets of $\mathbb{N}^{\widehat{P}}$; $oldY, Y$ subset of $\mathbb{N}^P$; t a transition
Output: *Closed* an array indexed by T_{ab} of upward closed sets of $\mathbb{N}^P$

1 $Y \leftarrow \emptyset$; **for** $t \in T_{ab}$ **do** $Closed[t] \leftarrow \emptyset$
2 **for** $t \in T_{el}$ **do**
3      $Grd_t \leftarrow \text{Proj}^{-1}(Grd_t)$
4   **for** $p \in P$ **do** $Upd_{t^-}(p) \leftarrow Upd_t(p) \circ \text{Proj}$
5   **for** $p_{t'} \in \widehat{P}$ **do** $Upd_t(p_{t'}) \leftarrow p_{t'}$
6 **end**
7 **for** $t \in T_{ab}$ **do**
8   **for** $p \in P$ **do** $Upd_{t^-}(p) \leftarrow Upd_t^-(p) \circ \text{Proj}$
9   **for** $p_{t'} \in \widehat{P}$ with $t' \neq t$ **do** $Upd_{t^-}(p_{t'}) \leftarrow p_{t'}$
10     $Upd_{t^-}(p_t) \leftarrow p_t + 1$; $Grd_{t^+} \leftarrow p_t > 0$
11   **for** $p \in P$ **do** $Upd_{t^+}(p) \leftarrow Upd_t^+(p) \circ \text{Proj}$
12   **for** $p_{t'} \in \widehat{P}$ with $t' \neq t$ **do** $Upd_{t^+}(p_{t'}) \leftarrow p_{t'}$
13     $Upd_{t^+}(p_t) \leftarrow p_t - 1$
14 **end**
15 **repeat**
16   **for** $t \in T_{ab}$ **do** $Grd_{t^-} \leftarrow \text{Proj}^{-1}(Grd_t \cap Closed[t])$
17     $oldY \leftarrow Y$; $X \leftarrow \text{Proj}^{-1}(End)$; $oldX \leftarrow \emptyset$
18   **while** $X \neq oldX$ **do**
19         $oldX \leftarrow X$; **for** $t \in \widehat{T}_{el}$ **do** $X \leftarrow X \cup (Upd_t^{-1}(X) \cap Grd_t)$
20   **end**
21     $Y \leftarrow \text{Proj}(X \cap \bigcap_{t \in T_{ab}} p_t = 0)$
22   **for** $t \in T_{ab}$ **do** $Closed[t] \leftarrow Beg_t^{-1}(Y)$
23 **until** $Y \neq oldY$
24 **return** $Y, Closed$

Proof. In the sequel $Closed[t]$ denotes the value of this variable at some execution point. Let us denote $\mathcal{N}'$ the version of $\widehat{N}_{el}$ built by the algorithm and updated at every iteration of loop of lines 15–23.

- **Termination.** We prove by induction that the sequence of sets Y_n and for all t, $Closed_n[t]$ at the beginning of iteration n of the **repeat** loop is an increasing sequence of upward closed sets of $\mathbb{N}^P$ and $\mathbb{N}^{\widehat{P}}$, respectively. So it must stabilize after a finite number of iterations. Since T is finite this will establish termination of the algorithm. First, for all t, $\emptyset = Closed_1[t] \subseteq Closed_2[t]$. Assume that for some $1 < n$, for all t, $Closed_{n-1}[t] \subseteq Closed_n[t]$. Then the nth iteration corresponds to the $n - 1^{th}$ iteration with $Closed_n[t]$ substituted to $Closed_{n-1}[t]$. Since the operations involving $Closed[t]$ are intersection, union

and projection, this immediately entails that $Closed_n[t] \subseteq Closed_{n+1}[t]$.
It remains to prove that for all n, Y_n and (for all t) $Closed_n[t]$ are upward closed and that the **while** loop terminates. We also prove it by induction on n. Consider the nth iteration of the **repeat** loop and let us prove the sequence of sets X_k at the beginning of iteration k of the **while** loop is an increasing sequence of upward closed sets. This will establish the termination of this loop. Observe that $X_1 = \texttt{Proj}^{-1}(End)$ is an upward closed set and that at every iteration X is increased because it is updated by union of some set with itself. Furthermore, X remains an upward closed set since (1) upward closed sets are closed by union, intersection, and inverse of non decreasing mappings. Finally while $X \cap \bigcap_{t \in T_{ab}} p_t = 0$ is not upward closed, this is the case for $Y_{n+1} = \texttt{Proj}(X \cap \bigcap_{t \in T_{ab}} p_t = 0)$. Thus for every t, $Closed_{n+1}[t]$ is upward closed.

- **Consistency.** We establish by induction on the iterations of the **repeat** loop that $Y \subseteq Endable(\mathcal{N})$ and for all $\mathbf{m} \in Closed[t]$, there is a sequence $s_{Beg(t)(\mathbf{m})} \xrightarrow{\sigma_{\mathbf{m}}} \perp$, implying $Closed[t] \subseteq Closed_t$. Consider an arbitrary iteration of the **repeat** loop. Thus the **while** loop computes X the set $Endable(\mathcal{N}')$. Since by induction, $Closed[t] \subseteq Closed_t$ one deduces that $X \subseteq Endable(\widehat{\mathcal{N}}_{el})$. Applying Assertion 4 of Lemma 1, one deduces that $Y \subseteq Endable(\mathcal{N})$ and so that at the end of the iteration $Closed[t] \subseteq Closed_t$.

- **Completeness.** Let $\mathbf{m} \in Closed_t$. Consider a sequence $s_{Beg(t)(\mathbf{m})} \xrightarrow{\sigma}_{\mathcal{N}} \perp$. Observe that if in σ, there exists a firing of and abstract transition creating some vertex v not later followed by a cut transition in v, then one can omit this firing and all firings in the subtree rooted at v and still reaches $\perp$. Thus we assume that every vertex v created by the firing of abstract transition is later deleted by a matching cut transition in v.
  We establish the completeness of the algorithm by recurrence on the depth of σ. If the depth is null, it means that σ only includes firing of elementary transitions in r ended by the cut transition. So $\widehat{s}_{Beg(t)(\mathbf{m})} \xrightarrow{\sigma}_{\mathcal{N}_{el}} \perp$. Furthermore since $\sigma \in (\{r\} \times T_{el})^*(r, \tau)$, $\widehat{s}_{Beg(t)(\mathbf{m})} \xrightarrow{\sigma}_{\mathcal{N}'} \perp$ for $\mathcal{N}'$ built at the beginning of the first iteration of the **repeat** loop. During every iteration of the **repeat** loop, the **while** loop computes in X the set of markings from which a sequence of transitions of $\mathcal{N}'$ leads to some marking in $\texttt{Proj}^{-1}(End)$. So $\widehat{Beg_t}(\mathbf{m}) \in X$ at the end of the iteration and after the **for** loop at line 22, $\mathbf{m} \in Closed[t]$.
  Assume that σ has depth $h > 0$. So every $\mathbf{m}'$ in the root from which there is a firing of an abstract transition t' belongs to $Closed[t']$ since the subsequence in the created vertex up to the cut transition has depth strictly less than h. Consider the last iteration of the **repeat** loop for which such t' is added to $Closed[t']$. Then either at this iteration $\mathbf{m}$ already belongs to $Closed[t]$ or it will be added at the next iteration (which exists since Y is enlarged) due to execution of the **while** loop. Indeed consider a closing subsequence of σ for a child of the root created by some transition t' and substitute the firing of t' by t'^-, delete the closing subsequence and substitute the cut step by the firing of t^+ in r. Doing this transformation (and ommitting the cut

step in r) one obtains a closing firing sequence in $\mathcal{N}'$ as described above from $\widehat{s}_{Beg(t)(\mathbf{m})}$. Thus at the beginning of last iteration of the **while** loop, $\mathcal{N}' = \widehat{\mathcal{N}}_{el}$. Using Assertion 4 of Lemma 1, one gets that at this end of this iteration, $Y = Endable(\mathcal{N})$. ∎

Theorem 1. *The restricted rooted coverability problem is decidable.*

Proof. Due to Assertion 1 of Lemma 1, we consider $\widehat{\mathcal{N}}$. Applying Assertion 3 of Lemma 1, the sequence we are looking for, does not create other vertices than the vertices of s since (1) the firing of abstract transitions matched by cut transitions are non necessary in $\widehat{\mathcal{N}}$ and (2) those that are not matched decrease the marking of a vertex and create a subtree useless for covering $s = (V, E, M, \Lambda)$.

Then for any vertex $v \in V$, one guesses an order of creation of its children along the sequence σ and for any transition t labelling an edge of E, one guesses a transition t' with $Upd^+_{t'} \geq Upd^+_t$. Observe that there are only a finite number of such guesses and so the algorithm enumerates them.

Afterwards the algorithm proceeds bottom-up from the leaves of s to the root. Let v be a leaf. Then the algorithm computes by backward exploration the upward closed set $Cov(v) = \{\mathbf{m} \in \mathbb{N}^{\widehat{P}} \mid \exists \mathbf{m}' \geq M(v) \; \exists \sigma \in \widehat{T}^*_{el} \; \mathbf{m} \xrightarrow{\sigma} \mathbf{m}'\}$.

Let v be an internal vertex with $v_1, \ldots, v_n$ its children enumerated in the guessed order and $t_1, \ldots, t_n$ the associated guessed transitions. Then the algorithm computes by backward exploration the upward closed sets $Cov_n(v), \ldots, Cov_0(v) = Cov(v)$ as follows: $Cov_n(v) = \{\mathbf{m} \in \mathbb{N}^{\widehat{P}} \mid \exists \mathbf{m}' \geq M(v) \; \exists \sigma \in \widehat{T}^*_{el} \; \mathbf{m} \xrightarrow{\sigma} \mathbf{m}'\}$ and for $i < n$,

$$Cov_i(v) = \{\mathbf{m} \in \mathbb{N}^{\widehat{P}} \mid$$
$$\exists \mathbf{m}' \in (\widehat{Upd^-_{t_{i+1}}})^{-1} \left(\widehat{Beg}^{-1}_{t_{i+1}} (Cov(v_{i+1})) \cap Cov_{i+1}(v) \right) \cap \widehat{Grd}_{t_{i+1}}$$
$$\exists \sigma \in \widehat{T}^*_{el} \; \mathbf{m} \xrightarrow{\sigma} \mathbf{m}'\}$$

By a straightforward induction, one establishes that $\mathbf{m} \in Cov_i(v)$ if and only if from vertex v marked by $\mathbf{m}$ there is a firing sequence $\sigma = \sigma_i t_{i+1} \sigma_{i+1} \ldots t_n \sigma_n$ with for all j, $\sigma_j \in \widehat{T}^*_{el}$ and such that (1) the marking of v reached by σ is greater or equal than $M(v)$, and (2) for all $j > i$, from the initial marking of v_j one can fire a sequence that builds a tree covering the subtree of s rooted at v_i using the guessed transitions and orders of creation. Finally the algorithm returns true if and only if for some guess $\widehat{\mathbf{m}}_0 \in Cov(r)$ where r is the root of s. ∎

Theorem 2. *The restricted coverability problem is decidable.*

Proof. Let us fix some net $\mathcal{N}$ and some state s. As above we substitute $\widehat{\mathcal{N}}$ to $\mathcal{N}$ but for sake of readability we omit the occurrences of '^'. Thus one observes that the state s' that should cover s can be chosen as a single branch, say Br, leading to a tree s^* isomorphic to s. Indeed the firings that would create other branches are useless since they only decrease the markings in Br or in s'.

Observe that in the previous proof, instead of answering the decision problem, one can compute the set $RRC(\mathcal{N}, s) = \{\mathbf{m} \mid \exists s_{\mathbf{m}} \xrightarrow{\sigma} s^*_{\mathbf{Id}} \succeq s\}$.

Let us define $RC(\mathcal{N}, s, k) = \{\mathbf{m} \mid \exists s_{\mathbf{m}} \xrightarrow{\sigma} s'_{\varphi} \succeq s \wedge |r \rightarrow_{s'} \varphi(r)| \leq k\}$ where $|x \rightarrow_{s'} y|$ denotes the length of the elementary path from x to y in s'. One immediately observes that $RC(\mathcal{N}, s, 0) = RRC(\mathcal{N}, s)$ that for all k, $RC(\mathcal{N}, s, k)$ is upward closed and $RC(\mathcal{N}, s, k) \subseteq RC(\mathcal{N}, s, k+1)$. Furthermore the answer to the restricted coverability problem is positive if and only if $\mathbf{m}_0 \in \bigcup_{k \in \mathbb{N}} RC(\mathcal{N}, s, k)$. So it only remains to show how to compute $RC(\mathcal{N}, s, k + 1)$ when one knows $RC(\mathcal{N}, s, k)$. Observe that:

$$RC(\mathcal{N}, s, k + 1) = RC(\mathcal{N}, s, k) \cup \{\mathbf{m} \mid$$
$$\exists \sigma \in \widehat{T}^*_{el} \, \exists \mathbf{m}' \in \bigcup_{t \in T_{ab}} Grd_t \cap Beg_t^{-1}(RC(\mathcal{N}, s, k))$$
$$\mathbf{m} \xrightarrow{\sigma} \mathbf{m}'\}$$

where the second term of this union can be computed by a backward exploration. ∎

Using the previous theorems, we are now in position to decide the coverability problem.

Theorem 3. *The coverability problem of DRPN is decidable.*

Proof. Let us fix some net $\mathcal{N}$ and states s_0 and s. As above we substitute $\widehat{\mathcal{N}}$ to $\mathcal{N}$ and omit the occurrences of '$\widehat{}$'. In order to decide the existence of a sequence $s_0 \xrightarrow{\sigma} s'_{\varphi} \succeq s$, we consider two cases depending on $\varphi(r)$

- $|\varphi(V_s) \cap V_{s_0}| \leq 1$. So one guesses a vertex $w \in V_{s_0}$ that is the deepest on the branch leading from r to $\varphi(r)$ in s'. Then we transform the net (whose current version is in the sequel denoted $\mathcal{N}'$) as follows. We examine bottom-up all the (proper) descendants of w in s_0 as follows. Let v be such a vertex, u its parent and t the abstract transition labelling the edge (u, v). If $M(v) \in Endable(\mathcal{N}')$ then one adds a place $p_{u,v}$ with a token in u and no token elsewhere and an elementary transition $t_{u,v}$ such that $Grd_{t_{u,v}} = p_{u,v} > 0$ and $Upd_{t_{u,v}} = Upd_t^+ - p_{u,v}$. By construction, there exists a sequence $s_0 \xrightarrow{\sigma}_{\mathcal{N}} s'_{\varphi} \succeq s$, if and only if there exists a sequence $s_{M(w)} \xrightarrow{\sigma}_{\mathcal{N}'} s'_{\varphi} \succeq s$. So this case is decidable by Theorem 2.
- $|\varphi(V_s) \cap V_{s_0}| > 1$. Then one builds a state s'_0 with root $\varphi(r)$ and simultaneously transform the net (whose current version is also denoted $\mathcal{N}'$) as follows. We eliminate all the vertices of s_0 which are not descendants of $\varphi(r)$ since there are irrelevant due to the choice of φ. Then we eliminate bottom-up all the descendants of $\varphi(r)$ in s_0 which do not belong to $\varphi(V_s)$ as follows. Let v be such a vertex, u its parent and t the abstract transition labelling the edge (u, v). When one examines v, it has become a leaf. If $M(v) \in Endable(\mathcal{N}')$ then one adds a place $p_{u,v}$ with a token in u and no token elsewhere and an elementary transition $t_{u,v}$ such that $Grd_{t_{u,v}} = p_{u,v} > 0$ and $Upd_{t_{u,v}} = Upd_t^+ - p_{u,v}$. Afterwards one deletes v and (u, v). By construction, there exists a sequence $s_0 \xrightarrow{\sigma}_{\mathcal{N}} s'_{\varphi} \succeq s$, if and only if there exists a sequence

$s'_0 \xrightarrow{\sigma}_{\mathcal{N}'} s'_{\varphi} \succeq s$.

For all vertices u of s'_0 denote s_u the subtree of s rooted at $\varphi(u)$ consisting of vertices whose deepest ancestor in s is $\varphi(u)$. Then there exists a sequence $s'_0 \xrightarrow{\sigma}_{\mathcal{N}'} s'_{\varphi} \succeq s$ if and only if the following conditions hold:

 - For all $(u,v) \in E_{s'_0}$, $Upd^+(\Lambda(u,v)) \geq Upd^+(\Lambda(\varphi^{-1}(u), \varphi^{-1}(v)))$;

 - For all $u \in V_{s'_0}$ there exists a sequence $s_{M(u)} \xrightarrow{\sigma_u}_{\mathcal{N}'} s'_{\mathrm{Id}} \succeq s_u$.

The first item is decidable by effectiveness of $\mathcal{N}$. The second item is decidable by Theorem 1.

Since there are only a finite number of possible φ (more precisely their restriction over s_0) the coverability problem is decidable. ∎

5   Conclusion

We introduced DRPN that extend recursive Petri nets in several directions. We established that w.r.t. coverability languages, this extension is strict and we have established that the coverability problem is still decidable.

We plan to define a restriction of DRPN for which reachability would be still decidable as in RPN. Moreover since our algorithm is based on backward explorations and our team has already developed an efficient tool for coverability in PN based on such explorations [1], we want to adapt it for DRPN.

References

1. Blondin, M., Finkel, A., Haase, C., Haddad, S.: The logical view on continuous Petri nets. ACM Trans. Comput. Log. **18**(3), 24:1–24:28 (2017)
2. Dufourd, C., Finkel, A., Schnoebelen, P.: Reset nets between decidability and undecidability. In: Larsen, K.G., Skyum, S., Winskel, G. (eds.) ICALP 1998. LNCS, vol. 1443, pp. 103–115. Springer, Heidelberg (1998). https://doi.org/10.1007/BFb0055044
3. Seghrouchni, A.E.F., Haddad, S.: A recursive model for distributed planning. In: ICMAS 1996, Kyoto, Japan, pp. 307–314 (1996)
4. Finkel, A., Haddad, S., Khmelnitsky, I.: Coverability and termination in recursive petri nets. In: Donatelli, S., Haar, S. (eds.) PETRI NETS 2019. LNCS, vol. 11522, pp. 429–448. Springer, Cham (2019). https://doi.org/10.1007/978-3-030-21571-2_23
5. Finkel, A., Haddad, S., Khmelnitsky, I.: Coverability and termination in recursive Petri nets (2019). https://hal.inria.fr/hal-02081019
6. Finkel, A., McKenzie, P., Picaronny, C.: A well-structured framework for analysing Petri net extensions. Inf. Comput. **195**, 1–29 (2004)
7. Finkel, A., Schnoebelen, P.: Well-structured transition systems everywhere!. Theor. Comput. Sci. **256**(1–2), 63–92 (2001)
8. Haddad, S., Khmelnitsky, I.: Dynamic recursive Petri nets (2020). https://hal.archives-ouvertes.fr/hal-02511321v1
9. Haddad, S., Poitrenaud, D.: Theoretical aspects of recursive petri nets. In: Donatelli, S., Kleijn, J. (eds.) ICATPN 1999. LNCS, vol. 1639, pp. 228–247. Springer, Heidelberg (1999). https://doi.org/10.1007/3-540-48745-X_14

10. Haddad, S., Poitrenaud, D.: Modelling and analyzing systems with recursive petri nets. In: WODES 2000, Ghent, Belgium, pp. 449–458. Kluwer Academic Publishers (2000)
11. Haddad, S., Poitrenaud, D.: Recursive Petri nets. Acta Informatica **44**(7–8), 463–508 (2007)
12. Leroux, J., Schnoebelen, P.: On functions weakly computable by petri nets and vector addition systems. In: Ouaknine, J., Potapov, I., Worrell, J. (eds.) RP 2014. LNCS, vol. 8762, pp. 190–202. Springer, Cham (2014). https://doi.org/10.1007/978-3-319-11439-2_15
13. Lomazova, I.A.: Nested petri nets: multi-level and recursive systems. Fundam. Inform. **47**(3–4), 283–293 (2001)
14. Mayr, R.: Process rewrite systems. Inf. Comput. **156**(1–2), 264–286 (2000)
15. Valk, R.: Self-modifying nets, a natural extension of Petri nets. In: Ausiello, G., Böhm, C. (eds.) ICALP 1978. LNCS, vol. 62, pp. 464–476. Springer, Heidelberg (1978). https://doi.org/10.1007/3-540-08860-1_35

Tools

Visualizing Token Flows Using Interactive Performance Spectra

Wil M. P. van der Aalst[1,2(✉)] ⓘ, Daniel Tacke Genannt Unterberg[1],
Vadim Denisov[2,3], and Dirk Fahland[2] ⓘ

[1] Process and Data Science (Informatik 9), RWTH Aachen University,
Aachen, Germany
wvdaalst@pads.rwth-aachen.de
[2] Process Analytics Group, Technische Universiteit Eindhoven, Eindhoven, The
Netherlands
[3] Vanderlande Industries, Veghel, The Netherlands

Abstract. Process mining techniques can be used to discover process models from event data and project performance and conformance
related diagnostics on such models. For example, it is possible to automatically discover Petri nets showing the bottlenecks in production,
administration, transport, and financial processes. Also basic statistics
(frequencies, average delays, standard deviations, etc.) can be projected
on the places and transitions of such nets to reveal performance and
compliance problems. However, real-life phenomena such as overtaking,
batching, queueing, concept drift, and partial blocking of multiple cases
remain invisible when considering basic statistics. This paper presents an
approach combining Petri-net-based discovery techniques and so-called
performance spectra based on token flows. Token production and consumption are visualized such that the true dynamics of the process are
revealed. Our ProM implementation supports a range of visual-analytics
features allowing the user to interact with the underlying event data
and Petri net. Event data related to the handling of orders are used to
demonstrate the functionality of our tool.

Keywords: Process mining · Visual analytics · Petri nets ·
Performance spectrum

1 Introduction

Process mining techniques can be used to automatically discover process models,
diagnose compliance and performance problems, predict behaviors, and recommend process improvements [1]. Currently, there are over 30 commercial process
mining tools (Celonis, ProcessGold, Disco, Everflow, Lana, Logpickr, Mehrwerk,
Minit, MyInvenio, PAFnow, PuzzleData, Timeline, QPR, etc.) and thousands
of organizations are using process mining to improve their processes.

In process mining, there are two main types of artifacts: *event logs* and *process models* [1]. Each *event* in an event log refers to an *activity* possibly executed by a *resource* at a particular *time* and for a particular *case*. An event may

© Springer Nature Switzerland AG 2020
R. Janicki et al. (Eds.): PETRI NETS 2020, LNCS 12152, pp. 369–380, 2020.
https://doi.org/10.1007/978-3-030-51831-8_18

have additional attributes such as transactional information, costs, customer, location, and unit. Process discovery techniques can be used to discover process models [1]. Although most of the commercial systems start by deriving a so-called *directly-follows graph* (a Markovian model with activities as states), there is consensus that *concurrency* should be uncovered to avoid Spaghetti-like underfitting process models [2]. Therefore, all of the more advanced tools provide ways of discovering *higher-level process models* such as Petri nets, process trees, statecharts, or BPMN models [1].

The approach presented in this paper assumes that we have an event log and a corresponding Petri net that was automatically discovered. Existing techniques are able to replay the event log on such a process model to show performance-related diagnostics by *aggregating* over all cases, e.g., the frequency of a transition and the average or median waiting time [1,10,11]. This is done by showing numeric values next to places, transitions, and arcs. It is also possible to use colors and line thickness, e.g., transitions with long waiting times are colored red and frequent paths are thicker than infrequent ones. Although existing tools also report minimum times, maximal times, and standard deviations, they have problems revealing the following phenomena that require *comparing all individual cases*:

- *Overtaking*: Some cases are handled much faster, thereby bypassing cases that started earlier. This can only be observed through pairwise comparison.
- *Batching*: Groups of events are combined into batched activities.
- *Queueing*: Cases are delayed waiting for shared resources.
- *Concept drift*: The dynamic behavior is changing over time (periodic, gradual, or sudden). This can only be observed by plotting the fine-grained behavior over time.
- *Partial blocking*: Specific subclasses of cases are delayed, whereas others are not.

For example, if the reported average delay in a part of the process is 8.5 days (computed over the past year), then we cannot say anything about phenomena such as overtaking, batching, queueing, concept drift, and partial blocking. These phenomena are caused by interactions between cases and resources that may change over time. Therefore, they cannot be identified by looking at aggregate data.

To address these limitations, we use so-called *performance spectra* based on *token flows*. A token flow refers to the production of a token for a place and the later consumption and also the corresponding two timestamps. Note that a token flow has a *duration*. Performance spectra were introduced in [6] to provide fine-grained visualizations of the performance of business processes and material handling systems along *sequences* of activities (rather than process models).

In this paper, we use performance spectra in the context of a Petri net and provide extensive tool support. We provide a tight integration between process models and performance spectra. By aligning model and log, we can visualize performance spectra for token flows in concurrent processes instead of along sequences of interleaved events, thereby overcoming a significant limitation of our

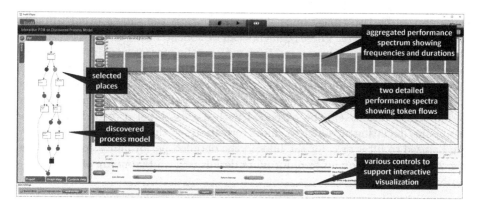

Fig. 1. Screenshot visualizing the token flows of the two selected places using three interactive performance spectra (based on an event log with 39.022 orders and 210.459 events). (Color figure online)

prior work [6]. We can now also handle concurrent processes, thereby exploiting the alignment of model and log.

The approach has been implemented as a ProM plug-in and can be downloaded from promtools.org (install the package "PerformanceSpectrumIntegration"). Figure 1 shows the tool while analyzing the detailed behavior of token flows of two places (left) over time (right). The performance spectra reveal concept drift, overtaking, and queueing. These phenomena would not be visible if one would consider basic place-related statistics such as average waiting times and standard deviations.

Our work can be seen as part of *visual analytics* [9], also called "the science of analytical reasoning facilitated by interactive visual interfaces", since we combine automated process discovery with interactive visualization. Therefore, we aim to apply best practices from visual analytics, e.g., interaction, supporting the identification of patterns, dealing with many data points, and the ability to drill down.

The remainder is organized as follows. Section 2 introduces performance spectra and Sect. 3 introduces token flows and how these correspond to performance spectra. Section 4 shows how token flows can be extracted from event logs, thus enabling the computation of performance spectra in the presence of concurrency. The implementation and its application are presented in Sect. 5. Section 6 concludes the paper.

2   Performance Spectra

Performance spectra were first introduced in [6]. They use the notion of *segments*. A segment is defined by a pair of activities (a, b), e.g., two subsequent sensors on a conveyor belt or two subsequent stations of an assembly line. Each object (e.g., a bag or product) flowing from a directly to b is an *observation* and is characterized by two timestamps (t_a, t_b) where t_a is the time a occurs directly followed by b at time t_b. These segments can be concatenated and it is possible to show the *detailed performance spectrum* with one line for every pair (t_a, t_b) or the *aggregate performance spectrum* aggregating results per time period (yielding a kind of bar chart with periods on the horizontal axis). The aggregate performance spectrum is based counting the number of lines (i.e., observations) per time period. Observations can be extended with a class label c (e.g., fast/slow or compliant/deviating) leading to characterizations of the form (t_a, t_b, c). Label c can be used to filter or color the lines in the detailed performance spectrum. See Fig. 2 for an example and [5,6] for more details (e.g., a taxonomy of patterns in performance spectra).

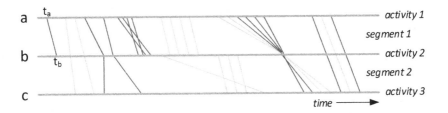

Fig. 2. Example of a detailed performance spectrum composed of two segments and three activities a, b, and c. Each observation (i.e., a line) connects two correlated events. The horizontal axis represents time. The color indicates the class. (Color figure online)

Figure 3 shows a few common patterns that should be self-explanatory. For example, crossing lines correspond to overtaking, e.g., a LIFO (Last-In-First-Out) queuing discipline. If the tokens in a place are always consumed in FIFO (First-In-First-Out) order there are no crossing lines.

Batching corresponds to converging lines. We can also see how patterns change over time, e.g., due to concept drift. Figure 3(g–i) shows three aggregate performance spectra. Here, information about frequencies, durations, etc. is aggregated over predefined time intervals. In all performance spectra, colors can be used to distinguish observations related to different classes.

Performance spectra have been used to analyze materials-handling systems such as the handling of luggage in airports and parcels in distribution centers. However, they can be applied to any operational process to gain novel insights regarding overtaking, batching, queueing, concept drift, and partial blocking that remain hidden when looking at traditional performance indicators like frequencies and average durations.

Performance spectra are created using *event data*. Each observation (i.e., a line) corresponds to two *correlated events*. In this paper, we use standard event data where each event has a timestamp and refers to at least an activity and a case.

Definition 1 (Log Events, Event Log). $\mathcal{U}_{LE}$ *is the universe of log events. A log event* $e \in \mathcal{U}_{LE}$ *can have any number of attributes.* $\pi_x(e)$ *is the value of* *attribute x for event e. $\pi_x(e) = \bot$ if there is no such value. We assume that each log event e has a timestamp $\pi_{time}(e) \neq \bot$ and refers to a case $\pi_{case}(e) \neq \bot$ and an activity $\pi_{act}(e) \neq \bot$. $L \subseteq \mathcal{U}_{LE}$ is an event log describing a collection of recorded events.*

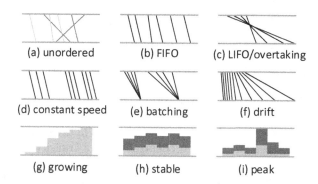

(a) unordered	(b) FIFO	(c) LIFO/overtaking
(d) constant speed	(e) batching	(f) drift
(g) growing	(h) stable	(i) peak

Fig. 3. Six detailed performance spectra (a–f) and three aggregate performance spectra (g–i). (Color figure online)

Note that events can have many more attributes than the mandatory ones, e.g., for $e \in L$, $\pi_{case}(e)$, $\pi_{act}(e)$, and $\pi_{time}(e)$ need to exist, but also $\pi_{resource}(e)$, $\pi_{costs}(e)$, and $\pi_{location}(e)$ may be defined.

The case identifier is used to correlate two subsequent events belong to the same case. The activities are used to build segments. Consider the leftmost observation (first blue line) in Fig. 3. t_a and t_b refer to the timestamps of the two corresponding events e_1 and e_2, i.e. $\pi_{act}(e_1) = a$, $\pi_{time}(e_1) = t_a$, $\pi_{act}(e_2) = b$, and $\pi_{time}(e_2) = t_b$. These events are correlated because they have the same case identifier (i.e., $\pi_{case}(e_1) = \pi_{case}(e_2)$). For example, events e_1 and e_2 refer to the same bag in a baggage handling system or to a particular car that is assembled. The class label c of an observation (t_a, t_b, c) may be based on the event attributes, e.g., $\pi_{resource}(e_1)$, $\pi_{costs}(e_1)$ or $\pi_{time}(e_2) - \pi_{time}(e_1)$.

Although the Performance Spectrum Miner (PSM) presented in [5,6] uses event data, the link to process models (e.g. Petri nets) was not supported thus far. Moreover, the PSM has difficulties dealing with concurrency since direct successions of *causally unrelated* activities are hiding the real causalities and durations. Consider, for example, a process where the traces $\langle a, b, c, d \rangle$ and $\langle a, c, b, d \rangle$ occur both 100 times. Although a is always followed by b (i.e., 200 times), without preprocessing, the PSM will only show this for half of the cases (and show misleading delays).

3   Token Flows

To combine process discovery with performance spectra, we introduce the notion of *token flows*. This way, we are able to relate performance spectra to process models and visualize concurrent processes.

Petri nets are defined as usual [7] and can be discovered using standard process mining techniques [1]. We assume that the reader is familiar with the basic concepts.

Definition 2 (Petri Net). *A Petri net is a tuple* $N = (P, T, F)$ *with* P *the set of places,* T *the set of transitions such that* $P \cap T = \emptyset$, *and* $F \subseteq (P \times T) \cup (T \times P)$ *the flow relation.* $\bullet x = \{y \mid (y, x) \in F\}$ *is the preset of a place or transition* $x \in P \cup T$, *and* $x\bullet = \{y \mid (x, y) \in F\}$ *its postset.*

In a *marking*, i.e., a state of the Petri net, places may contain *tokens*. A transition $t \in T$ is *enabled* if the input places are marked. A transition that *occurs* (i.e., fires) consumes one token from each output place and produces one token for each output place.

We need to relate transition occurrences to events in the event log. Therefore, we differentiate between the different occurrences of a transition and use *events* as a unifying notion. Events may refer to *transition occurrences* (cf. Definition 3) and/or events recorded in the *event log* (cf. $L \subseteq \mathcal{U}_{LE}$ in Definition 1).

Definition 3 (Transition Events). $\mathcal{U}_{TE}$ *is the universe of transition events, i.e., all events having a timestamp and referring to a model transition. A transition event* $e \in \mathcal{U}_{TE}$ *has a timestamp* $\pi_{time}(e) \neq \bot$ *and refers to a transition* $\pi_{trans}(e) \neq \bot$, *and can have any number of attributes. (as before,* $\pi_x(e)$ *is the value of attribute* x *for event* e *and* $\pi_x(e) = \bot$ *denotes that there is no such value).*

A *transition event* $e \in \mathcal{U}_{TE}$ refers to the occurrence of a transition $\pi_{trans}(e)$ at time $\pi_{time}(e)$. Note that transition events can have any number of attributes (just like log events). These attributes will be used to determine delays and to correlate transitions occurrences. Alignments will be used to relate log events and transition events (see Sect. 4). Therefore, we only consider the Petri net structure $N = (P, T, F)$ and abstract from the initial marking, token colors, guards, arc expressions, etc.

A *binary token flow* refers to the production of a particular token by some transition occurrence and the subsequent consumption by another transition occurrence. This relates to the notion of a place in an occurrence net and the notion of token flows defined in [8]. However, we adapt terminology to be able to relate binary token flows to pairs of events in event logs. Section 4 shows how this is done. Binary token flows correspond to observations in performance spectra (i.e., the individual lines in Fig. 2).

A binary token flow $tf = (e_{prod}, p, e_{cons})$ refers to a transition occurrence e_{prod} producing a specific token for place p that is later consumed by another transition occurrence e_{cons}.

Definition 4 (Binary Token Flow). *Let* $N = (P, T, F)$ *be a Petri net.* $\mathcal{U}_{BTF}^N = \{(e_{prod}, p, e_{cons}) \in \mathcal{U}_{TE} \times P \times \mathcal{U}_{TE} \mid \pi_{trans}(e_{prod}) \in \bullet p \wedge \pi_{trans}(e_{cons}) \in p \bullet \wedge \pi_{time}(e_{cons}) \geq \pi_{time}(e_{prod})\}$ *are the possible binary token flows of* N.

Note that a binary token flow corresponds to a *condition* in a *partially-ordered run* of the Petri net [7]. It is possible to describe an entire run of a net as a set of binary token flows. A necessary requirement is that the token flows in this set can only produce and/or consume one token for each place (of course, the same transition can consume or produce multiple tokens and a token cannot be consumed before it is produced). We call such a set a *valid token flow set*.

Definition 5 (Valid Token Flow Set). *Let $N = (P, T, F)$ be a Petri net structure. $TFS \subseteq \mathcal{U}_{BTF}^N$ is a valid token flow set if for any pair of token flows $(e_{prod}, p, e_{cons}) \in TFS$ and $(e'_{prod}, p, e'_{cons}) \in TFS$ sharing the same place p: $e_{prod} = e'_{prod}$ if and only if $e_{cons} = e'_{cons}$ (i.e., tokens are produced and consumed only once).*

A run of the Petri net N corresponds to a valid token flow set. Such a run may involve many different cases (e.g., all bags handled in a baggage handling system). Therefore, we do not define an initial marking or additional annotations (like in colored Petri nets). Section 4 explains how a valid token flow set can be extracted from an event log $L \subseteq \mathcal{U}_{LE}$.

Token flows abstract from tokens that are never produced or never consumed (e.g., tokens in the initial and final marking). If needed, it is always possible to add dummy source and sink transitions. Since an event log may contain information about many cases, this is more natural for process mining.

Definition 6 (Notations). *For any binary token flow $tf = (e_{prod}, p, e_{cons})$, we use the following shorthands: $\pi_{dur}(tf) = \pi_{time}(e_{cons}) - \pi_{time}(e_{prod})$ is the duration of a token flow, $\pi_{tp}(tf) = \pi_{trans}(e_{prod})$ is the transition starting the token flow, $\pi_{tp}^{time}(tf) = \pi_{time}(e_{prod})$ is the production time, $\pi_{tc}(tf) = \pi_{trans}(e_{cons})$ is the transition completing the token flow, $\pi_{tc}^{time}(tf) = \pi_{time}(e_{cons})$ is the consumption time, and $\pi_{pl}(tf) = p$ is the corresponding place. Let $TFS \subseteq \mathcal{U}_{BTF}^N$ be a valid token flow set. For any subset of places $Q \subseteq P$: $TFS\!\upharpoonright_Q = \{tf \in TFS \mid \pi_{pl}(tf) \in Q\}$ are the token flows through these places. We can also compute statistics over sets of token flows, e.g., $avgdur(TFS) = \sum_{tf \in TFS} \pi_{dur}(tf)/|TFS|$ is the average delay.*

The above notions facilitate computations over token flows, e.g., $avgdur(TFS\!\upharpoonright_{\{p\}})$ is the average time tokens spend in place p. Minimum, maximum, median, variance, and standard deviation can be computed in a similar way.

Each binary token flow $tf \in TFS$ needs to be converted into an observation of the form (t_a, t_b, c) where a represents the start of the segment (i.e., the top line), b represents the end of the segment (i.e., the bottom line), t_a is the start time of the observation, t_b is the end time of the observation, and c is the class. This can be done as follows. For a we can pick all the transitions producing tokens for place p or a specific transition in the preset of p, i.e., $a = \bullet p$ or $a \in \bullet p$. For b we can pick all the transitions consuming tokens from place p or a specific one, i.e., $b = p\bullet$ or $b \in p\bullet$. For $a = \bullet p$ and $b = p\bullet$, we create the *grouped multiset of observations* $Obs_p = [(t_a, t_b, c) \mid tf \in TFS\!\upharpoonright_{\{p\}} \wedge t_a = \pi_{tp}^{time}(tf) \wedge t_b =$

$\pi_{tc}^{time}(tf) \ \wedge \ c = class(tf)]$. For $a \in \bullet p$ and $b \in p\bullet$, we create the *ungrouped multiset of observations* $Obs_{a,b,p} = [(t_a, t_b, c) \mid tf \in TFS\restriction_{\{p\}} \wedge a = \pi_{tp}(tf) \wedge b = \pi_{tc}(tf) \ \wedge \ t_a = \pi_{tp}^{time}(tf) \ \wedge \ t_b = \pi_{tc}^{time}(tf) \ \wedge \ c = class(tf)]$. We use the notations from Definition 6 to define these two types of multisets. We leave the definition of the classification function *class* open. It may be based on properties of the case (e.g. the customer involved), events (e.g., the outcome), or some performance related computation (e.g., overall case duration). A simple example is $class(tf) = slow$ if $\pi_{dur}(tf) \geq 24$ h and $class(tf) = fast$ if $\pi_{dur}(tf) < 24$ h.

Using the above, we can create one grouped performance spectrum for place p based on Obs_p or $\mid \bullet p \mid \times \mid p\bullet \mid$ ungrouped spectra based on $Obs_{a,b,p}$ with $a \in \bullet p$ and $b \in p\bullet$.

4   Extracting Token Flows from Event Logs

Each binary token flow $tf = (e_{prod}, p, e_{cons})$ in a valid token flow set (i.e., a run of the Petri net) corresponds to an observation (i.e., a line) in the corresponding performance spectrum. Attributes and computations such as $\pi_{dur}(tf)$ can be used to add class labels (coloring of lines). Hence, given a valid token flow set $TFS \subseteq \mathcal{U}_{BTF}^N$, it is easy to construct performance spectra. In this section, we show how to obtain valid token flow sets from event data.

We start with an event log $L \subseteq \mathcal{U}_{LE}$ and first discover a Petri net $N = (P, T, F)$ using existing process discovery techniques (e.g., inductive mining, region-based discovery, etc.) [1]. The Petri net N may also be a discovered model that was modified or created manually. The event log L is replayed on the Petri net N by computing the so-called *alignments* [3,4]. Alignments make it possible to link each case in the event log (i.e., tokens having a common case identifier) to a path through the model. This may imply that events are ignored because they do not fit (called a "move on log") or that artificial events need to be introduced (called a "move on model"). In the latter case, also a timestamp needs to be added in-between the synchronous moves (i.e., fitting events before and after).

Alignments are typically computed per case and several approaches are available [3,4]. They all map each case in the log onto a path through the model such that the differences are minimized. Let c be a case in event log $L \subseteq \mathcal{U}_{LE}$. $\sigma_c = \{e \in L \mid \pi_{case}(e) = c\}$ are the events of c (often represented as sequence of activities). Replaying the firing sequence σ_c over N yields a collection of transition events $\gamma_c \subseteq \mathcal{U}_{TE}$. Note that events in σ_c may be ignored ("move on log") and events may be added ("move on model") due to silent transitions or deviating behavior. γ_c defines a partially-ordered run [7] for a particular case c. Each condition in such a run describes a token in place p with its producing event e_{prod} and its consuming event e_{cons}. This yields the binary token flows for case c. Taking the union over all cases yields the overall valid token flow set $TFS \subseteq \mathcal{U}_{BTF}^N$ for event log L. As shown in Sect. 3, such a TFS can be visualized in terms of performance spectra.

5   Implementation

The approach has been implemented as a ProM plug-in. Figure 1 already showed a screenshot analyzing a Petri net discovered from an event log with information about 39.022 orders and 7 activities (210.459 events in total). Figure 4 shows a larger visualization of the discovered process model.

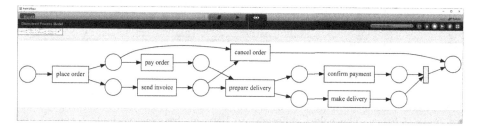

Fig. 4. The Petri net model automatically discovered based on the event log with 39.022 orders and 210.459 events.

To use the plug-in, install the package "PerformanceSpectrumIntegration" in combination with ProM's latest Nightly Build from promtools.org. The plug-in "Interactive Performance Spectrum" takes as input an event log and a Petri net.[1]. A possible scenario is to take an arbitrary XES event log, run the "Inductive Visual Miner", save the result as an "Accepting Petri Net", and then run the "Interactive Performance Spectrum" on the event log and accepting Petri net. Using these simple steps, one immediately gets the desired result and can now interactively show performance spectra side-by-side with the process model.

Alignments [3,4] play a key role in constructing token flows. The "Interactive Performance Spectrum" plug-in first applies the standard alignment techniques to the event log and the discovered Petri net. This is typically the biggest performance bottleneck and not specific to our approach (alignments are also needed to compute simple statistics for places such as the average waiting time). After this initialization step, the plug-in shows the Petri net and a variable number of performance spectra. Figure 1 shows three performance spectra. The top one is an aggregated performance spectrum with information about the token flow between activities *place order* and *send invoice* (the blue place). The two other performance spectra show details about the token flow of the green place, i.e., the place connecting activities *place order* and *pay order* and *cancel order*. The colors of the lines and bars are based on the overall flow time which is used as a class attribute.

Figure 5 shows another screenshot of the "Interactive Performance Spectrum" while analyzing the same process in a different period. Three performance spectra (marked ❶, ❷, and ❸) are shown. The first one (i.e., ❶) is a detailed performance

---

[1] Note that there are several variants depending on the type of Petri net, whether the alignment remains to be constructed, and to configure the class attribute.

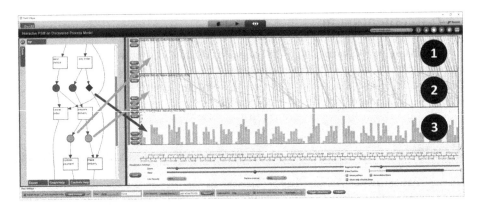

Fig. 5. Screenshot showing three performance spectra marked ❶, ❷, and ❸. The blue diamond was added to the process model to analyze the token flow between *pay order* and *make delivery*. The colors differentiate between frequent variants (orange) and infrequent variants (green). (Color figure online)

spectrum for the place connecting *prepare delivery* with *confirm payment*. As can be seen, work is sometimes queueing for *confirm payment* (see the gaps and diverging lines) and things are handled in FIFO (First-In-First-Out) order (there are no crossing lines). The second performance spectrum (i.e., ❷) shows the token flow for the place connecting *prepare delivery* with *make delivery*. Again we see queueing. However, now things are handled in a LIFO (Last-In-First-Out) order as illustrated by the crossing lines. The orange lines refer to token flows belonging to frequent variants and the green lines refer to token flows belonging to infrequent variants (i.e., cases following a less common path through the model). The third performance spectrum (i.e., ❸) refers to an artificially added place (indicated by the blue diamond in the Petri net). The virtual place connects *pay order* and *make delivery* and was added to analyze the token flow between these two activities. The tool facilitates the addition of such virtual places without recomputing alignments (the places do not need to be implicit and do influence the alignments; they are merely used to generate additional diagnostic token flows). Moreover, both detailed and aggregated performance spectra can be generated for such places.

By simply selecting places on the left-hand side, spectra are added on the right-hand side. It is also possible to add multiple virtual measurement places interactively. Given a place p, one can show one aggregated token-flow spectrum based on Obs_p or show individual spectra for all pairs of transitions in $\{(a, b) \mid a \in \bullet p \wedge b \in p \bullet \}$ based on $Obs_{a,b,p}$ (see Sect. 3). Moreover, there are many ways to adapt the view as illustrated by the bottom part of Fig. 5:

- It is possible to seamlessly rescale the time axis and to zoom-in on selected time periods using two sliders.
- The density of the lines can be adapted and it is possible to overlay the spectra with weeks, days, hours, and seconds.

- For smaller event logs, the views are immediately updated, for larger event logs it is possible to make multiple changes and then update the view.
- Performance spectra can be filtered based on attributes of events in the log or derived attributes such as duration or variant.
- Performance spectra use colors that may be based on event attributes or derived attributes, i.e., the *class*(*tf*) introduced before. This way one can, for example, see differences between products, customer groups, and departments. However, it can also be used to highlight exceptional or slow cases.

For aggregated performance spectra (like ❸ in Fig. 5) we need to select the time period used for binning the token flows. We can use years, quarters, months, weeks, days, days, hours, minutes or a customized time period. The aggregated performance spectrum in Fig. 5 uses day as a time period. Each stacked bar corresponds to one day. The height refers to the number of token flows and the colors refer to classes involved. One can clearly see the weekends and the varying load and changing mix of cases. Obviously, one cannot see such phenomena using traditional performance visualizations.

The "Interactive Performance Spectrum" plug-in has been applied to most of the publicly available real-life event logs from the 4TU Center for research data.[2] Experiments show that the biggest bottleneck is in the discovery (getting a reasonable Petri net) and alignment computation (which may be very time consuming).

6 Conclusion

This paper presented the novel "Interactive Performance Spectrum" plug-in (available as part of ProM's Nightly Build promtools.org) that visualizes token flows using performance spectra while providing a range of interaction mechanisms. This allows us to uncover real-life phenomena such as overtaking, batching, queueing, concept drift, and partial blocking that would otherwise remain invisible. By using a Petri net as the "lens" to look at event data, we overcome the limitations of our earlier work [5,6] where concurrency was not handled and we did not support the link between performance spectra and process models. This results in problems similar to using directly-follows graphs for process discovery [2]. Due to the "Interactive Performance Spectrum" plug-in we are now able to handle concurrency and tightly integrate process models and event data.

The applicability of the approach is mostly limited by the potential complexity of alignment computations. For large event logs and models, preprocessing may be time-consuming. Therefore, as future work, we would like to explore approximative alignments and apply various forms of sampling.

Acknowledgments. We thank the Alexander von Humboldt (AvH) Stiftung and Vanderlande Industries for supporting our research.

---

[2] The site https://data.4tu.nl/repository/collection:event_logs_synthetic provides 24 sets of real-life event data and 15 synthetic data sets.

References

1. van der Aalst, W.M.P.: Process Mining: Data Science in Action. Springer, Heidelberg (2016). https://doi.org/10.1007/978-3-662-49851-4
2. van der Aalst, W.M.P.: A practitioner's guide to process mining: limitations of the directly-follows graph. In: International Conference on Enterprise Information Systems (Centeris 2019), Procedia Computer Science, vol. 164, pp. 321–328. Elsevier (2019)
3. van der Aalst, W.M.P., Adriansyah, A., van Dongen, B.: Replaying history on process models for conformance checking and performance analysis. WIREs Data Mining Knowl. Discov. **2**(2), 182–192 (2012)
4. Carmona, J., van Dongen, B., Solti, A., Weidlich, M.: Conformance Checking: Relating Processes and Models. Springer, Heidelberg (2018). https://doi.org/10.1007/978-3-319-99414-7
5. Denisov, V., Belkina, E., Fahland, D., van der Aalst, W.M.P.: The performance spectrum miner: visual analytics for fine-grained performance analysis of processes. In: Proceedings of the BPM Demo Track, CEUR Workshop Proceedings, vol. 2196, pp. 96–100. CEUR-WS.org (2018)
6. Denisov, V., Fahland, D., van der Aalst, W.M.P.: Unbiased, fine-grained description of processes performance from event data. In: Weske, M., Montali, M., Weber, I., vom Brocke, J. (eds.) BPM 2018. LNCS, vol. 11080, pp. 139–157. Springer, Cham (2018). https://doi.org/10.1007/978-3-319-98648-7_9
7. Desel, J., Reisig, W.: Place/transition nets. In: Reisig, W., Rozenberg, G. (eds.) Lectures on Petri Nets I: Basic Models. LNCS, vol. 1491, pp. 122–173. Springer, Heidelberg (1998)
8. Juhás, G., Lorenz, R., Desel, J.: Unifying petri net semantics with token flows. In: Franceschinis, G., Wolf, K. (eds.) PETRI NETS 2009. LNCS, vol. 5606, pp. 2–21. Springer, Heidelberg (2009). https://doi.org/10.1007/978-3-642-02424-5_2
9. Keim, D., Kohlhammer, J., Ellis, G., Mansmann, F. (eds.) Mastering the Information Age: Solving Problems with Visual Analytics. VisMaster (2010). http://www.vismaster.eu/book/
10. Rogge-Solti, A., van der Aalst, W.M.P., Weske, M.: Discovering stochastic petri nets with arbitrary delay distributions from event logs. In: Lohmann, N., Song, M., Wohed, P. (eds.) BPM 2013. LNBIP, vol. 171, pp. 15–27. Springer, Cham (2014). https://doi.org/10.1007/978-3-319-06257-0_2
11. Rozinat, A., Mans, R.S., Song, M., van der Aalst, W.M.P.: Discovering colored petri nets from event logs. Int. J. Softw. Tools Technol. Transf. **10**(1), 57–74 (2008)

SNexpression: A Symbolic Calculator for Symmetric Net Expressions

Lorenzo Capra[1], Massimiliano De Pierro[2(✉)], and Giuliana Franceschinis[3]

[1] Dip. di Informatica, Università di Milano, Milan, Italy
lorenzo.capra@unimi.it
[2] Dip. di Informatica, Università di Torino, Turin, Italy
massimiliano.depierro@unito.it
[3] DISIT, Università del Piemonte Orientale, Alessandria, Italy
giuliana.franceschinis@uniupo.it

Abstract. The paper presents SNexpression: a tool for the symbolic structural analysis of Symmetric Nets (SN). It can operate at a low level, handling expressions required to compute the structural properties of interest, but features also a net-based way of interaction allowing to submit commands referring directly to the net structure avoiding error prone input of low level expressions. The User Interface implements a command line interpreter and provides also a multi-page notebook to keep track of the submitted commands and their result.

1 Introduction

The SNexpression tool has been developed with the aim of providing support to the structural analysis of Symmetric Nets (SN), a High-Level Petri Net (HLPN) formalism, without *unfolding* the net, allowing one to work at symbolic and parametric[1] level. A recently added feature is the possibility of deriving a set of *symbolic* ordinary differential equations (Symbolic ODE - SODE) from a Stochastic SN (SSN) model, making it possible an efficient computation of the average marking of colored tokens into places. A first version of SNexpression was presented in [6], but significant improvements/extensions have been implemented since then.

The theoretical work behind the tool has been published in a few papers defining the language for expressing the structural relations in symbolic form and the operators to be applied to the SN arc functions to derive several structural relations [5,7] or to generate a set of SODE from a model satisfying certain properties [3,4]. The basic idea consists of using a syntax similar to the SN arc expressions one, to symbolically represent structural relations useful for checking invariance properties, to deduce model behavioral properties, etc. Each symbolic structural relation is representative of several structural relations defined on the model unfolding: the latter can be derived from the former by instantiating it on specific colors. This approach has advantages: the compact representation,

---

[1] The method is parametric in the size of the *color classes*.

R. Janicki et al. (Eds.): PETRI NETS 2020, LNCS 12152, pp. 381–391, 2020.
https://doi.org/10.1007/978-3-030-51831-8_19

the similarity of the languages used to describe the model and that used to express the structural properties, and to some extent the possibility to apply it to models with parametric color class size, hence providing results that are valid for a family of similar models.

1.1 Definitions and Notation

SNs were introduced (with the name Well-Formed Nets) in [8]. It is a formalism, similar to Colored Petri Nets, featuring a syntax designed to naturally make symmetries explicit when the modelled system is symmetric (e.g. composed of several similarly behaving entities). A little SN model is depicted in Fig. 1, it is a small portion of a Distributed Memory fault tolerance mechanism model presented in [2]; the picture has been drawn with the GreatSPN GUI [1] and then (manually) translated into the textual format accepted by SNexpression (file with .sn extension). The automatic export from the GreatSPN GUI to the SNexpression net format is a planned future work. This is a natural choice since GreatSPN has been the first tool to support Symmetric Nets, moreover the GUI has been designed to allow extensions to the syntax (in SNexpression arc function terms may have both guards and filters) and handle several formalisms.

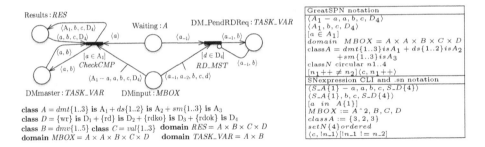

Fig. 1. A fragment of a distributed memory SSN model.

For the sake of space in this section we shall only describe in some detail the color structure of a SN, assuming that the reader is familiar with PN and HLPN formalisms and the definition of places, transitions, input, output and inhibitor arcs, marking. The color structure of a SN is built upon the basic color classes $\mathcal{C} = \{C_i, i = 1, \ldots, n\}$ which are finite and disjoint sets[2], may be (circularly) *ordered* or *partitioned into static subclasses* $C_{i,j}$. Transition and place *color domains* are defined as Cartesian products of classes : $D = \bigotimes C_i^{e_i}, e_i \geq 0, i = 1, \ldots, n$ (class C_i appears e_i times in the product). The color domain $cd(p)$ of place p defines the possible colors (tuples of color elements from $cd(p)$) of the tokens in its marking; the color domain $cd(t)$ of transition t defines its possible firing modes: these are

---

[2] In the tool color classes are denoted with a single capital letter: A, B, C, ... and the Cartesian product of classes is denoted as a comma separated list of classes.

tuples of color elements, and distinct typed variable names $(var(t))$ are used to refer to such elements in any tuple in $cd(t)$.

Let us consider the model in Fig. 1: $\mathcal{C} = \{A, B, C, D\}$, classes A and D are partitioned into static subclasses; the $cd(DMinput) = A^2, B, C, D$ and a tuple in this place could be $\langle dmt1, dmt2, dmv3, val2, rdok \rangle$; $cd(CheckCMP) = A, B, C$ $(var(CheckCMP) = \{a, b, c\})$ and one possible firing mode, also called instance, of this transition is $a = dmt1, b = dmv2, c = val1$. The enabling conditions of a transition instance and the effect of its firing depend on the functions on its input, inhibitor and output arcs. Guards can be associated with transitions, to constrain the set of valid instances. Transition $CheckCMP$ has several input and output places and its instances must satisfy predicate $a \in A_1$; the function on the arc from $DMinput$ is $\langle S\_A\{1\} - a, a, b, c, S\_D\{4\} \rangle$, while the functions on the arcs connecting it to place $DMmaster$ are $\langle a, b \rangle$. The domain of an arc function linking place p to transition t is $cd(t)$, whereas its codomain is $Bag[cd(p)]^3$. Its general form is: $\sum_i \lambda_i.T_i[g_i]$, $\lambda_i \in \mathbb{N}$, where T_i is a function-tuple $\langle f_1, \ldots, f_k \rangle$ denoting the Cartesian product of $class\ functions$ f_i. Each class-function f is a linear function defined on a subset of variables of $var(t)$ of the same type. Let $var_{C_i}(t)$ be the subset of $var(t)$ of type C_i, and $\widetilde{C_i}$ the set of static subclasses of C_i, then $f : cd(t) \to Bag[C_i]$ is so defined:

$$f = \sum_{v_k \in var_{C_i}(t)} \alpha_k.v_k + \sum_{v_k \in var_{C_i}(t)} \gamma_k.!v_k + \sum_{q \in \{1, \ldots, |\widetilde{C_i}|\}} \beta_q.S_{i,q} + \eta.S_i$$

where $\alpha_k, \gamma_k, \beta_q, \eta \in \mathbb{Z}$. $v_k \in var_{C_i}(t)$ in this context denotes the projection of a transition instance on the kth element of type C_i in its color domain; symbol $!$ denotes the $successor$ operator mapping the value of v_k to its successor (the type of v_k must be ordered). $S_{i,q}/S_i$ is a constant function mapping to the set $C_{i,q}/C_i$. Boolean expressions g_i (guards) on $var(t)$ may be associated with transitions or individual tuples; their terms are $standard\ predicates$ checking whether two variables hold the same value, or if a variable "belongs" to a given static subclass; if g_i is false for a given transition instance, the associated tuple evaluates to the empty-multiset. Scalars in class-functions must be such that no negative coefficients result from the evaluation of any color satisfying the guard possibly associated with the corresponding tuple/transition. Figure 1 contains examples of arc expressions involving several classes; in the table examples of functions operating on ordered classes (see Fig. 4) are also shown.

The calculus on which SNexpression operates, handles expressions of a language ($\mathcal{L}$) introduced in [5], very similar to arc functions but with additional constraints and a couple of extensions: the constraints are on the basic class functions (only $v_i, !^n v_i, S - v_i, S_i, S_{i,q}$ are allowed) and on their coefficients which cannot be negative, while the extensions are the use of intersection of basic class functions as tuple elements, and the possibility to use a predicate also as a $filter$ placed as a prefix in front of a tuple (filtering out the elements not satisfying

---

[3] A multiset on set D is a map $D \to \mathbb{N}$. $Bag[D]$ denotes the set of all multisets on D. If $m \in Bag[D]$, and $d \in D$, $m[d]$ is the multiplicity of d in m.

the filter predicate from the tuple evaluation). A number of unary and binary operators are defined on these expressions, which are useful when defining structural properties on SN models. The SNexpression tool implements a calculus on $\mathcal{L}$ and provides several off-the-shelf formulas to compute interesting structural properties of SN models.

To the best of our knowledge no other tools implement a general calculus for structural analysis of HLPNs. Even very advanced tools, e.g. Snoopy [9], take advantage of symmetry properties in the color structure to efficiently perform the net unfolding [10], but do not exploit it for structural analysis.

2   Tool Architecture and Functions

The architecture of SNexpression is organized in three layers, depicted in Fig. 2: the Library for Symbolic Calculus (LSC), the SN management framework (SNF), and the Command Line Interface (CLI). The LSC is a sort of Computer Algebra System that handles base-level SN expressions. The SNF middle layer manages more abstract objects, such as structural relation formulae, directly derived from a SN definition that may be loaded into the system; it also provides the algorithms needed to automatically derive the SODE for a given SSN model, based on a manipulation of SSN annotations.

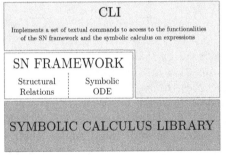

The CLI is a shell surrounding both the library and the SN framework, through which the user can operate directly on base-level expressions, using the CLI as a sort of symbolic calculator, or at a higher level, performing structural analysis of (S)SN models previously loaded.

To help the users operation during a session the CLI provides a multi-page textual notebook where it is possible to annotate and save formulae to be submitted, or results, or com-

Fig. 2. Architecture of SNexpression.

ments, in other words anything useful to support multi-step complex analysis. Since the format of the LSC output is pretty similar to the syntax of CLI input, copy-and-paste from the notebook to the command window and vice-versa may be conveniently used.

SNexpression is implemented in Java. The LSC is distributed as a standalone jar file, so that programmers can use it in other projects. Its API is available at URL: www.di.unito.it/~depierro/SNexpression/libAPI, we plan to make the LSC soon available as open-source project. At the current release, the CLI and the SNF are built as a unique executable, but we plan to make also the SNF accessible through a public API. Since the CLI reads from the standard input, it might be integrated in other tools. The following sections discuss the functions of the three layers of the tool architecture in more detail.

2.1   The LSC: A Computer Algebra System for SN Expressions

The major functionalities and the design of the LSC are summarized here. The library implements a (parametric) rewriting system: algebraic rules are used to rewrite symbolic expressions composed of terms of $\mathcal{L}$, and the associated set of operators: sum, difference, intersection, composition, transpose. Rewriting stops when no more rules apply, in which case the resulting term is considered in "normal form". Final expressions manipulated by the LSC match a sort of *disjunctive normal form*, where only SN functions, guards, filters and sum/intersection operators may occur.

With respect to earlier versions, the LSC currently supports both set- and bag-expressions (i.e. espressions returning multisets), with the only exception of composition, which is partially implemented for bag-expressions: its complete definition is work in progress. The support operator provides a convenient bridge between bag- and set-expressions.

Thanks to its modular layout and intuitive API, the LSC may also be used as a standalone component. As a direct consequence of its design, it is possible to directly build and manipulate objects (terms) at three different levels: class-functions, guards/filters, and function-tuples. Each level has its own set of operators, similar among the levels. Guards/Filters (standard predicates), class-functions, and single function-tuples have a canonical representative, which coincides with their normal form. There is no canonical form for sums (bags) of function-tuples: in general, however, equivalence between expressions may be syntactically checked by using the difference operator.

Library Architecture and API. The library consists of around one hundred Java classes/interfaces, divided in ten packages. The adopted design has many advantages. *Ease of extension/maintenance:* changes or updates (e.g., adding new language elements) are low-cost. *Modular testing/debugging:* every single element of the language can be managed in a uniform way. *Efficiency:* term normalization complexity is alleviated by a reduced use of recursion (the normalization times for many examples, some of which very complex, vary from msec. to sec.).

A code snippet illustrating the several steps needed to create and normalize a SN expression (the transpose of a tuple composition) is listed in Fig. 3. A (simplified) UML class-diagram describing the top of the LSC type hierarchy, and its connections to the lower levels, can be found in the tool home page, together with a small portion of the library's API concerning simplification methods.

2.2   The Symmetric Nets Management Framework

The SNF implements the method to load a SN description and those to compute some symbolic structural relations on it, listed in Table 1: some relations are functions on sets, others return multisets. For the structural relations computation it exploits the functions implemented in the library (difference, transpose, composition) on one hand, and the information on the loaded SN structure on the other hand: the model structure allows to select transition pairs that might be in

```
import wncalculus.color.ColorClass;
import wncalculus.classfunction.*;
import wncalculus.guard.*;
import wncalculus.tuple.*;
//...
ColorClass C = new ColorClass("C", true); //ordered class C s.t. |C| > 1
Interval i1 = new Interval(3,8), // [3,8] (constraint)
          i2 = new Interval(2,2); // [2,2]
ColorClass D = new ColorClass("D", new Interval[] {i1, i2}); // partitioned
       class D = D{1} \cup D{2}
Projection c_1= Projection.builder(1, C), // c_1
           c_2= Projection.builder(2, 1, C), // !c_2
           d_1= Projection.builder(1, D); // d_1
SetFunction comp_c_1, comp_d_1, sd2, inter;
comp_c_1 = ProjectionComp.factory(c_1); // S - c_1
comp_d_1 = ProjectionComp.factory(d_1); // S - d_1
sd2 = Subcl.factory(2, D); // constant D{2}
inter = Intersection.factory(comp_d_1,sd2);//D{2} \cap S -d1
Domain dom = new Domain(C,C,D); // domain C^2 x D
Guard g1 = Membership.build(d_1,sd2,true,dom); //d_1 \in D{2}
Tuple t1 , t2;
t1 = new Tuple(dom, c_1, comp_c_1, c_2, d_1); // <c_1,S-c_1,!c_2,d_1>
t2 = new Tuple(null, g1, dom, comp_c_1, c_2 , inter); // <S-c_1,!c_2,(S-d_1
     * S_D{2})>[d_1 in D{2}]
TupleComposition tcom = new TupleComposition(t1, t2);
TupleTranspose tcom_trans = new TupleTranspose(tcom);
List<LogicalExpr> result = tcom_trans.simplify();
System.out.println(result);
```

Fig. 3. Snippet showing creation and simplification of "function-tuple" expressions.

structural conflict or causal connection relation, then the arc functions labeling the involved arcs are processed through the symbolic calculus implemented by the LSC. For the mutual exclusion structural relation an ad-hoc computation algorithm [7] is applied after pre-processing the selected arc expressions through the library methods.

Finally, the SNF implements the algorithm to derive a set of Symbolic Ordinary Differential Equations from a (partially unfolded) SN [3,4]: it exploits the library to compute the (multiset) transpose of the arc expressions and to derive the *enabling degree* of homogeneous sets of transition instances. It operates with just one command *print_ode* after having loaded the SN to be translated. A file .ode is produced, including the set of SODE (one for each model place) ready to be solved using Rstudio.

2.3 The Command Line Interface

The CLI is the user interface of the tool: despite its simplicity it provides the essential commands to access the main functions of the LSC and of the SNF implementing four kinds of commands: definition of classes or language expressions; application of operators to expressions and simplification; loading a SN and computing some structural relations on it; derivation of a set of SODE from a SN, which in turn needs the computation of some structural relations. The syntax of all commands is described in the manual: Table 1 summarizes the main types of commands; a few detailed examples are described in Sect. 3. By convention the *color classes* are denoted with capital letters $(A, B, C, ...)$ while small letters, possibly indexed with an integer, denote *variables* whose type is the corresponding capital letter class (e.g. a or a_2 of type A). Classes may be partitioned into static subclasses denoted by the class capital letter followed by an integer index (e.g. $A\{1\}$ subclass of A). Classes have finite cardinality, but it can be defined to be parametric (by default any class has a parametric cardinality n greater than or equal to two; when a class is partitioned into static subclasses only one subclass may have parametric cardinality). Domains are Cartesian products of color classes, if one class appears more than once in a domain, it is listed once followed by the number of repetitions.

Table 1. Summary of the main commands implemented in the CLI.

Description	CLI syntax (examples)
Defining a class and showing its definition	
A (possibly ordered) class	*set C ordered*
and its subclasses	*set M := $\{10, 3, [4, n]\}$*
Show class definition	*class(A)*
Symbol definition	
Define domain symbol	*dom := $A \,\hat{}\, 2, B, C \,\hat{}\, 3$*
Define expression symbol	*exp := $@A \,\hat{}\, 3 < a\_1 * S - a\_2, (a\_1 + a\_2) * S - a\_3 >$*
(with domain prefix)	(*: intersection, a_i and $S - a\_j$: basic functions)
may include a *filter*	*$@D \,\hat{}\, 2[d\_1! = d\_2, d\_2 \text{ in } D\{1\}] < d\_2, S - d\_1 >$*
and a *guard*	*$@A \,\hat{}\, 3 < a\_1, a\_3 > [a\_3 \text{ in } M\{2\}]$*
Define multiset expression	*mexp := $@C, D \text{ mset} : 2 < d\_1, c\_1 > + < S\_D, c\_1 >$*
Operators application	
support (applies to bag-expressions)	*<<mset expression>>*
transpose	*expression'*
difference	*expression1 − expression2*
composition	*expression1.expression2*
	implementation for bag-expressions is not yet complete
simplify expression	*s(e)* rewrites an expression into a normalized form
simplify and fold	*sf(e)* merges terms or expressions ($\neq$ constraints)
Symmetric Nets Framework commands	
Nets management commands	
Load net	load "DistMem.sn"
Input/output/inhib. places and arc expr	I(t) / O(t) / H(t)
Symbols for arc (bag-)expressions	I(t,p) / O(t,p) / H(t,p)
Structural properties	
Conflict	$SC(t1, t2, p)$ or $SC(t1, t2)$
Self-Conflict	$SC(t, p)$ or $SC(t)$
Causal Connection	$SCC(t1, t2, p)$
Mutual Exclusion	$SME(t1, t2, p)$
Added By (set/ multiset)	$AB/ABmset(t, p)$
Removed By (set/multiset)	$RB/RBmset(t, p)$
Derivation of a set of SODE	*print_ode*

The expressions can be interpreted as functions mapping into multisets or functions mapping on sets, the latter case is the default. The prefix *mset:* indicates that an expression denotes a function mapping on multisets. The expressions syntax takes the form of a *sum of tuples*, each tuple may be prefixed with a *filter* and suffixed with a *guard* (both filter and guard take the form of SN *standard predicates*). The tuple elements are intersections (∗) of basic functions: projection, complement, successor, constant (whole class or one static subclass).

Operators can be applied to expressions: there are two unary operators, support and transpose, and two binary operators, difference and composition. The support operator can be applied to a multiset-expression to obtain the corresponding expression mapping on sets; the transpose operator is available both for expressions mapping on multisets and for those mapping on sets: the result of its application is an expression of the same type. The difference can be applied to any pair of expressions (of the same type) while the composition is completely implemented for expressions mapping on sets and on a significant subset of multiset expressions. These operators are the basis for the SN structural analysis implemented in the SN Management Framework. For instance the structural conflict between two transitions t_1 and t_2 sharing an input place p is computed as follows: $SC(t_1, t_2, p) = \ll I(t_1,p) - O(t_1,p) \gg'. \ll I(t_2,p) \gg +$ $\ll O(t_1,p) - I(t_1,p) \gg'. \ll H(t_2,p) \gg$ where $I(t,p)$ and $O(t,p)$ are respectively the expressions on the input and output arcs connecting p and t. This could be useful to identify the groups of immediate transitions that are potentially in conflict and define how such conflicts should be solved.

To support the user in performing experiments with the tool, the CLI embeds a multi-page notebook: it is possible to copy-and-paste commands annotated in the notebook to the command window and then copy-and-paste results from the command window back in the notebook. When the color classes involved in the expressions processed by the tool have parametric cardinality, the result of a computation is not a single expression but a list of expressions, each with associated a different range of possible values for the classes cardinality: indeed one of the strong points of the tool is its ability to handle expressions without necessarily fixing the color classes sizes, so that the obtained result is valid for a family of models differing only in the size of (some) color classes.

3   Use Cases: Exploiting SNexpression

The goal of this section is to show on a few practical examples some features of SNexpression. Let us consider the relay race model in Fig. 4 (described in [7]), representing a set of teams (class C), each composed of four athletes (ordered class N, $|N| = 4$), competing in a relay race. Some symbolic structural properties of interest are the *causal connection* and *structural conflict* involving transition *Run* and the immediate transitions *pass*, *Win* and *notWin*; these properties

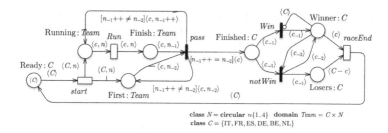

class N = **circular** $n\{1..4\}$ **domain** $Team = C \times N$
class $C = \{IT, FR, ES, DE, BE, NL\}$

Fig. 4. An SN model of a relay race.

are useful (among others) for model validation purposes, or to correctly define immediate transition weights. Let us consider the commands summary in Table 2

1) computes the instances $\langle c, n', n'' \rangle$ of $pass$ that may enable an instance $\langle c\_1, n\_1 \rangle$ of Run through $Running$; the result shows that such instances involve an athlete of the same team

Table 2. Examples of structural relation expressions.

1) $SCC(pass, Run, Running) = @C, N\langle c\_1, !3n\_1, S - n\_1 \rangle$
2) $SCC(pass, Win, Finished) = @C[n\_1 = !3n\_2]\langle c\_1, S\_N, S\_N \rangle$
3) $SC(Win) = @C\langle S - c\_1 \rangle$
4) $SME(Win, notWin, Winner) = @C, N\hat{\ }2\langle S\_C \rangle$
5) $s(fscc2.fsc3) = @C[n\_1 = !3n\_2]\langle S - c\_1, S\_N, S\_N \rangle$
where $fscc2 := @C[n\_1 = !3n\_2]\langle c\_1, S\_N, S\_N \rangle$, $fsc3 := @C\langle S - c \rangle$
6) $SC(RD\_MST, CheckCMP) = (\langle a\_1, S\_A, S\_B, S\_C, S\_D\{4\}\rangle$
$+ \langle (S\_A\{1\} * S - a\_1), a\_1, b\_1, c\_1, S\_D\{4\}\rangle)[a\_1\ in\ A\{1\}]$
7) $SC(CheckCMP, RD\_MST) = \langle a\_2, b\_1, c\_1 \rangle[a\_1! = a\_2, d\_1\ in\ D\{4\},$
$a\_1\ in\ A\{1\}, a\_2\ in\ A\{1\}] + \langle a\_1, S\_B, S\_C \rangle[d\_1\ in\ D\{4\}, a\_1\ in\ A\{1\}]$

(c_1) with sequence number n' equal to the predecessor[4] of n_1, provided that $n'' \neq n\_1$ (i.e. the team has not run the last section yet). 2) computes the instances $\langle c, n', n'' \rangle$ of $pass$ enabling instance $\langle c\_1 \rangle$ of Win: the result has a filter and denotes the instances involving the same team c_1, and the last section runner (the predecessor of n''). 3) computes the structural auto-conflicts among different instances of Win, while the result of 4) shows that Win and $notWin$ are mutually exclusive: indeed, $Winner$ is input place for $notWin$ and inhibitor place (with arc function $\langle S_C \rangle$) for Win. In SSNs with immediate transitions it is useful to check for confusion, i.e., situations where the model is underspecified (a situation solved by using priorities). In our example, this may arise due to the fact that different instances of Win are in conflict with each other. There would be confusion if an instance of $pass$ fired while an instance of Win is enabled: this could cause the enabling of another instance of Win in conflict with the former. 5) shows how to obtain the $confusing$ instances of $pass$ by composing the results of 2) and 3): in this case the SNF is not involved.

Other structural relations can be computed on the model in Fig. 1, whose arc functions are a bit more complex as illustrated by relations 6) and 7) in Table 2. The resulting expressions have the same domain as the 2nd parameter of SC, namely A, B, C for 6) and $A\hat{\ }2, B, C, D$ for 7). The terms are pair-wise disjoint: this enhances readability and interpretation.

---

[4] Since $|N| = 4$ the predecessor $!^{-1}n$ of n coincides with the third successor $!^3n$.

The tool can be used for other purposes. A recent implementation concerns the automatic generation of Symbolic ODE from an SSN model (command *print_ode*). The technique applies only to models whose underlying stochastic process is density dependent. One condition for an SSN to satisfy such property is the coverage of places by P-invariants. In [4] an application to a botnet model has been illustrated (this is one of the examples that can be downloaded from the tool's web page). SNexpression can be used to check if a given P-indexed vector of multiset expressions defines a set of *colored* P-invariants, and possibly prove the coverage of all place instances. The expressions in the P-indexed vector denote functions from the place color domains to the P-invariant's domain. An example of P-vector of expressions $C, L \to Bag[L]$ is: $pinv[NoConBot] := @C, L\ mset : \langle l \rangle [c\ in\ C\{1\} + c\ in\ C\{2\}];\ pinv[ConBot] := @C, L\ mset : \langle l \rangle [c\ in\ C\{2\}];\ pinv[InactBot] := @C, L\ mset : \langle l \rangle [c\ in\ C\{3\} + c\ in\ C\{4\}];\ pinv[ActBot] := @C, L\ mset : \langle l \rangle [c\ in\ C\{3\} + c\ in\ C\{4\}]$. In order to prove that this vector corresponds to a set of P-invariants we need to show, for each transition t, that the sum over all places of the compositions of P-invariant's function ($pinv[p]$) with the difference $O(t, p) - I(t, p)$ results in the null function. Due to space constraints we show only the result for transition $ConnectBot$: $s(pinv[NoConBot].f4 + pinv[ConBot].f5) = null$ where $f4 := @C, Lmset :< 0_C, 0_L > -I(ConnectBotNet, NoConnBot)$ and $f5 := O(ConnectBotNet, ConnBot)$. The same result holds for all transitions, thus $pinv[]$ is a P-invariant: it represents $|L|$ invariants, indicating that the number of tokens with second component $l \in L$ is constant, and have the correct label in C. In the tool web page other P-invariants for this model are available.

4   Conclusions and Installation Instructions

SNexpression implements a symbolic calculus useful for studying the structure of a Symmetric Net and deriving information on its behavioral properties. It has also been used to derive a set of SODE from an SSN model for performance analysis purposes. Several extensions are planned: completing the composition of bag-expressions, further automation of net structural calculus (e.g., checking P-invariants or building Extended Conflict Sets), automatizing the SN partial unfolding procedure which is a preliminary step for the generation of the SODE from a SN model, providing access to the different software layers with suitable APIs. Finally, we plan to build a bridge between GreatSPN and SNexpression.

A free version of the tool is available: download the archive SNEx.zip from the project homepage www.di.unito.it/~depierro/SNexpression, unzip its content into a folder. The extracted file structure contains the main program SnexCLI.jar and, in the folder lib, the library SNexLib.jar. To launch the tool in an OS shell run: java -jar <path to SNexCLI.jar> (JRE 1.8 or above is necessary). At the project's web page, the user can find a reference manual and some examples to immediately start using it.

References

1. Amparore, E.G., Balbo, G., Beccuti, M., Donatelli, S., Franceschinis, G.: 30 years of GreatSPN. In: Fiondella, L., Puliafito, A. (eds.) Principles of Performance and Reliability Modeling and Evaluation, pp. 227–254. Springer, Cham (2016). https://doi.org/10.1007/978-3-319-30599-8_9

2. Ballarini, P., Capra, L., Franceschinis, G., De Pierro, M.: Memory fault tolerance software mechanisms: design and configuration support through SWN models. In: 3rd International Conference on Application of Concurrency to System Design ACSD 2003, Guimaraes, Portugal, 18–20 June 2003, pp. 111–121 (2003)

3. Beccuti, M., Capra, L., De Pierro, M., Franceschinis, G., Follia, L., Pernice, S.: A tool for the automatic derivation of symbolic ODE from symmetric net models. In: 27th IEEE International Symposium MASCOTS 2019, Rennes, France, October 21–25 2019, pp. 36–48 (2019)

4. Beccuti, M., Capra, L., De Pierro, M., Franceschinis, G., Pernice, S.: Deriving symbolic ordinary differential equations from stochastic symmetric nets without unfolding. In: Bakhshi, R., Ballarini, P., Barbot, B., Castel-Taleb, H., Remke, A. (eds.) EPEW 2018. LNCS, vol. 11178, pp. 30–45. Springer, Cham (2018). https://doi.org/10.1007/978-3-030-02227-3_3

5. Capra, L., De Pierro, M., Franceschinis, G.: A high level language for structural relations in well-formed nets. In: Ciardo, G., Darondeau, P. (eds.) ICATPN 2005. a high level language for structural relations in well-formed nets, vol. 3536, pp. 168–187. Springer, Heidelberg (2005). https://doi.org/10.1007/11494744_11

6. Capra, L., De Pierro, M., Franceschinis, G.: A tool for symbolic manipulation of arc functions in symmetric net models. In: Proceedings of the 7th International Conference VALUETOOLS 2013, Torino, Italy, pp. 320–323, ICST. Belgium (2013)

7. Capra, L., De Pierro, M., Franceschinis, G.: Computing structural properties of symmetric nets. In: Proceedings of the 15th International Conference on Quantitative Evaluation of Systems, QEST 15, Madrid, ES. IEEE CS (2015)

8. Chiola, G., Dutheillet, C., Franceschinis, G., Haddad, S.: Stochastic well-formed coloured nets for symmetric modelling applications. IEEE Trans. Comput. **42**(11), 1343–1360 (1993)

9. Heiner, M., Herajy, M., Liu, F., Rohr, C., Schwarick, M.: Snoopy – a unifying petri net tool. In: Haddad, S., Pomello, L. (eds.) PETRI NETS 2012. LNCS, vol. 7347, pp. 398–407. Springer, Heidelberg (2012). https://doi.org/10.1007/978-3-642-31131-4_22

10. Liu, F., Heiner, M., Yang, M.: An efficient method for unfolding colored Petri nets. In: Winter Simulation Conference, WSC 2012, Berlin, Germany, 9–12 December 2012, pp. 295:1–295:12 (2012)

Cycl⟲n – A Tool for Determining Stop-Transitions of Petri Nets

Jörg Desel, Marc Finthammer[(✉)], and Andrea Frank

FernUniversität in Hagen, Hagen, Germany
{joerg.desel,marc.finthammer,andrea.frank}@fernuni-hagen.de

Abstract. This paper introduces the tool Cycl⟲n. The core functionality of Cycl⟲n is to determine for a transition t of an unbounded Petri net whether or not t stops the net. A transition t stops the net (and is called a stop-transition) if each reachable marking of the net enables only finite occurrence sequences without occurrences of t. Cycl⟲n provides a graphical user interface which also illustrates the graph structures leading to the computed result. This way, results are explained in a comprehensible manner, and the user gets a visual explanation of the cyclic behavior of the net causing these results.

Keywords: Unbounded Petri nets · Coverability graphs · Termination

1   Introduction

The core functionality of the tool Cycl⟲n is to provide a practical solution to the following problem, originally introduced in [1]: Assuming a Petri net[1] and a transition t of this net, can we eventually stop the behavior of the net by forbidding occurrences of t in occurrence sequences enabled at an arbitrary reachable marking? Or, equivalently: Does no reachable marking of the net enable an infinite occurrence sequence without occurrences of t? If this is the case, then we say that *transition t stops the net* or that *t is a stop-transition*. Consequently, if t does not stop the net, then some reachable marking enables an infinite occurrence sequence without occurrences of t.

As thoroughly explained in [1], determining if a transition t stops an *unbounded* net is a difficult task, since just checking a coverability graph of the net for cycles[2] does not provide a sufficient criterion for unbounded nets[3]. However, it is shown in [1] that t stops the net if and only if there is no *non-decreasing closed path* of the coverability graph without an arc labeled by t. A closed path of the coverability graph is called non-decreasing, if, for every place

---

[1] Cycl⟲n handles place/transition Petri nets without arc weights, capacity restrictions or inhibitor arcs (in accordance with the nets considered in [1]).

[2] A closed path is a *cycle* if no two distinct arcs of the path start at the same node.

[3] As also shown in [1], considering stop-transitions in *bounded* nets is a much simpler task, since indeed it merely requires to check cycles in the reachability graph.

© Springer Nature Switzerland AG 2020
R. Janicki et al. (Eds.): PETRI NETS 2020, LNCS 12152, pp. 392–402, 2020.
https://doi.org/10.1007/978-3-030-51831-8_20

of the net, at least as many tokens are added as are removed by the transitions at the arcs of the path, and hence the effect of these transitions is non-negative for each place. An algorithm has been developed in [1] which – very roughly speaking – identifies and processes non-decreasing closed paths in a coverability graph of the net and that way determines stop-transitions. The algorithm involves constructing a homogeneous system of linear inequalities over the arcs of the coverability graph, computing a basis of solutions and performing a divide-and-conquer approach on the solutions to finally determine if a transition t stops the net.

Cycl◯n implements this algorithm. It provides a graphical user interface and performs all necessary computations in the background, making it easy for the user to examine if a selected transition stops the net. Moreover, Cycl◯n does not only deliver a pure yes-or-no result, but also features a graphical representation of the involved graph structures to illustrate and explain the computed result. That way, Cycl◯n allows the user to comprehend the cyclic behavior of the net leading to the result. Furthermore, Cycl◯n offers additional features, e.g. considering several transitions simultaneously, automatically investigating each transition of the net in a batch-processing manner, focusing on a certain subset of the net, and visualizing cyclic behavior in a more general way (i.e. without a priori considering a certain transition).

In this contribution, we will illustrate all functionalities of Cycl◯n in detail by means of examples. Note that some of these functionalities are in close connection to the theoretical background explained in [1]. Hence, we encourage the reader to see [1] for a deeper understanding of some of the concepts, which we can only outline here very briefly.

2   User Interface and Functionalities

Figure 1 shows the graphical user interface of the Cycl◯n tool. Cycl◯n supports Petri net models specified in the PNML file format. After opening a PNML file, a graphical representation of the net is displayed in the upper left part of the GUI and a coverability graph (CG) of the net is constructed[4] and displayed in the upper right part. Each ω-marking m_ω in CG is represented as a tuple $(m_\omega(p_1), \ldots, m_\omega(p_n))$, i.e., each such tuple is implicitly based on the given order of places. Each arc of CG is labeled by $[a_i]$ t with a_i being a distinct arc identifier and t corresponding to the respective transition.

The output area in the lower left part informs about the results of the computation and the lower right part holds a selectable list of computed basis solutions (which we will explain in Sect. 2.2). The GUI has a toolbar which provides buttons for all functionalities of the Cycl◯n tool. Above the toolbar is another array of buttons which provides quick access to several preconfigured examples (i.e. PNML files with preselected transitions).

---

[4] The coverability graph is computed by a non-deterministic algorithm; consequently, the structure and arc numbering of the particular coverability graph may change with each computation.

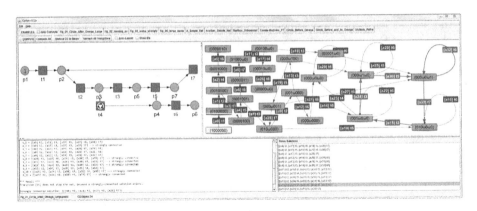

Fig. 1. Graphical user interface of the CyclOn tool

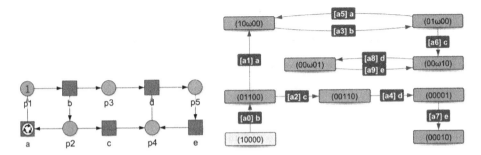

Fig. 2. An unbounded net with transition a selected as a forbidden transition and a coverability graph of the net

2.1 Core Functionalities

Figure 2 shows an unbounded[5] Petri net (which we will use as a running example throughout this section) and a coverability graph CG of this net. Note that simply taking a closer look at CG does not help in determining which transitions stops the net: Obviously, the coverability graph contains two[6] closed paths (even cycles), but we have to search for non-decreasing closed paths in order to determine stop-transitions. In the following, we will demonstrate how to employ the CyclOn tool for this task.

To examine whether or not a certain transition stops the net, e.g. transition a, the user selects the respective transition via mouse click (indicted by the

---

[5] Note that CyclOn also handles bounded Petri nets, but we focus on the much more challenging and interesting case of unbounded nets in this paper.

[6] Precisely, there are infinitely many closed paths, since e.g. not only $a5, a3$ is a closed path, but also $a5, a3, a5, a3$, and so on.

red-white sign[7] on transition a in Fig. 2). Roughly speaking, such a selected transition is considered to be forbidden to occur infinity often, therefore we also refer to a selected transition as a *forbidden* transition. Pressing the "Compute" button starts the algorithmic computation to determine if the selected transition a is a stop-transition. The following result is presented in the output area:

```
Forbidden Transition: T = [a]
*** Result ***
Transition [a] stops the net.
```

That is, if we forbid transition a to occur infinitely often, then the net eventually terminates. This result follows directly from the algorithmic computation which has shown that there is no non-decreasing closed path in CG without an arc labeled by transition a. Expressed the other way round: Every non-decreasing closed path in CG contains an arc labeled by transition a.

Next, we select transition e instead. Performing the computation yields:

```
Forbidden Transition: T = [e]
Basis Solutions:
  b_0 = [[a3] b, [a5] a] --> strongly connected
*** Result ***
Transition [e] does not stop the net,
because a strongly-connected solution exists: [[[a3] b, [a5] a]]
```

That is, even if we forbid transition e to occur infinitely often, the net does not terminate necessarily, since there exists an infinite occurrence sequence which does not rely on transition e. The output also gives reasons for the result by presenting a particular *strongly connected solution*, which – in the context of [1] – corresponds directly to a non-decreasing closed path without an arc labeled by transition e. The non-decreasing closed path indicated by the output [[a3] b, [a5] a] is also highlighted green in CG. So the green arcs a3 and a5 represent a non-decreasing closed path which has no arc labeled by transition e. Consequently, this non-decreasing closed path leads to an infinite occurrence sequence $b, a, b, a, \ldots$ without transition e and therefore transition e does not stop the net. The involved transitions a and b are also highlighted green in the graphical representation of the net. The graphical representation of CG makes it easy to see that the infinite occurrence sequence is enabled at marking $(10\omega00)$ and that the (finite) sequence b, a leads from the initial marking to that particular marking.

By clicking on a marking in CG, this particular marking becomes highlighted and is displayed in the graphical representation of the net, i.e. the respective number of tokens is displayed on each place of the net. This allows to click several markings in CG (e.g. the markings involved in a non-decreasing closed path) one by one and retrace the corresponding markings in the net, in this way e.g. studying the cyclic behavior of the net.

Cyclon also allows to select multiple transitions at once in order to examine if forbidding these transitions simultaneously stops the net. For example, selecting transitions c, d, and e gives the result:

---

[7] The red-white "traffic-circle-prohibition-sign" shall give an intuitive hint that this transition is not "absolutely" forbidden to occur, but it is forbidden to occur in "infinite cyclic activities" (very roughly speaking).

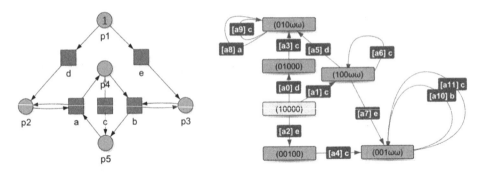

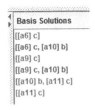

Fig. 3. An unbounded net which leads to several basis solutions; a coverability graph of the net

Basis Solutions
[[a6] c]
[[a6] c, [a10] b]
[[a9] c]
[[a9] c, [a10] b]
[[a10] b, [a11] c]
[[a11] c]

Fig. 4. Multi-selectable list of basis solutions in the GUI (green: strongly connected, red: not strongly connected) resulting from computation for transition a (Color figure online)

```
Forbidden Transitions: T = [c, d, e]
Basis Solutions:
  b_0 = [[a3] b, [a5] a] --> strongly connected
*** Result ***
Transitions [c, d, e] do not stop the net,
because a strongly-connected solution exists: [[[a3] b, [a5] a]]
```

Considering the previous result, this does not surprise, since the already known strongly connected solution [[a3] b, [a5] a] (non-decreasing closed path, respectively) does not contain any of these transitions.

Furthermore, CyclÖn features some sort of batch processing to check for each transition of the net automatically if it stops the net. Starting this batch processing via the "Compute All" button results in the following output:

```
*** Results for All Transitions ***
Transition [a] stops the net? true
Transition [b] stops the net? true
Transition [c] stops the net? false    strongly connected solution: [[[a3] b, [a5] a]]
Transition [d] stops the net? false    strongly connected solution: [[[a3] b, [a5] a]]
Transition [e] stops the net? false    strongly connected solution: [[[a3] b, [a5] a]]
```

2.2   Visualization of Basis Solutions

Now we consider the net and a coverability graph CG of this net depicted in Fig. 3. Selecting transition c yields:

```
Forbidden Transition: T = [c]
Basis Solutions:
  b_0 = [[a8] a, [a10] b]
*** Result ***
Transition [c] stops the net.
```

Note that the algorithm has determined a basis solution. However this basis solution is not strongly connected, i.e. the respective arcs a8 and a10 in CG do not constitute a strongly connected subgraph of CG; but being strongly connected is a necessary property of a closed path. Consequently, since no strongly connected solution exists, there is no non-decreasing closed path either and therefore transition c stops the net.

By performing the computation for transition a instead, we get a more comprehensive result:

```
Forbidden Transition: T = [a]
Basis Solutions:
  b_0 = [[a6]  c]              --> strongly connected
  b_1 = [[a6]  c, [a10] b]
  b_2 = [[a9]  c]              --> strongly connected
  b_3 = [[a9]  c, [a10] b]
  b_4 = [[a10] b, [a11] c] --> strongly connected
  b_5 = [[a11] c]              --> strongly connected
*** Result ***
Transition [a] does not stop the net,
because a strongly-connected solution exists: [[[a6] c]]
```

As we have already seen, the result that a transition stops the net is always justified by an appropriate strongly connected solution. However, this time the output also informs us about six basis solutions in total, four of them being strongly connected. All basis solutions are also listed in the lower-right area of the GUI (see Fig. 4). The color of each basis solution indicates whether it is strongly connected (green) or not (red). Selecting a solution from the list highlights the appropriate arcs in CG as well as the corresponding transitions in the net in the respective color (see Fig. 5a).

Moreover, multiple basis solutions can be selected at once and CyclOn determines on-the-fly for that particular set of basis solutions (i.e. for the induced subgraph) whether it is strongly connected or not. To indicate the result, all involved arcs in CG (and transitions in the net) are highlighted in green or red, respectively (see Fig. 5b). That way, the user can try out different combinations of basis solutions and receives a visual information of which arcs (and transitions) are involved and whether they are strongly connected or not.

2.3   Focusing on a Part of the Net

Next, we consider the net and coverability graph CG depicted in Fig. 6. CyclOn also supports to determine the answer to a more general version of our original question, namely: Does a transition t stop all transitions from *a certain part* of the net? That is, with U being a subset of transitions of the net, we ask: Does

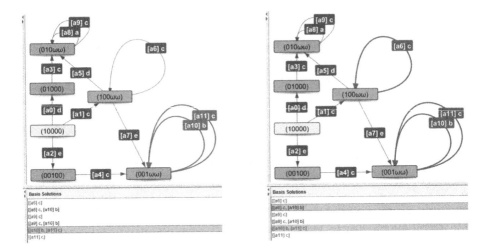

Fig. 5. a) A strongly connected basis solution is selected in the list, and the corresponding arcs in the coverability graph are highlighted green **b)** Two basis solutions (one strongly connected, the other not strongly connected) are selected in the list, and the corresponding arcs in the coverability graph are highlighted red, since together these basis solutions are not strongly connected. (Color figure online)

transition t stop all transitions in U? So we ask if there is no non-decreasing closed path which does not contain t but contains some transition from U. Note that if U contains all transitions of the net, then this generalized question coincides with our original question.

To define such a subset U of transitions, the user selects the respective transitions by holding the Shift-key and clicking them, that way labeling these transitions with yellow "or" lettering. In Fig. 6, we have selected transitions a and b that way, and additionally selected transition c as a forbidden transition. The computation gives the following result (note that we have abbreviated the output by leaving out some not strongly connected basis solutions; see Fig. 7a for the whole set of basis solutions):

```
Forbidden Transition: T = [c]
Part of the net: U = [a, b]
Basis Solutions:
  b_0  = [[a30] f, [a39] g] --> strongly connected
  b_1  = [[a32] a, [a47] b]
  ...
  b_8  = [[a58] a, [a67] b]
  ...
  b_11 = [[a59] f, [a68] g] --> strongly connected
  ...
  b_17 = [[a65] a, [a71] b]
*** Result ***
Transition [c] does not stop all transitions in U = [a, b], because a
strongly-connected solution exists which contains some of these transitions.
Strongly connected solution: [[[a58] a, [a67] b], [[a59] f, [a68] g]]
```

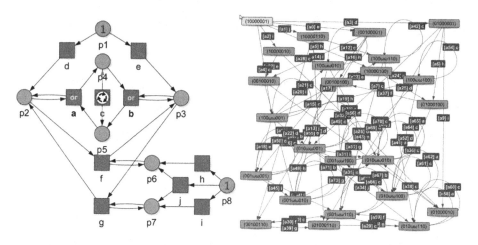

Fig. 6. An unbounded net with transition c selected as forbidden, and transitions a and b selected as the part of the net to focus on; a coverability graph of the net consisting of 74 arcs

Before discussing the above result, we quickly point out another feature of CyclOn which is very useful at this point: After performing a computation, CyclOn allows to reduce CG to the subgraph which is constituted by the determined set of basis solutions (see Fig. 7a). Pushing the "Reduce CG to Basis" button generates the subgraph shown in Fig. 7b, which offers much more clarity, by leaving out arcs and markings not relevant with respect to the set of basis solutions.

Next, we take a closer look at the above result: Due to the more generalized version of our question, finding a strongly connected solution which does not contain transition c is not sufficient, but we also require that transition a or b are contained in such a solution. Therefore, the divide-and-conquer part of the performed algorithm has to determine if there exists a combination of basis solutions which satisfies both requirements. The determined solution [[[a58] a, [a67] b], [[a59] f, [a68] g]], which is a combination of the strongly connected basis solution [[a59] f, [a68] g] and the not strongly connected basis solution [[[a58] a, [a67] b] (cf. the selected solutions in Fig. 7a), is strongly connected itself (indicated by the green arcs in Fig. 7b). Hence, it corresponds to a non-decreasing closed path not containing c but containing some transition from $U = \{a, b\}$ (it even contains both transitions).

2.4 Analyzing Basis Solutions Beyond Stop-Transitions

Although CyclOn has been developed for determining stop-transitions, it offers another useful feature to investigate the cyclic behavior of a net in a more general way. To illustrate this, we once again come back to the net from Fig. 3. However, this time we perform the computation without selecting any transition at all,

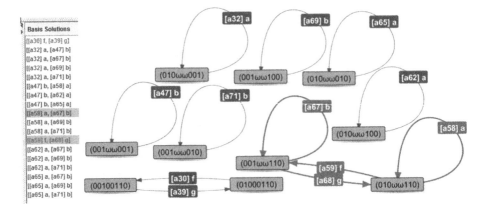

Fig. 7. a) List of basis solutions for the setting from Fig. 6 with a "green" and a "red" basis solution selected which together are strongly connected **b)** Subgraph of the coverability graph from Fig. 6 which is reduced to the arcs contained in the basis solutions (Color figure online)

i.e. we do not forbid any transition, and get the following result (again, we abbreviated the result; see Fig. 8a for the whole set of basis solutions):

```
Forbidden Transitions: T = []
Basis Solutions:
    b_0 = [[a6]  c]          --> strongly connected
    b_1 = [[a6]  c, [a8] a]
    ...
    b_8 = [[a10] b, [a11] c] --> strongly connected
    b_9 = [[a11] c]          --> strongly connected
*** Result ***
Transitions [] do not stop the net,
because a strongly-connected solution exists: [[[a6] c]]
```

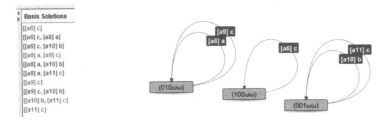

Fig. 8. a) List of basis solutions for the net from Fig. 3, if performing a "dry run" for this net (i.e. performing the computation without any forbidden transition) **b)** Subgraph of the coverability graph from Fig. 3 which is reduced to the arcs contained in the basis solutions

If we have an unbounded net and perform such a "dry run" (i.e. we do not forbid any transition), then there always exists a strongly connected solution corresponding to a non-decreasing closed path in CG. By performing the computation without forbidding any transition, we get an "unrestricted" set of basis solutions. We can use the functionalities of CyclƟn to take a closer look at these basis solutions, e.g. by investigating the involved arcs and transitions, and by checking if some combination of basis solutions is strongly connected. In this way we can already get some intuitive insights to the cyclic behavior of a net[8] in a practical way.

For example, a closer look at the determined basis solutions together with the reduced CG (see Fig. 8) makes it easy to see that every combination of basis solutions containing a "red" basis solution cannot be strongly connected, due to the structure of each "red" basis solution with regard to CG. Consequently, we can focus on the "green" basis solutions (in this example, not in general) for building combinations which are strongly connected and therefore correspond to a non-decreasing closed path. However, not every combination of strongly connected basis solutions is strongly connected itself, e.g. [[a6] c] combined with [[a9] c] is not strongly connected. Such considerations can help to identify and analyze non-decreasing closed paths in a net, even beyond the particular context of stop-transitions.

3   Architecture and Installation

CyclƟn is implemented in Java and employs the data structures provided by the APT^9 library for the internal representation of Petri nets. CyclƟn also uses an algorithm of the APT library to determine a coverability graph of the net. The $GraphStream^{10}$ library is employed to visualize the internal data structures of the net and the coverability graph in an illustrative way. CyclƟn also uses an implementation of Tarjan's algorithm provided by the GraphStream library to determine the strongly connected components of a graph. Since the algorithm from [1] implemented in CyclƟn requires solving a homogeneous system of linear inequalities, the *polco tool*[11] library is employed to compute a basis of solutions for such a system of inequalities. While CyclƟn does not provide any editing capabilities for Petri nets, it allows to process PNML files created with tools like e.g. WoPeD[12] (which has also been employed for all our examples).

The CyclƟn tool is freely available for download at our website: https://sttp1.fernuni-hagen.de/tools/cyclon/

The provided ZIP-archive file contains a runnable JAR file and several examples as PNML files (including all examples presented in this paper). Running CyclƟn merely requires an operation system with Java 8 (or higher) installed.

---

[8] considering a net of reasonable size, of course.

[9] APT – Analysis of Petri nets and labeled transition systems
   https://github.com/CvO-Theory/apt.

[10] GraphStream – A Dynamic Graph Library http://graphstream-project.org.

[11] polco Tool – Enumerate extreme rays of polyhedral cones
   https://csb.ethz.ch/tools/software/polco.html.

[12] WoPeD – Workflow Petri Net Designer http://www.woped.org.

4   Conclusion and Future Work

We presented the CyclOn tool and illustrated by several examples how it allows to determine stop-transitions of a Petri net in an easy way. We showed how the graphical representation and interactive features of CyclOn help to explain the results and make them comprehensible to the user.

In future work, we plan to extend the tool by editing capabilities for a Petri net under investigation, so that the effects of modifying certain elements of a net can be evaluated immediately without relying on external tools. Furthermore, we will work on replacing the current non-deterministic construction algorithm for a coverability graph by a more deterministic algorithm, so that every repeated coverability graph computation will result in the same graph structure (and the the same numbering of arcs).

References

1. Desel, J.: Can a single transition stop an entire net? In: International Workshop on Algorithms and Theories for the Analysis of Event Data 2019, ATAED@Petri Nets/ACSD 2019, Aachen, Germany, 25 June 2019, pp. 23–35 (2019). http://ceur-ws.org/Vol-2371/ATAED2019-23-35.pdf

A CTL* Model Checker for Petri Nets

Elvio Gilberto Amparore, Susanna Donatelli$^{(\boxtimes)}$, and Francesco Gallà

Dipartimento di Informatica, Università di Torino, Turin, Italy
{amparore,susi,galla}@di.unito.it

Abstract. This tool paper describes RGMEDD*, a CTL* model checker
that computes the set of states (sat-sets) of a Petri net that satisfy a
CTL* formula. The tool can be used as a stand-alone program or from
the GreatSPN graphical interface. The tool is based on the decision diagram library Meddly, it uses Spot to translate (sub)formulae into Büchi
automata and a variation of the Emerson-Lei algorithm to compute the
sat-sets. Correctness has been assessed based on the Model Checking
Context 2018 results (for LTL and CTL queries), the sat-set computation of GreatSPN (for CTL) and LTSmin (for LTL), and the μ-calculus
model checker of LTSmin for proper CTL* formulae (using a translator from CTL* to μ-calculus available in LTSmin). As far as we know,
RGMEDD* is the only available Büchi-based CTL* model checker.

1 Introduction

In recent years, the model checking of CTL [10] and LTL [22] temporal logics
for (colored) Petri nets (PN) has seen a boost in interest, and several tools and
methods have been developed. Efficiency has been a main driving force behind
this effort, also motivated by the lively Model Checking Competition (MCC) [19].
MCC models, formulae, and their evaluations are publicly available, making
MCC data a very valuable benchmark for (Petri net) tools. The MCC includes
LTL and CTL properties, but does not consider CTL* [15] ones. CTL* is a
temporal logic strictly more expressive than CTL and LTL. Although various
theoretical aspects of CTL* model checking have been extensively studied in the
past, very few CTL* model checkers exist, despite the fact that CTL* properties
are of practical interests, for example for modelling fairness constraints for CTL
properties. It is well known that various forms of fairness constraints are directly
expressible in LTL, while this is not the case for CTL.

Algorithms for CTL* model checking can either be based on Büchi automata,
as illustrated in [6](page 429), or they can rely on the translation from CTL*
into μ-calculus [20], as done in the work of Dam [11,12], using the standard fixed
point iteration of μ-calculus to compute the sat-set.

RGMEDD*, the CTL* model checker described in this paper, computes the
sat-sets using a Büchi-based model checking algorithm. Given a CTL* formula, the algorithm identifies the LTL sub-formulae of maximal length and uses
Spot [14] to translate each of them into a Büchi automaton. States are encoded
as Decision Diagrams (DD), using the Meddly [5] library. RGMEDD* can be run

© Springer Nature Switzerland AG 2020
R. Janicki et al. (Eds.): PETRI NETS 2020, LNCS 12152, pp. 403–413, 2020.
https://doi.org/10.1007/978-3-030-51831-8_21

as a stand-alone tool or as part of the GreatSPN [2] tool suite, called from a "check property" window of the GreatSPN graphical interface [1].

We could only find another Petri net tool that can deal directly with CTL*: LTSmin [18]. This tool translates CTL* into μ-calculus, using the procedures defined in [11,12]). Note that μ-calculus for Petri nets is available also in TINA [7], but no translator from CTL* to μ-calculus is provided. We could not find any CTL* tool based on Büchi automata. the set of available tools does not change even when looking to model checkers with input languages other than Petri Nets. There are papers on CTL* for Spin [17,25], but we could not find any implementation available. There is an implementation of μ-calculus for nuXmv [8], which could lead to a CTL* model checker but, for the time being, only CTL* formulae that are either LTL or CTL are actually processed.

Validation of RGMEDD* has been achieved taking advantage of the MCC2018 models, LTL and CTL formulae, and associated truth values for the models' initial state. Computations of the sat-sets of CTL* formulae have been checked in two ways: sat-sets of formulae that are plain CTL have been compared with the sat-set computed by RGMEDD3, the existing CTL model checker of GreatSPN, and LTSmin, while for CTL*-only formulae the comparison has been conducted against LTSmin, using its CTL* module.

We undertook the construction of a CTL* model checker to use it: 1) to test the efficiency of a DD-based implementation of LTL and CTL*; 2) to explore whether a Büchi automata approach can favour the formulation of counterexamples and witnesses; 3) to investigate the efficacy for CTL* of the variable ordering techniques developed in [3]; and 4) to support teaching: following the effort in [4] the users of GreatSPN to experiment (within the same window of the GUI) formulae of multiple logics: LTL, (*fair*)CTL and CTL*.

The paper is organized as follows: Sect. 2 introduces basic definitions, Sect. 3 reviews the algorithm for CTL* model checking, the symbolic data structures used to implement it, the tool architecture and the integration into the GreatSPN graphical interface. Section 4 describes the testing performed, while Sect. 5 concludes the paper.

2 Background

PT-nets. A place-transition (PT) Petri net M is defined [21] as a tuple $M = \langle P, T, A, W, m_0 \rangle$, where P is the set of places, T is the set of transitions, $A \subseteq (P \times T) \cup (T \times P)$ is the set of arcs, $W : A \to \mathbb{N}_{\geq 1}$ is the arc weight function, and $m_0 : P \to \mathbb{N}$ is the initial marking. Markings represent states of the system, i.e. assignments of tokens to places. A transition $t \in T$ is *enabled* if and only if all input places of t contain at least $W(p,t)$ tokens. The *firing* of t removes such tokens from the input places, and adds to each output place p' an amount of $W(t, p')$ new tokens. Notation $m \xrightarrow{t} m'$ indicates the firing of t from marking m to m'. The *reachability set* (RS) is the set of all markings reachable from m_0. Figure 1(**A**) shows a simple Petri net model M with 3 places and 4 transitions, with a RS of 4 markings.

GBA. A Generalized Büchi Automaton (GBA) is a tuple $A = \langle Q, \Sigma, \delta, Q_0, \mathcal{F} \rangle$, where Q is a finite set of locations, Σ is a set of atomic proposition labels, $\delta : Q \times \Sigma \to 2^Q$ is a total transition function, $q_0 \subseteq Q$ is the set of initial locations, and $\mathcal{F}$ is a subset of 2^Q and each element of $\mathcal{F}$ is called an acceptance set. Every LTL formula ϕ can be translated into an equivalent GBA [23]. The details of this translation are outside the scope of this paper. Figure 1(**B**) shows a GBA with 3 locations and a single atomic proposition $\#P2{=}1$, corresponding to a boolean formula on M. Location 1 is q_0, and $\mathcal{F} = \{\{0\}\}$.

CTL*. The language of CTL* formulae of RGMEDD* is inductively defined by:

$$\Psi ::= \top \mid \bot \mid \text{dead} \mid \text{en}(\tau) \mid \Theta \bowtie \Theta \mid \Psi \wedge \Psi \mid \Psi \vee \Psi \mid \neg\Psi \mid \exists\phi \mid \forall\phi$$
$$\phi ::= \Psi \mid \phi \wedge \phi \mid \phi \vee \phi \mid \neg\phi \mid X\phi \mid F\phi \mid G\phi \mid \phi U\phi$$
$$\Theta ::= n \mid \#\text{p} \mid \text{bounds}(\pi) \mid -\Theta \mid \Theta \circ \Theta$$

where $n \in \mathbb{N}$, $p \in P$, $\pi \subseteq P$, $\tau \subseteq T$, $\bowtie \in \{=, \neq, >, <, \geq, \leq\}$ is a comparison operator, and $\circ \in \{+, -, *, /\}$ is an arithmetic operator. The rules Ψ, ϕ and Θ are the rules for the *state*, *path* and *integer* formulae, respectively. The state formula *dead* is a special label for all RS states that do not enable any transition; $\text{en}(\tau)$ is satisfied in all RS states enabling at least one transition $t \in \tau$; $\#p$ evaluates to the cardinality of place p in the current marking; $\text{bounds}(\pi)$ is the maximum sum of token counts of all places in π in every reachable marking. A CTL* formula with no quantifiers but the initial one is an LTL formula.

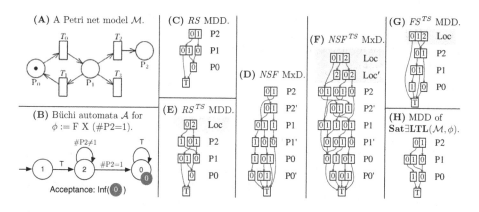

Fig. 1. An example of the MDDs and MxDs generated for CTL* evaluation.

Product $RS \otimes A$. Model checking of properties expressed using Büchi automata follows the schema of [24]. A Transition System (TS) is generated from the cross product $RS \otimes A$ between a path-formula GBA A and the RS of M. States of this TS are pairs $\langle m, q \rangle$, with m a Petri net marking and q a GBA location.

MDD and MxD. Decision Diagrams (DD) are a data structure used to encode large sets of structured data. GreatSPN uses the DD library Meddly [5], which

---

Algorithm 1. Sat-set generation of a maximal proper state subformula ϕ.

1: **procedure** SAT∃LTL$((M, \phi))$
2:     $A \leftarrow$ translate ϕ into a Büchi Automaton
3:     $\langle S_0^{TS}, NSF^{TS}, AS_{\mathcal{F}}^{TS} \rangle \leftarrow$ **BuildTransitionSystem**(M, A)
4:     $RS^{TS} \leftarrow Saturate(S_0^{TS}, NSF^{TS})$
5:     **switch** $type(A)$ **do**
6:         **case** StrongBüchi:
7:             $FS^{TS} \leftarrow RS^{TS} \models E_{fair}G(\textbf{true}, fair{=}AS_{\mathcal{F}}^{TS})$ ▷ Emerson Lei Algorithm
8:         **case** WeakBüchi:
9:             $FS^{TS} \leftarrow RS^{TS} \models EF\ EG(AS_{\mathcal{F}}^{TS}[0])$
10:        **case** TerminalBüchi:
11:            $FS^{TS} \leftarrow RS^{TS} \models EF(AS_{\mathcal{F}}^{TS}[0])$
12:    $Sat(\phi) \leftarrow relabel(FS^{TS} \cap S_0^{TS}, none)$
13:    **return** $Sat(\phi)$
14: **end procedure**

---

supports, among others, binary (BDD) and multivalued (MDD) DDs. We shall use MDD to encode both the RS of the Petri net, and the TS of the product $RS \otimes A$. MDDs encoding a relation function are called MxD, and are used by the tool to represent transitions and their union as a transition relation. An MxD has twice the levels of an MDD, and in each pair of levels it encodes the before/after relations of a variable (i.e. a place or a location).

Figure 1(C) and (D) shows the RS and the transition relation of M encoded as a MDD and as a MxD, respectively. A reachable state corresponds to a path in the MDD of (C): so, taking the rightmost path, we have that $(1, 0, 0)$ is a reachable state. The MDD are fully-reduce, so the leftmost path encodes both $(0, 0, 0)$ and $(0, 0, 1)$.

3 The RGMEDD* Architecture

Given a Petri net model M and a CTL* formula ψ, the solver first descends recursively through the syntax tree of ψ to extract each *maximal proper state subformula* [6, pp. 427], and then model checks each maximal proper state subformula ψ independently, in a bottom up process. Let ψ' be a maximal proper state subformula where all inner maximal proper state subformulae have been replaced with atomic propositions. For example if a and b are atomic propositions, then ∃XX($\forall a$UFb) has two maximal subformulas: $\forall a$UFb and ∃XXc, where c is the atomic proposition that stands for the sat-set of $\forall a$UFb. If ψ' only contain logical operations, it is trivially checked by applying the corresponding logical functions. Therefore, it remains to treat only the quantified cases $\psi' = \exists\phi$ and $\psi' = \forall\phi$. For the first case the model checker implements a single function to compute the set of states of M satisfying ϕ, denoted Sat∃LTL(M, ϕ). The method Sat∃LTL(M, ϕ) identifies the set of all markings m that have at least one path starting in m satisfying ϕ, which gives the semantics for the CTL* expression $\exists\phi$. The CTL* expression $\forall\phi$, corresponding to the more usual LTL semantics, is obtained from $\forall\phi = \neg(\exists\neg\phi)$, which is computed as $Sat(\forall\phi) = RS \setminus Sat\exists LTL(M, \neg\phi)$.

A pseudo-code of the $Sat\exists LTL(M, \phi)$ function is given in Algorithm 1. The function first translates the path formula ϕ into a GBA A using Spot (line 2). It then encodes (line 3) the transition system $RS \otimes A$: each TS state $\langle m, q \rangle$, is encoded in a MDD with $|P| + 1$ levels. The set of reachable states in $RS \otimes A$ is generated using *saturation* [9] (line 4). An infinite path meets the state-based acceptance condition of a GBA if it visits infinitely often at least one state in each acceptance set $F \in \mathcal{F}$. The work in [16] shows that the set of states originating accepting paths can be computed by the fair CTL formula $E_{fair}G(\mathbf{true}, \mathcal{F})$ on the TS. For some subclasses of GBAs (weak and terminal), a simplified procedure (lines 5–11) can be used [26]. The final set $Sat(\phi)$ is obtained by taking all the fair states (i.e. those that satisfy the acceptance condition of the GBA) that are also initial states of the RS (line 12).

Most of the complexity resides in the generation of $RS \otimes A$ with Decision Diagrams, summarized in Algorithm 2. Its generation requires both the MDD of the RS and the MxD of the transition relation, referred to as the *Next-State-Function* [9] (*NSF*) of the Petri net model. Each edge $q \xrightarrow{s} q'$ of A is encoded as a new MxD (line 5), by modifying the transition relation of the Petri net to reach only $Sat(s)$ markings, and at the same time by moving the GBA location from q to q'. To ensure that the generated transition system is a proper Kripke structure, all deadlock states of the Petri net model must be closed by a self-loop, which is built separately for each edge (lines 6–7). The transition relation NSF^{TS} of the TS is created from the union of all the edge's MxD (line 8). The function $relabel(d, q)$ takes a MDD d and sets the location level to q, while the $loc\_change(x, q, q')$ takes a MxD x and replaces for the location level the relation

---

Algorithm 2. Encoding of the $RS \otimes A$ transition system.

1: **procedure** BuildTransitionSystem$((M, A))$
2:     $S_0^{TS} \leftarrow \text{MDD}()$
3:     $NSF^{TS} \leftarrow \text{MxD}()$
4:     **for each** edge $e = q \xrightarrow{s} q'$ in δ: **do**
5:         $edgeMxD^e \leftarrow loc\_change(NSF \cap (RS \times Sat(s)), q, q')$
6:         $b \leftarrow Sat(s) \cap deadlock$ ▷ Self loops
7:         $selfLoopMxD^e \leftarrow loc\_change(b \times b, q, q')$
8:         $NSF^{TS} \leftarrow NSF^{TS} \cup (edgeMxD^e \cup selfLoopMxD^e)$
9:         **if** $q \in Q_0$ **then**
10:             $S_0^{TS} \leftarrow S_0^{TS} \cup relabel(Sat(s), q')$
11:         **end if**
12:     **end for**
13:     **for each** accepting set $F \in \mathcal{F}$: **do**
14:         $AS_F^S \leftarrow \bigcup\{edgeForVar(l) \mid \text{for each } l \in F\}$ ▷ MDD of all the locations in F
15:     **end for**
16:     **return** $\langle S_0^{TS}, NSF^{TS}, AS_\mathcal{F}^{TS}\rangle$
17: **end procedure**

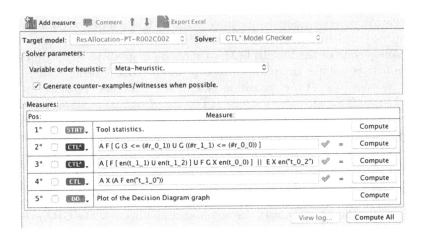

Fig. 2. A screenshot of the CTL/CTL* interface of RGMEDD* in GreatSPN.

$q \rightarrow q'$. The encoded TS is a tuple made by the transition relation NSF^{TS}, the set of initial states S_0^{TS} of the TS, which is the set of markings that are accepted by an edge leaving an initial location of the GBA (line 10), and the encoding $AS_{\mathcal{F}}^{TS}$ of the accepting sets $\mathcal{F}$ of the TS as MDDs (line 14).

Example. Figure 1 shows in **(E)** the MDD of RS^{TS}, generated as a fixed-point image of the MxD NSF^{TS} shown in **(F)**. The DDs of $RS \otimes A$ encode both the places of M and the locations of A, and therefore have an additional level of nodes. The set of fair states FS^{TS} visited infinitely often are also encoded as a MDD, shown in **(G)**. Finally **(H)** shows the MDD of the sat-set of the CTL* formula $\exists \phi$, encoding the 3 satisfying markings (all markings of RS except the one where all places have zero tokens).

Tool. The CTL* model checker is available both as a command line tool and inside the integrated GUI. Figure 2 shows the interface. For each model, a list of queries can be inserted, following the grammar given in Sect. 2. Queries are declared as either CTL or CTL*, which changes the model checking algorithm used. The algorithm described in Sect. 3 is used for CTL* queries only.

4   Testing

We have tested different aspects of RGMEDD*: the *correctness* of the results and the *performance* of the tool. MCC2018 models and relative model instances and formulae were used. We have considered only instances for which GreatSPN is able to build the RS, since the RS is required for sat-set computation.

The reported tests are the *final* ones, but the MCC instances have been extensively used also during the debugging phase allowing to discover both technical errors (in the implementation of the algorithm) and semantics ones (different behaviour for systems with deadlocks). Indeed Petri net models with deadlocks do not feature only infinite paths, as per the semantics of LTL and CTL. A standard way to turn around this problem is to "stutter" deadlock states by adding a self-loop. MCC assumes that deadlock states are stuttered only for LTL model checking, since this is required by the Büchi-based construction. RGMEDD* assumes deadlocks are stuttered, which may cause discrepancies when comparing RGMEDD* on CTL formulae. Therefore the MCC instances considered have been split in two sets: deadlock-free models (DFM), with 408 instances, and non deadlock-free models (DM), with 539 instances.

4.1   Testing of Truth Values for CTL and LTL Formulae

This test is based on the MCC2018 results for four categories of queries: *CTL-Cardinality*, *CTLFireability*, *LTLCardinality* and *LTLFireability*. Each category has 16 queries. A pair ⟨*model instance, category*⟩ is called an *examination*. For each examination an *expected result* is provided, that was used to check the correctness of RGMEDD*. To limit resource consumption we set a time limit of 300s and a memory limit of 2GB. With these limitations we could complete 360 examinations, that is to say 5760 ($=360 * 16$) queries. For LTL categories we have used instances of both DFM and DF models, while for CTL only DFM ones have been used. Over all the 5760 tests performed we have got a single mismatch, but this is query for which there is a mismatch also among the three tools that were able to evaluate the query in MCC2018.

4.2   Testing of Sat-Sets Computation of CTL Formulae

GreatSPN has a CTL model checker called RGMEDD3 that recursively computes the sat-set of CTL formulae. With a timeout of 60 seconds and 2GB of memory, RGMEDD3 completes 3544 queries from 168 different model instances from the DF set. RGMEDD* timeout-out on 10.05% of the queries. RGMEDD3 was faster than RGMEDD* in 92.65% of all queries completed by both tools.

Figure 3 (A) shows the execution times of the two tools, each query being a dot. In all completed tests the tools produce sat-sets of equal cardinality.

4.3   Assessment on CTL* Formulae

Here we report the assessment of RGMEDD* against LTSmin, run using the *pins2lts-sym* interface with the *−ctlstar* option which first converts CTL* into μ-calculus (which may incur an exponential cost). Also LTSmin is based on decision diagrams (of the Sylvan [13] library) and to make a more realistic comparison we have enforced the use of the same variable order by the two tools. Before checking CTL* formulae we have tested their behaviours on known results.

1. *Tool validation against MCC results.* We computed the sat-sets generated by RGMEDD* and pnml2lts-sym when provided with the same queries from the MCC examination of LTLCardinality and CTLCardinality (DF only models)

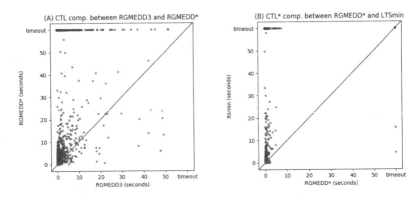

Fig. 3. Execution times of RGMEDD* vs. RGMEDD3 (A) and vs. LTSmin (B).

set for which the RS can be built. With the same resource limits as before, of the 1248 formulae considered (covering instances from 32 different models), RGMEDD* completed 1162 and LTSmin 702. Among these 702 formulae successfully completed by both model checkers, there were 7, from different models, over which the two tools did not agree: For these formulae MCC states that they are true in the initial marking, RGMEDD* returns sat-sets that include the initial marking, while LTSmin returns empty sat-sets, which seems a wrong answer.

2. *Relative performances of the two tools.* We checked the time performance of the two tools on the same instances as above, excluding the time for RS computation. Figure 3 (B) shows the comparison. RGMEDD* performs better in average and a closer look at LTSmin times reveals that a significant amount of time is spent converting CTL* formulae to μ-calculus.

3. *Correctness and performance on CTL* formulae.* Since there is no available set of CTL* formulae for the MCC models (and for any other models we could find) we have generated formulae algorithmically by parsing *CTL Cardinality* queries from MCC2018 and deleting each path quantifiers with a probability of 70%. The first quantifier is always kept, in order to preserve consistency with the CTL* grammar. For these tests, which are mainly aimed at checking correctness, we considered only 39 models whose RS are rather quick to generate. This gives a total of 624 queries, and on 218 both tools complete with the given resources. RGMEDD* timed out in 5.44% of the queries while LTSmin timed out in 65.06% of the queries. RGMEDD* was faster than LTSmin in 85.78% of all queries completed by both tools. More importantly the cardinality of the sat-sets coincides for both tools.

5 Conclusion

RGMEDD* is the new CTL* model checker of GreatSPN. It leverages two libraries: Spot for the translation from CTL* to Büchi automata and Meddly for decision diagram manipulation. RGMEDD* allows GreatSPN users to compute the set of reachable states that satify CTL* and LTL properties. The approach is fully integrated into the GreatSPN GUI, that already included the possibility of computing the sat-set of CTL formulae. RGMEDD* itself can be used to check also CTL properties, although our testing has confirmed that, as expected, the standard CTL model checking algorithm has superior performances.

Testing of RGMEDD* had to face a number of difficulties, due to the availability of a single CTL* tool, LTSmin. To be able to use LTSmin for our benchmark it was first necessary to fix a few syntactic problems in the LTSmin parsing of CTL* formulae: nevertheless it was an easy to use tool and we observed very limited discrepancies in the results. Although both tools are based on decision diagrams, RGMEDD* seems to perform significantly better than LTSmin, that suffers for the expensive translation from CTL* into μ-calculus.

Based on the experience gained in building RGMEDD* we plan to develop a model checker for the fair variant of CTL (reusing the available implementation of the Emerson-Lei algorithm for $E_{fair}G\phi$). We also plan to work on the generation of counterexamples and witnesses: these two topics are of paramount importance for the verification of distributed algorithms.

Finally, we shall work on improving memory and time performance by avoiding, as much as possible, the construction of a single monolithic decision diagram for the next state function of the Petri Net and of the $RS \otimes A$ transition system.

Availability. A virtual machine with the tool pre-installed can be downloaded from http://www.di.unito.it/~greatspn/VBox/RGMEDDstar-vm.ova. The source code of GreatSPN is available from https://github.com/greatspn/SOURCES.

Acknowledgements. We would like to thank Jaco van de Pol for the various insights given on LTSmin, and Yann Thierry Mieg for the discussion on finite paths and model checking.

References

1. Amparore, E.G.: A new GreatSPN GUI for GSPN editing and CSL^{TA} model checking. In: Norman, G., Sanders, W. (eds.) QEST 2014. LNCS, vol. 8657, pp. 170–173. Springer, Cham (2014). https://doi.org/10.1007/978-3-319-10696-0_13

2. Amparore, E.G., Balbo, G., Beccuti, M., Donatelli, S., Franceschinis, G.: 30 years of GreatSPN. In: Fiondella, L., Puliafito, A. (eds.) Principles of Performance and Reliability Modeling and Evaluation. SSRE, pp. 227–254. Springer, Cham (2016). https://doi.org/10.1007/978-3-319-30599-8_9

3. Amparore, E.G., Ciardo, G., Donatelli, S., Miner, A.: i_{Rank}: a variable order metric for DEDS subject to linear invariants. In: Vojnar, T., Zhang, L. (eds.) TACAS 2019. LNCS, vol. 11428, pp. 285–302. Springer, Cham (2019). https://doi.org/10.1007/978-3-030-17465-1_16

4. Amparore, E.G., Donatelli, S.: GreatTeach: a tool for teaching (stochastic) petri nets. In: Khomenko, V., Roux, O.H. (eds.) PETRI NETS 2018. LNCS, vol. 10877, pp. 416–425. Springer, Cham (2018). https://doi.org/10.1007/978-3-319-91268-4_24

5. Babar, J., Miner, A.: Meddly: multi-terminal and edge-valued decision diagram library. In: Proceedings of QEST Conf, pp. 195–196. IEEE (2010)
6. Baier, C., Katoen, J.: Principles of Model Checking. MIT Press, Cambridge (2008)
7. Berthomieu, B., Ribet, P.O., Vernadat, F.: The tool TINA - construction of abstract state spaces for Petri nets and time Petri nets. Int. J. Prod. Res. **42**, 2741–2756 (2004)
8. Cavada, R., et al.: The NUXMV symbolic model checker. In: Biere, A., Bloem, R. (eds.) CAV 2014. LNCS, vol. 8559, pp. 334–342. Springer, Cham (2014). https://doi.org/10.1007/978-3-319-08867-9_22
9. Ciardo, G., Lüttgen, G., Siminiceanu, R.: Saturation: an efficient iteration strategy for symbolic state—space generation. In: Margaria, T., Yi, W. (eds.) TACAS 2001. LNCS, vol. 2031, pp. 328–342. Springer, Heidelberg (2001). https://doi.org/10.1007/3-540-45319-9_23
10. Clarke, E.M., Emerson, E.A.: Design and synthesis of synchronization skeletons using branching time temporal logic. In: Kozen, D. (ed.) Logic of Programs 1981. LNCS, vol. 131, pp. 52–71. Springer, Heidelberg (1982). https://doi.org/10.1007/BFb0025774
11. Dam, M.: Translating CTL* into the modal μ-calculus. Technical report ECS-LFCS-90-123, University of Edinburgh (1990)
12. Dam, M.: CTL* and ECTL* as fragments of the modal μ-calculus. Theoret. Comput. Sci. **126**(1), 77–96 (1994)
13. van Dijk, T., van de Pol, J.: Sylvan: multi-core framework for decision diagrams. STTT **19**(6), 675–696 (2017)
14. Duret-Lutz, A., Lewkowicz, A., Fauchille, A., Michaud, T., Renault, É., Xu, L.: Spot 2.0—a framework for LTL and ω-automata manipulation. In: Artho, C., Legay, A., Peled, D. (eds.) ATVA 2016. LNCS, vol. 9938, pp. 122–129. Springer, Cham (2016). https://doi.org/10.1007/978-3-319-46520-3_8
15. Emerson, E.A., Halpern, J.: "Sometimes" and "not never" revisited: on branching versus linear time temporal logic. JACM **33**(1), 151–178 (1986)
16. Emerson, E.A., Lei, C.: Efficient model checking in fragments of the propositional mu-calculus. In: Logic in Computer Science (LICS 1986), pp. 267–278. IEEE (1986)
17. Holzmann, G.: The Spin Model Checker: Primer and Reference Manual. Addison-Wesley, Boston (2004)
18. Kant, G., Laarman, A., Meijer, J., van de Pol, J., Blom, S., van Dijk, T.: LTSmin: high-performance language-independent model checking. In: Baier, C., Tinelli, C. (eds.) TACAS 2015. LNCS, vol. 9035, pp. 692–707. Springer, Heidelberg (2015). https://doi.org/10.1007/978-3-662-46681-0_61
19. Amparore, E., et al.: Presentation of the 9th edition of the model checking contest. In: Beyer, D., Huisman, M., Kordon, F., Steffen, B. (eds.) TACAS 2019. LNCS, vol. 11429, pp. 50–68. Springer, Cham (2019). https://doi.org/10.1007/978-3-030-17502-3_4
20. Kozen, D.: Results on the propositional μ-calculus. Theoret. Comput. Sci. **27**(3), 333–354 (1983)
21. Murata, T.: Petri nets: properties, analysis and applications. Proc. IEEE **77**(4), 541–580 (1989)
22. Pnueli, A.: The temporal logic of programs. In: FOCS, pp. 46–57. IEEE Computer Society (1977)
23. Vardi, M.Y.: An automata-theoretic approach to linear temporal logic. In: Moller, F., Birtwistle, G. (eds.) Logics for Concurrency. LNCS, vol. 1043, pp. 238–266. Springer, Heidelberg (1996). https://doi.org/10.1007/3-540-60915-6_6

24. Vardi, M.Y., Wolper, P.: Reasoning about infinite computations. Inf. Comput. **115**(1), 1–37 (1994)
25. Visser, W., Barringer, H.: CTL* model checking for spin. In: The 4th International Spin Workshop. ENST, France, November 1998
26. Wang, C., Hachtel, G.D., Somenzi, F.: Abstraction Refinement for Large Scale Model Checking. Springer, Heidelberg (2006). https://doi.org/10.1007/0-387-34600-7

The Information Systems Modeling Suite
Modeling the Interplay Between Information and Processes

Jan Martijn E. M. van der Werf[1(✉)] and Artem Polyvyanyy[2]

[1] Department of Information and Computing Science,
Utrecht University, Utrecht, The Netherlands
`j.m.e.m.vanderwerf@uu.nl`
[2] School of Computing and Information Systems,
The University of Melbourne,
Parkville, VIC 3010, Australia
`artem.polyvyanyy@unimelb.edu.au`

Abstract. According to our recent proposal, an information system is a combination of a process model captured as a Petri Net with Identifiers, an information model specified in the first-order logic over finite sets with equality, and a specification of how the transitions in the net manipulate information facts. The Information Systems Modeling (ISM) Suite is an integrated environment for developing, simulating, and analyzing models of information systems, released under an open-source license. This paper presents the basic features of the ISM Suite.

Keywords: Information systems · Modeling · Simulating · Tools

1 Introduction

An *information system* (IS) is an integrated system of components that aim to collect, store, organize, manipulate, process, and disseminate data, information, and knowledge, often in the form of digital products. Finding the right balance between static and dynamic aspects is essential when designing an IS. As shown in [8], existing modeling languages often focus only on one of the two aspects. Therefore, we introduced in [8] the Information Systems Modeling Language (ISML), which focuses on expressing both these aspects and their interplay. An ISML model captures information aspects of using first-order logic with finite sets and equality and describes dynamic aspects that govern the information using Petri nets with Identifiers (PNIDs). Due to this symbiosis, ISML models feature:

1. CRUD operations over information facts and arbitrary information constraints;
2. Process dependencies that extend to infinite-state processes; and
3. Formal foundation which enables automated verification, e.g., one can decide if the system can evolve from its initial state into some other state of interest.

R. Janicki et al. (Eds.): PETRI NETS 2020, LNCS 12152, pp. 414–425, 2020.
https://doi.org/10.1007/978-3-030-51831-8_22

This approach to modeling and analyzing information systems promises advantages in educating future Information Systems professionals. Some benefits reported by students include the immediate experience of the consequences of design decisions and traceability between abstract and implemented concepts [11].

This paper focuses on presenting the Information Systems Modeling Suite (ISM Suite), an integrated collection of programs and tools for designing, executing, testing, simulating, and analyzing models of information systems captured in ISML. The next section describes a motivating scenario of a small yet comprehensive information system. Then, Sect. 3 exemplifies ISML by giving an in-depth presentation of an ISML model that captures the motivating scenario. Section 4 presents the ISM Suite and its features. The paper closes with conclusions.

2   Motivating Example

Consider the following running example, which we employ throughout the paper. A small organization uses a process-aware information system to support the management of its purchase orders. The organization works with a bidding system for its suppliers. If the organization wants to order a Product, it requests a bid from each of its Suppliers. At least two Suppliers need to respond with their bids, before the organization can select the best bid, and order the Product at the chosen Supplier.

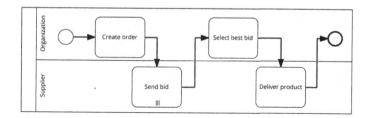

Fig. 1. BPMN model of the Purchase Order system.

The BPMN model that describes the dynamics of the IS is depicted in Fig. 1. First, an employee at the organization creates an Order. Next, several Suppliers *bid* for the Order, implemented via a "multi-instance activity" [2]. Finally, the best Supplier is selected and *receives* the Order, after which the Supplier *delivers* the ordered Product.

The information model of the IS is shown as an Object-Role Model (ORM) [3] in Fig. 2. According to this model, Suppliers *supply* Products. An Order always *contains* exactly one Product (c_1, c_4). Supplier can *bid* on an Order. From the *bids*, a Supplier is selected, and *receives* the Order (c_5). At most one Supplier

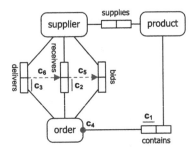

Fig. 2. Information model of the Purchase Order system.

receives the Order (c_2). Eventually, the selected Supplier can *deliver* the Order (c_6). Note that at most one Supplier can *deliver* the Order (c_3).

The organization faces a problem with the IS: *no progress can be made for many orders*. Both the process model and the information model do not expose any error: the process model is sound and the constraints of the information model do not contradict. However, it is the interplay of these models that causes the problem: if at most one Supplier *supplies* a requested Product, the process cannot progress. In the remainder of this paper, we show how our tool, the Information Systems Modeling Suite (ISM Suite), can assist the modeler to analyze the interplay between the process and data.

3   Information Systems Modeling Language

Modeling a process-aware information system requires at least three aspects (cf. [1,7,8]): an *information model* for the structure of the information, a *process model* describing the possible activities and their order, and a *specification* that defines how activities manipulate the information model. To model such systems, we proposed the Information Systems Modeling Language (ISML) in [8]. In the remainder of this section, we exemplify ISML using the motivating scenario from Sect. 2.

3.1   Information Modeling

An *information model* consists of a set of entity types, relations – characterized by finite sequences of entity types – and a set of constraints, specified in first-order logic. Given a sequence or tuple σ, by $\sigma(i)$, $i \in \mathbb{N}$, we denote the element at position i in σ. Let $\mathcal{I}$ and Λ be a universe of *identifiers* and a universe of *labels*, respectively.

Definition 1 (Information model)
An *information model* is a 5-tuple (E, R, ρ, η, Ψ), where:

– $E \subseteq \mathcal{P}(\mathcal{I})$ is a finite set of *entity types*;

- $R \subseteq \Lambda$ is a finite set of *relation types*;
- $\rho : R \rightarrow E^*$ is a *relation definition function* that maps every relation type onto a finite sequence of entity types;
- $\eta : E \rightarrow R$ is an *entity relation definition function* that maps every entity type onto a relation type such that η is injective and for every $e \in E$ it holds that $\rho(\eta(e)) = \langle e \rangle$, $\eta(e)$ is called the *entity relation* of e;
- Ψ is a collection of statements in first-order logic that for every $r \in R$, $n = |\rho(r)|$, (i) can use predicate r over the domain $\mathcal{I}^n$, called the *relation predicate* of r, and (ii) contains the statement:

$$\forall\, i \in \mathcal{I}^n : \left(r(i(1),\ldots,i(n)) \Rightarrow \left(\bigwedge_{k=1}^{n} \eta(\rho(r)(k))(i(k)) \right) \right),$$

called the *relation predicate statement* of r.

Definition 1 *refines* the corresponding definition presented in [8]. It introduces function η that explicitly defines entity relations. In addition, it requires that the predicates induced by the relation types, i.e., relation predicates, are named after the corresponding relation types, i.e., for every relation $r \in R$ statements in Ψ can use predicate r of the corresponding arity. Finally, for every relation type, it introduces one relation predicate statement to the theory that establishes dependencies between the truth values of the relation predicates. Note that a relation predicate statement of an entity relation is a tautology. The information model of the Purchase Order system is shown in Example 1.

Example 1 (Information model of the Purchase Order system)
The information model of the Purchase Order system as depicted in Fig. 2 is captured as the 5-tuple (E, R, ρ, η, Ψ), where:

- $E = \{E_s, E_p, E_o\}$;
- $R = \{supplier, product, order, supplies, receives, bids, delivers, contains\}$;
- $\rho = \{(supplier, \langle E_s \rangle), (product, \langle E_p \rangle), (order, \langle E_o \rangle), (supplies, \langle E_s, E_p \rangle),$
  $(receives, \langle E_s, E_o \rangle), (bids, \langle E_s, E_o \rangle), (delivers, \langle E_s, E_o \rangle), (contains, \langle E_o, E_p \rangle)\}$;
- $\eta = \{(E_s, supplier), (E_p, product), (E_o, order)\}$; and
- $\{\psi_1, \ldots, \psi_{11}\} \subset \Psi$, such that:
  - $\psi_1 \Leftrightarrow \forall i \in \mathcal{I} : (supplier(i) \Rightarrow (\neg product(i) \wedge \neg order(i)))$;
  - $\psi_2 \Leftrightarrow \forall i \in \mathcal{I} : (product(i) \Rightarrow (\neg supplier(i) \wedge \neg order(i)))$;
  - $\psi_3 \Leftrightarrow \forall i \in \mathcal{I} : (order(i) \Rightarrow (\neg supplier(i) \wedge \neg product(i)))$;
  - $\psi_4 \Leftrightarrow \forall o \in E_o \forall p_1 \in E_p \forall p_2 \in E_p : ((contains(o, p_1) \wedge contains(o, p_2)) \Rightarrow (p_1 = p_2))$;
  - $\psi_5 \Leftrightarrow \forall s_1 \in E_s \forall s_2 \in E_s \forall o \in E_o : ((receives(s_1, o) \wedge receives(s_2, o)) \Rightarrow (s_1 = s_2))$;
  - $\psi_6 \Leftrightarrow \forall s_1 \in E_s \forall s_2 \in E_s \forall o \in E_o : ((delivers(s_1, o) \wedge delivers(s_2, o)) \Rightarrow (s_1 = s_2))$;
  - $\psi_7 \Leftrightarrow \forall o \in E_o \exists p \in E_p : contains(o, p)$;
  - $\psi_8 \Leftrightarrow \forall s \in E_s \forall o \in E_o : (receives(s, o) \Rightarrow bids(s, o))$;
  - $\psi_9 \Leftrightarrow \forall s \in E_s \forall o \in E_o : (delivers(s, o) \Rightarrow receives(s, o))$;
  - $\psi_{10} \Leftrightarrow \forall s \in E_s \forall o \in E_o : (\exists p \in E_p : (bids(s, o) \Rightarrow (supplies(s, p) \wedge contains(o, p))))$;
  - $\psi_{11} \Leftrightarrow \forall s_1 \in E_s \forall o \in E_o : (bids(s_1, o) \Rightarrow (\exists s_2 \in E_s : (s_1 \neq s_2 \wedge bids(s_2, o))))$.

To visualize an information model, we use the Object-Role Modeling (ORM) notation. For each Entity Type $e \in E$, we draw an ORM Entity Type, for each relation types that are not an entity relation, we draw an ORM Fact Type with r as its caption. Some constraints are supported in ORM notation. For example, uniqueness constraints (e.g. ψ_4, visualized as c_1 in Fig. 2), mandatory constraints (e.g. ψ_7 visualized as c_4), and subset constraints (e.g. ψ_8 visualized as c_5).

An information model can be populated with facts. If a population satisfies all constraints, it is called valid.

Definition 2 ((Valid) Population, Fact)
A *population* of information model (E, R, ρ, η, Ψ) is a function $\pi : R \to \mathcal{P}(\bigcup_{n \in \mathbb{N}} \mathcal{I}^n)$ such that every element in the population is correctly typed, i.e., for every $r \in R$ it holds that $\pi(r) \in \mathcal{P}\left(\prod_{i=1}^{|\rho(r)|} \rho(r)(i)\right)$. An element in $\pi(r)$ is called a *fact*. The population is *valid*, denoted by $\pi \models \Delta$, only if $\pi \models \Psi$, given that for every $r \in R$, $n = |\rho(r)|$, it holds that $\forall (i_1, \ldots, i_n) \in \mathcal{I}^n : (r(i_1, \ldots, i_n) \Leftrightarrow (i_1, \ldots, i_n) \in \pi(r))$; otherwise π is *invalid*, denoted by $\pi \not\models \Delta$.

To manipulate populations, we define two operations for inserting and removing a fact from a relation. Operations can be concatenated, resulting in a transaction.

Definition 3 (Transaction)
Let $\mathcal{D} = (E, R, \rho, \eta, \Psi)$ be an information model and let π be a population of $\mathcal{D}$. An *operation* is a tuple $o \in \mathcal{O}(\mathcal{D})$, where $\mathcal{O}(\mathcal{D}) = (R \times \{\oplus, \ominus\} \times \bigcup_{n \in \mathbb{N}} \mathcal{I}^n)$.

- Operation $o = (r, \oplus, v)$ *inserts* fact v of type r into π iff $\pi' = (\pi \smallsetminus \{(r, \pi(r))\}) \cup \{(r, \pi(r) \cup \{v\})\}$ is a valid population of $\mathcal{D}$, denoted by $(\mathcal{D} : \pi \xrightarrow{r \oplus v} \pi')$.
- Operation $o = (r, \ominus, v)$ *removes* fact v of type r from π iff $\pi' = (\pi \smallsetminus \{(r, \pi(r))\}) \cup \{(r, \pi(r) \smallsetminus \{v\})\}$ is a valid population of $\mathcal{D}$, denoted by $(\mathcal{D} : \pi \xrightarrow{r \ominus v} \pi')$.

A *transaction* $s \in (\mathcal{O}(\mathcal{D}))^*$ is a finite sequence of operations such that every subsequent operation is performed over a population resulting from the previous operation. A transaction is *valid* if the initial and resulting populations are both valid.

Example 2 (Transaction in the Purchase Order system)
A transaction in the Purchase Order system to add some Order o_1 for Product p can be expressed as follows: $\langle (\text{order}, \oplus, (o_1)); (\text{contains}, \oplus, (o_1, p)) \rangle$. For any valid population π such that $o_1 \notin \pi(\text{order})$, this transaction is valid.

3.2 Process Modeling with Petri Nets with Identifiers

For modeling the activities and their order we use Petri nets with Identifiers (PNID) [4,8]. PNIDs can be seen as an extension of ν-Nets [9]. In a PNID, tokens carry a vector of identifiers. These vectors have the advantage that a single token can represent multiple entities at the same time. In this way, a token may represent a (composed) fact from a population of an information model. Each place is typed with a vector of identifiers, i.e., all tokens in a place have the same vector length, called the *cardinality*. Arcs are annotated with vectors of variables. Its size is implied by the cardinality of the place it is connected to. Let Σ denote the universe of variables.

Definition 4 (Petri net with Identifiers)

A *Petri net with Identifiers* (PNID) N is a 5-tuple (P, T, F, α, β), where:

- P and T are two disjoint sets of *places* and *transitions*, resp., i.e., $P \cap T = \varnothing$;
- $F : ((P \times T) \cup (T \times P)) \to \mathbb{N}^0$ is the *flow function*;
- $\alpha : P \to \mathbb{N}^0$ defines the *cardinality* of a place, i.e., the length of the vector of identifiers carried on the tokens residing at that place; its color is defined by $C(p) = \mathcal{I}^{\alpha(p)}$;
- β defines the *variable vector* for each arc, i.e., $\beta \in \prod_{\{f|F(f)>0\}} V_f$, where $V_{(p,t)} = V_{(t,p)} = \Sigma^{\alpha(p)}$ for $p \in P, t \in T$.

Its set of all possible markings is defined as $\mathcal{M}(N) = \prod_{p \in P}(C(p) \to \mathbb{N}^0)$. The pair (N, m) is a *marked PNID* if $m \in \mathcal{M}(N)$.

To fire a transition, the variables on its arcs need to be *valuated* to match identifiers the tokens carry. A valuation maps each variable to an identifier. New identifiers can be created if a transition contains variables that only occur on outgoing arcs. The valuation guarantees that variables occurring on an outgoing arc each receive a new, fresh identifier. For the full semantics of PNIDs, we refer the reader to [4,8].

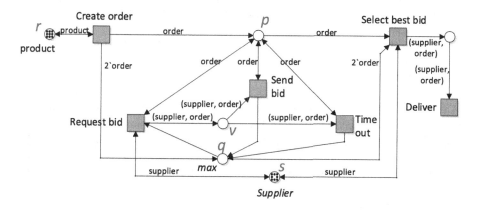

Fig. 3. PNID of the Purchase Order system.

Example 3 (PNID of the Purchase Order system)

The PNID of the Purchase Order system is depicted in Fig. 3. The multiple-instance activity is translated in a standard pattern in which two suppliers can be asked simultaneously. The net starts in the marking with three tokens in place r, and three tokens in place s. These tokens resemble products and suppliers, resp., i.e., $m_0 = \{r \mapsto [p, q, r], s \mapsto [s, t, v]\}$. In this marking, transition Create order is enabled with valuation $\nu_1 = \{\text{order} \mapsto o_1, \text{product} \mapsto p\}$, where o_1 is a fresh identifier. Firing this transition results in marking $m_1 = \{p \mapsto [o_1], q \mapsto [o_1{}^2], r \mapsto [p, q, r], s \mapsto [s, t, v]\}$.

3.3   Semantics of the Information Systems Modeling Language

Transitions in the process model often resemble events that manipulate the information model. For example, in the Purchase Order system, the transition Create order resembles adding a new order fact to the information model. In ISML, each transition is specified with an *abstract transaction* that describes how the transition manipulates the information model. Similar to transition firing, abstract transactions rely on valuations to be instantiated. The set of all abstract transactions over some information model $\mathcal{D}$ is denoted by $\mathcal{T}(\mathcal{D})$, and extends normal operations by allowing variables in the facts.

Example 4 (Specification of the Purchase Order process)
The Purchase Order system has four transitions with a non-empty abstract transaction. The variables used in the transactions coincide with the variables used on the arcs.

- Transition Create order ensures an order contains exactly one product:
  $S(\text{Create order}) = \langle(\text{order}, \oplus, (order)); (\text{contains}, \oplus, (order, product))\rangle$.
- Transition Send bid resembles the activity that a supplier responds to an order:
  $S(\text{Send bid}) = \langle(\text{bids}, \oplus, (supplier, order))\rangle$
- Transition Select best bid sends the order to the best supplier:
  $S(\text{Select best bid}) = \langle(\text{receives}, \oplus, (supplier, order))\rangle$
- Transition Deliver models that the order is delivered by the supplier:
  $S(\text{Deliver}) = \langle(\text{deliver}, \oplus, (supplier, order))\rangle$

An information system model has three constituents: an information model $\mathcal{D}$, a PNID N, and a specification S. In an information system, executing a transaction should not result in an invalid population. Therefore, given a state (π, m), a transition can only fire if (i) it is enabled in (N, m), and (ii), its transaction results again in a valid population. For an overview of ISML and its semantics, we refer the reader to [8]. The information system model of the Purchase Order system is given in Example 5.

Example 5 (ISM of the Purchase Order System)
The information system model for the Purchase Order system uses the information model of Example 1, the PNID depicted in Fig. 3, and the specification in Example 4. Consider the following initial population, with three products and three suppliers:

$$\begin{array}{llll}
\text{product}(p) & \text{supplier}(s) & \text{supplies}(s,p) & \text{supplies}(t,r) \\
\text{product}(q) & \text{supplier}(t) & \text{supplies}(s,q) & \text{supplies}(v,r) \\
\text{product}(r) & \text{supplier}(v) & \text{supplies}(t,p) &
\end{array}$$

Hence, the tokens in place r resemble the products, and the tokens in place s suppliers. Transition **Create order** is enabled in the PNID with three valuations, one for each product. None of the valuations result in an invalid transaction on the current population. Hence, the transition is enabled in the ISM with all three valuations.

Selecting valuation ν_1 results in transaction: $\langle(\text{order}, \oplus, (o_1));$ $(\text{contains}, \oplus, (o_1, p))\rangle$. Firing the transition with this valuation results in marking m_1 (see Example 4), and the valid population:

$$\begin{array}{lllll}
\text{product}(p) & \text{supplier}(s) & \text{supplies}(s,p) & \text{supplies}(t,r) & \text{order}(o_1) \\
\text{product}(q) & \text{supplier}(t) & \text{supplies}(s,q) & \text{supplies}(v,r) & \text{contains}(o_1,p) \\
\text{product}(r) & \text{supplier}(v) & \text{supplies}(t,p) & &
\end{array}$$

In marking m_1, two transitions are enabled in the PNID: **Request bid** and **Select best bid**. Both are enabled with three valuations, one for each supplier. Consider valuation $\nu_2 = \{\text{supplier} \mapsto v, \text{order} \mapsto o_1\}$. For transition **Select best bid**, this valuation results in the transaction $\langle(\text{receives}, \oplus, (v, o_1))\rangle$. However, this transaction invalidates the constraints ψ_8 and ψ_{11}, as the population contains no bids for order o_1. Hence, this transition is not enabled in the ISM for any of the valuations. Only transition **Request bid** is enabled. Firing it with valuation ν_2 results in the marking $m_2 = \{p \mapsto [o_1], q \mapsto [o_1], v \mapsto [o_1], r \mapsto [p,q,r], s \mapsto [s,t,v]\}$. Now, transitions **Time out** and **Send bid** are enabled in the PNID, with valuation $\nu_3 = \{\text{order} \mapsto o_1\}$. However, the transaction induced by **Send bid** yields an invalid population, as $\text{supplies}(v,p)$ is not a fact in our population. Hence, only transition **Time out** is enabled in the current state.

4 ISM Suite

The ISM Suite is implemented as a plug-in for Eclipse and can be installed via a publicly available Update Site.[1] The ISM Suite consists of an editor and simulator for PNIDs, and a simulator for ISMs. It heavily builds upon ePNK [6] and is fully PNML compliant [5]. The ISM Suite has its own perspective in Eclipse. It uses its own prover and simulation libraries, which, together with the source code of the ISM Suite, can be found on Github.[2]

4.1 Editing and Simulating PNIDs

As the ISM Suite process editor is based on ePNK, it opens with a process explorer, which displays the content of the PNML file. Initially, a net is created of the correct type, together with a page. As the editor is PNML compliant, all Petri net content has to be created on pages. The editor supports reference

---

[1] See: http://informationsystem.org/ismsuite/.
[2] See: https://github.com/information-systems/ISMSuite.

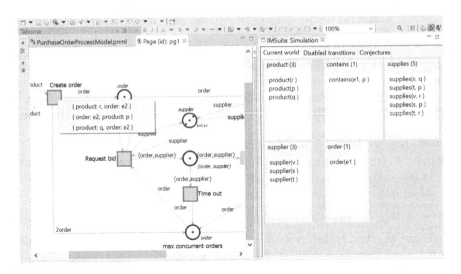

Fig. 4. The ISM Suite. The view on the left depicts the PNID, a popup menu with enabled valuations is shown. The panel on the right shows the current population of the ISM.

places and reference transitions to "divide" the net over several pages. By double-clicking on a page, the graphical editor is started. Places, transitions, and arcs can be inserted on the canvas from the pallet.

By default, all places have the cardinality of 0, i.e., can hold "black" tokens. Instead of working with cardinalities, the ISM Suite is structurally typed, i.e., each constituent has its type, and variables need to be consistently typed. The cardinality is derived from this type. A place can be typed by adding a label to the canvas and connecting it with a "Link label". The type is specified as a bracketed list of entity types. Each entity type has to start with a lower case. Similarly, the inscription of arcs can be added. Places with a non-empty type are colored yellow in the editor. Tokens can be added by creating another label. Each token is identified by its vector. For example, $3`()$ denotes three black tokens, while $2`(a,b)+(c,d)$ resembles three "colored" tokens: two tokens carry the vector (a,b) and one token carries the vector (c,d).

To simulate the PNID, one can start the simulator via the menu "ISM Suite". Before the engine is started, the net is validated and type-checked. If the net is invalid, a warning is shown, together with the violations. The simulator is shown in the left panel of Fig. 4. Once the simulator is started, enabled transitions are marked with a thick red outline. On clicking on such transition, a menu is opened with the possible valuations. After selecting a valuation, the transition fires. The tokens residing in a place can be checked by clicking on that place. A menu is opened with a list of all tokens.

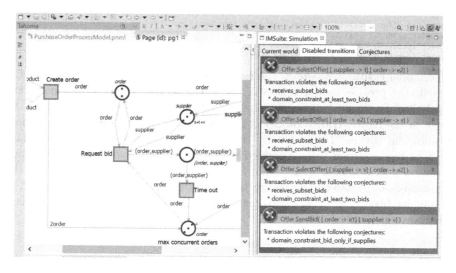

Fig. 5. The ISM Suite. The panel on the right explains why certain transitions are not enabled.

4.2 Simulating ISMs

To simulate an information system model, choose the "Start ISM Simulator" option in the menu "ISM Suite". A dialog is opened that asks for two files: an information model and a specification. The information model needs to be specified in the Typed First-Order Formulae (TFF) format [10]. For example, constraint ψ_{11}, refer to Example 1, at least two bids of different suppliers are required, can be written in TFF as:

```
tff(domain_constraint_at_least_two_bids, conjecture,
  ! [S1: universe, O: universe] :
    ( receives(S1,O) => ( ? [S2: universe] : ( (S1 != S2) & bids(S2,O) ) ) ) ).
```

A transaction is specified by a name and a set of typed variables and consists of a sequence of operations. Different from the information model, the ISM Suite does not "know" which entities are present in each entity type. These need to be added explicitly in the transaction. To this end, there are four operations:

- Operation `register p;` adds element p to its entity type set;
- Operation `deregister p;` removes element p from its entity type set;
- Operation `insert (a,...,z) into relation;` inserts fact $(a, \ldots, z)$ to 'relation'; and
- Operation `remove (a,...,z) from relation;` removes fact $(a, \ldots, z)$ from 'relation'.

Transactions are matched with the transitions of the PNID by their identifiers. To create an initial population, upon starting the simulator, all places that contain tokens are also matched with transactions. For places, the variable names have to match the identity types as specified in the place type. For example, to specify the initial population, as presented in Example 5, a place is added with five tokens, and the place is matched with the following transaction:

```
transaction Offer.provides(supplier: universe, product: universe)    {
  register product; register supplier;
  insert (product) into product; insert (supplier) into supplier;
  insert (supplier, product) into supplies;                          }
```

If the model is fully consistent, the simulator is started, as shown in Fig. 4. On the left, the process model is shown, the panel on the right shows the current population. By firing transitions, the population is updated automatically. As shown in the figure, transition Select best bid is not enabled, even though there are sufficient tokens. To study why the transition is not enabled, the panel on the right has a second tab, "Disabled transitions", as shown in Fig. 5. For each transition and valuation that is not enabled, an explanation is given why the transition is not enabled. As shown in this figure, transition Select best bid, is not enabled because it violates two constraints "receives subset bids" and domain constraint "at least two bids".

5   Conclusion

As the interplay between data and processes can be very subtle, it is not sufficient to only study information and process models in isolation. In this paper, we present the Information System Modeling Suite (ISM Suite). It is a tool that helps to study how processes and data are related in an information system. The tool provides editing and simulation facilities for modeling Petri Nets with Identifiers. Furthermore, it allows simulating information system models. It combines process models with an information model in terms of first-order logic constraints, and a simple specification language to define how transitions manipulate the information model. The tool provides visual aids to assist the modeler by explaining why certain transitions are disabled in the ISM.

We envision the ISM Suite as a tool for learning modeling of information systems. In the future, we want to perform several experiments with students to validate the modeling approach, and how it is experienced, similar to [11]. As the ISM Suite currently only supports visual modeling of processes, we plan to extend it with visual facilities for information modeling, together with simulation and analysis options.

Acknowledgment. Artem Polyvyanyy was partly supported by the Australian Research Council Discovery Project DP180102839.

References

1. De Masellis, R., Di Francescomarino, C., Ghidini, C., Montali, M., Tessaris, S.: Add data into business process verification: bridging the gap between theory and practice. In: AAAI, pp. 1091–1099. AAAI Press (2017)
2. Dumas, M., La Rosa, M., Mendling, J., Reijers, H.: Fundamentals of Business Process Management. Springer, Heidelberg (2018). https://doi.org/10.1007/978-3-642-33143-5

3. Halpin, T.A., Morgan, T.: Information Modeling and Relational Databases, 2nd edn. Morgan Kaufmann Publishers, Burlington (2008)
4. van Hee, K.M., Sidorova, N., Voorhoeve, M., van der Werf, J.M.E.M.: Generation of database transactions with Petri nets. Fundamenta Informatica **93**(1–3), 171–184 (2009)
5. Hillah, L.M., Kindler, E., Kordon, F., Petrucci, L., Tréves, N.: A primer on the Petri net markup language and ISO/IEC 15909-2. Petri Net Newslett. **76**, 9–28 (2009)
6. Kindler, E.: The ePNK: a generic PNML tool - users' and developers' guide for version 1.0.0. Technical Report IMM-Technical Report-2012-14, DTU Informatics (2012)
7. Montali, M., Rivkin, A.: DB-Nets: on the marriage of colored Petri nets and relational databases. In: Koutny, M., Kleijn, J., Penczek, W. (eds.) Transactions on Petri Nets and Other Models of Concurrency XII. LNCS, vol. 10470, pp. 91–118. Springer, Heidelberg (2017). https://doi.org/10.1007/978-3-662-55862-1_5
8. Polyvyanyy, A., van der Werf, J.M.E.M., Overbeek, S., Brouwers, R.: Information systems modeling: language, verification, and tool support. In: Giorgini, P., Weber, B. (eds.) CAiSE 2019. LNCS, vol. 11483, pp. 194–212. Springer, Cham (2019). https://doi.org/10.1007/978-3-030-21290-2_13
9. Rosa-Velardo, F., de Frutos-Escrig, D.: Decidability and complexity of Petri nets with unordered data. Theoret. Comput. Sci. **412**, 4439–4451 (2011)
10. Sutcliffe, G.: The TPTP problem library and associated infrastructure. From CNF to TH0, TPTP v6.4.0. J. Autom. Reason. **59**(4), 483–502 (2017)
11. van der Werf, J.M.E.M., Polyvyanyy, A.: An assignment on information system modeling. In: Daniel, F., Sheng, Q.Z., Motahari, H. (eds.) BPM 2018. LNBIP, vol. 342, pp. 553–566. Springer, Cham (2019). https://doi.org/10.1007/978-3-030-11641-5_44

MCC: A Tool for Unfolding Colored Petri Nets in PNML Format

Silvano Dal Zilio[(✉)] [ID]

LAAS-CNRS, Université de Toulouse, CNRS, Toulouse, France
`dalzilio@laas.fr`

Abstract. MCC is a tool designed for a very specific task: to transform the models of High-Level Petri nets, given in the PNML syntax, into equivalent Place/Transition nets. The name of the tool derives from the annual Model-Checking Contest, a competition of model-checking tools that provides a large and diverse collection of PNML models. This choice in naming serves to underline the main focus of the tool, which is to provide an open and efficient solution that lowers the access cost for developers wanting to engage in this competition. We describe the architecture and functionalities of our tool and show how it compares with other existing solutions. Despite the fact that the problem we target is abundantly covered in the literature, we show that it is still possible to innovate. To substantiate this assertion, we put a particular emphasis on two distinctive features of MCC that have proved useful when dealing with some of the most challenging colored models in the contest.

Keywords: Tools · PNML · High-Level Petri Nets · Colored Petri nets

1 Introduction

The Petri Net Markup Language (PNML) [4] is an XML-based interchange format for representing Petri nets and their extensions. One of its main goal is to provide developers of Petri net tools with a convenient, open and standardized format to exchange and store models. While its focus is on openness and extensibility, the PNML spotlights two main categories of models: standard Place/Transition nets (P/T nets), and a class of Colored Petri nets, called High-Level Petri Nets (HLPN), where all types have finite domains and expressions are limited to a restricted set of operators [6,12].

In this paper we present `mcc`, a tool designed for the single task of *unfolding* the models of High-Level Petri nets, given in the PNML syntax, into equivalent Place/Transition nets. The name of the tool derives from the annual Model-Checking Contest (MCC) [1], a competition of "Petri tools" that makes an extensive use of PNML and that provides a large and diverse collection of PNML models, some of which are colored. Our choice when naming `mcc` was to underline the main focus of the tool, which is to provide an open and efficient solution that lowers the access cost for developers wanting to engage in the MCC.

© Springer Nature Switzerland AG 2020
R. Janicki et al. (Eds.): PETRI NETS 2020, LNCS 12152, pp. 426–435, 2020.
https://doi.org/10.1007/978-3-030-51831-8_23

We seek to follow the open philosophy of PNML by providing a software that can be easily extended to add new output formats. Until recently, the tool supported the generation of Petri nets in both the TINA [3] (.net) and LOLA [16] formats; but it has been designed with the goal to easily support new tools. To support this claim, we have very recently added a new command to print the resulting P/T net in PNML format. This extension to the code serves as a guideline for developers that would like to extend mcc for their need.

The rest of the paper is organized as follows. In Sect. 2, we describe the basic functionalities of mcc and give an overview of the PNML elements supported by our tool, we also propose three new classes of colored models that are representative of use cases found in the MCC repository [11]. Next, we describe the architecture of mcc and discuss possible applications of its libraries. Before concluding, we compare mcc with other existing solutions. Despite the fact that the problem we target is abundantly covered in the literature, we show that it is still possible to innovate. We describe two particular examples of optimizations that have proved useful when dealing with some of the most challenging colored models in the contest, namely the use of a restricted notion of "higher-order invariant", and the support of a Petri net scripting language.

2   Installation, Usage and Supported PNML Elements

The source code of mcc is made freely available on GitHub and is released as open-software under the CECILL-B license; see https://github.com/dalzilio/mcc. The code repository also provides binaries for Windows, Linux and MacOS at https://github.com/dalzilio/mcc/releases. The tool can also be easily compiled, from source, on any computer that provides a recent distribution of the Go programming language.

Basic Usage. Tool mcc is a command-line application that accepts three primary subcommands: hlnet, lola and pnml. In this paper, we focus on the mcc hlnet command, that generates a Petri net file in the TINA *net* format [3]. Similarly, commands lola and pnml generate an equivalent output but targeting, respectively, the LoLa [16] and PNML formats for P/T nets.

We follow the UNIX philosophy and provide a small program, tailored for a precise task, that can be composed using files, pipes and shell script commands to build more complex solutions. As it is customary, option -h prints a usage message listing the parameters and options accepted by the command.

The typical usage scenario is to provide a path to a PNML file, say model.pnml, and invoke the tool with a command such as "mcc hlnet -i model.pnml". By default, the result is written in file model.net, unless option -o or --name is used. We discuss some of the other options of mcc in the sections that follow.

PNML Elements Supported by MCC. The input format supported by mcc covers most of the PNML syntax defined in the ISO/IEC 15909-2 standard, which corresponds to the definition of HLPN.

High-Level Petri nets form a subset of colored nets defined by a restriction on the types and expressions that are allowed in a net [4,10]. The core action language of HLPN is a simple, first-order declarative language organized into categories for types, values and expressions. Essentially, HLPN is built around a nominal type system where possible ground types include a constant for "plain tokens" (dot), and three different methods for declaring finite, ordered enumeration types (finite, cyclic, and integer range). The type system also includes "product types", used for tuples of values, and a notion of *partition elements*, which are (named) subsets of constants belonging to the same type.

Expressions are built from values and operations and describe multisets of colors, which act as the marking of places. For instance, the language include operators add and subtract, that correspond to multiset union and difference. The language also includes a notion of *patterns*, which are expressions that includes variables (in a linear way), and of *conditions*, which are boolean expressions derived from a few comparison operators. A simple way to describe the subset of the PNML standard supported in mcc is to list the XML elements supported in each of these categories (most of the element names are self-explanatory):

```
types        ::= dot | cyclicenumeration | finiteenumeration
               | finiteintrange | productsort
               | partition | partitionelement
values       ::= dotconstant | feconstant | finiteintrangeconstant
expressions ::= variable | successor | predecessor | tuple
               | all | add | subtract
conditions   ::= or | and | equality | inequality
               | lessthan | greaterthan
               | greaterthanorequal | lessthanorequal
```

The mcc tool, in its latest version, supports all the operators used in models of the Model-Checking Contest. To better understand this fragment, we give three examples of HLPN that can be expressed using these constructs, see Fig. 1 to 3. Each of these examples illustrate an interesting class of parametric models found in the MCC and will be useful later to discuss the strengths and weaknesses of our approach. None of these models are part of the MCC repository (yet), but their PNML specification can be found in the mcc source code repository.

Three Representative Examples. Our first example, Fig. 1, illustrates the use of colors to model a complex network topology. While Diffusion is not part of the MCC repository, it is the colored equivalent of model Grid2d; it is also the main benchmark in [14]. In this model, values in the place Grid are of the form (x, y), with $x, y \in 0..4$. Hence we can interpret colors as cells on a 5×5 grid and values as "tokens" in these cells. Tokens can move to an adjacent cell by firing transition t_1 but cannot cross borders. (In our diagrams we use ++ for successor, > for greaterthan, and + for add.) All the behavior is concentrated on the condition associated with t_1. Since the expression contains four variables; we potentially have $|CD|^4$ different ways to enable t_1.

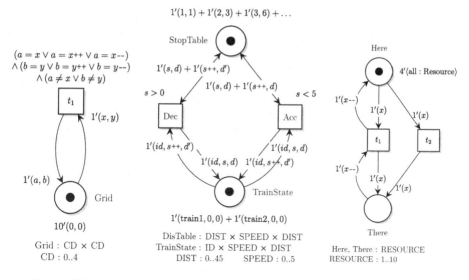

Fig. 1. Diffusion   **Fig. 2.** TrainTable   **Fig. 3.** Resource swap

TrainTable, is an example where colors are used to simulate complex relations between data values. Place StopTable is initialized with a list of pairs associating, to each (integer) speed in 0...5, the safety distance needed for a train to stop. Hence TrainTable tabulates a non-linear constraint between speed and distance. Place TrainState stores the current state of two different trains. Each time a train accelerate (Acc), or decelerate (Dec), the safety distance is updated. TrainTable is a simplified version of the BART model. We can make this model more complex by storing the distance traveled instead of the safety distance (Traintable-Dist); or even more complex by storing both values (TrainTable-Stop+Dist).

Our last example, Swap, is typical of systems built from the composition of multiple copies of the same component and where interactions are limited to "neighbors". The model obeys some interesting syntactical restrictions: it does not use conditions on the transitions and inscriptions on arcs are limited to two patterns, x or $x-$. This is representative of many models, such as the celebrated *Dining Philosophers* example (known as Philosopher in the MCC).

3   Architecture of MCC

The mcc tool is a standalone Go program built from three main software components[1] (called *packages* in Go): pnml, hlnet, and corenet. Basically, the architecture of mcc is designed to resemble that of a compiler that translates high-level code (HLPN) into low-level instructions (P/T net). We follow a traditional structure with three stages where: pnml corresponds to the front-end (responsible for syntax and semantics analysis); hlnet provides the intermediate representation;

---

[1] See the documentation at https://godoc.org/github.com/dalzilio/mcc.

and `corenet` is the back-end, which includes functions for unfolding an `hlnet` and for "code generation". A last package, `cmd`, contains boilerplate code for parsing command-line parameters and manage inputs/outputs.

Each of these packages is interesting taken separately and can be reused in other applications. Package `pnml`, for instance, includes all the types and functions necessary for parsing a PNML file: it defines a `pnml.Decoder`, which encapsulates an efficient, UTF-8 compatible XML parser and can provide meaningful error messages in case of problems. The `hlnet` package, for its part, defines the equivalent of an Abstract Syntax Tree data structure for PNML files. Both of these packages can be easily reused in programs that need to consume PNML data. In particular, they can help build a standalone PNML parser with good error handling.

Finally, package `corenet` contains the code for unfolding an `hlnet.Net` value into a `corenet.Net`, which is a simple, graph-like data structure representing a P/T net. The package also contains the functions for marshalling a core net structure into other formats; see function `corenet.LolaWrite` for an example. A tool developer that would like to adopt `mcc` to generate a "core net", using his own format, only needs to provide a similar `Write` function. In the case of the `pnml` subcommand, that was added on the last release of the tool, one hundred line of codes were enough to add the ability to generate PNML files. A figure that is similar to what we observed with the `lola` subcommand.

Using Package hlnet for Drawing Colored Nets. Package `hlnet` also includes a function to output a textual representation of an AST that is compatible with TINA's net syntax. It generates a net that includes all the places and transitions in a colored model as if it was a P/T net and uses labels to display the expressions associated with transitions and the initial marking of places. The net also includes "nodes" (comments similar to sticky notes) for information about types, variables and arc inscriptions. The result can be displayed and modified with **nd**, the *NetDraw* graphical editor distributed with TINA. We show such an example in the screen capture of Fig. 4, which is obtained by using `mcc` with option `--debug` on the HLPN model *TrainTable* of Fig. 2.

While modifications cannot be saved back into PNML, this capability is still useful to inspect colored model (and is often more accurate than the graphical information included in the "cover flow" provided with every model). We can also use the export function included in **nd** to generate a LaTeX(*tikz*) representation of the net. This is what we used to generate an initial version of the diagrams that appear in Fig. 1 to 3 of this paper.

4   Comparison with Other Tools

The problem of (efficiently) unfolding colored models has been abundantly covered in the literature and many of the proposed algorithms have been implemented. We can cite the works of Mäkelä, with his tool MARIA [15]; of Heiner et al. with Marcie [9,14]; or the work of Kordon et al. [13], that makes a clever use of decision diagrams in order to compute results for very large instances.

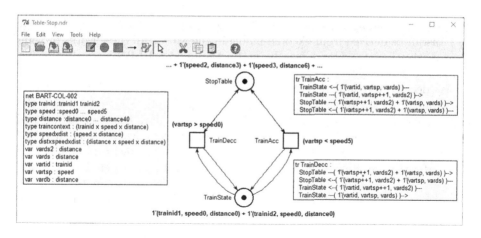

Fig. 4. Result of option debug on model TrainTable, displayed in **nd**

Table 1. Execution time (in s) when unfolding complex PNML instances

MODEL	PLACES	TRANS.	MCC	TAPAAL	MARCIE	GSPN
GlobalResAllocation-07	133	291 067	**1.7**	3	14.4	22.3
GlobalResAllocation-11	297	2.10⁶	**15.1**	29.3	144.6	—
DrinkVendingMachine-16	192	10⁶	15.5	10.7	52.8	108.1
DrinkVendingMachine-24	288	8.10⁶	97.1	95.9	—	—
PhilosophersDyn-50	2 850	255 150	1	2.1	11.1	15.7
PhilosophersDyn-80	6 960	10⁶	**4.1**	9.9	55.9	61.0
Diffusion-D050	2 500	8 109	14.5	**0.6**	4.1	—
Diffusion-D100	10 000	31 209	243.3	**8.6**	31.3	—
TokenRing-100	10 201	10⁶	**4**	8.2	33.5	49.3
TokenRing-200	40 401	8.10⁶	**67.4**	166.1	—	—
SafeBus-50	5 606	140 251	14.2	**1.4**	6.2	25.1
SafeBus-80	13 766	550 801	89.5	**7**	20.6	133.1
TrainTable-Dist	722	602	**1.4**	12.6	59.5	69.4
TrainTable-Stop+Dist	728	602	**2.1**	—	—	—
BART-002	764	646	**3.1**	—	—	—
BART-060	15 032	19 380	**3.2**	—	—	—
SharedMemory-000200	40 801	80 400	**0.3**	1.7	2.6	5.1
SharedMemory-001000	10⁶	2.10⁶	**8.9**	—	60.3	160.2
SharedMemory-002000	4.10⁶	8.10⁶	**55.3**	—	—	—
FamilyReunion-L800	2.10⁶	2.10⁶	**5.5**	—	84.8	143.0
FamilyReunion-L3000	28.10⁶	27.10⁶	**89.5**	—	—	—
Swap-P010000	20 000	20 000	0.1/ 0.6	0.4	0.9	5.0
Swap-P100000	200 000	200 000	0.4/ 4.8	26.1	15.7	—

This approach is implemented in CPN-AMI [8] and provides the reference for P/T instances derived from Colored model in the MCC. All these works provide good motivations for why it may be useful to unfold a HLPN instead of trying to analyze it directly.

We decided to compare mcc with three tools that participated in the Model-Checking Contest: Tapaal [7] (with its verifypn tool); Marcie [9] (with the andl_converter); and GreatSPN [2] (that includes a Java based unfolding tool in its editor). Since each tool is tailored for a different toolchain—and therefore generate very different results—it is difficult to make a precise comparison of the performances. Hence these results should only be interpreted as a rough estimate. For instance, Tapaal is the only tool in this list that do not output the unfolded net on disk. This means that its computation time do not include the time spent marshalling the result and printing it on file.

Unfolding Algorithm. We follow a very basic strategy. For each place p, of type say T, we create one instance of p for any value that inhabits T. (This part is common to most of the existing unfolding algorithms.) For each transition, t, we consider the set of variables occurring in the inscription of arcs attached to it (its *environment*). Then we enumerate all possible valuations of the environment and keep only those that satisfy the conditions associated with t.

Our main optimization is to follow a "constraint solving" approach where we can avoid enumerating a large part of the possible assignments when we know that the condition cannot be satisfied. For instance a subexpression in a conjunction is falsified. This is a less sophisticated approach than those described in existing works [14,15]. For instance, we do not try to detect particular kind of expressions where an unification-based approach could have better performances.

Typically, we should perform badly with instances similar to Diffusion, where we may fail to cut down the size of our search space. On the other hand, our approach is not hindered when we need to deal with complex expressions, such as with TrainTable, that involve at the same time tuples, successor, and add. Actually, our approach may also work with nonlinear patterns, where the same variable is reused in the same expression. Finally, all the algorithms should work equally well on examples like Swap, because of its simplicity.

Even if our approach is quite rustic, our experiments shows that this does not hinder our performances. This may be because few of the colored instances in the MCC fall in the category where clever algorithms shine the most.

Benchmarks. We selected instances, with a processing time of over a second, from different models listed in the MCC repository [11] and from the three examples in Sect. 3. We give the results of our experiment in Table 1. Computations were performed with a time limit of 5 min and a limit of 16 GB of RAM. In each case we give the number of places and transitions in the unfolded net and highlight the best time (when there is a significant difference). An absence of values (—) means a timeout. Tapaal shows very good performances on many instances and significantly outperforms mcc for models SafeBus and Diffusion. On the opposite, we see that many instances can only be processed with mcc. This is the case with model BART (even with a time limit of 1 h). Other interesting examples are models SharedMemory and FamilyReunion. This suggest that we could further improve our tool by including some of the optimizations used in `verifytpn` that seems to be orthogonal to what we have implemented so far. We describe two of the optimizations performed by `mcc` below.

Actually, sheer performance is not our main goal. We rather seek to return a result for all the colored instances used in the MCC in a sensible time. (Who needs to unfold a model too big to be analyzed anyways?) At present, there are 193 *instances* of Colored nets in the MCC repository, organized into 23 different classes, simply referred to as *models*. We can return a result for 184 of these instances, with the condition of the competition. Moreover, to the best of our knowledge, `mcc` is the only tool able to return results (for at least one instance) in all the models. But some instances, like DrinkVendingMachine-48, should stay out of reach for a long time, mostly due to memory space limitations.

Use of Colored Invariants. The first "additional" optimization added to `mcc` explains our good result on models such as BART. The idea is to identify invariant places; meaning places whose marking cannot be changed by firing a transition. A sufficient condition for place p to be invariant is if, for every transition t, there is an arc with inscription e from p to t iff there is an arc from t to p with inscription e' equivalent to e. (Syntactical equality between e and e' is enough for our purpose.) We say that such places are *stable*; a concept equivalent to "test arcs" for an HLPN. This is the case, for example, for place StopTable in model TrainTable. When a place is stable, we know that its marking is fixed. This can significantly reduce the set of assignments that need to be enumerated.

Use of a Petri Scripting Language. The effect of our second improvement can be observed in model Swap (Fig. 3). In this case, like with model Philosopher of the MCC, it is possible to detect that the unfolded net is the composition of n copies of the same component; where $n = |\text{Resource}|$. Each component x (with $x \in$ Resource) is a net with a local copy of the places. As for the transitions, we need to keep one copy for each "local interactions" (such as t_2) and two copies for distant interactions (t_1): one for the pair of components $(x-, x)$; the other for the pair $(x, x++)$. Since type Resource is a cyclic enumeration—this is basically a "scalar set"—the composition of all these components form a ring architecture.

Our tool is able to recognize this situation automatically. In such a case we output a result that uses the *TPN format*, a scripting language for Petri net supported by the TINA toolchain. This scripting language includes operators for make copies of net; add and rename places and transitions; compute the product or chaining of nets; ... It also provides higher-order composition patterns, such as pools or rings of components. We use the latter for model Swap.

Our benchmarks of Table 1 include the results on two instances of model Swap. The computation time for `mcc` (the first value) is mostly independent from the size of the instance (the only difference is in parsing the PNML file.) This result conceals a much more complex realty. Indeed, a tool that consumes a TPN script still needs to "expand it". This is why we added a second value in Table 1, which is the time taken to generate the result using the `mcc pnml` command. For information, the size of the PNML result for model Swap-P100000 is 99 MB, while it is only 200 bytes for the TPN version.

5   Conclusion

Tool mcc is a new solution to an old problem. It is also an unassuming tool, that focuses on a single, very narrow task. Nonetheless, we believe that it can still be of interest for the Petri net community, and beyond, by enriching the PNML ecosystem. As a matter of fact, there has been a total of 26 verification tools to participate to the MCC since its beginning [1], not all "Petri tools". Many of these tools could benefit from using mcc.

Development on mcc started in 2017, as a pet project for studying the suitability of the Go programming language to develop formal verification tools. Our assessment in this regard is very positive: performances are competitive with regards to C++, with good code productivity and mature software libraries; building executables for multiple platforms and distributing code is easy; ... Since then, work has progressed steadily in-between each edition of the MCC, with a focus on stability of the tool and on compliance with the PNML standard. Three iterations later, mcc is now sufficiently mature to gain more exposure and provides a good showcase for an efficient PNML parser written in Go. But mcc is more than that. First, mcc was designed to lower the work needed by developers wanting to engage in the Model-Checking Contest. It also provides new features, such as the ability to display an interactive (read-only), graphical view of a PNML model; see Fig. 4. Finally, it provides a testbed for evaluating new unfolding algorithms (we show two of these ideas in Sect. 4).

In the future, we plan to enrich mcc by computing interesting properties of the models during unfolding. For example by computing invariants or by finding sets of places that can be clustered together. In that respect, the possibility to identify HLPN that can be expressed using a "Petri net scripting language" could potentially leads to new advances. For example to simplify the detection of symmetries, something that we have been working on recently in the context of Time Petri nets [5].

References

1. Amparore, E., et al.: Presentation of the 9th edition of the model checking contest. In: Beyer, D., Huisman, M., Kordon, F., Steffen, B. (eds.) TACAS 2019. LNCS, vol. 11429, pp. 50–68. Springer, Cham (2019). https://doi.org/10.1007/978-3-030-17502-3_4
2. Amparore, E.G., Balbo, G., Beccuti, M., Donatelli, S., Franceschinis, G.: 30 years of GreatSPN. In: Fiondella, L., Puliafito, A. (eds.) Principles of Performance and Reliability Modeling and Evaluation. SSRE, pp. 227–254. Springer, Cham (2016). https://doi.org/10.1007/978-3-319-30599-8_9
3. Berthomieu, B., Ribet, P.O., Vernadat, F.: The tool TINA-construction of abstract state spaces for Petri nets and time Petri nets. Int. J. Prod. Res. **42**(14), 2741–2756 (2004). https://doi.org/10.1080/00207540412331312688

4. Billington, J., et al.: The Petri Net Markup Language: concepts, technology, and tools. In: van der Aalst, W.M.P., Best, E. (eds.) ICATPN 2003. LNCS, vol. 2679, pp. 483–505. Springer, Heidelberg (2003). https://doi.org/10.1007/3-540-44919-1_31

5. Bourdil, P.A., Berthomieu, B., Dal Zilio, S., Vernadat, F.: Symmetry reduction for time Petri net state classes. Sci. Comput. Program. **132**(2), 209–225 (2016)

6. Chiola, G., Dutheillet, C., Franceschinis, G., Haddad, S.: On well-formed coloured nets and their symbolic reachability graph. In: Jensen, K., Rozenberg, G. (eds.) High-level Petri nets, pp. 373–396. Springer, Heidelberg (1991). https://doi.org/10.1007/978-3-642-84524-6_13

7. David, A., Jacobsen, L., Jacobsen, M., Jørgensen, K.Y., Møller, M.H., Srba, J.: TAPAAL 2.0: integrated development environment for timed-arc petri nets. In: Flanagan, C., König, B. (eds.) TACAS 2012. LNCS, vol. 7214, pp. 492–497. Springer, Heidelberg (2012). https://doi.org/10.1007/978-3-642-28756-5_36

8. Hamez, A., et al.: New features in CPN-AMI 3: focusing on the analysis of complex distributed systems. In: Sixth International Conference on Application of Concurrency to System Design (ACSD 2006). IEEE (2006). https://doi.org/10.1109/ACSD.2006.15

9. Heiner, M., Rohr, C., Schwarick, M.: MARCIE – model checking and reachability analysis done efficiently. In: Colom, J.-M., Desel, J. (eds.) PETRI NETS 2013. LNCS, vol. 7927, pp. 389–399. Springer, Heidelberg (2013). https://doi.org/10.1007/978-3-642-38697-8_21

10. Hillah, L., Kordon, F., Petrucci, L., Trèves, N.: PN standardisation: a survey. In: Najm, E., Pradat-Peyre, J.-F., Donzeau-Gouge, V.V. (eds.) FORTE 2006. LNCS, vol. 4229, pp. 307–322. Springer, Heidelberg (2006). https://doi.org/10.1007/11888116_23

11. Hillah, L.M., Kordon, F.: Petri nets repository: a tool to benchmark and debug petri net tools. In: van der Aalst, W., Best, E. (eds.) PETRI NETS 2017. LNCS, vol. 10258, pp. 125–135. Springer, Cham (2017). https://doi.org/10.1007/978-3-319-57861-3_9

12. Jensen, K.: Coloured petri nets. In: Brauer, W., Reisig, W., Rozenberg, G. (eds.) Petri Nets: Central Models and Their Properties, vol. 254, pp. 248–299. Springer, Heidelberg (1987). https://doi.org/10.1007/BFb0046842

13. Kordon, F., Linard, A., Paviot-Adet, E.: Optimized colored nets unfolding. In: Najm, E., Pradat-Peyre, J.-F., Donzeau-Gouge, V.V. (eds.) FORTE 2006. LNCS, vol. 4229, pp. 339–355. Springer, Heidelberg (2006). https://doi.org/10.1007/11888116_25

14. Liu, F., Heiner, M., Yang, M.: An efficient method for unfolding colored Petri nets. In: Winter Simulation Conference (WSC). IEEE (2012). https://doi.org/10.1109/WSC.2012.6465203

15. Mäkelä, M.: Optimising enabling tests and unfoldings of algebraic system nets. In: Colom, J.-M., Koutny, M. (eds.) ICATPN 2001. LNCS, vol. 2075, pp. 283–302. Springer, Heidelberg (2001). https://doi.org/10.1007/3-540-45740-2_17

16. Schmidt, K.: LoLA a low level analyser. In: Nielsen, M., Simpson, D. (eds.) ICATPN 2000. LNCS, vol. 1825, pp. 465–474. Springer, Heidelberg (2000). https://doi.org/10.1007/3-540-44988-4_27

Author Index

Printed in the United States
By Bookmasters